Platelets, Thrombosis and the Vessel Wall

Advances in Vascular Biology
A series of books bringing together important advances and reviewing all areas of vascular biology.
Edited by *Mathew A. Vadas, The Hanson Centre for Cancer Research, Adelaide, South Australia* and *John Harlan, Division of Hematology, University of Washington, Seattle, USA.*

This book is part of a series. The publisher will accept continuation orders which may be cancelled at any time and which provide for automatic billing and shipping of each title in the series upon publication. Please write for details.

Platelets, Thrombosis and the Vessel Wall

edited by

Michael C. Berndt

Baker Medical Research Institute
Prahran, Victoria
Australia

harwood academic publishers
Australia • Canada • France • Germany • India
Japan • Luxembourg • Malaysia • The Netherlands
Russia • Singapore • Switzerland

Amsteldijk 166
1st Floor
1079 LH Amsterdam
The Netherlands

British Library Cataloguing in Publication Data

Platelets, thrombosis and the vessel wall. – (Advances in
   vascular biology; v. 6)
   1. Blood platelets   2. Thrombosis   3. Hemostasis
   I. Berndt, Michael C.
   612.1'17

   ISBN: 90-5702-369-5
   ISSN: 1072-0618

CONTENTS

Contents

SERIES PREFACE

It is our privilege to live at a time when scientific discoveries are providing insights into human biology at an unprecedented rate. It is also a time when the sheer quantity of information tends to obscure underlying principles, and when hypotheses or insights that simplify and unify may be relegated to the shadow of hard data.

The driving force for editing a series of books on Vascular Biology was to partially redress this balance. In inviting editors of excellence and experience, it is our aim to draw together important facts, in particular areas of vascular biology, and to allow the generation of hypotheses and principles that unite an area and define newer horizons. We also anticipate that, as is often the case in biology, the formulation and application of these principles will interrelate with other disciplines.

Vascular biology is a frontier that has been recognised since at least the time of Cohnheim and Metchikoff, but has really come into prominence over the last 10–15 years, once the molecules that mediate the essential functions of the blood vessel started to be defined. The boundaries of this discipline are, however, not clear. There are intersections, for example, with hypertension and atherogenesis that bring in, respectively, neuroendocrine control of vessel tone and lipid biochemistry which exist as separate bodies of knowledge. Moreover, it would be surprising if some regional vascular biology (for example, pulmonary, renal, etc.) were not to emerge as subgroups in the future. Our aims for the moment, however, are to concentrate on areas of vascular biology that have a wide impact.

It is our hope to publish two books each year for the next 3–4 years. Indeed the first five books have been commissioned and address areas primarily in endothelial biology (hemostasis and thrombosis), immunology, leukocyte adhesion molecules, platelet adhesion molecules, adhesion molecules that mediate cell-cell contact. Subsequent volumes will cover the physiology and pathology of other vascular cells as well as developmental vascular biology.

We thank the editors and contributors for their very hard work.

Mathew VADAS   John HARLAN

PREFACE

Unstable angina, myocardial infarction and stroke, precipitated by thrombosis, are the leading combined causes of death in the Western world. In response to vascular trauma, platelets rapidly adhere to the exposed subendothelial matrix. At high shear flow rates, as would occur at sites of stenosis, the initial platelet adhesion is dependent upon von Willebrand factor and a specific von Willebrand factor receptor on platelets, the GP Ib-IX-V complex. Platelet activation through agonist and adhesion receptors leads to a cascade of signaling events, cytoskeletal re-organization and secretion of the contents of platelet dense bodies and alpha-granules. Substances released from platelets include ADP that acts in the recruitment of additional platelets to the developing thrombus. A variety of proteins involved in the regulation of coagulation and fibrinolysis are also released as well as proteins such as thrombospondin and platelet-derived growth factor that are involved in subsequent vascular remodeling and wound repair. Concomitant with platelet activation and release, the platelet integrin, GP-IIb-IIIa, is converted from a low affinity to a high affinity receptor for fibrinogen and von Willebrand factor leading to platelet aggregation and subsequent thrombus. Activated platelets also express adhesion receptors for neutrophils and endothelial cells and therefore there is the potential for cross-talk between these cell types in the developing thrombus.

In recent years, there have been major advances in our understanding of platelets and their role within the vasculature. The purpose of this text is not to provide an exhaustive overview of this subject, but to highlight key aspects of platelet function in the regulation of thrombosis and haemostasis. The various chapters cover the regulation of megakaryocytopoiesis and platelet production, platelet cell surface receptors and ligands involved in platelet adhesion, aggregation, and the interaction of platelets with leukocytes and endothelial cells, the role of the platelet cytoskeleton in agonist-dependent platelet activation, the role of platelets in regulation of fibrinolysis, and mechanisms of cross-talk between platelets, leukocytes and endothelium. Finally, there are two chapters highlighting two major clinical causes of thrombosis, circulating anti-phospholipid antibodies and genetic predisposition.

LIST OF CONTRIBUTORS

Andrews, Robert K.
Hazel and Pip Appel Vascular Biology
   Laboratory
Baker Medical Research Institute
PO Box 348, Commercial Road
Prahran, VIC 3181
Australia

Baker, Ross
Clinical Thrombosis Unit
Haematology Department
Royal Perth Hospital
University of Western Australia
GPO Box X2213
Perth, WA 6001
Australia

Begley, C. Glenn
The Walter and Eliza Hall Institute for
   Medical Research
PO Royal Melbourne Hospital
Melbourne, VIC 3050
Australia

Berndt, Michael C.
Hazel and Pip Appel Vascular Biology
   Laboratory
Baker Medical Research Institute
PO Box 348, Commercial Road
Prahran, VIC 3181
Australia

Brown, Susan
Department of Medicine
Monash Medical School
Box Hill Hospital
Nelson Road
Box Hill, VIC 3128
Australia

Burns, Gordon F.
Cancer Research Unit, Level 5
David Maddison Clinical Sciences
   Building
Royal Newcastle Hospital
Newcastle, NSW 2300
Australia

Campbell, Janine K.
Department of Medicine
Monash Medical School
Box Hill Hospital
Nelson Road
Box Hill, VIC 3128
Australia

Chesterman, Colin N.
Centre for Thrombosis and Vascular
   Research
School of Pathology
University of New South Wales
Sydney, NSW 2052
Australia

Chignard, Michel
Unité de Pharmacologie Cellulaire
Institut Pasteur
25 rue de Dr Roux
75724 Paris Cedex 15
France

Collins, Tucker
Vascular Research Division
Brigham and Women's Hospital and
   Harvard Medical School
Boston, MA 02115
USA

Crossno, Jr., Joseph T.
Departments of Medicine and
   Biochemistry
University of Tennessee
Coleman Building H. 335
956 Court Avenue
Memphis, TN 38117
USA

Dong, Jing-Fei
Veterans Affairs Medical Center
Hematology/Oncology 111H
2002 Holcombe Boulevard
Houston, TX 77030-4298
USA

Dorahy, Douglas J.
Cancer Research Unit, Level 5
David Maddison Clinical
   Sciences Building
Royal Newcastle Hospital
Newcastle, NSW 2308
Australia

Du, Xiaoping
Department of Pharmacology
University of Illinois, Chicago
835 S. Wolcott Avenue
Chicago, IL 60612
USA

Eikelboom, John
Clinical Thrombosis Unit
Haematology Department
Royal Perth Hospital
University of Western Australia
GPO Box X2213
Perth, WA 6001
Australia

Fox, Joan E.B.
Department of Molecular Cardiology
Joseph J. Jacobs Center for Thrombosis
   and Vascular Biology
Cleveland Clinic Foundation
9500 Euclid Avenue
Cleveland, OH 44195
USA

Hogg, Philip J.
Centre for Thrombosis and Vascular
   Research
School of Pathology
University of New South Wales
Sydney, NSW 2052
Australia

Hotchkiss, Kylie A.
Centre for Thrombosis and Vascular
   Research
School of Pathology
University of New South Wales
Sydney, NSW 2052
Australia

Jennings, Lisa K.
Departments of Medicine and
   Biochemistry
University of Tennessee
Coleman Building H. 335
956 Court Avenue
Memphis, TN 38117
USA

Khachigian, Levon M.
Centre for Thrombosis and Vascular
   Research
School of Pathology
University of New South Wales
Sydney, NSW 2052
Australia

Leopold, Jane A.
Evans Department of Medicine and
   Whitaker Cardiovascular Institute
Boston University School of
Medicine
80 E. Concord Street, W-507
Boston, MA 02118
USA

Lindner, Volkhard
Department of Surgery
Maine Medical Research Institute
South Portland, ME 04106
USA

López, José A.
Veterans Affairs Medical Center
Hematology/Oncology 111H
2002 Holcombe Boulevard
Houston, TX 77030–4298
USA

Loscalzo, Joseph
Evans Department of Medicine and
    Whitaker Cardiovascular Institute
Boston University School of Medicine
80 E. Concord Street, W-507
Boston, MA 02118
USA

McEver, Rodger P.
W.K. Warren Medical Research
    Institute
University of Oklahoma Health
    Sciences Center
825 N.E. 13th Street
Oklahoma City, OK 73104
USA

McNally, Tracy
Hazel and Pip Appel Vascular Biology
    Laboratory
Baker Medical Research Institute
PO Box 348, Commercial Road
Prahran, VIC 3181
Australia

Meyer, Sylvie C.
Department of Molecular Cardiology
Joseph J. Jacobs Center for Thrombosis
    and Vascular Biology
Cleveland Clinic Foundation
9500 Euclid Avenue
Cleveland, OH 44195
USA

Mitchell, Christina A.
Department of Medicine
Monash Medical School
Box Hill Hospital
Nelson Road
Box Hill, VIC 3128
Australia

Munday, Adam D.
Department of Medicine
Monash Medical School
Box Hill Hospital
Nelson Road
Box Hill, VIC 3128
Australia

Pidard, Dominique
Unité de Pharmacologie Cellulaire
Institut Pasteur
25 rue de Dr Roux
75724 Paris Cedex 15
France

Rasko, John E.J.
Centenary Institute of Cancer Medicine
    and Cell Biology
Locked Bag No 6
Newtown, NSW 2042
Australia

Si-Tahar, Mustapha
Unité de Pharmacologie Cellulaire
Institut Pasteur
25 rue de Dr Roux
75724 Paris Cedex 15
France

Silverman, Eric S.
Vascular Research Division
Brigham and Women's Hospital and
    Harvard Medical School
Boston, MA 02115
USA

Smith, David R.
Veterans Affairs Medical Center
Hematology/Oncology 111H
2002 Holcombe Boulevard
Houston, TX 77030-4298
USA

Thorne, Rick F.
Cancer Research Unit, Level 5
David Maddison Clinical Sciences Building
Royal Newcastle Hospital
Newcastle, NSW 2300
Australia

Ward, Christopher M.
Department of Haematology and
    Transfussion Medicine
Royal North Shore Hospital
Pacific Highway
St. Leonards, NSW 2065
Australia

White, Melanie M.
Departments of Medicine and Biochemistry
University of Tennessee
Coleman Building H. 335
956 Court Avenue
Memphis, TN 38117
USA

Wilkinson, Robyn M.
Cancer Research Unit, Level 5
David Maddison Clinical Sciences
    Building
Royal Newcastle Hospital
Newcastle, NSW 2308
Australia

Williams, Amy J.
Vascular Research Division
Brigham and Women's Hospital
    and Harvard Medical
    School
Boston, MA 02115
USA

1 Thrombopoietin

John E.J. Rasko[1] and C. Glenn Begley[2]

[1] *Gene Therapy Research Unit, Centenary Institute of Cancer Medicine & Cell Biology, Locked Bag No 6, Newtown, NSW 2042, Australia*
Tel: +61 2 9565 6156, Fax: +61 2 9565 6101, email: j.rasko@centenary.usyd.edu.au
[2] *The Walter and Eliza Hall Institute for Medical Research, PO Royal Melbourne Hospital, Victoria, 3050, Australia*

INTRODUCTION

Following a prolonged gestation, thrombopoietin was cloned in 1994 and now commands center stage as the preeminent growth factor affecting megakaryopoiesis. In this chapter, following a brief introduction to the megakaryocytic lineage, the history and discovery of thrombopoietin is described. Building on this background, a detailed account of the biochemistry, molecular biology and genetics of thrombopoietin and its receptor c-mpl is presented. Although available for only a short time, this recombinant protein has facilitated many new insights concerning the biology of platelet production and homeostasis. Furthermore, thrombopoietin has been applied in the clinical context of high dose chemoradiotherapy and bone marrow transplantation. The current phase of thrombopoietin's clinical development represents an exciting culmination of intense basic research and a promising début for future therapies.

THE MEGAKARYOCYTIC LINEAGE

The haemopoietic compartment is comprised of multiple parallel pathways of differentiation over which a hierarchy of stem cell, progenitors and end cells may be superimposed (Metcalf, 1988). Pluripotential stem cells differentiate to become megakaryocytes by the process of megakaryopoiesis. Functional platelets are the end cells of the process of thrombopoiesis (see Figure 1.1). In order to orientate the reader and prepare a foundation for subsequent chapters, a brief outline of the stages of the megakaryocytic lineage will be presented.

The first stage of differentiation commitment solely toward the megakaryocyte lineage is the megakaryocyte burst-forming unit (BFU-meg) (Metcalf *et al.*, 1975; Long *et al.*, 1985). It has been suggested that a high proliferative potential cell found only in fetal bone marrow may represent an even more primitive lineage-restricted megakaryocyte (Bruno *et al.*, 1996; Murray *et al.*, 1996). The subsequent stage of development following BFU-meg is the megakaryocyte colony-forming unit (CFU-meg) (Vainchenker *et al.*, 1979; Long *et al.*, 1986). None of these early stages may be distinguished by morphology. After the CFU-meg stage, expression of the CD34 marker of primitive haemopoiesis is gradually lost (Briddell *et al.*, 1989; Debili *et al.*, 1992). In concert with a loss of proliferative capacity,

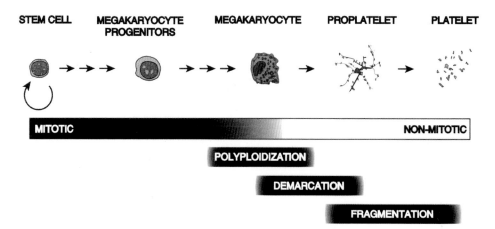

Figure 1.1   Stages of megakaryocyte and platelet development. The self-reflexive arrow beneath the haemopoietic stem cell indicates its potential for self-renewal. The shaded region separating mitotic (black) from non-mitotic (white) phases of megakaryocyte maturation represents the "transitional" stage (Jackson, 1973; Young and Weiss, 1987). The approximate stages of development during which polyploidization, demarcation and fragmentation occur are indicated.

mature megakaryocyte progenitors progressively forgo their capacity to form colonies (Thean *et al.*, 1983; Hoffman, 1989).

Following the mitotic phase of differentiation, megakaryocytes commit to the process of nuclear endoreduplication, otherwise called polyploidization (Jackson, 1990). This poorly understood phenomenon is characterised by successive rounds of DNA replication and endomitosis in the absence of cytokinesis. Within the haemopoietic compartment, polyploidization is believed to be unique to the megakaryocytic lineage (Long, 1995). Commitment to endomitosis occurs at the "promegakaryoblast" stage, during which cells may be identified ultrastructurally by the presence of platelet peroxidase in the endoplasmic reticulum and perinuclear space (Breton-Gorius *et al.*, 1976). Promegakaryoblasts are characterised by a diameter of 7–15µm and expression of surface glycoprotein (GP) IIb/IIIa, cytoplasmic platelet factor 4 and von Willebrand factor (vWF) (Rabellino *et al.*, 1979; Vainchenker *et al.*, 1982). In rodent cells, acetylcholine esterase activity is the most frequently used marker of the megakaryocytic lineage (Jackson, 1973). Subsequent morphological and phenotypic stages of megakaryocyte maturation have been defined (Williams *et al.*, 1982; Debili *et al.*, 1992).

Mature megakaryocytes develop a filigree of demarcation membranes in the cytoplasm by presumed invagination of the plasma membrane (Behnke, 1968; Breton-Gorius *et al.*, 1976). These membranes are thought to define platelet fields from which formed platelets may be released into the circulation following fragmentation (Tavassoli, 1980; Zucker-Franklin *et al.*, 1984). An alternative (although not necessarily exclusive) hypothesis concerning platelet release involves the intermediary "proplatelet" structure (Choi *et al.*, 1995b). Time-lapse video, and electron- and light-microscopic images indicate that these delicate structures represent nascent platelets budding off from fine stalks (Thiery *et al.*, 1956; Behnke, 1969; Choi *et al.*, 1995a). It has been established that megakaryocytes of 8N ploidy and above can produce platelets (Paulus, 1970).

It is just over ninety years since Wright first proposed that megakaryocytes function as the proximal cell from which platelets are shed (Wright, 1906). Platelets are small, anucleate effector cells essential for maintenance of normal haemostasis. In addition, platelets contribute to the fibrinolytic cascade and mechanisms maintaining vascular integrity/tone as well as providing factors involved in angiogenesis and wound recovery. The absence of normal platelet function either from quantitative or qualitative deficits leads to a bleeding diathesis.

HISTORY AND DISCOVERY OF THROMBOPOIETIN

Haemopoietic Growth Factors (HGFs), including thrombopoietin, are typically glyco-proteins which transmit instructions to cells within the haemopoietic compartment. The resulting panoply of biological effects in target cells include lineage commitment, differentiation induction, survival, proliferation, functional activation and physiological cell death (apoptosis). The production of approximately 35 million platelets per litre per day in a normal male is daunting, but the system must also allow for at least an eight-fold increase in production following emergency demands in the event of haemorrhage or other threats to haemostasis (Harker *et al.*, 1969). Thus a fundamental question in the study of haemopoiesis is: what mechanism(s) control the differentiation of stem cells to lineage-specific progenitor cells whereby mature cell numbers are maintained within strict limits? In seeking an answer to this question, the HGFs were identified and characterised (Metcalf *et al.*, 1995).

Around the time of a bitter civil uprising in Hungary, Keleman and his co-workers in Budapest performed experiments suggesting the presence of a natural factor which could promote the production of platelets (Kelemen *et al.*, 1958). Yamamoto in Japan drew similar conclusions (Yamamoto, 1957). These investigators showed that serum or plasma from thrombocytopenic patients could cause a rapid elevation in the platelet count in recipient mice (Kelemen, 1995). This activity, also present in the serum or plasma from several animal models, was termed "thrombopoietin" to parallel erythropoietin (Spector, 1961; de Gabriele *et al.*, 1967; Ebbe, 1976).

Although the identity of the thrombopoietic activity was not to be cloned for many years, from pioneering investigations it was highly likely that it could bind platelets. Evidence for this was presented in 1967 when Pennington showed that platelet transfusions could reduce the megakaryocyte ploidy-promoting activity present in thrombocytopenic rat plasma (de Gabriele *et al.*, 1967). Similar conclusions have been drawn in rabbits, humans and mice (Mazur *et al.*, 1981; Kuter *et al.*, 1995; Fielder *et al.*, 1996; Stoffel *et al.*, 1996). The following brief summary of thrombopoietin's eventual cloning in 1994 by several biotechnology companies is a strong argument for the importance of both basic research and serendipity in science.

Prior to the cloning of thrombopoietin, its receptor, c-mpl, was cloned by homology to a viral oncogene, v-mpl. Indeed many oncogenes have been unveiled as the transforming gene of oncogenic retroviruses (Courtois *et al.*, 1995; Onishi *et al.*, 1996). Adopting a well-established approach, Francoise Wendling and colleagues infected over two hundred neonatal DBA/2 mice with Friend Murine Leukemia Virus. By seven months, only one mouse developed hepato-splenomegaly and polycythaemia (Wendling *et al.*, 1986). The replication-defective virus (MPLV) responsible for the myeloproliferative leukemia was eventually cloned and found to be rapidly transforming for multiple lineages *in vivo* (Wendling *et al.*, 1989). Genetic analysis of the new virus showed that part of a gene, consequently named v-mpl, had rearranged in the *env* region of the virus (Souyri *et al.*, 1990). Interest immediately shifted from an exclusively virological audience to the experimental hematologists when it was shown that c-mpl, the normal cellular homologue of v-mpl, encoded a portion of a previously unknown member of the cytokine receptor superfamily (Vigon *et al.*, 1992).

Parenthetically, the "receptor first" approach to cloning cytokines has met with incredible success over the last decade. Since many cytokine receptors share regions of marked homology, it is much more likely that other members of large receptor families will be identified using molecular similarity (Bazan, 1990; Wilks, 1993). In contrast, ligands tend not to exhibit such regions of homology and so are less likely to be cloned by the same venture (Minasian *et al.*, 1992). This has led to the unusual situation in which a number of HGF receptors have been cloned prior to their biological characterization. These so-called "orphan receptors" (which, to be literal, lack a partner — not a parent!) have been cloned often years before their cognate ligand. Nevertheless, identification of orphan receptors provide molecular probes for putative ligands.

By 1993, the attention of several biotechnology companies and research laboratories was determinedly focused. Five lines of evidence led to the compelling conclusion that c-mpl performed a critical role in the positive control of megakaryopoiesis. First, c-mpl had been found to be expressed throughout the megakaryocytic lineage and on many cell lines known to have megakaryocytic potential (Methia *et al.*, 1993; Vigon *et al.*, 1993). Second, antisense oligonucleotides used to block c-mpl expression in stimulated human CD34$^+$ cells led to a reduction in megakaryocytic, but not myeloid or erythroid, colony formation (Methia *et al.*, 1993). Third, soluble c-mpl could bind to, and remove, the thrombopoietic, but not myelopoietic or erythropoietic, activity present in aplastic serum from several species (Bartley *et al.*, 1994; de Sauvage *et al.*, 1994; Wendling *et al.*, 1994). Fourth, by using extracellular domains of either G-CSF or IL-4 receptors fused to make chimeric c-mpl receptors, it was shown that the cytoplasmic domain of the latter could transduce a proliferative signal (Skoda *et al.*, 1993; Vigon *et al.*, 1993). Finally, mice rendered nullizygous for c-mpl by gene targeting technology exhibited platelet concentrations less than 15% of those present in normal littermates (see below) (Gurney *et al.*, 1994).

Thrombopoietin was cloned in 1994 by five independent laboratories using several approaches. The first ("c-mpl affinity") approach, adopted succesfully by teams at Amgen, Inc. and Genetech, Inc. in California, depended on the specific, high affinity of c-mpl for its ligand (Bartley *et al.*, 1994; de Sauvage *et al.*, 1994). Beginning with litres of plasma from aplastic pigs or dogs, various chromatographic techniques, including c-mpl affinity, were used to enrich for Mpl-ligand activity (Hunt *et al.*, 1995). Cells stably expressing c-mpl were established as a reliable proliferation assay for Mpl-ligand to monitor the approximately 10^7-fold purification.

The second ("brute force") approach performed by biochemists both at Kirin Brewery Company in Tokyo and Massachusetts Institute of Technology in Boston relied primarily on a robust bioassay for thrombopoietin (Kuter *et al.*, 1994; Kato *et al.*, 1995; Kuter *et al.*, 1997). These dedicated researchers started purifications using eight litres of plasma from 1100 irradiated, thrombocytopenic rats or twenty-one litres of plasma from busulfan-treated, thrombocytopenic sheep, respectively. Following an 11-step purification, the resulting material yielded a 10^7- to 10^8-fold increase in specific activity, thereby enabling amino acid sequencing and subsequent cloning.

The third, and perhaps most elegant, method by which the c-mpl ligand was cloned utilised a novel "autocrine approach" (Kaushansky *et al.*, 1994; Lok *et al.*, 1994). HGF-dependent cells expressing c-mpl were chemically mutagenised to become HGF-independent. The resulting eighteen cell lines were screened for secretion of self-stimulating (autocrine) HGF and one was shown to produce a ligand for c-mpl. Having established a reliable cellular source of Mpl-ligand, construction and screening of a conventional cDNA expression library yielded molecular clones. Our attempts using a similar retroviral mutagenesis approach proved less productive (Metcalf, 1994; Rasko *et al.*, 1995).

MOLECULAR BIOLOGY OF THROMBOPOIETIN

The ligand for c-mpl is variously known as thrombopoietin (Tpo), megakaryocyte growth and development factor (MGDF), megapoietin and c-mpl ligand conforming to the usage adopted by the laboratory in which it was realised. These differing nomenclatures also

Table 1.1A Thrombopoietin — molecular details

Feature	Human	Mouse
Chromosomal locus	3q26.3-27	proximal 16
Transcript size	1.8 kb	1.8 kb
Sites of expression	adult and fetal liver, kidney, muscle	adult and fetal liver, kidney, muscle
Protein size	353 amino acids	356 amino acids
	68–85 kDa	68–85 kDa

Table 1.1B c-mpl — molecular details

Feature	Human	Mouse
Chromosomal locus	1p34	4 band D
Transcript size	3.7 (major) and 2.8 kb (minor)	3.0 kb
Sites of expression	Haemopoietic Stem Cells, CD34$^+$ cells, megakaryocytes, platelets	Haemopoietic Stem Cells, megakaryocytes, platelets
Protein size	635 amino acids approximately 80 kDa	625 amino acids approximately 80 kDa

reflect the recognition that the activities of the newly-isolated Mpl-ligand were partly discrepant with those previously attributed to "thrombopoietin". The cDNA molecules identified by all groups appear to be identical. Aside from splice variants, there is no current evidence for any alternate protein capable of high affinity binding to c-mpl, so the "thrombopoietin" designation has been adopted throughout this chapter. However, the growth-promoting activity described by other investigators termed thrombocytopoietic-stimulating factor (McDonald *et al.*, 1981), megakaryocyte stimulatory factor (Tayrien *et al.*, 1987) and megakaryocyte colony-stimulating factor (Erickson-Miller *et al.*, 1993) may also, in retrospect, have contained c-mpl ligand. Table 1.1 summarises some important details pertaining to thrombopoietin and its receptor.

Protein Structure

Partial sequencing of purified thrombopoietin polypeptides from human, murine, porcine, canine and rat species was required for their cloning. However, other than for human thrombopoietin which has been fully analysed using liquid chromatography-electrospray mass spectrometry (Hoffman *et al.*, 1996), the complete sequence was deduced from cDNA clones obtained from various libraries. There is approximately 70% overall identity between the thrombopoietin sequences shown in Figure 1.2. Indeed, thrombopoietin from several different species will bind c-mpl from diverse species. For example, both human and murine thrombopoietin induce proliferative responses in cells containing either human or murine c-mpl (Gurney *et al.*, 1995a). Thrombopoietin is a glycoprotein ranging in size between 326 (in rat) and 356 (in mouse) residues, including a 21 amino acid signal peptide. Four wholly conserved cysteine residues are located at positions 7, 29, 85 and 151 of the mature protein. As with erythropoietin, the first and last cysteines form an essential disulphide

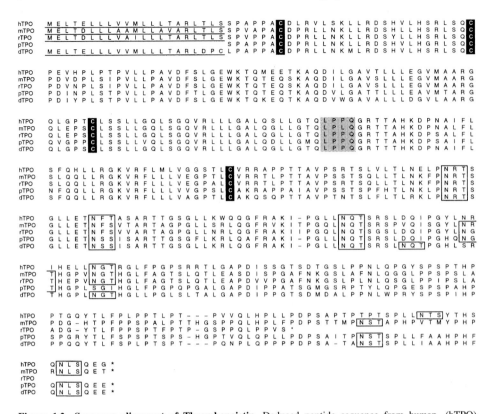

Figure 1.2 Sequence alignment of Thrombopoietin. Deduced peptide sequence from human- (hTPO), mouse- (mTPO), rat- (rTPO), pig- (pTPO) and dog-Thrombopoietin (dTPO) were obtained from various databases or publications and aligned using the pileup command in the Wisconsin software package. The four conserved cysteine residues are highlighted using black backgrounds. The putative signal sequence is underlined and stop codons are indicated by an asterix. Predicted N-linked glycosylation sites are boxed. An alternate thrombopoietin isoform (Tpo-2) results from a four amino acid deletion indicated by stippling at positions 133-136. Each row is 50 peptides long and gaps (–) have been inserted to optimise alignment.

bond (Wada *et al.*, 1995; Hoffman *et al.*, 1996). Although the crystal structure of thrombopoietin has not been solved, it is likely that the amino-terminal domain comprises four α-helices, as observed in many other HGFs (Bazan, 1990; Minasian *et al.*, 1992).

There are two distinct regions in thrombopoietin separated by a potential Arg-Arg proteolytic cleavage site at amino acid 153. Whether native thrombopoietin circulates as the full-length or cleaved peptide remains an open question. The amino-terminal domain retains full biological activity and displays homology to erythropoietin, several neurotrophins and interferon-α (Lok *et al.*, 1994; Li *et al.*, 1995). Indeed the amino-terminal domain of thrombopoietin exhibits 21–23% amino acid identity to erythropoietin, and they share approximately 50% similarity. In contrast, the carboxy-terminal domain diverges between species and does not share homology with any other known protein. However the native protein is definitely glycosylated at the conserved N- and O-linked glycosylation sites in the carboxy-terminal domain (see Figure 1.2) (Gurney *et al.*, 1995a; Hunt *et al.*, 1995; Hoffman *et al.*, 1996). Since the carboxy-terminal domain may be deleted without apparent loss of activity, its function is unclear — although hyperglycosylation is likely to facilitate molecular stability (Narhi *et al.*, 1991).

Genomic Structure and Chromosomal Localization

Thrombopoietin is encoded by a single copy gene which spans approximately 7 kb (Foster *et al.*, 1994; Gurney *et al.*, 1995a). The human gene comprises 6 exons with one or two noncoding 5′ exons. The final exon contains all the coding sequence for the carboxy-terminal domain as well as about one third of the amino-terminal domain. The thrombopoietin gene shares homologous intron/exon boundaries with the erythropoietin gene, except for the non-conserved carboxy-terminus (Chang *et al.*, 1995b). It is therefore suggested that thrombopoietin and erythropoietin share a common genetic ancestry which may reflect common cellular origins and ontogeny of the megakaryocytic and erythroid lineages (Romeo *et al.*, 1990; Tronik-Le Roux *et al.*, 1995).

The genomic locus for thrombopoietin, designated THPO, has been mapped by several groups to human chromosome 3, at 3q26.3-q27 (Foster *et al.*, 1994; Chang *et al.*, 1995b; Gurney *et al.*, 1995a). Rearrangements, translocations and inversions of this general region have been associated with many clinical examples of increased megakaryopoiesis and thrombopoiesis in the context of leukemia or myelodysplasia (Carbonell *et al.*, 1982; Pintado *et al.*, 1985; Pinto *et al.*, 1985; Bernstein *et al.*, 1986; Jenkins *et al.*, 1989). However, despite this tantalising association, several groups have shown that increased expression of thrombopoietin is not the likely cause of thrombocytosis in the 3q21-q26 syndrome (Schnittger *et al.*, 1996). A syntenic region in the mouse genome, *Thpo*, is located on the proximal part of chromosome 16 (Chang *et al.*, 1995a).

Expression and Regulation of Thrombopoietin

There are four isoforms of thrombopoietin present in humans, pigs, dogs and mice (Chang *et al.*, 1995b; Gurney *et al.*, 1995a; Stoffel *et al.*, 1996). The alternative isoforms in each species differ by deletion of four amino acids, a large deletion/frameshift, or both. In tissues expressing thrombopoietin, mRNAs coding for the four isoforms are equally abundant. However, the main form of thrombopoietin is responsible for essentially all *in vivo* biological activity, since other isoforms are not secreted from cells (Gurney *et al.*, 1995a). The original suggestion that differential expression of isoforms may provide a further level of control appears not to be so (Stoffel *et al.*, 1996).

S1 nuclease mapping and 5′-rapid amplification of cDNA ends by PCR have demonstrated two different, functional thrombopoietin promoters (Sohma *et al.*, 1994; Chang *et al.*, 1995b). To date transcription factors that control thrombopoietin expression have not been thoroughly characterised, although Ets and GATA binding site motifs are present in upstream regions (see below).

Human thrombopoietin is produced predominately by cells of the proximal convoluted tubules in the kidney and peri-portal hepatocytes (Sungaran *et al.*, 1997). Considerably less thrombopoietin is produced by bone marrow stromal cells and splenic cells. In contrast to the predominately renal source of erythropoietin, the wider distribution of sites capable of producing thrombopoietin may explain why anephric patients typically suffer significant anemia but less thrombocytopenia (Erslev *et al.*, 1980; Michalak *et al.*, 1991).

As mentioned above, it has been known for thirty years, and recently confirmed, that platelets can remove Tpo from the circulation (de Gabriele *et al.*, 1967; Kuter *et al.*, 1995). Consequently, the combination of soluble c-mpl, platelet and megakaryocyte mass probably determines the serum thrombopoietin concentration by means of a direct negative feedback loop (Shivdasani *et al.*, 1995; Emmons *et al.*, 1996; Stoffel *et al.*, 1996). In such a system, thrombopoietin would be constitutively produced. A similar regulatory mechanism may occur in the myeloid lineage during states of leucocytosis when polymorphonuclear leucocytes and macrophages may sequester HGF's important for their production: such as GM-CSF, G-CSF and M-CSF (Bartocci *et al.*, 1987; Layton *et al.*, 1989; Metcalf *et al.*, 1993). However, in contrast to the case with thrombopoietin, this negative feedback mechanism for GM-CSF is apparently irrelevant in steady state myelopoiesis (Robb *et al.*, 1995). Still, a trivial element of bone marrow microenvironmental control may be through modulation of thrombopoietin mRNA levels (McCarty *et al.*, 1995; Fielder *et al.*, 1996; Sungaran *et al.*, 1997). Most convincing is the observation that mice heterozygous for the Tpo gene exhibit a considerable reduction in circulating platelet mass — rather than compensating for reduced gene dosage by means of increased expression (de Sauvage *et al.*, 1996). Further, Tpo transcription is not increased in thrombocytopenic, c-mpl nullizygous mice. Thus, in the simplest model of platelet homeostasis, thrombopoietin levels (which stimulate megakaryopoiesis) are inversely proportional to the total body, functional c-mpl load (Penny *et al.*, 1966; Kuter, 1997).

Receptor Binding, Signaling Cascade and Target Genes

Thrombopoietin stimulation of cells leads to probable homodimerization of c-mpl mediated through the conserved haemopoietin receptor dimer interface domain (Alexander *et al.*, 1995). There is no current evidence for alternate receptor chains in the thrombopoietin-c-mpl complex (Kaushansky *et al.*, 1995). Thrombopoietin binding to platelets is saturable and Scatchard analysis reflects only a single class of receptors. The reported binding affinity (K_d) of thrombopoietin for c-mpl ranges between 100 and 750pM, depending on the assay method (Broudy *et al.*, 1995; Fielder *et al.*, 1996). Each platelet expresses between 25 and 220 c-mpl receptors and is capable of internalising and degrading bound ligand (Kuter *et al.*, 1995).

A number of known signal transduction molecules are phosphorylated following stimulation of cells with thrombopoietin (Dorsch *et al.*, 1995; Drachman *et al.*, 1995; Mu *et al.*, 1995). Beyond autophosphorylation of c-mpl itself, these include, SHC, Grb2, p145[SHIP], SOS, Vav, p120[c-cbl], p39[crkl], JAK2, Signal Transducer and Activator of Transcription (STAT)1, STAT3 and MGF-STAT5 and probably TYK2 (Bacon *et al.*, 1995; Ezumi *et al.*, 1995; Gurney *et al.*, 1995b; Sattler *et al.*, 1995; Tortolani *et al.*, 1995; Miyakawa *et al.*, 1996; Oda *et al.*, 1996a; Oda *et al.*, 1996c). The tyrosine residue 599 of c-mpl has been shown to play a decisive role in initiating differentiation signals mediated by SHC binding and activation of the Ras signal transduction cascade (Alexander *et al.*, 1996a; Morita *et al.*, 1996).

Stimulation of cells through the thrombopoietin-c-mpl axis and abovementioned signal transduction molecules will eventually lead to activation of transcription factors. Such factors will thereby drive the cellular response within the megakaryocytic lineage — although no single transcription factor is known to be restricted to this lineage. As such, a complex interplay of factors promote specific responses: for example, expression of platelet glycoproteins IIb, Ib and IX described in detail in later chapters. Transcription factors known to be expressed in megakaryocytes include GATA family members (Martin *et al.*, 1990; Romeo *et al.*, 1990; Dorfman *et al.*, 1992), Ets family members (Hromas *et al.*, 1993; Lemarchandel *et al.*, 1993), SCL/TAL-1 (Green *et al.*, 1992; Kallianpur *et al.*, 1994), rbtn2 (Warren *et al.*, 1994), c-fes, c-myb and c-myc (Emilia *et al.*, 1986).

The c-mpl promoter itself binds several transcription factors at low affinity: for example, GATA-1 and members of the Ets family including Ets-1, Fli-1 and Elf-1 (Alexander *et al.*, 1995; Deveaux *et al.*, 1996). The 200bp promoter fragment capable of conferring high-level and specific transcription contains one GATA-1 binding motif at −70, two Ets binding motifs at −25 and −65 and four Sp1 binding motifs at −175, −100, −70 and −15 (Mignotte *et al.*, 1994). The close proximity of GATA and Ets binding motifs in the promoters of the GpIIb and c-mpl genes suggests that such regions are important for regulation of mega-karyocytic differentiation.

BIOLOGICAL FUNCTION

Thrombopoietin is the most important physiological regulator of platelet homeostasis (Kaushansky, 1995). Earlier models of megakaryocyte development included two separate regulatory activities: one to promote the proliferation of committed megakaryocyte pro-genitors, and the other to increase megakaryocyte maturation. However the activity of thrombopoietin appears to subsume both functions (Debili *et al.*, 1995). Whereas scores of papers have been published, constraints of space limit the presentation of thrombopoietin's biological function to the more decisive *in vitro* and *in vivo* studies.

Action of Thrombopoietin *In Vitro*

Thrombopoietin acts at all stages of megakaryopoiesis and thrombopoiesis, consistent with the distribution of its receptor, c-mpl, throughout platelet production from repopulating stem cells and CD34$^+$ cells to megakaryocytes and platelets (Vigon *et al.*, 1992; Zeigler *et al.*, 1994; Berardi *et al.*, 1995; Sitnicka *et al.*, 1996). Thrombopoietin acts directly on single human CD34$^+$, CD34$^+$Thy−1$^+$Lin$^-$ progenitors, CD34$^+$CD61$^+$ and CD34$^+$CD41$^+$ megakaryocyte progenitors to stimulate their survival, proliferation, differentiation and maturation (Debili *et al.*, 1995; Angchaisuksiri *et al.*, 1996; Rasko *et al.*, 1996; Young *et al.*, 1996). In bulk culture of human bone marrow, peripheral or cord blood CD34$^+$ cells, CD34$^+$CD41$^+$ and CD34$^+$CD38$^-$ cells, thrombopoietin stimulates increases in megakaryo-cyte ploidy and expression of megakaryocyte surface markers (Bartley *et al.*, 1994; de Sauvage *et al.*, 1994; Kaushansky *et al.*, 1994).

Although originally described as a "lineage-restricted" megakaryopoietic HGF, thrombo-poietin in fact stimulates multi-potential progenitors and cells of the erythroid lineage. Stimulation of human CD34$^+$ bone marrow or cord blood cells with thrombopoietin and erythropoietin (but not thrombopoietin alone) promoted proliferation of primitive (BFU-E) and mature (CFU-E) erythroid progenitor cells (Kobayashi *et al.*, 1995; Papayannopoulou *et al.*, 1996). Also, thrombopoietin stimulated the proliferation and differentiation of erythroid-colony forming cells obtained from murine yolk sac and embryonic stem cells (Era *et al.*, 1997). Perhaps the best evidence for thrombopoietin's erythroid lineage promoting activity comes from experiments in which the combination of thrombopoietin and Stem Cell Factor (SCF) was shown to stimulate the proliferation and maturation of embryonic liver or yolk-sac cells obtained from Epo-receptor nullizygous mice (Kieran *et al.*, 1996). That is, thrombopoietin could sustain and stimulate primitive erythroid progenitor cells incapable of responding to erythropoietin.

Despite c-mpl expression on the target cells, thrombopoietin does not act as a platelet release factor and is even inhibitory to proplatelet formation (Choi *et al.*, 1996). Still, exquisite images of previously neglected, proplatelet-displaying megakaryocytes were obtained by Choi and colleagues following liquid culture of CD34$^+$ cells stimulated by thrombopoietin (Choi *et al.*, 1995b; Choi *et al.*, 1995c). Indeed, these culture conditions led to release of functional platelets into the culture medium (Choi *et al.*, 1995a).

Thrombopoietin does however act to "prime" platelets *ex vivo* for aggregation induced by shear stress and several agonists (Toombs *et al.*, 1995; Harker *et al.*, 1996a; Montrucchio *et al.*, 1996; Oda *et al.*, 1996b; Rasko *et al.*, 1997).

In addition to direct actions, thrombopoietin cooperates with other HGFs in additive and synergistic activities. For example, thrombopoietin acts with SCF, but not IL-6, IL-11, or G-CSF, to support clonal proliferation of highly enriched lineage-marker negative, sca-1$^+$c-kit$^+$ primitive murine cells in serum free conditions (Broudy, *et al.*, 1995; Kobayashi *et al.*, 1996; Ku *et al.*, 1996). In serum containing cultures, thrombopoietin acts synergystically with SCF and/or IL-3 on human CD34$^+$c-kit^{low}CD38^{low} cells to stimulate granulocyte-macrophage, erythroid and, more primitive, mixed colony formation (Kobayashi *et al.*, 1995). Similar results have been demonstrated on quiescent murine cells enriched for long term haemopoietic repopulating ability (Sitnicka *et al.*, 1996). As well as proliferative effects, the survival of Meg-CFC, GM-CFC and BFU-E arising from human CD34$^+$CD61$^+$ cells was prolonged by poly[ethylene-glycol]-conjugated recombinant human MGDF alone and in combination with SCF (Rasko *et al.*, 1996).

"Knockout" Mouse Models and Overexpression

Substantial information regarding the actions of thrombopoietin has been obtained through detailed analysis of nullizygous mouse models. Tpo- and c-mpl-nullizygous mice exhibit very similar phenotypes. Both nullizygous mouse strains exhibit 10–15% of the normal peripheral platelet concentration, suggesting that other molecules cannot substitute for, or complement, the thrombopoietin/c-mpl axis (de Sauvage *et al.*, 1996). Principal control of megakaryopoiesis therefore is non-redundant. Nevertheless, HGFs with known megakaryopoietic and thrombopoietic activity (IL-6, IL-11, and SCF) can stimulate a minor increment in platelet concentration in c-mpl nullizygous mice (Gurney *et al.*, 1994). The 10–15% population of functional platelets in the Tpo- and c-mpl-nullizygous mice are likely produced in response to any one or more of IL-1β, IL-3, IL-6, IL-11, Epo, GM-CSF, Leukemia Inhibitory Factor, Flk2/Flt3 ligand and SCF. Mice rendered nullizygous for combinations of these HGFs or their receptors will hopefully settle such questions.

It is important to note that the 10–15% platelet concentration is sufficient to protect these nullizygous mice against lethal haemorrhage during delivery and throughout life (as is usually the case in humans with platelet concentrations of less than 15% of normal (Bussel *et al.*, 1991)). It remains to be seen whether these mice are able to respond adequately to conditions which stress megakaryopoiesis — such as acute haemorrhage. Other changes in both Tpo- and c-mpl- nullizygous mice include a 1.5–2-fold increase in platelet volume and reduced numbers, size and ploidy of megakaryocytes in the spleen and bone marrow. Indeed, increases in mean platelet volume have been previously noted in response to marked thrombocytopenic states (Martin *et al.*, 1983; Corash *et al.*, 1990).

Myeloid and erythroid peripheral blood cells are within normal ranges in both thrombopoietin and c-mpl "knockout" models (Alexander *et al.*, 1996b; Carver-Moore *et al.*, 1996). Surprisingly, erythroid, myeloid and multi-potential progenitors were reduced in the bone marrow and spleen of adult and neonatal "knockout" mice, as well as the expected lack of megakaryocytic precursors. Indeed, repopulating stem cells and their immediate progeny were profoundly reduced in c-mpl "knockout" mice (Kimura *et al.*,). No differences were detected in the progenitor levels of multiple lineages from fetal liver cells of c-mpl nullizygous mice compared to wild-type and heterozygous littermates (Alexander *et al.*, 1996b). Accordingly, c-mpl is not required for early fetal haemopoiesis, but is essential for the maintenance of normal progenitor levels by birth. These observations nevertheless attest to thrombopoietin's role in stimulating primitive haemopoiesis and raise the prospect of therapeutic efficacy beyond the megakaryocytic compartment (see below).

Abnormalities were generally concordant in Tpo- and c-mpl-nullizygous mice compared to their normal littermates. However some changes were discordant in mice lacking only one allele. Heterozygous Tpo (+/−) mice exhibited certain intermediate aberrations in contrast to the, otherwise normal, heterozygous c-mpl (+/−) mice. For example, heterozygous Tpo (+/−) mice had approximately two thirds of normal platelet and bone marrow megakaryocyte levels (de Sauvage *et al.*, 1996). The implications of this partial dosage phenomenon are discussed above in the section "Expression and Regulation of Thrombopoietin".

A model of constitutive thrombopoietin overexpression has been developed. Lethally irradiated recipient mice were reconstituted with syngeneic bone marrow cells which had been infected with a retrovirus engineered to produce excess thrombopoietin (Yan *et al.*, 1995). Platelet concentration was elevated at least three-fold above control-transplanted mice for over 17 months. Other changes in thrombopoietin-overexpressing mice include: myelofibrosis and osteosclerosis, probably due to elevation of fibrosis-inducing HGFs; decreased bone marrow haemopoietic function, with resulting extramedullary haemopoiesis; and elevated peripheral blood progenitor cells (Yan *et al.*, 1996). The general significance of this model of myelofibrosis remains to be demonstrated, perhaps through creation of a transgenic inducible-thrombopoietin mouse strain.

Therapeutic Potential and *In Vivo* Actions of Thrombopoietin

Two proprietary preparations of thrombopoietin are currently being evaluated for their clinical utility. Scientists at Amgen Inc. generated a recombinant protein that includes the receptor-binding domain but lacks the carbohydrate-rich carboxy-terminus. This truncated mpl-ligand has been conjugated to poly[ethylene-glycol] and is known as pegylated recombinant human megakaryocyte growth and development factor (hereafter "PEG-rHuMGDF"). Derivitisation with PEG greatly improves stability *in vivo* and potency (Hokom *et al.*, 1995; Foster *et al.*, 1997). Scientists at Genentech Inc. generated a full-length recombinant human thrombopoietin (hereafter "rhTpo"). The actions of PEG-rhuMGDF and/or rhTpo have been examined in normal mice (Bartley *et al.*, 1994; de Sauvage *et al.*, 1994; Lok *et al.*, 1994), dogs (Peng *et al.*, 1996), rhesus monkeys (Farese *et al.*, 1995), baboons (Harker *et al.*, 1996b) and humans (Basser *et al.*, 1996; Fanucchi *et al.*, 1997). Although the kinetics of the platelet increment were quite different, all species demonstrated a gratifying dose-response to injections of clinical formulations of thrombopoietin. Indeed, a single intravenous injection into mice of PEG-recombinant murine MGDF resulted in increases in platelet count (3–5-fold), megakaryocyte frequency (10-fold), megakaryocyte size (1.3-fold) and megakaryocyte modal ploidy (from 16N to 64N) (Arnold *et al.*, 1997).

In the bone marrow of nonhuman primates and patients with cancer, increases were demonstrated in megakaryocyte number (mean 1.8–fold in the highest human dose cohort), size, nuclear lobulation and ploidy (Farese *et al.*, 1995; Rasko *et al.*, 1997). Other lineages were unchanged in the bone marrow. An example of bone marrow trephines taken prior to and after 8 doses of PEG-rHuMGDF is shown in Figure 1.3. At supra-therapeutic doses of PEG-rHuMGDF, a rapidly reversible increase in marrow reticulin has been shown in mice (Ulich *et al.*, 1996).

The *in vivo* effects of thrombopoietins are lineage-dominant. Administration of pegylated- or unpegylated-rHuMGDF in several species induced negligible, if any, alterations in peripheral myeloid or erythroid cell numbers (Farese *et al.*, 1995; Basser *et al.*, 1996; Ulich

day 0   **day 8**

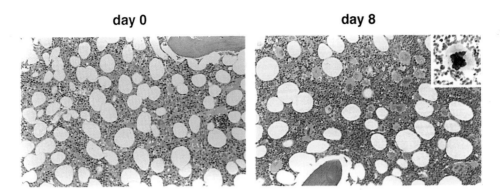

Figure 1.3   Bone Marrow Changes Following PEG-rHuMGDF. A trephine biopsy of bone marrow was obtained from a patient prior to (day 0) and following 8 daily doses (day 8) of PEG-rhuMGDF (0.3μg/kg/day). Samples were decalcified and stained using haematoxylin and eosin. An increase in megakaryocyte number was demonstrated. The inset shows an enlarged megakaryocyte exhibiting hyperlobulation of the nucleus. Original magnification X10 objective.

et al., 1996). However, in contrast to results in pre-clinical models, PEG-rHuMGDF mobilised haemopoietic progenitors of erythroid, myeloid and megakaryocytic lineages into the peripheral blood of untreated cancer patients (Rasko *et al.*, 1997). In this way, primitive cells may be harvested from the peripheral blood for use in transplantation settings. Indeed, if the initial report is supported by further favorable results, the combination of PEG-rHuMGDF and G-CSF (Filgrastim) may replace G-CSF alone as the preferred means by which stem and progenitor cells are mobilised into the peripheral blood for collection (Basser *et al.*, 1997). Such a prediction is strengthened by the reported safety profile of the combined agents, their different mobilisation kinetics (providing a rationale for potential synergy), and the 100- to 1000-fold increase in peripheral blood progenitor cells when used after chemotherapy (Basser *et al.*, 1997).

Two groups have reported the results of Phase I/II randomised trials of PEG-rHuMGDF in cancer patients before and after chemotherapy (Basser *et al.*, 1997; Fanucchi *et al.*, 1997). A dose-related recovery from thrombocytopenia after intensive chemotherapy was demonstrated (Basser *et al.*, 1997). The drug was well tolerated and there were no significant differences in the frequencies of adverse events between the PEG-rHuMGDF-treated and placebo arms. Patients were tested for the development of antibodies to PEG-rHuMGDF, but none had seroconverted. Both in humans and non-human primates, there was no evidence of platelet activation following administration of PEG-rHuMGDF (Harker *et al.*, 1996b; O'Malley *et al.*, 1996; Rasko *et al.*, 1997). However, the two cases each of superficial thrombophlebitis and deep venous thrombosis complicated by pulmonary embolism which occurred in patients receiving PEG-rhuMGDF were probably attributable to a paraneoplastic state. As such, clinical use of thrombopoietins is being undertaken with appropriate caution (Basser *et al.*, 1996; Fanucchi *et al.*, 1997).

A current major impasse in the development of dose-intensive therapies for cancer is the problem of thrombocytopenia (Schick, 1994; Gibbs *et al.*, 1995; Kaushansky, 1996). Consequently, most attention has been focused on the use of rhTpo and PEG-rHuMGDF in the context of myelosupression following chemo-radiotherapy. In both mouse and non-human primate models of dose-intensive chemo-radiotherapy, rhTpo and PEG-rHuMGDF ameliorated or abrogated the duration and/or nadir of neutropenia, anemia and thrombocytopenia as well as reducing mortality (Hokom *et al.*, 1995; Farese *et al.*, 1996; Grossmann *et al.*, 1996b; Ulich *et al.*, 1996). Other agents with thrombopoietic activity in the context of myeloablation, including IL-3, IL-6 and LIF, are not as efficacious in primates as thrombo-poietins and their use in humans is complicated by undesirable side effects (Farese *et al.*, 1994; MacVittie *et al.*, 1994). Clinical Phase II/III trials of PEG-rHuMGDF are currently being performed and early indications are encouraging (Nimer, 1997). In addition there are promising data to support the co-administration of G-CSF and PEG-rhuMGDF or rhTpo alone to ameliorate or protect against myeloablation (Farese *et al.*, 1996; Grossmann *et al.*, 1996a; Basser *et al.*, 1997).

Beyond therapy-induced thrombocytopenias, a large number of diverse thrombocyto-penic states may benefit from the clinical application of thrombopoietins. For example, the serum concentration of thrombopoietin in a patient with cyclic thrombocytopenia has been shown to inversely fluctuate with her platelet count (Oh *et al.*, 1996). If other related con-genital thrombocytopenias exhibit a similar reciprocal relationship, then therapeutic trials will be warranted. Indeed, a minor elevation of endogenous serum thrombopoietin concen-trations in patients with immune thrombocytopenic purpura (ITP) compared to those with aplastic anemia suggests the possibility that pharmacological doses of thrombopoietin may be efficacious in treating a relative thrombopoietin deficiency in ITP (Kosugi *et al.*, 1996). The possible need for long-term treatments in these diseases may benefit from current developments in "gene therapy" delivery systems (Thompson, 1995). Finally, if the safety profile of PEG-rHuMGDF and rhTpo are confirmed, it may be that normal platelet donors may receive these agents in the future to increase the frequency, quantity and quality of their platelet donations.

Acknowledgements

The authors are grateful for continued support from colleagues at The Walter and Eliza Hall Institue of Medical Research, The Rotary Bone Marrow Research Laboratory, The Centre for Developmental Cancer Therapeutics, The Royal Melbourne Hospital and Amgen, (Kew, Australia and Thousand Oaks, USA).

References

Alexander, W.S. and Dunn, A.R. (1995). Structure and transcription of the genomic locus encoding murine c-Mpl, a receptor for thrombopoietin. *Oncogene*, **10**:795–803.

Alexander, W.S., Maurer, A.B., Novak, U. and Harrison-Smith, M. (1996a). Tyrosine-599 Of The C-Mpl Recep-tor Is Required For Shc Phosphorylation And The Induction Of Cellular Differentiation. *EMBO J.*, **15**:6531–6540.

Alexander, W.S., Metcalf, D. and Dunn, A.R. (1995). Point mutations within a dimer interface homology domain of c-Mpl induce constitutive receptor activity and tumorigenicity. *EMBO J.*, **14**:5569–5578.

Alexander, W.S., Roberts, A.W., Nicola, N.A., Li, R. and Metcalf, D. (1996b). Deficiencies in progenitor cells of multiple hematopoietic lineages and defective megakaryocytopoiesis in mice lacking the thrombopoietic receptor c-Mpl. *Blood*, **87**:2162–2170.

Angchaisuksiri, P., Carlson, P.L. and Dessypris, E.N. (1996). Effects of recombinant human thrombopoietin on megakaryocyte colony formation and megakaryocyte ploidy by human CD34[+] cells in a serum-free system. *Br. J. Haematol.*, **93**:13–17.

Arnold, J.T., Daw, N.C., Stenberg, P.E., Jayawardene, D., Srivastava, D.K. and Jackson, C.W. (1997). A Single Injection of Pegylated Megakaryocyte Growth and Development Factor (MGDF) Into Mice is Suf-ficient to Produce a Profound Stimulation of Megakaryocyte Frequency, Size and Ploidization. *Blood*, **89**:823–833.

Bacon, C.M., Tortolani, P.J., Shimosaka, A., Rees, R.C., Longo, D.L. and O'Shea, J.J. (1995). Thrombopoietin (TPO) induces tyrosine phosphorylation and activation of STAT5 and STAT3. *FEBS Lett.*, **370**:63–68.

Bartley, T.D., Bogenberger, J., Hunt, P., Li, Y.S., Lu, H.S. and Martin, F. (1994). Identification and cloning of a megakaryocyte growth and development factor that is a ligand for the cytokine receptor Mpl. *Cell*, **77**:1117–1124.

Bartocci, A., Mastrogiannis, D.S., Migliorati, G., Stockert, R.J., Wolkoff, A.W. and Stanley, E.R. (1987). Macrophages specifically regulate the concentration of their own growth factor in the circulation. *Proc. Natl. Acad. Sci., USA*, **84**:6179–6183.

Basser, R.L., Rasko, J.E.J., Clarke, K., Cebon, J., Green, M.D., Grigg, A.P. *et al.* (1997). Randomised, Blinded, Placebo-Controlled Phase 1 Trial of Pegylated Recombinant Human Megakaryocyte Growth and Development Factor (PEG-rHuMGDF) with Filgrastim after Dose Intensive Chemotherapy in Patients with Advanced Cancer. *Blood*, **89**:3118–28.

Basser, R.L., Rasko, J.E.J., Clarke, K., Cebon, J., Green, M.D., Hussein, S. *et al.* (1996). Thrombopoietic Effects Of Pegylated Recombinant Human Megakaryocyte Growth And Development Factor (PEG-rhuMGDF) In Patients With Advanced Cancer. *Lancet*, **348**:1279–1281.

Bazan, J.F. (1990). Haemopoietic receptors and helical cytokines. *Immunol. Today*, **11**:350–354.

Behnke, O. (1968). An electron microscope study of the megacaryocyte of the rat bone marrow. I. The development of the demarcation membrane system and the platelet surface coat. *J. Ultrastruct. Res.*, **24**:412–433.

Behnke, O. (1969). An electron microscope study of the rat megacaryocyte. II. Some aspects of platelet release and microtubules. *J. Ultrastruct. Res.*, **26**:111–129.

Berardi, A.C., Wang, A.L., Levine, J.D., Lopez, P. and Scadden, D.T. (1995). Functional isolation and characterization of human hematopoietic stem cells. *Science*, **267**:104–108.

Bernstein, R., Bagg, A., Pinto, M., Lewis, D. and Mendelow, B. (1986). Chromosome 3q21 abnormalities associated with hyperactive thrombopoiesis in acute blastic transformation of chronic myeloid leukemia. *Blood*, **68**:652–657.

Breton-Gorius, J. and Reyes, F. (1976). Ultrastructure of human bone marrow cell maturation. *Int. Rev. Cytol.*, **46**:251–321.

Briddell, R.A., Brandt, J.E., Straneva, J.E., Srour, E.F. and Hoffman, R. (1989). Characterization of the human burst-forming unit-megakaryocyte. *Blood*, **74**:145–151.

Broudy, V.C., Lin, N., Fox, N., Atkins, H., Iscove, N. and Kaushansky, K. (1995). Hematopoietic cells display high affinity receptors for thrombopoietin. *Blood*, **86**:593a (abstract 2361).

Broudy, V.C., Lin, N.L. and Kaushansky, K. (1995). Thrombopoietin (c-mpl ligand) acts synergistically with erythropoietin, stem cell factor, and interleukin-11 to enhance murine megakaryocyte colony growth and increases megakaryocyte ploidy *in vitro*. *Blood*, **85**:1719–1726.

Bruno, E., Murray, L.J., DiGiusto, R., Mandich, D., Tsukamoto, A. and Hoffman, R. (1996). Detection of a primitive megakaryocyte progenitor cell in human fetal bone marrow. *Exp. Hematol.*, **24**:552–558.

Bussel, J., Kaplan, C. and McFarland, J. (1991). Recommendations for the evaluation and treatment of neonatal autoimmune and alloimmune thrombocytopenia. *Thromb. Haemost.*, **65**:631–634.

Carbonell, F., Hoelzer, D., Thiel, E. and Bartl, R. (1982). Ph1-positive CML associated with megakaryocytic hyperplasia and thrombocythemia and an abnormality of chromosome no. 3. *Cancer Genet. Cytogenet.*, **6**:153–161.

Carver-Moore, K., Broxmeyer, H.E., Luoh, S.M., Cooper, S., Peng, J., Burstein, S.A. *et al.* (1996). Low levels of erythroid and myeloid progenitors in thrombopoietin-and c-mpl-deficient mice. *Blood*, **88**:803–808.

Chang, M.S., Hsu, R.Y., McNinch, J., Copeland, N.G. and Jenkins, N.A. (1995a). The gene for murine megakaryocyte growth and development factor (thrombopoietin, Thpo) is located on mouse chromosome 16. *Genomics*, **26**:636–637.

Chang, M.S., McNinch, J., Basu, R., Shutter, J., Hsu, R.Y., Perkins, C. *et al.* (1995b). Cloning and characterization of the human megakaryocyte growth and development factor (MGDF) gene. *J. Biol. Chem.*, **270**:511–514.

Choi, E.S., Hokom, M., Bartley, T., Li, Y.S., Ohashi, H., Kato, T. *et al.* (1995a). Recombinant human megakaryocyte growth and development factor (rHuMGDF), a ligand for c-Mpl, produces functional human platelets *in vitro*. *Stem Cells*, **13**:317–322.

Choi, E.S., Hokom, M.M., Chen, J.L., Skrine, J., Faust, J., Nichol, J. *et al.* (1996). The Role Of Megakaryocyte Growth And Development Factor In Terminal Stages Of Thrombopoiesis. *Brit. J. Haematol.*, **95**:227–233.

Choi, E.S., Hokom, M.M., Nichol, J.L., Hornkohl, A. and Hunt, P. (1995b). Functional human platelet generation *in vitro* and regulation of cytoplasmic process formation. *Comptes Rendus de l Academie des Sciences — Serie Iii, Sciences de la Vie*, **318**:387–393.

Choi, E.S., Nichol, J.L., Hokom, M.M., Hornkohl, A.C. and Hunt, P. (1995c). Platelets generated *in vitro* from proplatelet-displaying human megakaryocytes are functional. *Blood*, **85**:402–413.

Corash, L., Mok, Y., Levin, J. and Baker, G. (1990). Regulation of platelet heterogeneity: effects of thrombocytopenia on platelet volume and density. *Exp. Hematol.*, **18**:205–212.

Courtois, G., Benit, L., Mikaeloff, Y., Pauchard, M., Charon, M., Varlet, P. *et al.* (1995). Constitutive activation of a variant of the env-mpl oncogene product by disulfide-linked homodimerization. *J. Virol.*, **69**:2794–2800.

de Gabriele, G. and Pennington, D.G. (1967). Regulation of platelet production: Thrombopoietin. *Br. J. Haematol.*, **13**:210–215.

de Sauvage, F.J., Carver-Moore, K., Luoh, S.M., Ryan, A., Dowd, M., Eaton, D.L. *et al.* (1996). Physiological regulation of early and late stages of megakaryocytopoiesis by thrombopoietin. *J. Exp. Med.*, **183**:651–656.

de Sauvage, F.J., Hass, P.E., Spencer, S.D., Malloy, B.E., Gurney, A.L., Spencer, S.A. *et al.* (1994). Stimulation of megakaryocytopoiesis and thrombopoeisis by the c-Mpl ligand. *Nature*, **369**:533–538.

Debili, N., Issaad, C., Masse, J.M., Guichard, J., Katz, A., Breton-Gorius, J. *et al.* (1992). Expression of CD34 and platelet glycoproteins during human megakaryocytic differentiation. *Blood*, **80**:3022–3035.

Debili, N., Wendling, F., Katz, A., Guichard, J., Breton-Gorius, J., Hunt, P. *et al.* (1995). The Mpl-ligand or thrombopoietin or megakaryocyte growth and differentiative factor has both direct proliferative and differentiative activities on human megakaryocyte progenitors. *Blood*, **86**:2516–2525.

Deveaux, S., Filipe, A., Lemarchandel, V., Ghysdael, J., Romeo, P.H. and Mignotte, V. (1996). Analysis of the thrombopoietin receptor (MPL) promoter implicates GATA and Ets proteins in the coregulation of megakaryocyte-specific genes. *Blood*, **87**:4678–4685.

Dorfman, D.M., Wilson, D.B., Bruns, G.A. and Orkin, S.H. (1992). Human transcription factor GATA-2. Evidence for regulation of preproendothelin-1 gene expression in endothelial cells. *J. Biol. Chem.*, **267**:1279–1285.

Dorsch, M., Fan, P.D., Bogenberger, J. and Goff, S.P. (1995). TPO and IL-3 induce overlapping but distinct protein tyrosine phosphorylation in a myeloid precursor cell line. *Biochem. Biophys. Res. Comm.*, **214**:424–431.

Drachman, J.G., Griffin, J.D. and Kaushansky, K. (1995). The c-Mpl ligand (thrombopoietin) stimulates tyrosine phosphorylation of Jak2, Shc, and c-Mpl. *Journal of Biological Chemistry*, **270**:4979–4982.

Ebbe, S. (1976). Biology of megakaryocytes. *Prog. Hemostasis Thromb.*, **3**:211.

Emilia, G., Donelli, A., Ferrari, S., Torelli, U., Selleri, L., Zucchini, P. *et al.* (1986). Cellular levels of mRNA from c-myc, c-myb and c-fes onc-genes in normal myeloid and erythroid precursors of human bone marrow: an *in situ* hybridization study. *Br. J. Haematol.*, **62**:287–292.

Emmons, R.V., Reid, D.M., Cohen, R.L., Meng, G., Young, N.S., Dunbar, C.E. *et al.* (1996). Human thrombopoietin levels are high when thrombocytopenia is due to megakaryocyte deficiency and low when due to increased platelet destruction. *Blood*, **87**:4068–4071.

Era, T., Takahashi, T., Sakai, K., Kawamura, K. and Nakano, T. (1997). Thrombopoietin Enhances Proliferation and Differentiation of Murine Yolk Sac Erythroid Progenitors. *Blood*, **89**:1207–1213.

Erickson-Miller, C.L., Ji, H., Parchment, R.E. and Murphy, M. J., Jr. (1993). Megakaryocyte colony-stimulating factor (Meg-CSF) is a unique cytokine specific for the megakaryocyte lineage. *Brit. J. Haematol.*, **84**:197–203.

Erslev, A.J., Caro, J., Kansu, E. and Silver, R. (1980). Renal and extrarenal erythropoietin production in anaemic rats. *Br. J. Haematol.*, **45**:65–72.

Ezumi, Y., Takayama, H. and Okuma, M. (1995). Thrombopoietin, c-Mpl ligand, induces tyrosine phosphorylation of Tyk2, JAK2, and STAT3, and enhances agonists-induced aggregation in platelets *in vitro*. *FEBS Lett.*, **374**:48–52.

Fanucchi, M., Glaspy, J., Crawford, J., Garst, J., Figlin, R., Sheridan, W. *et al.* (1997). Effects of Polyethylene Glycol-Conjugated Recombinant Megakaryocyte Growth And Development Factor On Platelet Counts After Chemotherapy For Lung Cancer. *New Eng. J. Med.*, **336**:404–409.

Farese, A.M., Hunt, P., Boone, T. and MacVittie, T.J. (1995). Recombinant human megakaryocyte growth and development factor stimulates thrombocytopoiesis in normal nonhuman primates. *Blood*, **86**:54–59.

Farese, A.M., Hunt, P., Grab, L.B. and MacVittie, T.J. (1996). Combined administration of recombinant human megakaryocyte growth and development factor and granulocyte colony-stimulating factor enhances multilineage hematopoietic reconstitution in nonhuman primates after radiation-induced marrow aplasia. *J. Clin. Invest.*, **97**:2145–2151.

Farese, A.M., Myers, L.A. and MacVittie, T.J. (1994). Therapeutic efficacy of recombinant human leukemia inhibitory factor in a primate model of radiation-induced marrow aplasia. *Blood*, **84**:3675–3678.

Fielder, P.J., Gurney, A.L., Stefanich, E., Marian, M., Moore, M.W., Carver-Moore, K. *et al.* (1996). Regulation of thrombopoietin levels by c-mpl-mediated binding to platelets. *Blood*, **87**:2154–2161.

Foster, D. and Hunt, P. (1997) In *Thrombopoiesis and Thrombopoietins: Molecular, Cellular, Preclinical, and Clinical Biology*, eds. D.J. Kuter, P. Hunt, W. Sheridan and D. Zucker-Franklin, pp. 203–214. Totowa, NJ: Humana Press.

Foster, D.C., Sprecher, C.A., Grant, F.J., Kramer, J.M., Kuijper, J.L., Holly, R.D. *et al.* (1994). Human thrombopoietin: gene structure, cDNA sequence, expression, and chromosomal localization. *Proc. Natl. Acad. Sci., USA*, **91**:13023–13027.

Gibbs, J., Green, M. and Basser, R. (1995). Thrombocytopenia: New Approaches To Resolve A Major Problem In Chemotherapy. *The Cancer J.*, **8**:255–259.

Green, A.R., Lints, T., Visvader, J., Harvey, R. and Begley, C.G. (1992). SCL is coexpressed with GATA-1 in hemopoietic cells but is also expressed in developing brain. *Oncogene*, **7**:653–660.

Grossmann, A., Lenox, J., Deisher, T.A., Ren, H.P., Humes, J.M., Kaushansky, K. *et al.* (1996a). Synergistic effects of thrombopoietin and granulocyte colony-stimulating factor on neutrophil recovery in myelosuppressed mice. *Blood*, **88**:3363–3370.

Grossmann, A., Lenox, J., Ren, H.P., Humes, J.M., Forstrom, J.W., Kaushansky, K. *et al.* (1996b). Thrombopoietin accelerates platelet, red blood cell, and neutrophil recovery in myelosuppressed mice. *Exp. Hematol.*, **24**:1238–1246.

Gurney, A.L., Carver-Moore, K., de Sauvage, F.J. and Moore, M.W. (1994). Thrombocytopenia in c-mpl-deficient mice. *Science*, **265**:1445–1447.

Gurney, A.L., Kuang, W.J., Xie, M.H., Malloy, B.E., Eaton, D.L. and de Sauvage, F.J. (1995a). Genomic structure, chromosomal localization, and conserved alternative splice forms of thrombopoietin. *Blood*, **85**:981–988.

Gurney, A.L., Wong, S.C., Henzel, W.J. and de Sauvage, F.J. (1995b). Distinct regions of c-Mpl cytoplasmic domain are coupled to the JAK-STAT signal transduction pathway and Shc phosphorylation. *Proc. Natl Acad. Sci., USA*, **92**:5292–5296.

Harker, L.A. and Finch, C.A. (1969). Thrombokinetics in man. *J. Clin. Invest.*, **48**:963–974.

Harker, L.A., Hunt, P., Marzec, U.M., Kelly, A.B., Tomer, A., Hanson, S.R. *et al.* (1996a). Regulation of platelet production and function by megakaryocyte growth and development factor in nonhuman primates. *Blood*, **87**:1833–1844.

Harker, L.A., Marzec, U.M., Hunt, P., Kelly, A.B., Tomer, A., Cheung, E. *et al.* (1996b). Dose-response effects of pegylated human megakaryocyte growth and development factor on platelet production and function in nonhuman primates. *Blood*, **88**:511–521.

Hoffman, R. (1989). Regulation of megakaryocytopoiesis. *Blood*, **74**:1196–1212.

Hoffman, R.C., Andersen, H., Walker, K., Krakover, J.D., Patel, S., Stamm, M.R. *et al.* (1996). Peptide, Disulfide, And Glycosylation Mapping Of Recombinant Human Thrombopoietin From Ser1 To Arg246. *Biochem.*, **35**:14849–14861.

Hokom, M.M., Lacey, D., Kinstler, O.B., Choi, E., Kaufman, S., Faust, J. *et al.* (1995). Pegylated megakaryocyte growth and development factor abrogates the lethal thrombocytopenia associated with carboplatin and irradiation in mice. *Blood*, **86**:4486–4492.

Hromas, R., Orazi, A., Neiman, R.S., Maki, R., Van Beveran, C., Moore, J. *et al.* (1993). Hematopoietic lineage- and stage-restricted expression of the ETS oncogene family member PU.1. *Blood*, **82**:2998–3004.

Hunt, P., Li, Y.S., Nichol, J.L., Hokom, M.M., Bogenberger, J.M., Swift, S.E. *et al.* (1995). Purification and biologic characterization of plasma-derived megakaryocyte growth and development factor. *Blood*, **86**:540–547.

Jackson, C.W. (1973). Cholinesterase as a possible marker for early cells of the megakaryocytic series. *Blood*, **42**:413–421.

Jackson, C.W. (1990). Megakaryocyte endomitosis: a review. *Int. J. Cell Clon.*, **8**:224–226.

Jenkins, R.B., Tefferi, A., Solberg, L.A., Jr. and Dewald, G.W. (1989). Acute leukemia with abnormal thrombo-poiesis and inversions of chromosome 3. *Cancer Genet. Cytogenet.*, **39**:167–179.

Kallianpur, A.R., Jordan, J.E. and Brandt, S.J. (1994). The SCL/TAL-1 gene is expressed in progenitors of both the hematopoietic and vascular systems during embryogenesis. *Blood*, **83**:1200–1208.

Kato, T., Ogami, K., Shimada, Y., Iwamatsu, A., Sohma, Y., Akahori, H. *et al.* (1995). Purification and characterization of thrombopoietin. *J. Biochem.*, **118**:229–236.

Kaushansky, K. (1995). Thrombopoietin: the primary regulator of platelet production. *Blood*, **86**:419–431.

Kaushansky, K. (1996). The thrombocytopenia of cancer. Prospects for effective cytokine therapy. *Hematology — Oncology Clinics of North America*, **10**:431–455.

Kaushansky, K., Broudy, V.C., Lin, N., Jorgensen, M.J., McCarty, J., Fox, N. *et al.* (1995). Thrombopoietin, the Mpl ligand, is essential for full megakaryocyte development. *Proc. Natl. Acad. Sci., USA*, **92**:3234–3238.

Kaushansky, K., Lok, S., Holly, R.D., Broudy, V.C., Lin, N., Bailey, M.C. *et al.* (1994). Promotion of mega-karyocyte progenitor expansion and differentiation by the c-Mpl ligand thrombopoietin. *Nature*, **369**:568–571.

Kelemen, E. (1995). Specific thrombopoietin cloned and sequenced — with personal retrospect and clinical prospects. *Leukemia*, **9**:1–2.

Kelemen, E., Cserháti, I. and Tanos, B. (1958). Demonstration of some properties of human thrombopoietin in thrombocythaemic serum. *Acta Haematol.*, **20**:350–355.

Kieran, M.W., Perkins, A.C., Orkin, S.H. and Zon, L.I. (1996). Thrombopoietin rescues *in vitro* erythroid colony formation from mouse embryos lacking the erythropoietin receptor. *Proc. Natl. Acad. Sci., USA*, **93**:9126–9131.

Kimura, S., Roberts, A.W., Metcalf, D. and Alexander, W.S. (1998). Hematopoietic stem cell deficiencies in mice lacking c-mpl, the receptor for thrombopoietin. *Proc. Natl. Acad. Sci., USA*, **95**: 1195–1200.

Kobayashi, M., Laver, J.H., Kato, T., Miyazaki, H. and Ogawa, M. (1995). Recombinant human thrombopoietin (Mpl ligand) enhances proliferation of erythroid progenitors. *Blood*, **86**:2494–2499.

Kobayashi, M., Laver, J.H., Kato, T., Miyazaki, H. and Ogawa, M. (1996). Thrombopoietin supports proliferation of human primitive hematopoietic cells in synergy with steel factor and/or interleukin-3. *Blood*, **88**:429–436.

Kosugi, S., Kurata, Y., Tomiyama, Y., Tahara, T., Kato, T., Tadokoro, S. *et al.* (1996). Circulating thrombopoietin level in chronic immune thrombocytopenic purpura. *Br. J. Haematol.*, **93**:704–706.

Ku, H., Yonemura, Y., Kaushansky, K. and Ogawa, M. (1996). Thrombopoietin, the ligand for the Mpl receptor, synergizes with steel factor and other early acting cytokines in supporting proliferation of primitive hemato-poietic progenitors of mice. *Blood*, **87**:4544–4551.

Kuter, D.J. (1997) In *Thrombopoiesis and Thrombopoietins: Molecular, Cellular, Preclinical, and Clinical Biology*, eds. D.J. Kuter, P. Hunt, W. Sheridan and D. Zucker-Franklin, pp. 377–395. Totowa, NJ: Humana Press.

Kuter, D.J., Beeler, D.L. and Rosenberg, R.D. (1994). The purification of megapoietin: a physiological regulator of megakaryocyte growth and platelet production. *Proc. Natl. Acad. Sci., USA*, **91**:11104–11108.

Kuter, D.J., Miyazaki, H. and Kato, T. (1997) In *Thrombopoiesis and Thrombopoietins: Molecular, Cellular, Preclinical, and Clinical Biology*, eds. D.J. Kuter, P. Hunt, W. Sheridan and D. Zucker-Franklin, pp. 143–164. Totowa, NJ: Humana Press.

Kuter, D.J. and Rosenberg, R.D. (1995). The reciprocal relationship of thrombopoietin (c-Mpl ligand) to changes in the platelet mass during busulfan-induced thrombocytopenia in the rabbit. *Blood*, **85**:2720–2730.

Layton, J.E., Hockman, H., Sheridan, W.P. and Morstyn, G. (1989). Evidence for a novel *in vivo* control mechanism of granulopoiesis: mature cell-related control of a regulatory growth factor. *Blood*, **74**:1303–1307.

Lemarchandel, V., Ghysdael, J., Mignotte, V., Rahuel, C. and Romeo, P.H. (1993). GATA and Ets cis-acting sequences mediate megakaryocyte-specific expression. *Mol. Cell. Biol.*, **13**:668–676.

Li, B. and Dai, W. (1995). Thrombopoietin and neurotrophins share a common domain [letter]. *Blood*, **86**:1643–1644.

Lok, S., Kaushansky, K., Holly, R.D., Kuijper, J.L., Lofton-Day, C.E., Oort, P.J. *et al.* (1994). Cloning and expression of murine thrombopoietin cDNA and stimulation of platelet production *in vivo*. *Nature*, **369**:565–568.

Long, M.W. (1995). Cyclins and cell division kinases in megakaryocytic endomitosis. *Comptes. Rend. Acad. Sci.*, **318**:649–654.

Long, M.W., Gragowski, L.L., Heffner, C.H. and Boxer, L.A. (1985). Phorbol diesters stimulate the development of an early murine progenitor cell. The burst-forming unit-megakaryocyte. *J. Clin. Invest.*, **76**:431–438.

Long, M.W., Heffner, C.H. and Gragowski, L.L. (1986). *In vitro* differences in responsiveness of early (BFU-Mk) and late (CFU-Mk) murine megakaryocyte progenitor cells. *Prog. Clin. Biol. Res.*, **215**:179–186.

MacVittie, T.J., Farese, A.M., Patchen, M.L. and Myers, L.A. (1994). Therapeutic efficacy of recombinant interleukin-6 (IL-6) alone and combined with recombinant human IL-3 in a nonhuman primate model of high-dose, sublethal radiation-induced marrow aplasia. *Blood*, **84**:2515–2522.

Martin, D.I., Zon, L.I., Mutter, G. and Orkin, S.H. (1990). Expression of an erythroid transcription factor in megakaryocytic and mast cell lineages. *Nature*, **344**:444–447.

Martin, J.F., Trowbridge, E.A., Salmon, G. and Plumb, J. (1983). The biological significance of platelet volume: its relationship to bleeding time, platelet thromboxane B2 production and megakaryocyte nuclear DNA concentration. *Thromb. Res.*, **32**:443–460.

Mazur, E.M., Hoffman, R. and Bruno, E. (1981). Regulation of human megakaryocytopoiesis. An *in vito* analysis. *J. Clin. Invest.*, **68**:733–741.

McCarty, J.M., Sprugel, K.H., Fox, N.E., Sabath, D.E. and Kaushansky, K. (1995). Murine thrombopoietin mRNA levels are modulated by platelet count. *Blood*, **86**:3668–3675.

McDonald, T.P., Andrews, R.B., Clift, R. and Cottrell, M. (1981). Characterization of a thrombocytopoietic-stimulating factor from kidney cell culture medium. *Exp. Hematol.*, **9**:288–296.

Metcalf, D. (1988). *The Molecular Control of Blood Cells*, Harvard University Press, Cambridge, Mass.

Metcalf, D. (1994). Thrombopoietin — at last. *Nature*, **369**:519–520.

Metcalf, D., MacDonald, H.R., Odartchenko, N. and Sordat, B. (1975). Growth of mouse megakaryocyte colonies *in vitro*. *Proc. Natl. Acad, Sci., USA*, **72**:1744–1748.

Metcalf, D. and Nicola, N.A. (1995). *The Hemopoietic Colony-Stimulating Factors: From Biology to Clinical Applications*, Cambridge University Press, Cambridge.

Metcalf, D. and Rasko, J.E.J. (1993). Leukemic transformation of immortalised FDC-P1 cells engrafted in GM-CSF transgenic mice. *Leukemia*, **7**:878–886.

Methia, N., Louache, F., Vainchenker, W. and Wendling, F. (1993). Oligodeoxynucleotides antisense to the proto-oncogene c-*mpl* specifically inhibit *in vitro* megakaryocytopoeisis. *Blood*, **82**:1395–1401.

Michalak, E., Walkowiak, B., Paradowski, M. and Cierniewski, C.S. (1991). The decreased circulating platelet mass and its relation to bleeding time in chronic renal failure. *Thromb. Haemost.*, **65**:11–14.

Mignotte, V., Vigon, I., Boucher de Crevecoeur, E., Romeo, P.H., Lemarchandel, V. and Chretien, S. (1994). Structure and transcription of the human c-mpl gene (MPL). *Genomics*, **20**:5–12.

Minasian, E. and Nicola, N.A. (1992). A review of cytokine structures. *Protein Sequences & Data Analysis*, **5**:57–64.

Miyakawa, Y., Oda, A., Druker, B.J., Miyazaki, H., Handa, M., Ohashi, H. *et al.* (1996). Thrombopoietin induces tyrosine phosphorylation of Stat3 and Stat5 in human blood platelets. *Blood*, **87**:439–446.

Montrucchio, G., Brizzi, M.F., Calosso, G., Marengo, S., Pegoraro, L. and Camussi, G. (1996). Effects of recombinant human megakaryocyte growth and development factor on platelet activation. *Blood*, **87**:2762–2768.

Morita, H., Tahara, T., Matsumoto, A., Kato, T., Miyazaki, H. and Ohashi, H. (1996). Functional Analysis Of The Cytoplasmic Domain Of The Human Mpl Receptor For Tyrosine-Phosphorylation Of The Signaling Molecules, Proliferation And. *FEBS Let.*, **395**:228–234.

Mu, S.X., Xia, M., Elliott, G., Bogenberger, J., Swift, S., Bennett, L. *et al.* (1995). Megakaryocyte growth and development factor and interleukin-3 induce patterns of protein-tyrosine phosphorylation that correlate with dominant differentiation over proliferation of mpl-transfected 32D cells. *Blood*, **86**:4532–4543.

Murray, L.J., Mandich, D., Bruno, E., DiGiusto, R.K., Fu, W.C., Sutherland, D.R. *et al.* (1996). Fetal bone marrow CD34$^+$CD41$^+$ cells are enriched for multipotent hematopoietic progenitors, but not for pluripotent stem cells. *Exp. Hematol.*, **24**:236–245.

Narhi, L.O., Arakawa, T., Aoki, K.H., Elmore, R., Rohde, M.F., Boone, T. *et al.* (1991). The effect of carbohydrate on the structure and stability of erythropoietin. *J. Biol. Chem.*, **266**:23022–23026.

Nimer, S.D. (1997). Platelet Stimulating Agents — Off The Launching Pad. *Nature Medicine*, **3**:154–155.

O'Malley, C.J., Rasko, J.E.J., Basser, R.L., Mcgrath, K.M., Cebon, J., Grigg, A.P. *et al.* (1996). Administration Of Pegylated Recombinant Human Megakaryocyte Growth And Development Factor To Humans Stimulates The Production Of Functional Platelets That Show No Evidence Of *In Vivo* Activation. *Blood*, **88**:3288–3298.

Oda, A., Miyakawa, Y., Druker, B.J., Ishida, A., Ozaki, K., Ohashi, H. *et al.* (1996a). Crkl Is Constitutively Tyrosine Phosphorylated In Platelets From Chronic Myelogenous Leukemia Patients And Inducibly Phosphorylated In Normal Platelets Stimulated By Thrombopoietin. *Blood*, **88**:4304–4313.

Oda, A., Miyakawa, Y., Druker, B.J., Ozaki, K., Yabusaki, K., Shirasawa, Y. *et al.* (1996b). Thrombopoietin primes human platelet aggregation induced by shear stress and by multiple agonists. *Blood*, **87**:4664–4670.

Oda, A., Ozaki, K., Druker, B.J., Miyakawa, Y., Miyazaki, H., Handa, M. *et al.* (1996c). p120c-cbl is present in human blood platelets and is differentially involved in signaling by thrombopoietin and thrombin. *Blood*, **88**:1330–1338.

Oh, H., Nakamura, H., Yokota, A., Asai, T. and Saito, Y. (1996). Serum thrombopoietin levels in cyclic thrombo-cytopenia. *Blood*, **87**:4918.

Onishi, M., Mui, A.L., Morikawa, Y., Cho, L., Kinoshita, S., Nolan, G.P. *et al.* (1996). Identification of an oncogenic form of the thrombopoietin receptor MPL using retrovirus-mediated gene transfer. *Blood*, **88**:1399–1406.

Papayannopoulou, T., Brice, M., Farrer, D. and Kaushansky, K. (1996). Insights into the cellular mechanisms of erythropoietin-thrombopoietin synergy. *Exp. Hematol.*, **24**:660–669.

Paulus, J.M. (1970). DNA metabolism and development of organelles in guinea-pig megakaryocytes: a combined ultrastructural, autoradiographic and cytophotometric study. *Blood*, **35**:298–311.

Peng, J., Friese, P., Wolf, R.F., Harrison, P., Downs, T., Lok, S. *et al.* (1996). Relative reactivity of platelets from thrombopoietin-and interleukin-6-treated dogs. *Blood*, **87**:4158–4163.

Penny, R., Rozenberg, M.C. and Firkin, B.G. (1966). The splenic platelet pool. *Blood*, **27**:1–16.

Pintado, T., Ferro, M.T., San Roman, C., Mayayo, M. and Larana, J.G. (1985). Clinical correlations of the 3q21;q26 cytogenetic anomaly. A leukemic or myelodysplastic syndrome with preserved or increased platelet production and lack of response to cytotoxic drug therapy. *Cancer*, **55**:535–541.

Pinto, M.R., King, M.A., Goss, G.D., Bezwoda, W.R., Fernandes-Costa, F., Mendelow, B. *et al.* (1985). Acute megakaryoblastic leukaemia with 3q inversion and elevated thrombopoietin (TSF): an autocrine role for TSF? *Br. J. Haematol*, **61**:687–694.

Rabellino, E.M., Nachman, R.L., Williams, N., Winchester, R.J. and Ross, G.D. (1979). Human megakaryocytes. I. Characterization of the membrane and cytoplasmic components of isolated marrow megakaryocytes. *J. Exp. Med.*, **149**:1273–1287.

Rasko, J.E.J., Basser, R.L., Boyd, J., Mansfield, R., O'Malley, C., Hussein, S. *et al.* (1997). Multilineage Mobi-lisation of Peripheral Blood Progenitor Cells in Humans Following Administration of PEG-rHuMGDF. *Br. J. Haematol.*, **97**:871–80.

Rasko, J.E.J., Metcalf, D., Gough, N.M. and Begley, C.G. (1995). The cytokine receptor repertoire specifies auto-crine growth factor production in factor-dependent cells. *Exp. Hematol.*, **23**:453–460.

Rasko, J.E.J., O'Flaherty, E. and Begley, C.G. (1997). Thrombopoietin (MGDF) Alone Or In Combination With SCF Promotes The Proliferation And Survival Of Megakaryocyte, Erythroid And Granulocyte/Macrophage Progenitors. *Stem Cells*, **15**:33–42.

Robb, L., Drinkwater, C.C., Metcalf, D., Li, R.L., Kontgen, F., Nicola, N.A. *et al.* (1995). Hematopoietic and lung abnormalities in mice with a null mutation of the common beta subunit of the receptors for granulocyte-macrophage colony-stimulating factor and interleukins 3 and 5. *Proc. Natl. Acad. Sci., USA*, **92**:9565–9569.

Romeo, P.H., Prandini, M.H., Joulin, V., Mignotte, V., Prenant, M., Vainchenker, W. *et al.* (1990). Megakaryo-cytic and erythrocytic lineages share specific transcription factors. *Nature*, **344**:447–449.

Sattler, M., Durstin, M.A., Frank, D.A., Okuda, K., Kaushansky, K., Salgia, R. *et al.* (1995). The thrombopoietin receptor c-MPL activates JAK2 and TYK2 tyrosine kinases. *Experimental Hematology*, **23**:1040–1048.

Schick, B.P. (1994). Hope for treatment of thrombocytopenia. *N. Eng. J. Med.*, **331**:875–876.

Schnittger, S., De Sauvage, F.J., Lepaslier, D. and Fonatsch, C. (1996). Refined Chromosomal Localization Of The Human Thrombopoietin Gene To 3q27-Q28 And Exclusion As The Responsible Gene For Thrombo-cytosis In Patients With Rearrangements Of 3q21 And 3q26. *Leukemia*, **10**:1891–1896.

Shivdasani, R.A., Rosenblatt, M.F., Zucker-Franklin, D., Jackson, C.W., Hunt, P., Saris, C.J. *et al.* (1995). Trans-cription factor NF-E2 is required for platelet formation independent of the actions of thrombopoietin/MGDF in megakaryocyte development. *Cell*, **81**:695–704.

Sitnicka, E., Lin, N., Priestley, G.V., Fox, N., Broudy, V.C., Wolf, N.S. *et al.* (1996). The effect of thrombopoietin on the proliferation and differentiation of murine hematopoietic stem cells. *Blood*, **87**:4998–5005.

Skoda, R.C., Seldin, D.C., Chiang, M.K., Peichel, C.L., Vogt, T.F. and Leder, P. (1993). Murine c-mpl: a member of the hematopoietic growth factor receptor superfamily that transduces a proliferative signal. *EMBO J.*, **12**:2645–2653.

Sohma, Y., Akahori, H., Seki, N., Hori, T., Ogami, K., Kato, T. *et al.* (1994). Molecular cloning and chromosomal localization of the human thrombopoietin gene. *FEBS Letters*, **353**:57–61.

Souyri, M., Vigon, I., Penciolelli, J.F., Heard, J.M., Tambourin, P. and Wendling, F. (1990). A putative truncated cytokine receptor gene transduced by the myeloproliferative leukemia virus immortalizes hematopoietic progenitors. *Cell*, **63**:1137–1147.

Spector, B. (1961). *In vivo* transfer of a thrombopoietic factor. *Proc. Soc. Exp. Biol. Med.*, **108**:146–149.

Stoffel, R., Wiestner, A. and Skoda, R.C. (1996). Thrombopoietin in thrombocytopenic mice: evidence against regulation at the mRNA level and for a direct regulatory role of platelets. *Blood*, **87**:567–573.

Sungaran, R., Markovic, B. and Chong, B.H. (1997). Localization And Regulation Of Thrombopoietin mRNA Expression In Human Kidney, Liver, Bone Marrow, And Spleen Using *In Situ* Hybridization. *Blood*, **89**:101–107.

Tavassoli, M. (1980). Megakaryocyte-platelet axis and the process of platelet formation and release. *Blood*, **55**:537–545.

Tayrien, G. and Rosenberg, R.D. (1987). Purification and properties of a megakaryocyte stimulatory factor present both in the serum-free conditioned medium of human embryonic kidney cells and in thrombocytopenic plasma. *J. Biol. Chem.*, **262**:3262–3268.

Thean, L.E., Hodgson, G.S., Bertoncello, I. and Radley, J.M. (1983). Characterization of megakaryocyte spleen colony-forming units by response to 5-fluorouracil and by unit gravity sedimentation. *Blood*, **62**:896–901.

Thiery, J.P. and Bessis, M. (1956). Mécanisme de la plaquettogénèse: étude *in vitro* par la microcinématographie. *Rev. Hematol.*, **11**:162.

Thompson, A.R. (1995). Progress towards gene therapy for the hemophilias. *Thromb. Haemostas.*, **74**:45–51.

Toombs, C.F., Young, C.H., Glaspy, J.A. and Varnum, B.C. (1995). Megakaryocyte growth and development factor (MGDF) moderately enhances *in-vitro* platelet aggregation. *Thromb. Res.*, **80**:23–33.

Tortolani, P.J., Johnston, J.A., Bacon, C.M., McVicar, D.W., Shimosaka, A., Linnekin, D. *et al.* (1995). Thrombopoietin induces tyrosine phosphorylation and activation of the Janus kinase, JAK2. *Blood*, **85**:3444–3451.

Tronik-Le Roux, D., Roullot, V., Schweitzer, A., Berthier, R. and Marguerie, G. (1995). Suppression of erythromegakaryocytopoiesis and the induction of reversible thrombocytopenia in mice transgenic for the thymidine kinase gene targeted by the platelet glycoprotein alpha IIb promoter. *J. Exp. Med.*, **181**:2141–2151.

Ulich, T.R., del Castillo, J., Senaldi, G., Kinstler, O., Yin, S., Kaufman, S. *et al.* (1996). Systemic hematologic effects of PEG-rHuMGDF-induced megakaryocyte hyperplasia in mice. *Blood*, **87**:5006–5015.

Vainchenker, W., Bouguet, J., Guichard, J. and Breton-Gorius, J. (1979). Megakaryocyte colony formation from human bone marrow precursors. *Blood*, **54**:940–945.

Vainchenker, W., Deschamps, J.F., Bastin, J.M., Guichard, J., Titeux, M., Breton-Gorius, J. *et al.* (1982). Two monoclonal antiplatelet antibodies as markers of human megakaryocyte maturation: immunofluorescent staining and platelet peroxidase detection in megakaryocyte colonies and in *in vivo* cells from normal and leukemic patients. *Blood*, **59**:514–521.

Vigon, I., Florindo, C., Fichelson, S., Guenet, J.L., Mattei, M.G., Souyri, M. *et al.* (1993). Characterization of the murine Mpl proto-oncogene, a member of the hematopoietic cytokine receptor family: molecular cloning, chromosomal location and evidence for a function in cell growth. *Oncogene*, **8**:2607–2615.

Vigon, I., Mornon, J.P., Cocault, L., Mitjavila, M.T., Tambourin, P., Gisselbrecht, S. *et al.* (1992). Molecular cloning and characterization of MPL, the human homolog of the v-mpl oncogene: identification of a member of the hematopoietic growth factor receptor superfamily. *Proc. Natl. Acad. Sci., USA*, **89**:5640–5644.

Wada, T., Nagata, Y., Nagahisa, H., Okutomi, K., Ha, S.H., Ohnuki, T. *et al.* (1995). Characterization of the truncated thrombopoietin variants. *Biochem. Biophys. Res. Commun.*, **213**:1091–1098.

Warren, A.J., Colledge, W.H., Carlton, M.B., Evans, M.J., Smith, A.J. and Rabbitts, T.H. (1994). The oncogenic cysteine-rich LIM domain protein rbtn2 is essential for erythroid development. *Cell*, **78**:45–57.

Wendling, F., Maraskovsky, E., Debili, N., Florindo, C., Teepe, M., *et al.* (1994). c-Mpl ligand is a humoral regulator of megakaryocytopoiesis. *Nature*, **369**:571–574.

Wendling, F., Varlet, P., Charon, M. and Tambourin, P. (1986). MPLV: a retrovirus complex inducing an acute myeloproliferative leukemic disorder in adult mice. *Virology*, **149**:242–246.

Wendling, F., Vigon, I., Souyri, M. and Tambourin, P. (1989). Myeloid progenitor cells transformed by the myeloproliferative leukemia virus proliferate and differentiate *in vitro* without the addition of growth factors. *Leukemia*, **3**:475–480.

Wilks, A.F. (1993). Protein tyrosine kinase growth factor receptors and their ligands in development, differentiation and cancer. *Adv. Cancer Res.*, **60**:43–73.

Williams, N. and Levine, R.F. (1982). The origin, development and regulation of megakaryocytes. *Brit. J. Haematol.*, **52**:173–180.

Wright, J.H. (1906). The origin and nature of blood plates. *Boston Med. Surg J.*, **154**:643–645.

Yamamoto, S. (1957). Mechanisms of the Development of Thrombocytosis due to Bleeding. *Acta Haematol. Jpn*, **20**:163.

Yan, X.Q., Lacey, D., Fletcher, F., Hartley, C., McElroy, P., Sun, Y. *et al.* (1995). Chronic exposure to retroviral vector encoded MGDF (mpl-ligand) induces lineage-specific growth and differentiation of megakaryocytes in mice. *Blood*, **86**:4025–4033.

Yan, X.Q., Lacey, D., Hill, D., Chen, Y., Fletcher, F., Hawley, R.G. *et al.* (1996). A model of myelofibrosis and osteosclerosis in mice induced by overexpressing thrombopoietin (mpl ligand): reversal of disease by bone marrow transplantation. *Blood*, **88**:402–409.

Young, J.C., Bruno, E., Luens, K.M., Wu, S., Backer, M. and Murray, L.J. (1996). Thrombopoietin stimulates megakaryocytopoiesis, myelopoiesis, and expansion of CD34$^+$ progenitor cells from single CD34$^+$Thy-1$^+$Lin$^-$ primitive progenitor cells. *Blood*, **88**:1619–1631.

Young, K.M. and Weiss, L. (1987). Megakaryocytopoiesis: incorporation of tritiated thymidine by small acetylcholinesterase-positive cells in murine bone marrow during antibody-induced thrombocytopenia. *Blood*, **69**:290–295.

Zeigler, F.C., de Sauvage, F., Widmer, H.R., Keller, G.A., Donahue, C., Schreiber, R.D. *et al.* (1994). *In vitro* megakaryocytopoietic and thrombopoietic activity of c-mpl ligand (TPO) on purified murine hematopoietic stem cells. *Blood*, **84**:4045–4052.

Zucker-Franklin, D. and Petursson, S. (1984). Thrombocytopoiesis — analysis by membrane tracer and freeze-fracture studies on fresh human and cultured mouse megakaryocytes. *J. Cell Biol.*, **99**:390–402.

2 Platelets and Fibrinolysis

Jane A. Leopold and Joseph Loscalzo

Evans Department of Medicine and Whitaker Cardiovascular Institute, Boston University School of Medicine, Boston, MA 02118, USA

THE ROLE OF PLATELETS IN THE HEMOSTATIC RESPONSE

The hemostatic response to endothelial damage is contingent upon the coordinated interaction between platelets and vessel wall constituents. Under basal conditions, platelets circulate in the resting state and are not avid for the endothelium. The endothelium further maintains this non-thrombogenic milieu by synthesizing cell products and expressing surface molecules that inhibit platelet activation. Endothelial cells metabolize endoperoxides to prostacyclin, which inhibits platelet aggregation by activating platelet adenylyl cyclase and augmenting levels of cAMP (Leopold and Loscalzo, 1995). The nitrosovasodilators, endothelium-derived-relaxing factor and nitric oxide, diffuse through platelet membranes to increase cGMP levels and, thereby, prevent platelet adhesion (Leopold and Loscalzo, 1995). Endothelin additionally enhances these mechanisms by stimulating prostacyclin synthetase and nitric oxide synthase (Thiemermann *et al.*, 1988; Herman *et al.*, 1989; Lidbury *et al.*, 1990). Endothelial cells constitutively express thrombomodulin, a surface receptor which binds thrombin with high affinity and facilitates the protease's activation of protein C; and the enzyme ecto-ADPase, which converts ADP to AMP (Leopold and Loscalzo, 1995). Endothelial cell dysfunction perturbs these homeostatic mechanisms and promotes platelet adhesion and activation.

Vascular injury exposes the subendothelial matrix to platelets, which adhere to collagen via surface receptors glycoprotein Ia/IIa and glycoprotein Ib/IX (Figure 2.1); the former binds collagen directly, the latter does so through von Willebrand factor (Leopold and Loscalzo, 1995). In areas of pathologic shear stress (>30 dynes/cm^2), platelet activation and aggregation may occur independently of adhesion (Leopold and Loscalzo, 1995). When collagen and thrombin bind to their specific receptors, phospholipase C-dependent hydrolysis of membrane phospholipids is stimulated to generate inositol 1,4,5 bisphosphate (IP$_3$) and diacylglycerol (DAG) (Leopold and Loscalzo, 1995). IP$_3$ elevates levels of intracellular calcium resulting in phosphorylation of myosin light chain (Leopold and Loscalzo, 1995), while DAG activates protein kinase C with subsequent phosphorylation of pleckstrin, a 47-kDA platelet protein, and secretion of granule ADP and serotonin (Leopold and Loscalzo, 1995). Thrombin-induced platelet aggregation is accompanied by calpain-mediated cleavage of aggregin, an ADP-receptor on the platelet surface (Leopold and Loscalzo, 1995). Weak platelet agonists, such as ADP and epinephrine, activate phospholipase A$_2$ to release membrane arachidonic acid, which is metabolized to form prostaglandin endoperoxides (PGG$_2$/PGH$_2$) and thromboxane A$_2$. These metabolites then feedback to activate phospholipase C (Leopold and Loscalzo, 1995).

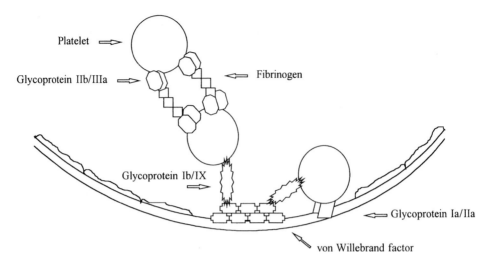

Figure 2.1 The role of platelets in primary hemostasis. At sites of vessel wall injury, circulating platelets adhere to the exposed subendothelial matrix via surface receptor integrins glycoprotein Ia/IIa, which binds collagen directly, and glycoprotein Ib/IX, which binds to von Willebrand factor and thereby to collagen. Additional platelets are recruited to the site of injury and bind plasma fibrinogen via the surface receptor integrin glycoprotein IIb/IIIa to form the primary hemostatic plug.

Once activated, platelets lose their characteristic discoid shape and extend finger-like projections (pseudopodia) by altering the proportion of cytoskeletal actin in the filamentous form (Lind, 1994). Myosin becomes increasingly incorporated into the cytoskeleton and participates in contraction of the inner core microfilaments (Carroll *et al.*, 1982; Leopold and Loscalzo, 1995). Actin-binding protein and α-actinin are concentrated at the periphery and interact with tropomyosin, vinculin, and talin to form focal adhesion sites (Pho *et al.*, 1984; Beckerle *et al.*, 1989; Asijee *et al.*, 1990). Through these linkages glycoprotein Ib/IX, glycoprotein IIb/IIIa, fibrinogen, fibronectin, and thrombospondin become associated with the cytoskeleton (Tuszynski *et al.*, 1985). Granule secretion occurs concomitant with platelet shape change. Platelet α-granules are secreted by a novel mechanism that involves central movement of the secretory granules, which then fuse with the open canalicular system to discharge their contents (White and Krumwiede, 1987). Secreted factors aide in the recruitment of additional platelets, which then undergo shape change and form a hemostatic plug.

FIBRINOLYSIS

Plasminogen

The fibrinolytic system is activated in response to clot or thrombus formation. The proenzyme plasminogen is converted to plasmin, which proteolytically degrades fibrin within the clot. Plasminogen is a 92-kDa single-chain glycoprotein which contains 791 amino acids with an amino-terminal glutamic acid residue (Robbins, 1987). The inactive precursor contains five homologous 80-amino-acid triple-loop structures or kringles, K1 through K5, which are secured by three disulfide bonds. Plasminogen binds to fibrin via lysine binding sites located in the plasmin (A) chain on K1 and K4. The kringle domains are involved in the interaction between plasmin(ogen) and α₂-antiplasmin as well as the binding of plasmin(ogen) to the platelet surface (Forsgren *et al.*, 1987; Handin and Loscalzo, 1992).

Plasminogen is synthesized primarily in the liver (Raum *et al.*, 1980). The plasma concentration of plasminogen is approximately 2μmol/l and its circulating half-life is between 1.75 and 2.65 days. Plasminogen has been found to exist in several isoforms (Collin *et al.*, 1972). The two molecular forms of fibrinolytic importance, Lys-plasminogen and Glu-plasminogen, can be readily distinguished by their reactivity with plasminogen activators and their affinity for fibrin (Christensen, 1977; Wohl *et al.*, 1978). Lys-plasminogen, which contains an amino-terminal lysine residue, is activated three- to ten-fold more efficiently than Glu-plasminogen by plasminogen activators. Lys-plasminogen also has a two-fold higher affinity for fibrin than does Glu-plasminogen and, therefore, is more efficiently metabolized to plasmin (Collen and Linjen, 1986; Rakoczi *et al.*, 1978).

A plasminogen receptor has been detected on platelets that binds plasminogen with a relatively high capacity, 10^4–10^7 sites per cell, and a $K_d=1\,\mu$mol/l. It is speculated that α-enolase, a 54-kDa protein with a carboxy-terminal lysine, is the plasminogen receptor (Miles *et al.*, 1991). The plasminogen receptor may uniquely participate in the regulation of fibrinolysis by promoting plasmin formation at the cell surface while offering a protected site of action from serpin inhibitors (Plow *et al.*, 1991).

Gangliosides, sialic acid containing glycosphingolipids, are present in the outer leaflet of plasma membranes and have been shown to bind plasminogen and urokinase-type plasminogen activator(s) directly. Plasminogen binding to both resting and stimulated platelets is inhibited by gangliosides, suggesting that they may modulate ligand-receptor interactions (Miles *et al.*, 1989).

Plasminogen binding sites are generated during fibrinolysis which ultimately concentrate plasminogen at the surface of a dynamically lysing clot. Utilizing model thrombi and fluorescein-labelled plasminogen, Sakharov and Rijken showed that a 50-micron superficial layer of plasmin-treated fibrin clot exposes approximately 2.5 plasminogen-binding sites per fibrin monomer with a $K_d=2.2\,\mu$mol/l. The presence of additional receptor sites allows plasminogen to be concentrated ten-fold in the superficial layer (Sakharov and Rijken, 1995).

Furthermore, confocal microscopy has demonstrated that plasminogen is preferentially concentrated in zones of lysis within a decomposing fibrin network. The fibrin network is dismantled in two phases: a phase of prelysis, during which the fibrin network remains fixed and plasminogen accumulates only moderately on the immobile fibers, and a phase of final lysis, during which fibers are loosened and plasminogen is concentrated up to 30-fold. The two phases occurs simultaneously at different regions of the clot; however, the zone of prelysis is located deep within the clot, while the zone of final lysis is detected within a 5–8 micron superficial layer. Therefore, the highest concentrations of plasminogen are found in the dynamically changing superficial layer, or zone of final lysis, coincident with significant decomposition of the fibrin network (Sakharov *et al.*, 1996).

Plasminogen Activators

Plasminogen activators are specific proteolytic enzymes that convert the inactive pro-enzyme plasminogen to plasmin by hydrolysis of the Arg_{560}-Val_{561} peptide bond. Conversion of Glu-plasminogen to plasmin requires initial cleavage of the preactivation peptide, while Lys-plasminogen is activated by cleavage of the internal Arg_{560}-Val_{561} bond alone (Castellino and Violand, 1979). *In vivo*, plasminogen activation is stimulated by tissue-type and urokinase-type plasminogen activators. This response may be pharmacologically enhanced to promote fibrinolysis with the exogenous plasminogen activators, streptokinase, anisoylated plasminogen-streptokinase activator complex (APSAC), and staphylokinase;

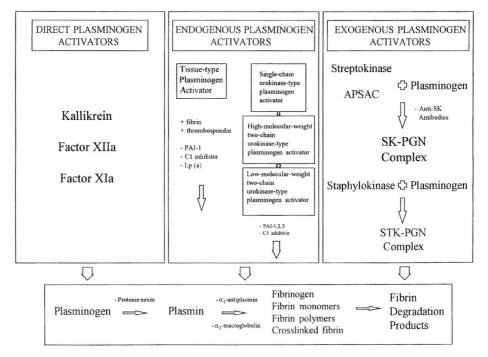

Figure 2.2   Plasminogen activators and the fibrinolytic system. The fibrinolytic system is activated in response to thrombus formation at sites of vessel wall injury. This response may be further enhanced pharmacologically to augment thrombus dissolution. Kallikrein, Factor XIIa, and Factor XIa cleave plasminogen directly to form plasmin. The endogenous plasminogen activators, tissue-type plasminogen activator and urokinase-type plasminogen activator(s), bind to specific platelet receptors in the presence of fibrin and thrombospondin and subsequently cleave plasminogen. Urokinase-type plasminogen activator is synthesized as a single chain precursor which undergoes proteolytic modification to form the high- and low- molecular-weight forms which have equivalent fibrinolytic activity. Platelet plasminogen activator inhibitor type-1 (PAI-1), C1 inhibitor, and surface bound lipoprotein (a) [Lp(a)] inhibit tissue-type plasminogen activator-mediated plasminogen activation while plasminogen activator inhibitor type(s)-1,2, and 3 (PAI-1,2,3) and C1 inhibitor inhibit urokinase-type plasminogen activator-mediated plasminogen activation. The exogenous plasminogen activators, streptokinase and staphylokinase, possess no enzymatic activity until they form a complex with plasminogen, which induces a conformational change in plasminogen and thereby confers enzymatic activity to the complex. Once plasminogen is cleaved to plasmin, fibrin degradation is initiated. The platelets products, protease nexin I and α_2-antiplasmin, as well as plasma α_2-macroglobulin, diminish this response.

and the endogenous plasminogen activators, recombinant tissue-type plasminogen activator and the urokinase-type plasminogen activators (Figure 2.2).

Kallikrein, factor XIa, and factor XIIa are capable of activating plasminogen directly; however, these factors appear to contribute only 15% of the total plasminogen activator activity in plasma (Kluft *et al.*, 1987; Coleman, 1968; Mandle and Daplan, 1979; Goldsmith *et al.*, 1978). Kallikrein can also convert single-chain tissue-type plasminogen activator to a two-chain form which has a greater catalytic efficiency in the absence of fibrin.

Native tissue-type plasminogen activator is synthesized predominantly by the endothelium. It is a 68-kDa serine protease with a circulating half-life of six minutes owing to rapid hepatic metabolism. Endothelial cells release tissue-type plasminogen activator in the presence of thrombin, bradykinin, interleukin-1β, epinephrine, and elevated levels of shear stress (Lucas and Miller, 1988; Collen *et al.*, 1989). Tissue-type plasminogen activator is converted by plasmin, factor Xa, and kallikrein to a two-chain form by cleavage of the Arg_{275}-Ile_{276} peptide bond. The two-chain form has greater catalytic activity towards plasminogen than its single-chain precursor. Binding to fibrin enhances the activity of tissue-type plasminogen activator several hundred-fold (Hoylaerts *et al.*, 1982; Loscalzo, 1988; Leopold and Loscalzo, 1995).

Urokinase is a serine protease that is initially synthesized as a single-chain precursor. Following hydrolysis of the Lys_{158}-Ile_{159} bond by plasmin and kallikrein, it is converted to a high-molecular-weight two-chain form (Leopold and Loscalzo, 1995). Further cleavage of the Lys_{135}-Lys_{136} bond yields a low-molecular-weight form that provides no additional catalytic benefit and is presently utilized as a thrombolytic agent. Single-chain urokinase-type plasminogen activator remains inactive in plasma unless fibrin is present, whereupon it initiates nonselective plasminogen activation. This activation occurs with a lag phase owing to the need for conversion to the high-molecular-weight two-chain form (Zamarron *et al.*, 1984). Fibrinolytic activity is conferred by plasmin-mediated proteolysis of fibrin to expose carboxy-terminal lysine binding sites. Once exposed, Glu-plasminogen binds to these sites, which induces a conformational change in plasminogen that enhances single-chain urokinase-type plasminogen activator-mediated generation of plasmin (Leopold and Loscalzo, 1995). The plasma half-life of single-chain urokinase-type plasminogen activator is 26.5 minutes and that of high-molecular-weight urokinase-type plasminogen activator is 12 minutes. Urokinase-type plasminogen activator is cleared by the liver.

Streptokinase is synthesized by group C β-hemolytic streptococci and purified for use as a pharmacologic agent. Streptokinase has neither a fibrin binding domain nor enzymatic activity, and exerts its fibrinolytic effects indirectly by forming a complex with plasminogen. Complex formation induces a conformational change in plasminogen and thereby confers enzymatic activity, without peptide bond cleavage, to the proenzyme (Leopold and Loscalzo, 1995). Streptokinase has a plasma half-life of 25 minutes. Anisoylated plasminogen-streptokinase activator complex (APSAC) is a noncovalent complex of modified streptokinase and plasminogen. The addition of a *p*-anisoyl group offers protection of the plasminogen active site and increases the plasma half-life of the compound to approximately 120 minutes. APSAC binds fibrin with similar affinity as streptokinase and is cleared by hepatic metabolism (Leopold and Loscalzo, 1995).

Staphylokinase is a 15.5-kDa protein that is produced by *Staphylococcus aureus* and possesses fibrinolytic properties (Lack, 1948). Staphylokinase alone also has no enzymatic activity until it forms a stoichiometric complex with plasminogen. The complex then activates plasminogen with a $K_m = 7.0 \mu mol/l$ (Lijnen *et al.*, 1991). In the absence of fibrin, the plasminogen-staphylokinase complex is readily neutralized by circulating α_2-antiplasmin; however, in the presence of fibrin, the lysine binding sites of the complex are occupied and α_2-antiplasmin-mediated inhibition is diminished, favoring plasminogen activation at the fibrin surface (Lijnen *et al.*, 1991).

MODULATION OF FIBRINOLYSIS BY PLATELETS (TABLE 2.1)

Platelet Content of Thrombi and Clot Retraction

Platelets are critical for the formation of occlusive thrombi, especially under conditions of high shear rates ($>30 dynes/cm^2$) (Leopold and Loscalzo, 1995). Platelet content of thrombi confers sensitivity to thrombolysis: platelet-poor thrombi are more sensitive to autologous lysis than platelet-rich thrombi, and on addition of some exogenous plasminogen activators (Serizawa *et al.*, 1993), platelet-poor thrombi are lysed slower than platelet-rich thrombi (Carroll *et al.*, 1982). Reocclusive thrombi that develop following treatment with plasminogen activators or coronary angioplasty are richer in platelets than thrombi that form without these therapeutic interventions (Leopold and Loscalzo, 1995). The relative resistance of platelet-rich thrombi to thrombolytic therapy suggests that the 20% of patients who fail to lyse thrombi in response to the administration of a plasminogen activator may possess thrombi containing a higher platelet content compared with those who successfully lyse (Jang *et al.*, 1989).

Table 2.1   Effects of platelets on fibrinolysis

Platelet Content of Thrombi
    Determines sensitivity to thrombolysis
    Clot retraction diminishes plasminogen activator binding

Stimulation of Plasminogen Activation
    Platelet surface provides an assembly site for components of the fibrinolytic system
    Autocatalytic increase in stimulation following plasmin exposure
    Insulin-mediated release of platelet plasminogen activator
    Release of osteonectin which binds plasminogen
    Enhance endothelial cell production of platelet-endothelial cell activated protease
    Binds prekallikrein

Inhibition of Plasminogen Activation
    Release of plasminogen activator inhibitor type-1
    Release of α_2-antiplasmin
    Release of C1-inhibitor
    Release of α_2-macroglobulin
    Release of protease nexin I
    Binds lipoprotein (a)

Inhibition of clot lysis and diminished binding of plasminogen activators occur as a consequence of clot retraction. *Ex vivo* thrombus preparations labeled with [^{125}I]-fibrin demonstrated a 47% decrease in radioactivity six hours after the addition of tissue-type plasminogen activator when formed in platelet-rich compared with platelet-poor plasma. This difference could not be attributed to the release of plasminogen activator inhibitor type-1. A similar platelet-dependent reduction in clot lysis was observed utilizing a tissue-type plasminogen activator mutant resistant to inhibition by plasminogen activator inhibitor type-1. The detected reduction in tissue-type plasminogen activator activity paralleled the decrease in binding to the platelet-enriched clot. Pretreatment with cytochalasin D, which inhibits clot retraction, abolished platelet-mediated inhibition of tissue-type plasminogen activator binding and thrombolysis, suggesting that platelets may inhibit clot lysis at therapeutic concentrations of tissue-type plasminogen activator as a consequence of clot retraction and decreased availability of fibrinolytic proteins (Kunitada *et al.*, 1992).

The role of platelet content and clot retraction in resistance to thrombolysis may be differentially affected by the choice of thrombolytic agent. In contrast to tissue-type plasminogen activator, staphylokinase demonstrates a reduction in clot lysis time and enhanced rates of plasminogen activation in the presence of platelet-rich plasma. This effect may be due to the existence of an intact platelet membrane surface which serves as a catalytic site for fibrinolysis (Suehiro *et al.*, 1995). A plasma clot lysis system with the ability to distinguish between fibrin-specific and non-fibrin specific plasminogen activators was developed by utilizing [^{125}I]-fibrinogen labeled human plasma clots formed on needles and mechanically compressed after spontaneous retraction. Recombinant staphylokinase produced high rates of clot lysis without significantly influencing fibrinogen, plasminogen or α_2-antiplasmin levels. In contrast, equimolar concentrations of streptokinase induced low rates of clot lysis but markedly depleted these factors. Tissue-type plasminogen activator demonstrated concentration-dependent rates of clot lysis with relatively high lysis rates at low concentrations, but once plasminogen levels were depleted, the rate of clot lysis was reduced (Hauptmann and Glusa, 1995).

Stimulation of Plasminogen Activation by Platelets

Platelets enhance the catalytic efficiency of plasminogen activation and thereby facilitate fibrinolysis (Miles and Plow, 1985; Miles *et al.*, 1986; Gao *et al.*, 1990). The platelet surface serves as an assembly site for the component molecules of the plasminogen activator system. Glu-plasminogen binds specifically and saturably to approximately 37 000 sites on the resting platelet surface with a K_d of approximately 1.9 μm (Miles *et al.*, 1986). Tissue-type plasminogen activator and urokinase-type plasminogen activator (Leopold and Loscalzo, 1995) similarly bind to the platelet surface. As a consequence of this coordinated assembly of plasminogen and plasminogen activators on the platelet surface, the catalytic efficiency of plasminogen activation is increased up to eight-fold compared to solution phase. This increase appears to be related to a decrease in the K_m of the plasminogen activator for plasminogen (Gao *et al.*, 1990).

Plasminogen activation on the platelet surface is also an autocatalytic process. Platelets exposed to plasmin (1 caseinolytic unit/ml for 1 hour at 37°C) demonstrate an increase in the number of glu-plasminogen binding sites by approximately 78% without altering tissue-type plasminogen activator binding. The dissociation constants for each of these binding processes are not significantly altered by plasmin treatment. Platelets that were activated with adenosine 5′-diphosphate and then exposed to plasmin showed an increase in the number of plasminogen binding sites by 41% (Ouimet *et al.*, 1994a; Leopold, 1995; Loscalzo *et al.*, 1995).

Human platelets have been shown to release a plasminogen activator in the presence of insulin. *Ex vivo* studies utilizing platelet membranes exposed to insulin demonstrate the release of a platelet plasminogen activator. This insulin-dependent release was inhibited in the presence of elevated levels of adenosine 3′,5′-cyclic monophosphate. Washed platelets incubated with insulin express a threefold increase in plasminogen activator activity over basal levels. The substrate specificity and ability to inhibit enzymatic activity suggest that the platelet plasminogen activator is similar to tissue-type plasminogen activator. Fibrinolytic activity of the platelet plasminogen activator and its insulin-dependent release were confirmed by observing clot lysis times of platelet-rich plasma from both normal and insulin-dependent diabetic patients (Kahn *et al.*, 1995).

Osteonectin is an adhesive glycoprotein that is synthesized constitutively by megakaryocytes. Platelet-derived osteonectin binds to plasminogen with an approximate K_d of 1.2×10^{-7} M and enhances the rate of plasmin generation by tissue-type plasminogen activator to a similar extent as fibrinogen. The osteonectin-plasminogen interaction is inhibited by α_2-antiplasmin and ε-aminocaproic acid, suggesting that this interaction is mediated through the kringle 1 domain of plasminogen (Kelm *et al.*, 1994).

In addition, fibrinolysis-potentiating activity following platelet-endothelial cell interaction has been observed *in vitro*. Porcine endothelial cells incubated in the presence of human platelets generate a 90-kDa neutral protease activity, platelet-endothelial cell-activated protease or PECAP, with physiochemical properties similar to glu-plasmin. This activity was not detected in either platelet extract or serum-free endothelial cell conditioned medium, indicating that platelet-endothelial cell interaction is necessary for its production (Flores-Delgado *et al.*, 1994).

Platelet-bound prekallikrein promotes pro-urokinase-induced plasminogen activation and subsequent clot lysis. Platelet-rich thrombi were found to be more readily lysed by low concentrations of single-chain urokinase-type plasminogen activator. This observation was not attributable to clot retraction, thereby implicating a platelet-mediated enzymatic mechanism. Casein autoradiography of isolated platelets revealed a 90-kDa band of activity that comigrated with plasma prekallikrein/kallikrein, a known activator of single-chain urokinase-type plasminogen activator. The activation of single-chain urokinase-type plasminogen activator by platelets pretreated with plasma factor XIIa, an activator of prekallikrein, was dose-dependent and inhibited by soybean trypsin inhibitor but not by an inhibitor of factor XIIa. Kinetic analysis of the single-chain urokinase-type plasminogen activator activation revealed that platelets reduced the K_m by seven-fold. The activation of single-chain urokinase-type plasminogen activator by platelet-bound kallikrein provides an explanation for the observed platelet-mediated promotion of single-chain urokinase-type plasminogen activator-induced clot lysis (Loza *et al.*, 1994).

Fibrinolytic Activity and Platelets

Platelets, once activated, may further modulate the fibrinolytic response by releasing peptides that confer resistance to thrombolysis (Figure 2.3). Several of these peptides belong to the serine protease inhibitor, or serpin, superfamily. These proteins are characterized by a mechanism that involves both unusual flexibility and ligand-receptor interactions. The principal feature is a refolding step, during which a disordered or helical strand is inserted into the center of a β-sheet. This transition, which is essential for inhibition, can be induced by heating, proteolytic cleavage of the serpin, or allowing the serpin to complex with the protease target (Engh *et al.*, 1995).

Plasminogen activator inhibitor type-1 is a 52-kDa single-chain glycoprotein that accounts for 60% of the plasminogen activator inhibitory activity in plasma (Sprengers and Kluft, 1987). Although plasminogen activator inhibitor type-1 is synthesized by endothelial cells (Loskutoff and Edington, 1977), approximately 90% of the total plasminogen activator inhibitor type-1 present in blood is contained in platelet α-granules (Kruithof *et al.*, 1986). Plasminogen activator inhibitor type-1 is synthesized in an active form but is converted to a latent form (Hekman and Loskutoff, 1985). It may be maintained in the active conformation bound to extracellular matrix or to vitronectin; however, free plasminogen activator inhibitor type-1 in the fluid phase spontaneously decays into the latent form as a consequence of a conformational change that masks its reactive center. The reactive center can be transiently exposed by treating latent plasminogen activator inhibitor type-1 with sodium dodecyl sulfate and other denaturants, but it eventually decays into the latent form (Loskutoff *et al.*, 1988).

Plasminogen activator inhibitor type-1 irreversibly inactivates both tissue-type and urokinase-type plasminogen activators (Wiman *et al.*, 1984)), but not streptokinase, by forming a stoichiometric, covalent complex. The formation of a covalent complex between tissue-type plasminogen activator and plasminogen activator inhibitor type-1 involves an interaction between the active site of tissue-type plasminogen activator and residues Arg_{346}-Met_{347} of plasminogen activator inhibitor type-1 (Madison *et al.*, 1990). Aggregating platelets release plasminogen activator inhibitor type-1 resulting in a two- to five-fold increase in plasminogen activator inhibitor type-1 activity level and a six- to 10-fold increase in plasminogen activator inhibitor type-1 antigen levels (Kruithof *et al.*, 1987; Declerck *et al.*, 1988).

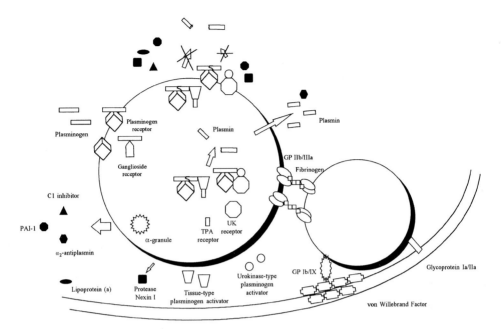

Figure 2.3   Platelets and the fibrinolytic system. Platelets modulate the fibrinolytic response by enhancing plasminogen activation on the platelet surface and secreting inhibitors to limit the extent of the fibrinolytic response. Circulating plasminogen binds to plasminogen and ganglioside receptors on the platelet surface. Binding places plasminogen in close proximity to surface bound tissue-type plasminogen activator and urokinase-type plasminogen activator(s), which are similarly bound to the platelet membrane. Following cleavage of plasminogen, plasmin is liberated. Platelets regulate fibrinolysis by releasing plasminogen activator inhibitor type-1 (PAI-1), C1 inhibitor, and α_2-antiplasmin from α-granules, secreting protease nexin I, and binding lipoprotein (a), which inhibit tissue-type plasminogen activator and urokinase-type plasminogen activator(s) and prevent further plasmin formation. (TPA, tissue-type plasminogen activator; UK, urokinase-type plasminogen activator(s); GP Ib/IX, glycoprotein Ib/IX; GP IIb/IIIa, glycoprotein IIb/IIIa).

Plasminogen activator inhibitor type-1 has been detected bound to the fibrin network of residual clot late in lysis, representing a pool of platelet-derived inhibitor complexed to fibrin. It is believed that fibrin binding enhances the autofibrinolytic effects of plasminogen activator inhibitor type-1 by protecting the serpin from normal degradation mechanisms (Braaten *et al.*, 1993). Plasminogen activator inhibitor type-1 is a relatively specific inhibitor of plasminogen activators; however, it has been shown to have thrombin-inhibitory properties once bound to vitronectin (Ehrlich *et al.*, 1988). Plasminogen activator inhibitor type-1 also binds to activated protein C, which, in turn, inactivates the protease and prevents its release from endothelial cells (Ouimet and Loscalzo, 1994b).

Ex vivo studies of clot lysis have established that both the concentration and activity of plasminogen activator inhibitor type-1 elaborated from platelets peak approximately 15 minutes after induction of clotting (Torr-Brown and Sobel, 1993); however, only one third of thrombus-associated plasminogen activator inhibitor type-1 exhibited significant activity (Fay *et al.*, 1996). Clot lysis initiated by tissue-type plasminogen activator is inhibited in the presence of platelet-rich, but not platelet-poor, plasma. This inhibition was reversed with the addition of an anti-plasminogen activator inhibitor type-1 monoclonal antibody, suggesting that plasminogen activator inhibitor type-1 mediates this effect (Torr-Brown and Sobel, 1993). Laser light scattering kinetic measurements revealed that clot lysis was significantly delayed both by thrombin-stimulated platelets and their cell-free lysate but returned to baseline in the presence of an anti-plasminogen activator inhibitor type-1 monoclonal antibody (Braaten *et al.*, 1993). The "Chandler loop" model, which creates thrombi that are polarized with a platelet-rich "white head" and red blood cell-rich "red tail," has been utilized to demonstrate that resistance to tissue-type plasminogen activator-mediated thrombolysis parallels the presence of activated platelets. This observation is attributed to the presence of plasminogen activator inhibitor type-1 that is released by activated platelets and preferentially retained within the head of the thrombus (Leopold and Loscalzo, 1995).

The contribution of plasminogen activator inhibitor type-1 to the antifibrinolytic effect of platelets was further investigated employing a murine plasminogen activator inhibitor type-1 deficient animal model subjected to arterial injury. Compared to wild-type animals, plasminogen activator inhibitor type-1 deficient mice demonstrated a significantly greater thrombus burden by planimetric analysis twenty-four hours following injury (Farrehi *et al.*, 1998). In addition, platelets isolated from a patient with a homozygous mutation in the plasminogen activator inhibitor type-1 gene resulting in complete loss of protein expression confirm these findings. *Ex vivo* clot lysis assays utilizing these platelets revealed a moderate inhibition of tissue-type plasminogen activator-mediated fibrinolysis while platelets from normal patients showed marked inhibition. When streptokinase or plasminogen activator inhibitor type-1-resistant mutant tissue-type plasminogen activator was used to initiate fibrinolysis, no difference was detected between plasminogen activator inhibitor type-1-deficient and normal platelets (Fay *et al.*, 1994).

α_2-antiplasmin is a 67-kDa glycoprotein peptide involved in the inactivation of plasmin. The protein is comprised of 452 amino acid residues linked by two disulfide bridges (Lijnen *et al.*, 1987), with Arg_{346}-Met_{365} placed at the active center. α_2-antiplasmin is synthesized and secreted by hepatocytes, present in platelet α-granules (Plow and Collen, 1981; Gogstad *et al.*, 1983), and has a circulating half-life of approximately 2.4 days (Collen and Wiman, 1979). α_2-antiplasmin in plasma is heterogeneous existing in both functionally active and inactive states (Kluft and Los, 1981), with approximately 30% of the α_2-antiplasmin functionally inactive (Muellertz and Clemmensen, 1976). The inactive form retains the Arg_{346}-Met_{365} active center, but lacks 26 residues that presumably contain the plasminogen-binding site (Sasaki *et al.*, 1983). This truncated form thus reacts slowly with plasmin and does not bind plasminogen (Kluft and Los, 1981; Sasaki *et al.*, 1983).

Platelet-derived α_2-antiplasmin is believed to contribute significantly as a regulator of fibrinolysis in the platelet-rich thrombus. In addition, the plasma contribution of α_2-antiplasmin has been demonstrated by *in vitro* studies of fibrin clot lysis utilizing a microtiter plate system. In the presence of platelet-rich or platelet-poor plasma, purified α_2-antiplasmin inhibits clot lysis in a dose-dependent manner (Robbie *et al.*, 1993). Once released in plasma, α_2-antiplasmin reacts very rapidly with the plasmin light chain and irreversibly forms a 1:1 stoichiometric complex, which is lacking protease or esterase activity (Moroi and Aoki, 1976; Wiman and Collen, 1977). α_2-antiplasmin also interferes competitively with the binding of plasminogen to fibrin and can be crosslinked to fibrin by factor XIIIa, resulting in resistance to fibrinolysis by plasminogen activators (Reed *et al.*, 1992). *Ex vivo* human clot preparations demonstrated approximately 20% of the plasma α_2-antiplasmin exists crosslinked to fibrin. The addition of Factor XIII inhibitors L722151 and mono-dansylcadaverine inhibited crosslinking to fibrin by more than 80% (van Giezen *et al.*, 1993). It has also been demonstrated that ligands such as lysine, 6-aminohexanoic acid, fibrinogen, and lipoprotein (a) inhibit the interaction between α_2-antiplasmin and plasmin (Edelberg and Pizzo, 1992).

C1-inhibitor is a 105-kDa peptide member of the serpin superfamily that is secreted from platelet α-granules. C1-inhibitor messenger RNA has been detected in megakaryocytes, suggesting that platelet α-granule content is determined by megakaryocyte production (Schmaier *et al.*, 1993). The amino terminal of the C1-inhibitor is heavily glycosylated and it is speculated that this confers protease inhibitory activity; however, truncated C1-inhibitor molecules with physiochemical profiles similar to wild type effectively diminish C1-inhibitor activity in hemolytic assays, thereby demonstrating that the amino-terminal domain does not influence complex formation with target proteases (Coutinho *et al.*, 1994). Indirect antibody consumption assays have demonstrated that 0.55 ± 0.4 ng C1-inhibitor/10^8 platelets is expressed on the surface of unstimulated platelets. Platelets activated with thrombin secreted 38% and externalized another 23% of the total platelet C1-inhibitor to their membrane (Schmaier *et al.*, 1993). Purified C1-inhibitor reacts with both single-chain and two-chain tissue-type plasminogen activator. Tissue-type plasminogen activator/C-1 inhibitor complex measured in plasma taken from patients during a recombinant tissue-type plasminogen activator infusion revealed that approximately 8% of the infused dose of recombinant tissue-type plasminogen activator was inhibited by C1-inhibitor at peak level (Huisman *et al.*, 1995).

Protease nexin I is a 47-kDa glycoprotein that binds to single-chain tissue-type plasminogen activator, two-chain tissue-type plasminogen activator, urokinase, as well as plasminogen, plasmin, trypsin, factor Xa, and thrombin in the presence of heparin (Van Nostrandt, 1987) and modulates fibrinolytic activity. Protease nexin I has recently been identified in and isolated from platelets (Chen and Essex, 1995).

α_2-macroglobulin is a 725-kDa dimeric glycoprotein that is synthesized by endothelial cells and macrophages and secreted by platelet α-granules. Although α_2-macroglobulin is not a member of the serpin superfamily, it serves as a ubiquitous protease inhibitor and binding protein for locally released platelet-derived growth factor-BB (Bonner and Orsonio-Vargas, 1995). The plasma level of α_2-macroglobulin varies with age and disease states, but is approximately 2.5 g/l in adults. α_2-macroglobulin is approximately 10% as efficacious as α_2-antiplasmin and participates in the inactivation of plasmin, primarily when α_2-antiplasmin levels are depleted (Aoki *et al.*, 1978).

Platelets further modulate fibrinolysis by binding lipoprotein(a) which has been shown to inhibit plasminogen activation by tissue-type plasminogen activator. Lipoprotein(a) binds specifically, saturably, and reversibly to platelets with a K_d of 0.20 μm. Scatchard analysis revealed a single class of binding sites with approximately 81 000 particles of lipoprotein(a) bound at saturation. Activated platelets demonstrated a 50% reduction in capacity to bind lipoprotein(a), with no demonstrable change in affinity. Lipoprotein(a) also inhibited the binding of plasminogen to platelets with an IC_{50} of approximately 0.23 μm. Kinetic analysis demonstrated that lipoprotein(a) behaved as a competitive inhibitor of tissue-type plasminogen activator-mediated plasminogen activation on the platelet surface with an estimated K_i of 0.49 μm (Ezratty *et al.*, 1993).

MODULATION OF PLATELET FUNCTION BY THE FIBRINOLYTIC SYSTEM

Plasmin has been noted to exert a variety of effects on platelet function *in vitro* (Table 2.2) that are influenced by concentration, incubation medium, incubation time, and source of both platelets and enzyme (Pasche and Loscalzo, 1991). Plasmin may either activate or inhibit platelet function depending primarily upon the concentration of enzyme to which the platelet is exposed and the incubation time. High concentrations of plasmin (>1.0 caseinolytic unit/ml) directly induce platelet aggregation, while lower concentrations (<0.5 caseinolytic units/ml) inhibit aggregation in the presence of agonists (Schafer *et al.*, 1986).

Table 2.2   Effects of plasmin on platelets

Inhibition of Platelets by Plasmin

Direct
  Degrades glycoprotein Ib/IX and glycoprotein IIIa
  Induces redistribution of glycoprotein Ib
  Stimulates expression of irreversibly active fibrinogen receptors
  Enhances signal transduction
  Release of transforming growth factor-β_1

Indirect
  Degrades von Willebrand factor, thrombospondin, and fibronectin
  Produces fibrin(ogen) degradation products
  Influences effects of diffusive and convective transport mechanisms
  Inhibits binding of Factor XIII

Activation of Platelets by Plasmin
  Occurs at a higher concentration of plasmin
  Streptokinase-dependent immune complex-mediated platelet aggregation
  Thrombin generated during fibrinolysis
  Facilitates elaboration of prothrombinase activity

Direct Effects of Plasmin on Platelets

Plasmin can directly disrupt two glycoproteins on the platelet surface that participate in normal adhesion and aggregation. Glycoprotein Ib/IX, a surface receptor for von Willebrand factor, is particularly sensitive to proteolysis by plasmin, and this effect appears to be mediated by the lysine-binding domain of the enzyme. Plasmin degrades the amino-terminal domain of the glycoprotein Ib α-subunit, which contains the binding site for von Willebrand factor (Adelman *et al.*, 1985; Adelman *et al.*, 1986). In the presence of elevated levels of shear stress (120 dynes/cm^2), plasmin similarly inhibits aggregation of washed platelets as a consequence of surface glycoprotein Ib proteolysis (Kamat *et al.*, 1995).

Plasmin also limits availability of the glycoprotein Ib/IX complex by inducing a redistribution of glycoprotein Ib molecules from the platelet surface to the lining of the open canalicular system. Once plasmin is neutralized, basal levels of glycoprotein Ib are restored to the platelet surface and functional recovery of the receptor occurs within 30 minutes, as measured by binding of both monoclonal anti-glycoprotein Ib antibody SZ-2 and [^{125}]I-labeled von Willebrand factor. Cytochalasin D does not inhibit this process, suggesting that passive actin depolymerization may play a role in glycoprotein Ib redistribution. Immunoelectron microscopy further demonstrates that the observed recovery results from recycling the glycoprotein Ib molecules from the open canalicular system back to the plasma membrane (Lu *et al.*, 1993).

Plasmin exerts opposing effects on the glycoprotein IIb/IIIa complex, resulting in impaired or enhanced fibrinogen binding and platelet aggregation. Plasmin degrades glycoprotein IIIa by cleaving a proteolytically susceptible loop domain. This cleavage can be detected only by subsequent proteolysis with Glu-C endoprotease followed by Western blotting with a polyclonal antibody. This cleavage event may be prevented by fibrinogen binding to glycoprotein IIb/IIIa (Pasche *et al.*, 1990).

Platelets exposed to plasmin have also been shown to express irreversibly active fibrinogen receptors. Plasmin-induced platelet aggregability is associated with the expression of approximately 2300 molecules of the integrin $\alpha_{IIb}\beta_{IIIa}$ on the platelet surface, which appears to depend partially on granule exocytosis at the time of incubation with plasmin. Plasmin-induced activation of glycoprotein IIb/IIIa is sustained and cannot be reversed by exposure of platelets to prostaglandin E_1. Immunoblotting of the receptor subunits shows no evidence of extensive proteolytic modification by plasmin, and only reveals a limited proteo-lysis of the amino-terminal domain of the α_{IIb} subunit. In addition to their capacity to aggregate in the presence of fibrinogen alone, plasmin-treated platelets also show potenti-ated aggregability in response to low doses of ADP. Therefore, plasmin is capable of activ-ating the fibrinogen receptor such that it remains present on the surface of nonaggregated platelets (Rabhi-Sabile and Pidard, 1995).

Plasmin mediates further platelet aggregation by inhibiting thrombin-induced platelet activation via cleavage of a functional thrombin receptor. Platelets exposed to plasmin demonstrated diminished binding of the anti-thrombin receptor antibody anti-TR-(33-46) in both dose- and time-dependent manners. This response was not associated with attenuated cytosolic calcium levels (Kimura *et al.*, 1996).

Elements of the signal transduction system in platelets may be modulated further by plasmin. Plasmin increases membrane-bound adenylyl cyclase activity in platelets, and diminishes the inhibitory effects of agonists like epinephrine, thereby attenuating the aggregation response (Adnot *et al.*, 1987).

Platelets exposed to physiological levels of plasmin become activated and exhibit an early, concentration-dependent increase in cytoplasmic calcium. Further challenge with thrombin or the endoperoxide analog U44069 inhibits the observed calcium response in proportion to the degree of initial activation. Adenosine 5'-diphosphate scavengers impair the cytoplasmic calcium increase, plasmin-mediated platelet aggregation, and their inhibit-ory effects on thrombin-induced activation, thereby suggesting that adenosine 5'-diphos-phate plays an integral role in both the initial platelet activation and subsequent inhibition by plasmin (Penny and Ware, 1992).

The observed increase in platelet cytosolic calcium following plasmin exposure is achieved by enhancing calcium mobilization and external calcium entry, which is associated with accelerated calcium extrusion and protein tyrosine phosphorylation. Plasmin-mediated external calcium influx and efflux, but not calcium mobilization, are attenuated by the tyro-sine kinase inhibitor, genistein, suggesting multiple mechanisms for intracellular calcium recruitment. In addition, plasmin inhibits calcium flux in the presence of thrombin. These observations imply a dual effect of plasmin on calcium mobilization. Plasmin initiates an increase in platelet cytosolic calcium and a tyrosine kinase-dependent enhancement of calcium turnover, calcium influx, and efflux. Plasmin also attenuates the cytosolic calcium response to thrombin by blocking calcium mobilization and slowing the rate of external calcium influx, resulting in a plasmin-induced inhibition of clot formation (Nakamura *et al.*, 1995).

Plasmin induces release of transforming growth factor-β_1 from human platelets. *Ex vivo* studies have demonstrated that platelets contain two reservoirs of transforming growth factor-β_1. A large pool composed of transforming growth factor-β serum binding protein, the latency-associated peptide, and the 25-kD transforming growth factor β_1 dimer is released during clot formation. The second pool of transforming growth factor β_1, which contains only the latency-associated peptide, is retained within the thrombus but may be activated and released by plasmin (Grainger *et al.*, 1995).

Indirect Effects of Plasmin on Platelets

Shear stress-induced platelet aggregation is mediated by von Willebrand factor binding to platelet membrane glycoproteins Ib and GP IIb/IIIa. Von Willebrand factor is susceptible to pharmacologically achievable levels of plasmin activity, and the resulting proteolytic degradation products impair platelet agglutination (Andersen *et al.*, 1980; Leopold, and

Loscalzo, 1995) Tissue-type plasminogen activator, which induces thrombolysis through the local generation of plasmin, has been shown to disrupt platelet aggregation by inducing proteolysis of both glycoprotein Ib (Hoffman and Janssen, 1992) and plasma von Willebrand factor. At concentrations of 2000 IU/ml, tissue-type plasminogen activator significantly inhibits shear stress-induced platelet aggregation without disrupting platelet glycoprotein Ib. Purified von Willebrand factor multimer degradation by tissue-type plasminogen activator was associated with the inability of von Willebrand factor to support shear stress-induced platelet aggregation. These results demonstrate that selective proteolysis of von Willebrand factor is one mechanism by which tissue-type plasminogen activator inhibits shear stress-mediated platelet aggregation in response to pathological shear stress (Kamat *et al.*, 1995).

Plasmin has been shown to influence the effects of diffusive and convective transport on fibrinolysis of whole blood clots. Thrombolytic agents were delivered into whole blood clots utilizing a constant pressure gradient ($\Delta P/L$) from 0 to 3.7 mmHg/cm-clot to drive fluid permeation. The velocity at which a lysis front moved across a whole blood clot was markedly enhanced by increasing the $\Delta P/L$. Without permeation, a thrombolytic agent placed adjacent to a clot boundary created a reaction front that moved at a velocity dependent on the concentration of plasmin. Therefore, local plasmin concentration modulates transport phenomena within the clot and thereby may influence the time required for reperfusion during fibrinolysis (Wu *et al.*, 1994).

Plasmin modulates platelet adhesion and spreading by cleaving fibrin at or near sites involved in platelet recognition. Scanning electron microscopy demonstrated that exposure to plasmin transformed the tight fibrillar fibrin surface to a less dense structure with irregular and broken fibers. A gradient of proteolytic degradation through the fibrin clot was detected with the most extensive degradation at the surface, resulting in a rapid and progressive decrease in platelet adhesion. Five minutes of plasmin exposure resulted in only 6% solubilization of the fibrin but a 56% decrease in platelet adhesion, while after 30 minutes of plasmin exposure, spreading of adherent platelets on fibrin was diminished to 35% of baseline. Platelet adhesion to undegraded fibrin involved residues that contain the RGDS site, the carboxyl terminal dodecapeptide of the gamma chain, and the aminoterminus of the beta chain. The monoclonal antibody 7E3 inhibited platelet adhesion to fibrinogen by approximately 90%. Plasmin exposure to fibrin markedly decreases platelet adhesion and spreading, indicating that plasmin degradation may play a role in modulating cellular response to fibrin (Hamaguchi *et al.*, 1994).

Fibrin(ogen) degradation products produced by plasmin also impair platelet aggregation. The proteolytic derivatives compete with fibrinogen for binding to platelet glycoprotein IIb/IIIa and, thereby attenuate the aggregation response. Fibrinogen A_α fragments are generated at plasmin cleavage sites and at novel cleavage sites involving hydrophobic and basic amino acid residues. *In vitro* studies have demonstrated that the fragment that contains the cell attachment site of fibrinogen A_α efficiently inhibits fibrinogen binding and platelet aggregation with an IC_{50} of 20–50 μm (Standker *et al.*, 1995). Electrophoresis of plasmin-digested fibrin(ogen) revealed that fragment E was the only inhibitor of plasmin-mediated hydrolysis of fibrinogen, with the antifibrinolytic activity of the fragments being increased in a fragment D light < fragment E < fragment D heavy series. Elastic light scattering and analytical ultracentrifugation showed that the fragments were able to form a complex with plasminogen in a fragment E > fragment D light > fragment D heavy series. Therefore, complex formation did not contribute significantly to the mechanism of fibrinolysis inhibition. An antifibrinolytic effect of fragment DH is conferred by its antipolymerization activity (Khavkina *et al.*, 1995). Fibrin(ogen) degradation products and glycoprotein

IIIa cleavage induced by plasmin act cumulatively to impair platelet aggregation during fibrinolysis (Gouin *et al.*, 1992).

Thrombospondin and fibronectin (Leopold and Loscalzo, 1995) are susceptible to proteolysis by plasmin. In addition, thrombospondin may also serve as a slow tight-binding inhibitor of plasmin, by forming a 1:1 stoichiometric complex. Plasmin complexed to streptokinase or ε-aminocaproic acid is protected from inhibition by thrombospondin, implicating the lysine-binding kringle domains of plasmin in the inhibition process. Thrombospondin similarly inhibits urokinase-type plasminogen activator, but less efficiently than plasmin; however, it also stimulates the amidolytic activity of tissue-type plasminogen activator, but exerts no effect on the amidolytic activity of α-thrombin or factor Xa (Hogg, 1992). Cytofluorometric analysis has demonstrated that plasmin inhibits the binding of Factor XIII to platelets, thereby potentially impairing the stability of the forming thrombus conferred by this transamidinase (Leopold and Loscalzo, 1995).

Plasmin Mediated Platelet Activation

Evidence for platelet activation appears early during incubation with plasmin, and is associated with an increase in cytoplasmic calcium levels (Penny and Ware, 1992) and cleavage of aggregin with accompanying activation of platelet calpain (Leopold and Loscalzo, 1995). Streptokinase has been noted to activate platelets independent of plasminogen activation, an effect observed in approximately 10% of the population (Leopold and Loscalzo, 1995). This effect involves the formation of antibody-antigen complexes with non-neutralizing antistreptokinase antibodies on the platelet surface (Vaughan *et al.*, 1988).

Fibrinolysis generates thrombin (Owen *et al.*, 1988; Eisenberg *et al.*, 1987), which binds to and is incorporated in fibrin thrombi where it remains active (Francis *et al.*, 1983). Thrombin activation of platelets amplifies the procoagulant response (Aronson *et al.*, 1992), leading to further elaboration of thrombin through the direct release of procoagulants from platelets and the interaction of these procoagulants with the platelet surface (Miletich *et al.*, 1977). Plasmin also appears to facilitate the elaboration of prothrombinase activity (Leopold and Loscalzo, 1995), which results in additional platelet activation and attenuated rates of fibrinolysis.

Platelets exposed to plasmin following the administration of fibrinolytic agents paradoxically demonstrate a diminished aggregatory response in the presence of thrombin yet are potentiated in the presence of other agonists such as ADP, platelet-activating factor, collagen and the calcium ionophore A23187. These observations suggest that platelets released from a dissolving thrombus may express increased responsiveness to agonists in the local milieu and thereby promote platelet activation (Kinlough-Rathbone *et al.*, 1997).

CONCLUSIONS

The dynamic role of platelets in hemostasis and fibrinolysis is dependent upon the coordinated interplay of activating and inhibiting mechanisms within a tightly regulated system. Platelet function is modulated by interactions with components of the fibrinolytic system in complex ways that may enhance or diminish net fibrinolytic efficiency. In turn, platelets modify constituents of the fibrinolytic system to manipulate the fibrinolytic response further. Pharmacologic perturbations of this complex system may be associated with functional responses that may not be readily predictable from first principles. Thus, all antiplatelet and fibrinolytic therapies must be evaluated carefully and thoroughly with respect to each component of the hemostatic response.

References

Adelman, B., Michelson, A.D., Loscalzo, J., Greenberg, J. and Handin, R.I. (1985). Plasmin effect on glycoprotein Ib-von Willebrand factor interactions. *Blood*, **65**:32–40.

Adelman, B., Michelson, A.D., Greenberg, J. and Handin, R.I. (1986). Proteolysis of platelet glycoprotein Ib by plasmin is facilitated by plasmin lysine-binding regions. *Blood*, **68**:1280–1284.

Adnot, S., Ferry, N., Hanoune, J. and Lacombe, M.L. (1987). Plasmin: a possible physiological modulator of the human platelet adenylate cyclase system. *Clinical Science*, **72**:467–473.

Anderson, J.C., Switzer, M.E. and McKee, P. (1980). Support of ristocetin-induced platelet aggregation by procoagulant-inactive and plasmin-cleaved forms of human factor VIII/von Willebrand factor multimers. Degradation *in vitro* and stimulation and release *in vivo*. *J. Clin. Invest.* **76**:261–267.

Aoki, N., Moroi, M. and Tachiya, K. (1978). Effects of α_2-plasmin inhibitor on fibrin clot lysis. Its comparison with α_2-macroglobulin. *Thromb. Haemost.*, **39**:22–31.

Aronson, D.L., Chang, P. and Kessler, C.M. (1992). Platelet-dependent thrombin generation after *in vitro* fibrinolytic treatment. *Circulation*, **85**:1706–1712.

Asijee, G.M., Sturk, A., Bruin, T., Wilkinson, J.M. and Ten Cate, J.W. (1990). Vinculin is a permanent component of the membrane cytoskeleton and is incorporated into the (re)organising cytoskeleton upon platelet activation. *Eur. J. Biochem.*, **189**:131–136.

Beckerle, M.C., Miller, D.E., Bertagnolli, M.E. and Locke, S.J. (1989). Activation-dependent redistribution of the adhesion plaque protein, talin, in intact human platelets. *J. Cell. Biol.*, **109**:3333–3346.

Bonner, J.C. and Osornio-Vargas, A.R. (1995). Differential binding and regulation of platelet-derived growth factor A and B chain isoforms by α_2-macroglobulin. *J. Biol. Chem.*, **270**:16236–16242.

Braaten, J.V., Handt, S., Jerome, W.G., Kirkpatrick, J., Lewis, J.C. and Hantgan, R.R. (1993). Regulation of fibrinolysis by platelet-released plasminogen activator inhibitor 1: light scattering and ultrastructural examination of lysis of a model platelet-fibrin thrombus. *Blood*, **81**:1290–1299.

Carroll, R.C., Butler, R.G., Morris, P.A. and Gerrard, J.M. (1982). Separable assembly of platelet pseudopodal and contractile cytoskeletons, *Cell*, **30**:385–393.

Castellino, F.J. and Violand, B.N. (1979). The fibrinolytic system-basic considerations. *Progr. Cardiovasc. Dis.*, **21**:241–254.

Chen, K. and Essex, D. (1995). Purification of secreted platelet protease nexin I. *Thrombosis Research*, **79**, 527–529.

Christensen, U. (1977). Kinetic studies of the urokinase-catalyzed conversion of NH_2-terminal glutamic acid plasminogen to plasmin. *Biochim. Biophys. Acta.*, **481**:638–647.

Coleman, R.W. (1968). Activation of plasminogen by human plasma kallikrein. *Biochem. Biophys. Res. Commun.*, **35**:273–279.

Collen, D. and Wiman, B. (1979). Turnover of antiplasmin, the fast-acting plasmin inhibitor of plasma. *Blood*, **53**:313–324.

Collen, D. and Lijnen, H.R. (1986). The fibrinolytic system in man. *CRC Crit. Rev. Oncol. Hematol.*, **4**:249–301.

Collen, D., Lijnen, H.R., Todd, P.A. and Goa, K.L. (1989). Tissue-type plasminogen activator. A review of its pharmacology and therapeutic use as a thrombolytic agent. *Drugs*, **38**:346–388.

Collin, D., Tytgot, G., Claeys, H., Verstraete, M. and Wallen, P. (1972). Metabolism of plasminogen in healthy volunteers. *J. Clin. Invest.*, **51**:1310–1318.

Coutinho, M., Aulak, K.S. and Davies, A.E. 3rd. (1994). Functional analysis of the serpin domain of C1 inhibitor. *J. Immunol.*, **153**:3648–3654.

Declerck, P.J., Alessi, M.C., Verstreken, M., Kruithof, E.K.O., Juhan-Vague, I. and Collen, D. (1988). Measurement of plasminogen activator-1 in biologic fluids with a murine monoclonal antibody-based enzyme-linked immunosorbent assay. *Blood*, **71**:220–225.

Edelberg, J.M. and Pizzo, S.V. (1992). Lipoprotein(a) promotes plasmin inhibition by α_2-antiplasmin. *Biochem. J.* **286**(Pt 1):79–84.

Ehrlich, H.J., Gebbink, R.K., Keijer, J., Linders, M., Preissner, K.T. and Pannkoek, H. (1990). Alteration of serpin specificity by a protein cofactor. Vitronectin endows plasminogen activator inhibitor-1 with thrombin inhibitory properties. *J. Biol. Chem.*, **265**:13029–13035.

Eisenberg, P.R., Sherman, L.A. and Jaffe, A.S. (1987). Paradoxic elevation of fibrinopeptide A after streptokinase: evidence for continued thrombosis despite intense fibrinolysis. *J. Amer. Coll. Card.*, **10**:527–529.

Engh, R.A., Huber, R., Bode, W. and Schulze, A.J. (1995). Divining the serpin inhibition mechanism: a suicide substrate "spring"? *Trends in Biotechnology* **13** (12):503–510.

Ezratty, A., Simon, D.I. and Loscalzo, J. (1993). Lipoprotein(a) binds to human platelets and attenuates plasminogen binding and activation. *Biochemistry*, **32**:4628–4633.

Farrehi P.M., Ozaki, C.K., Carmeliet, P. and Fay, W.P. (1998). Regulation of arterial thrombolysis by plasminogen activator inhibitor-1 in mice. *Circulation*, **97**:1002–1008.

Fay, W.P., Eitzman, D.T., Shapiro, A.D., Madison, E.L. and Ginsburg, D. (1994). Platelets inhibit fibrinolysis *in vitro* by both plasminogen activator inhibitor-1-dependent and -independent mechanisms. *Blood*, **83**:351–356.

Fay, W.P., Murphy, J.G. and Owen W.G. (1996). High concentrations of active plasminogen activator inhibitor-1 in porcine coronary artery thrombi. *Arterioscler. Thromb. Vasc. Biol.*, **16**:1277–1284.

Flores-Delgado, G., Hornebeck, W., Legrand, Y. and Menashi, S. (1994). Fibrinolysis potentiating activity following endothelial cells-platelet interaction. *Thrombosis Research*, **76**:245–252.

Forsgren, M., Raden, B., Israelsson, M., Larsson, K. and Heden, L.O. (1987). Molecular cloning and characterization of a full-length cDNA clone for human plasminogen. *FEBS Letts.*, **213**:254–260.

Francis, C.W., Markham, R.E., Barlow, G.H., Florack, T.M., Dobrzynski, D.M. and Marder, V.J. (1983). Thrombin activity of fibrin thrombi and soluble plasmin derivatives. *J. Lab. Clin. Med.*, **102**, 220–230.

Gao, S-W., Morser, J., McLean, K. and Shuman, M.A. (1990). Differential effect of platelets on plasminogen activation by tissue plasminogen activator, urokinase, and streptokinase. *Thromb. Res.*, **58**:421–423.

Gogstad, G.O., Stormorken, H. and Solum, N.O. (1983). Platelet α_2-antiplasmin is isolated in the platelet α-granules. *Thromb. Res.*, **31**:387–390.

Goldsmith, G.H., Saito, H. and Ratnoff, O.D. (1978). The activation of plasminogen by Hageman factor (factor XII) and Hageman factor fragments. *J. Clin. Invest.*, **62**, 54–60.

Gouin, I., Lecompte, T., Morel, M.C., Lebrazi, J., Modderman, P.W., Kaplan, C. and Samama, M.M. (1992). *In vitro* effect of plasmin on human platelet function in plasma. Inhibition of aggregation caused by fibrinogenolysis. *Circulation*, **85**, 935–941.

Grainger, D.J., Wakefield, L., Bethell, H.W., Farndale, R.W. and Metcalfe, J.C. (1995). Release and activation of platelet latent TGF-β in blood clots during dissolution with plasmin. *Nature Medicine*, **1**:932–7.

Hamaguchi, M., Bunce, L.A., Sporn, L.A. and Francis, C.W. (1994). Plasmic degradation of fibrin rapidly decreases platelet adhesion and spreading. *Blood*, **84**:1143–1150.

Handin, R.I. and Loscalzo, J. (1992). Hemostasis, thrombosis, fibrinolysis and cardiovascular disease. In *Heart Disease — A Textbook of Cardiovascular Medicine*, 4th ed., edited by E. Braunwald, pp. 1767–1789. Philadelphia: Saunders.

Hauptmann, J. and Glusa, E. (1995). Differential effects of staphylokinase, streptokinase and tissue-type plasminogen activator on the lysis of retracted human plasma clots and fibrinolytic plasma parameters *in vitro*. *Blood Coag. & Fibrinolysis*, **6**:578–583.

Hekman, C.M. and Loskutoff, D.J. (1985). Endothelial cells produce a latent inhibitor of plasminogen activators that can be activated by denaturants. *J. Biol. Chem.*, **260**:11581–11587.

Herman, F., Magyar, K., Chabrier, P-E., Braquet, P. and Filep, J. (1989). Prostacyclin mediates antiaggregatory and hypotensive actions of endothelin in anesthetized beagle dogs. *Br. J. Pharmacol.*, **98**:38–40.

Hoffman, J.J. and Janssen, W.C. (1992). Interactions between thrombolytic agents and platelets: effects of plasmin on platelet glycoproteins Ib and IIb/IIIa. *Thromb. Res.*, **67**:711–719.

Hogg, P.J., Stenflo, J. and Moshser, D.F. (1992). Thrombospondin is a slow tight-binding inhibitor of plasmin. *Biochemistry*, **31**:265–269.

Hoylaerts, M., Rijken, D., Lijnen, H.R. and Collen, D. (1982). Kinetics of the activation of plasminogen by human tissue plasminogen activator, role of fibrin. *J. Biol. Chem.*, **259**:2080–2083.

Huisman, L.G., van Griensven, J.M. and Kluft, C. (1995). On the role of C1-inhibitor as inhibitor of tissue-type plasminogen activator in human plasma. *Thromb. Haemost.*, **73** (3):466–471.

Jang, I-K., Gold, H.K., Yaoita, H., Ziskind, A.A., Fallon, J.T., Holt, R.E., Leinbach, R.C., May, J.W. and Collen, D. (1989). Differential sensitivity of erythrocyte-rich arterial thrombi to lysis with recombinant tissue-type plasminogen activator: a possible explanation for resistance to coronary thrombolysis. *Circulation*, **79**:920–928.

Kahn, N.N., Bauman, W.A. and Sinha, A.K. (1995). Insulin-induced release of plasminogen activator from human blood platelets. *Am. J. Phys.*, **268**:H117–124.

Kamat, S.G., Michelson, A.D., Benoit, S.E., Moake, J.L., Rajasekhar, D., Hellums, J.D., Kroll, M.H. and Schafer, A.I. (1995). Fibrinolysis inhibits shear stress-induced platelet aggregation. *Circulation*, **92**:1399–1407.

Kelm, R.J., Swords, N.A., Orfeo, T. and Mann, K.G. (1994). Osteonectin in matrix remodeling. A plasminogen-osteonectin-collagen complex. *J. Biol. Chem.*, **269**:30147–30153.

Khavkina, L.S., Rozenfeld, M.A. and Leonova, V.B. (1995). Mechanism of inhibition of fibrinolysis and fibrinogenolysis by the end fibrinogen degradation products. *Thromb. Res.*, **78**:173–187.

Kimura, M., Andersen, T.T., Fenton, J.W. 2nd, Bahou, W.F. and Aviv, A. (1996). Plasmin-platelet interaction involves cleavage of functional thrombin receptor. *Am. J. Physiol.*, **271**:C54–C60.

Kinlough-Rathbone, R.L., Perry, D.W., Rand, M.L. and Packham, M.A. (1997). Pretreatment of human platelets with plasmin inhibits responses to thrombin, but potentiates reponses to low concentrations of aggregating agents, including the thrombin receptor activating peptide, SFLLRN. *Thromb. Haemost.*, **77**:741–747.

Kluft, C. and Los, N. (1981). Demonstration of two forms of α_2-antiplasmin in plasma by modified crossed electrophoresis. *Thromb. Res.*, **21**:65–71.

Kluft, C., Dooijewaard, G. and Emeis, J.J. (1987). Role of the contact system in fibrinolysis. *Semin. Thromb. Haemost.*, **13**:50–68.

Kruithof, E.K.O., Tran-thang, D. and Bachmann, F. (1986). Studies on the release of plasminogen activator inhibitor by human platelets. *Thromb. Haemost.*, **55**:201–205.

Kruithof, E.K.O., Nicoloso, G. and Bachmann, F. (1987). Plasminogen activator inhibitor-1: development of a radioimmunoassay and observations on its plasma concentration during various occlusion and after platelet aggregation. *Blood*, **70**:1645–1653.

Kunitada, S., FitzGerald, G.A. and FitzGerald, D.J. (1992). Inhibition of clot lysis and decreased binding of tissue-type plasminogen activator as a consequence of clot retraction. *Blood*, **79**:1420–1427.

Lack, C.H. (1948). Staphylokinase: an activator of plasma protein. *Nature*, **161**:559–560.

Leopold, J.A. and Loscalzo, J. (1995). Platelet activation by fibrinolytic agents: A potential mechanism for resistance to thrombolysis and reocclusion after successful thrombolysis. *Coronary Artery Disease*, **6**:923–929.

Lidbury, P.S., Thiemermann, D., Korbut, R. and Vane, J.R. (1990). Endothelins release tissue plasminogen activator and prostanoids. *Eur. J. Pharmacol.*, **186**:205–212.

Lijnen, H.R., Holmes, W.E., Van Hoef, B., Rodriguez, H. and Collen, D. (1987). Amino acid sequence of human α_2-antiplasmin. *Eur. J. Biochem.*, **166**:565–574.

Lijnen, H.R., Van Hoef, B., DeCock, F., Okada, K., Ueshima, S., Matsuo, O. and Collen, D. (1991). On the mechanism of fibrin-specific plasminogen activation by staphylokinase. *J. Biol. Chem.*, **266**:11826–11832.

Loscalzo, J. and Vaughan, D.E. (1987). Tissue plasminogen activator promotes platelet disaggregation in plasma. *J. Clin. Invest.*, **79**:1749–1755.

Loscalzo, J. (1988). A structural and kinetic comparison of recombinant human single and two-chain tissue type plasminogen activator. *J. Clin. Invest.*, **82**:1391–1397.

Loscalzo, J., Pasche, B., Ouimet, H. and Freedman, J.E. (1995). Platelets and plasminogen activation. *Thromb. Haemost.*, **74**:291–293.

Loskutoff, D. and Edgington, T. (1977). Synthesis of a fibrinolytic activator and inhibitor by endothelial cells. *Proc. Natl. Acad. Sci., USA*, **74**:3903–3907.

Loskutoff, D.J., Sawdey, M. and Mimuro, J. (1988). Type 1 plasminogen activator inhibitor. *Prog. Hemost. Thromb.*, **9**:87–115.

Loza, J.P., Gurewich, V., Johnstone, M. and Pannell, R. (1994). Platelet-bound prekallikrein promotes pro-urokinase-induced clot lysis: a mechanism for targeting the factor XII dependent intrinsic pathway of fibrinolysis. *Thromb. Haemost.*, **71**:347–352.

Lu, H., Soria, C., Soria, J., De Romeuf, C., Perrot, J.Y., Tenza, D., Garcia, I., Caen, J.P. and Cramer, E.M. (1993). Reversible translocation of glycoprotein Ib in plasmin-treated platelets: consequences for platelet function. *Eur. J. Clin. Invest.*, **23**:785–793.

Lucas, F.V. and Miller, M.L. (1988). The fibrinolytic system. Recent advances. *Cleveland Clin. J. Med.*, **55**:531–541.

Madison, E.L., Goldsmith, E.J., Gerard, R.D., Gething, M-J.H., Sambrook, J.F. and Bassel-Duby, R.S. (1990). Amino acid residues that affect interaction of tissue-type plasminogen activator with plasminogen activator inhibitor 1. *Proc. Natl. Acad. Sci., USA*, **87**:3530–3533.

Mandle, R.J. and Daplan, A.P. (1979). Hageman factor-dependent fibrinolysis: generation of fibrinolytic activity by the interaction of human activated factor XI and plasminogen. *Blood*, **54**:850–861.

Miles, L.A. and Plow, E.F. (1985). Binding and activation of plasminogen on the platelet surface. *J. Biol. Chem.*, **260**:4304–4311.

Miles, L.A., Ginsberg, M.H., White, J.G. and Plow, E.F. (1986). Plasminogen interacts with human platelets through two distinct mechanisms. *J. Biol. Chem.*, **77**:2001–2009.

Miles, L.A., Dahlberg, C.M., Levin, E.G. and Plow, E.F. (1989). Gangliosides interact directly with plasminogen and urokinase and may mediate binding of these fibrinolytic components to cells. *Biochem.*, **28**:9337–9343.

Miles, L.A., Dahlberg, C.M., Plescia, J., Felez, J., Kato, K. and Plow, E.F. (1991). Role of cell surface lysines in plasminogen binding to cells: identification of α-enolase as a candidate plasminogen receptor. *Biochemistry*, **30**:1682–1691.

Miletich, J.P., Jackson, C.M. and Majerus, P.M. (1977). Interaction of coagulation factor Xa with human platelets. *Proc. Nat. Acad. Sci., USA*, **74**:4033–

Moroi, M. and Aoki, N. (1976). Isolation and characterization of α_2-plasmin inhibitor from human plasma. A novel proteinase inhibitor which inhibits activator-induced clot lysis. *J. Biol. Chem.*, **251**:5956–5965.

Muellertz, S. and Clemmensen, I. (1976). The primary inhibitor of plasmin in human plasma. *Biochem. J.*, **159**:545–553.

Nakamura, K., Kimura, M., Fenton, J.W. 2nd., Anderson, T.T. and Aviv, A. (1995). Duality of plasmin effect on cytosolic free calcium in human platelets. *Amer. J. Physiol.*, **268**:C958–967.

Ouimet, H., Freedman, J.E. and Loscalzo, J. (1994a). Kinetics and mechanism of platelet-surface plasminogen activation by tissue-type plasminogen activator. *Biochemistry*, **33**:2970–2976.

Ouimet, H. and Loscalzo, J. (1994b). Fibrinolysis. In *Thrombosis and Hemorrhage*, edited by J. Loscalzo and A.I. Schafer, pp. 127–144. Cambridge, MA: Blackwell Scientific.

Owen, J., Friedman, K.D., Grossman, B.A., Wilkins, C., Berke, A.D. and Powers, E.R. (1988). Thrombolytic therapy with tissue plasminogen activator or streptokinase induces transient thrombin activity. *Blood*, **72**:616–620.

Pasche, B., Collins, L., Ouimet, H., Francis, S. and Loscalzo, J. (1990). Modulation of platelet function by plasmin and fibrin(ogen) degradation products during thrombolysis. *Blood*, **78**:142a.

Pasche, B. and Loscalzo, J. (1991). Platelets and fibrinolysis. *Platelets*, **2**:125–130.

Peerschke, E.I. (1995). Bound fibrinogen distribution on stimulated platelets. Examination by confocal scanning laser microscopy. *Amer. J. Pathol.*, **147**:678–687.

Penny, W.F. and Ware, J.A. (1992). Platelet activation and subsequent inhibition by plasmin and recombinant tissue-type plasminogen activator. *Blood*, **79**:91–98.

Pho, D.B., Vasseur, C., Desbruyeres, E. and Olomucki, A. (1984). Evidence for the presence of tropomyosin in the cytoskeletons of ADP- and thrombin-stimulated platelets. *FEBS Lett*, **173**:164–168.

Plow, E.F. and Collen, D. (1981). The presence and release of α_2-antiplasmin from human platelets. *Blood*, **58**:1069–1074.

Plow, E.F., Felez, J. and Miles, L.A. (1991). Cellular regulation of fibrinolysis. *Thromb, Haemost.*, **66**:32–36.

Rabhi-Sabile, S. and Pidard, D. (1995). Exposure of human platelets to plasmin results in the expression of irreversibly active fibrinogen receptors. *Thromb. Haemost.*, **73**:693–701.

Rakoczi, I., Wiman, B. and Collen, D. (1978). On the biological significance of the specific interaction between fibrin, plasminogen and antiplasmin. *Biochim. Biophys. Acta.*, **540**:295–300.

Raum, D., Lever, R., Taylor, P.O. and Starzl, T.E. (1980). Synthesis of human plasminogen by the liver. *Science*, **208**:1036–1037.

Reed, G.L., Matsueda, G.R. and Haber, E. (1992). Platelet factor XIII increases the fibrinolytic resistance of platelet-rich clots by accelerating the crosslinking of α_2-antiplasmin to fibrin. *Thromb. Haemost.*, **68** (3):315–320.

Robbie, L.A., Booth, N.A., Croll, A.M. and Bennett, B. (1993). The roles of α_2-antiplasmin and plasminogen activator inhibitor-1 (PAI-1) in the inhibition of clot lysis. *Thromb. Haemost.*, **70** (2):301–306.

Robbins, K.C. (1987). The plasminogen-plasmin enzyme system. In *Hemostasis and Thrombosis*, 2nd ed., edited by R.W. Coleman, J. Hirsh, V.J., Marder and E. Salzmann, pp. 340–357. Philadelphia: Lippincott.

Sakharov, D.V. and Rijken, D.C. (1995). Superficial accumulation of plasminogen during plasma clot lysis. *Circulation*, **92**:1883–1990.

Sakharov, D.V., Nagelkerke, J.F. and Rijken, D.C. (1996). Rearrangements of the fibrin network and spatial distribution of fibrinolytic components during plasma clot lysis. Study with confocal microscopy. *J. Biol. Chem.*, **271**:2133–2138.

Sasaki, T., Morita, T. and Iwanaga, S. (1983). Identification of the plasminogen-binding site of human α_2-antiplasmin inhibitor. *Thromb. Haemost.*, **50**:170.

Schafer, A.I., Maas, A.K., Ware, A. and Johnson, P.C. (1986). Phosphorylation, elevation of cytosolic calcium, and inositol phospholipid breakdown in platelet activation induced by plasmin. *J. Clin. Invest.*, **78**:73–79.

Schmaier, A.H., Amenta, S., Xiong, T., Heda, G.D. and Gewirtz, A.M. (1993). Expression of platelet C1 inhibitor. *Blood*, **82**:465–474.

Serizawa, K., Urano, T., Kozima, Y., Takada, Y. and Takada, A. (1993). The potential role of platelet PAI-1 in t-PA mediated clot lysis of platelet rich plasma. *Thromb. Res.*, **71**:289–300.

Sprengers, E.D. and Kluft, D. (1987). Plasminogen activator inhibitors. *Blood*, **69**:381–387.

Standker, L., Sillard, R., Bensch, K.W., Ruf, A., Raida, M., Schulz-Knappe, P., Schepky, A.G., Patscheke, H. and Forssmann, W.G. (1995). *In vivo* degradation of human fibrinogen A α: detection of cleavage sites and release of antithrombotic peptides. *Biochem. Biophys. Res. Comm.*, **215**:896–902.

Suehiro, A., Tsujioka, H., Yoshimoto, H., Ueda, M., Higasa, S. and Kakishita, E. (1995). Enhancing effect of platelets on staphylokinase-mediated clot lysis and plasminogen activation. *Thromb. Res.*, **80**:135–142.

Thiemermann, C., Lidbury, P., Thomas, R. and Vane, J. (1988). Endothelin inhibits *ex vivo* platelet aggregation in the rabbit. *Eur. J. Pharmacol.*, **158**:181–182.

Torr-Brown, S.R. and Sobel, B.E. (1993). Attenuation of thrombolysis by release of plasminogen activator inhibitor type-1 from platelets. *Thromb. Res.*, **72** (5):413–421.

Tuszynski, G.P., Daniel, J.L. and Stewart, G. (1985). Association of proteins with the platelet cytoskeleton. *Semin. Hematol.*, **22**:303–312.

van Giezen, J.J., Minkema, J., Bouma, B.N. and Jansen, J.W. (1993). Cross-linking of α_2-antiplasmin to fibrin is a key factor in regulating blood clot lysis: species differences. *Blood Coag. & Fibrino.*, **4** (6):869–875.

Van Nostrandt, W. and Cunningham, D. (1987). Purification of protease nexin II from human fibroblasts *J. Biol. Chem.*, **262**:8508–8514.

Vaughan, D.E., Kirschenbaum, J. and Loscalzo, J. (1988). Streptokinase-induced, antibody-mediated platelet aggregation — a potential cause of clot propagation *in vivo*. *J. Amer. Coll. Card.*, **11**:1343–1354.

White, J.G. and Krumwiede, M. (1987). Further studies of the secretory pathway in thrombin-stimulated human platelets. *Blood*, **69**:1196–1203.

Wiman, B. and Collen, D. (1977). Purification and characterization of human antiplasmin, the fast-acting plasmin inhibitor in plasma. *Eur. J. Biochem.*, **78**:19–26.

Wiman, B., Chmielewska, J. and Randy, M. (1984). Inactivation of tissue plasminogen activator in plasma. Demonstration of a complex with a new rapid inhibitor. *J. Biol. Chem.*, **259**:3644–3647.

Wohl, R., Summaria, L., Arzadon, L. and Robins, K. (1978). Steady state kinetics of activation of human and bovine plasminogens by streptokinase and its equimolar complexes with various activated forms of human plasminogen. *J. Biol. Chem.*, **253**:1402–1407.

Wu, J.H., Siddiqui, K. and Diamond, S.L. (1994). Transport phenomena and clot dissolving therapy; an experimental investigation of diffusion-controlled and permeation-enhanced fibrinolysis. *Thromb. Haemost.*, **72**:105–112.

Zamarron, C., Lijnen, H.R. and Collen, D. (1982). Kinetics of the activation of plasminogen by natural and recombinant tissue-type plasminogen activator. *J. Biol. Chem.*, **259**:2080–2083.

3 von Willebrand Factor and Platelet Adhesion

Christopher M. Ward[1,†] and Michael C. Berndt[2]

[1]*Platelet Molecular Biology Laboratory, Blood Research Institute, The Blood Center of Southeastern Wisconsin, Milwaukee, USA*
[2]*Hazel and Pip Appel Vascular Biology Laboratory, Baker Medical Research Institute, Prahran, Victoria 3181, Australia*

VON WILLEBRAND DISEASE

In 1924, a family from the remote Åland Islands between Sweden and Finland brought their 5-year old daughter to a Helsinki physician, Professor Erik von Willebrand, for investigation of a severe bleeding disorder. This girl suffered from recurrent mucosal bleeding and was later to die at age 13 from uncontrollable menstrual haemorrhage. Her parents were third cousins and both had bleeding symptoms, as did 10 of her 11 siblings. von Willebrand began an extensive study of this family and concluded that they suffered from a previously undescribed bleeding disorder (von Willebrand, 1926). Later studies found that these patients were deficient in a plasma factor, now termed von Willebrand Factor (vWF).

Quantitative and qualitative abnormalities of vWF, collectively termed von Willebrand disease (vWD), are the commonest inherited bleeding disorder, with an estimated prevalence of up to 1% of the general population. vWD is generally inherited as an autosomal dominant disorder, with the exception of a more severe, autosomal recessive form (Eikenboom *et al.*, 1995). The clinical manifestations of vWD are muco-cutaneous bruising and bleeding, excessive menstrual bleeding and bleeding following trauma or surgery (Ginsburg, 1992). This reflects defective primary haemostasis, with a failure of platelet adhesion at the site of vessel injury. A rare form of vWD (type 2N) produces an autosomally inherited haemophilia, due to an abnormal vWF which fails to bind Factor VIII (Ginsburg, 1992). Many other variants of vWD have been described and a simplified classification of vWD (Table 3.1) based on pathophysiology has recently been proposed by Sadler (1994). Acquired vWD occurs rarely, secondary to the development of an anti-vWF

Table 3.1 Classification of von Willebrand disease (vWD)

Type 1	Partial quantitative deficiency of vWF
Type 2	Qualitative deficiency of vWF
	2A Decreased platelet-dependent function with loss of high-molecular weight multimers
	2B Increased affinity for platelet GPIb
	2M Decreased platelet-dependent function without loss of high molecular weight multimers
	2N Decreased affinity for Factor VIII
Type 3	Severe/complete deficiency of vWF

†current address Department of Haematology and Transfussion Medicine, Royal North Shore Hospital, St. Leonards, NSW 2065, Australia.

inhibitor, usually as a complication of lymphoproliferative disorders (Jakway *et al.*, 1992). Many groups have identified the genotypes of vWD cases and a large amount of data has been compiled (Sadler *et al.*, 1995). This review will discuss only those variants which provide an insight into the structure-function of vWF, particularly the gain-of-function mutations of type 2B vWD.

vWF STRUCTURE

vWF Synthesis and Secretion

The molecular structure of vWF has been extensively investigated during the last ten years (reviewed by Lyons *et al.*, 1994; Girma *et al.*, 1995). The vWF gene is ~180 kilobases in length, contains 52 exons and is located on the short arm of chromosome 12. A partial pseudogene, corresponding to exons 23–24 of vWF has also been described on chromosome 22 (Mancuso *et al.*, 1989). The vWF gene is expressed exclusively in endothelial cells and megakaryocytes (Bonthron *et al.*, 1986). Positive and negative regulatory domains have recently been described for endothelial cell-specific vWF gene expression (Jahroudi *et al*, 1994). vWF is synthesized as a 2813 amino acid precursor, pre-pro-vWF (Bonthron *et al.*, 1986), which consists of a 22 amino-acid signal peptide, a 741 amino-acid propeptide and a 2050 amino-acid mature vWF protein (Titani *et al.*, 1986) (Figure 3.1). The mature vWF protein consists of 19% carbohydrate by weight, with 12 N-linked and 10 O-linked oligosaccharide chains and the amino- and carboxyl-terminal regions are rich in cystine residues (Titani *et al.*, 1986).

The post-translational processing of the vWF precursor involves dimerization, glycosylation, sulfation, multimerization and cleavage of the prosequence, in a complex and ordered sequence (Lyons *et al.*, 1994) (Figure 3.1). Within the endoplasmic reticulum, the signal peptide is cleaved during translation and two molecules of pro-vWF are disulfide-linked through their C-terminal domains to form vWF dimers. Transport of vWF from the endoplasmic reticulum to the Golgi requires both dimerization and the addition of high-mannose N-linked oligosaccharide chains (Wagner *et al.*, 1991). The Golgi apparatus is the site of further glycosylation and sulfation (Lyons *et al.*, 1994) and in the acidic environment of the *trans*-Golgi, pro-vWF dimers are linked by a second set of interchain disulfide bonds to form multimers. The assembly of vWF dimers into multimers requires the presence of both D domains of the pro-sequence (Journet *et al.*, 1993).

Mature vWF is secreted into the plasma *via* the luminal surface of the endothelial cell and also abluminally, into the subendothelial matrix (Bloom *et al.*, 1973). Two pathways of vWF secretion from the endothelium have been described: small multimers are secreted constitutively by endothelial cells in culture (Wagner *et al.*, 1990), whereas high-molecular-weight multimers (10–20,000 kDa) are stored in a specialized storage organelle, the Weibel-Palade body, and released following stimulation with secretagogues such as thrombin, histamine and oxygen radicals (Wagner *et al.*, 1993). Proteolytic cleavage and multimerization of vWF precede the formation of Weibel-Palade bodies, which contain both the propolypeptide and mature vWF, noncovalently associated in a 1:1 ratio (Vischer *et al.*, 1994). The polymerized vWF within the Weibel-Palade body forms organized tubular arrays visible by electron microscopy (Wagner *et al.*, 1993). The only other documented components of the Weibel-Palade body are the membrane glycoproteins P-selectin and the lysosomal protein CD63 (Wagner *et al.*, 1993). In megakaryocytes and platelets, vWF multimers are stored with many other components in the α-granules (Lind *et al.*, 1994). The

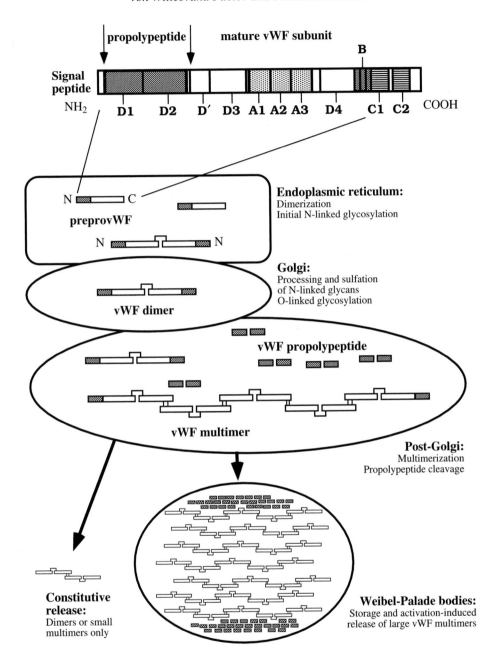

Figure 3.1 Biosynthesis of vWF multimers in endothelial cells.

release of vWF from platelet α-granules follows platelet activation by a variety of agonists (Kroll *et al.*, 1994), analogous to the release of Weibel-Palade body vWF by stimulated endothelium.

Once secreted into plasma, vWF is subjected to further proteolysis in normal subjects with a probable cleavage site between Tyr842 and Met843 (Dent *et al.*, 1990). Two recent reports describe the partial purification of a cation-dependent plasma protease which digests purified vWF into fragments similar to those seen *in vivo* (Furlan *et al.*, 1996; Tsai,

1996). Mutations in vWF leading to an increased susceptibility to proteolysis, with rapid degradation of high molecular-weight multimers, are one cause of Type 2A vWD (Lyons *et al.*, 1994).

vWF Multimerization

Plasma vWF consists of multimers with molecular weights varying from 500 kDa (dimers) to more than 10,000 kDa (Meyer *et al.*, 1993). Electron microscopic studies show that vWF multimers can form linear molecules almost 2 μm long, making them the largest soluble protein known (Slayter *et al.*, 1985). The largest multimers, however, are found primarily in storage organelles and only appear transiently in the plasma at sites of active release (Ruggeri *et al.*, 1992). Functionally, the multimerization of vWF appears to maximize ligand binding. Large vWF multimers bind platelets (Gralnick *et al.*, 1981) and extracellular matrix (Sporn *et al.*, 1987) more avidly than small multimers and in a perfusion study, platelets adhered more strongly to immobilized multimeric vWF than to vWF dimers (Wu *et al.*, 1996). This suggests that high molecular-weight vWF released locally at sites of tissue damage is particularly effective in mediating platelet adhesion. In conditions where the affinity of vWF for the platelet GPIb receptor is increased, such as Type 2B vWD (see below), these high-molecular-weight multimers are lost through adhesion to the platelet surface, leading to spontaneous platelet aggregation and a secondary thrombocytopenia (Ginsburg, 1992).

vWF Domain Structure

Analysis of the amino-acid sequence of vWF (Bonthron *et al.*, 1986; Shelton-Inloes *et al.*, 1986; Titani *et al.*, 1986) demonstrated that almost 80% of the sequence comprised homologous repeats of four domains (Figure 3.1). The most abundant amino acid in pre-pro-vWF is Cys (8.4% of residues), and the ~280 amino-acid D repeats are particularly Cys-rich with conservation of these Cys residues (Bonthron *et al.*, 1986). The D repeats and the smaller B and C domains are not homologous to other known molecules, with the exception of an Arg-Gly-Asp sequence in domain C1 (Shelton-Inloes *et al.*, 1986). In contrast, the three A domains of mature vWF contain a structural motif common to many adhesive proteins (Colombatti *et al.*, 1991). Each A domain of ~200 residues contains only two Cys residues: in the first and third A domains these form intrachain disulfide bonds (between Cys509-Cys695, and Cys923-Cys1109, respectively) , whereas the second A domain has a disulfide bond between adjacent residues Cys906-Cys907 (Meyer *et al.*, 1993). Regions homologous to the vWF A domains have been described in a variety of extracellular matrix proteins, complement components and in several adhesion receptors of the integrin family (Colombatti *et al.*, 1993) (Figure 3.2). In integrins of the β_1 and β_2 families, the A-like domains are "imbedded" within the α subunit and named "I-domains" (Larson *et al.*, 1989). Crystal structures for I domains from $\alpha_L\beta_2$ (Qu and Leahy, 1995) and $\alpha_M\beta_2$ (Lee *et al.*, 1995) have recently been reported and provide new insights into vWF structure and function (see below). The A domain loops of vWF also bind a wide variety of ligands, such as platelet GPIb, heparin and collagen (Figure 3.3).

vWF-Factor VIII Binding

vWF plays an important role in the coagulation pathway by protecting Factor VIII from proteolytic inactivation by activated protein C (APC). In plasma, the approximate molar

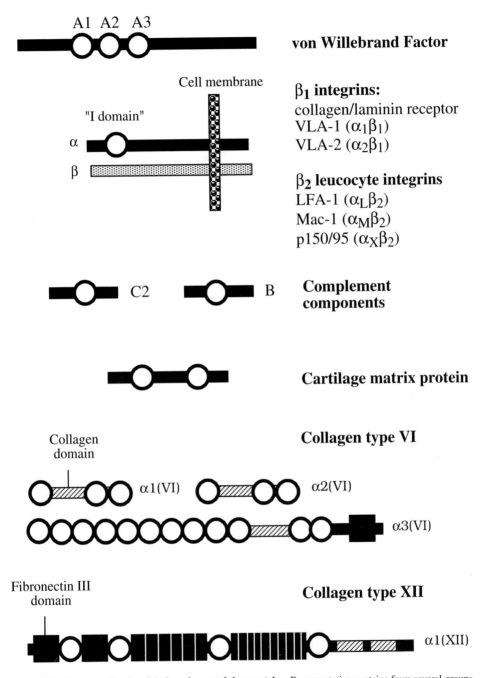

Figure 3.2 The superfamily of A domain-containing proteins. Representative proteins from several groups are shown with the A-type domains represented by open circles.

ratio of Factor VIII to vWF is 1:50 (Kaufmann *et al.*, 1992), but *in vitro* studies suggest that each vWF subunit contains an accessible binding site for Factor VIII (Vlot *et al.*, 1996). Several lines of evidence have localized the Factor VIII binding site to the N-terminal region of the mature vWF subunit (Wise *et al.*, 1991). Firstly, Factor VIII bound to a N-terminal proteolytic fragment of vWF (residues 1-272), but not to a reduced and alkylated

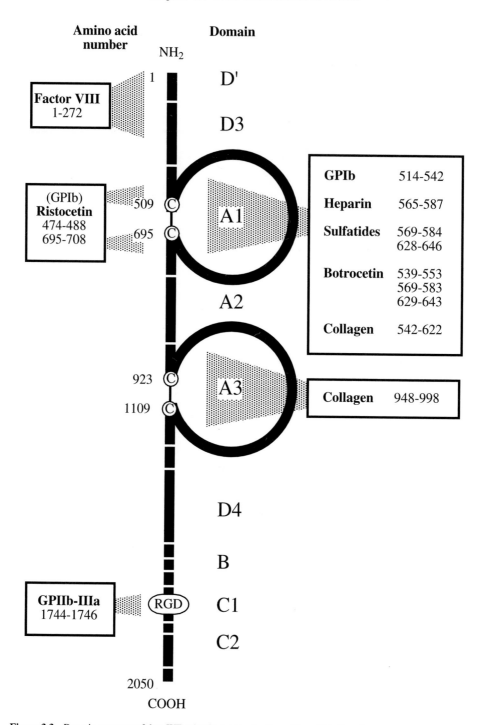

Figure 3.3   Domain structure of the vWF subunit and localization of ligand binding sites.

fragment, suggesting that the binding site was conformation-dependent (Foster *et al.*, 1987). A monomeric vWF fragment of residues 1-272 prevented APC cleavage of immobilized Factor VIII, but did not protect Factor VIII in solution, perhaps due to a lower binding affinity than that of multimeric vWF (Koppelman *et al.*, 1996). It has been proposed that regions of vWF other than residues 1-272 may be needed to protect Factor VIII from degradation (Layet *et al.*, 1992); however, vWF mutants lacking either the A1, A2, A3 and D4 domains still protected Factor VIII against APC (Koppelman *et al.*, 1996). Secondly, anti-vWF monoclonal antibodies which inhibit Factor VIII binding have epitopes which have been mapped to the N-terminal region, at vWF residues 2-53, 35-81 (Jorieux *et al.*, 1994; Piétu *et al.*, 1994) and 78-96 (Bahou *et al.*, 1989; Ginsburg *et al.*, 1992). Finally, point mutations have been identified in several cases of the rare type 2N ("Normandy") vWD, where a vWF defect of Factor VIII binding causes a haemophilia-like syndrome: these mutations (Arg19Trp, Thr28Met, Arg53Trp, His54Gln and Arg91Gln) are located within the first 100 residues (Ginsburg and Sadler, 1993). Therefore, multiple residues within the N-terminal 272 residues of the mature vWF subunit appear to make up the Factor VIII binding site.

vWF-GPIb-IX-V Binding

vWF mediates platelet adhesion through the binding of matrix-bound vWF to the platelet surface receptor GPIb-IX-V complex (Booth *et al.*, 1990). The structure and function of the leucine-rich GPIb-IX-V complex has been recently reviewed (López, 1994) and is discussed elsewhere in this volume (Chapter 4). Native vWF does not bind to platelet GPIb *in vitro*, however, and studies of the vWF-GPIb interaction have employed modulators which trigger vWF binding, such as the antibiotic, ristocetin (Berndt *et al.*, 1988), or the snake venom protein, botrocetin (Andrews *et al.*, 1989a). vWF stripped of sialic acid residues (asialo-vWF) binds spontaneously to human GPIb (De Marco and Shapiro, 1981), as do bovine and porcine vWF. More recently, high rates of shear stress have been found also to activate vWF-GPIb binding (Ikeda *et al.*, 1991). These modulators of vWF-platelet binding appear to induce a change in vWF conformation, exposing a cryptic binding site for GPIb (Berndt *et al.*, 1992).

Early studies with protease-derived fragments of vWF convincingly localized the GPIb binding site to the first A domain of vWF. An N-terminal dimeric vWF fragment including domains A1-A3, generated by *Staphylococcus aureus* V8 protease, bound platelets in the presence of ristocetin (Girma *et al.*, 1986). A monoclonal antibody which inhibited ristocetin-dependent vWF binding to platelets was epitope-mapped to a 116-kDa tryptic vWF fragment, comprising two A1 domains linked by interchain disulfide bonds (Sixma *et al.*, 1984). A monomeric 52/48-kDa form of this tryptic A1 domain fragment (Val449-Lys728) inhibited vWF-dependent platelet agglutination induced by ristocetin, asialo-vWF (Fujimura *et al.*, 1986) and botrocetin (Fujimura *et al.*, 1987a). Finally, a monomeric, 39/34-kDa dispase fragment of vWF (Leu480/Val481-Gly718) which inhibited asialo-vWF and bovine vWF binding to GPIb-IX-V, and bound GPIb in the presence of botrocetin, was shown by chemical cross-linking to associate directly with the N-terminal peptide domain of GPIbα (Andrews *et al.*, 1989b). Together, these studies localize the GPIb-binding site to a vWF fragment (480/481-718) that is essentially equivalent to the first A domain.

Several groups have attempted to define specific GPIb-binding residues using synthetic peptides corresponding to linear sequences from the A1 domain. Two peptide sequences adjacent to the 509-695 bond, Cys474-Pro488 and Leu694-Pro708, were found to inhibit platelet binding of vWF in the presence of ristocetin, and of asialo-vWF (Mohri *et al.*, 1988). This group proposed that the GPIb-binding site consisted of the two discontinuous regions, residues 474-488 and 694-708, brought into proximity by the intrachain 509-695 disulfide bond. This hypothesis was examined by Girma *et al.* (1990), who reported that vWF peptides 474-488 and 694-708 inhibited ristocetin-dependent vWF binding to plate-lets but had no effect on vWF-platelet binding in the presence of botrocetin, a result con-firmed by our laboratory (Berndt *et al.*, 1992). The combined data suggest instead that residues 474-488 and 694-708 are necessary for the ristocetin-specific modulation of vWF-GPIb binding but not for modulation by agents such as botrocetin and raise the pos-sibility that there is a GPIb-binding site elsewhere in the A1 domain. We have identified a peptide sequence, Asp514-Glu542 which is an effective inhibitor of all four *in vitro* mod-ulators, inhibiting ristocetin- and botrocetin-dependent vWF binding to platelets, as well as asialo-vWF- and bovine vWF-induced platelet agglutination (Berndt *et al.*, 1992). The Asp514-Glu542 sequence of vWF shows a high degree of homology with A domains from other proteins (Colombatti and Bonaldo, 1991) and with the A1 domain sequence of porcine (Bahnak *et al.*, 1992) and bovine vWF (Bakhshi *et al.*, 1992). This evidence led our group to propose a model of vWF-GPIb binding in which residues 514-542 form a specific GPIb-binding site normally hidden within the flexible A1 loop, but which can be exposed through a conformational shift induced by modulators, such as ristocetin and botrocetin (Berndt *et al.*, 1992; Andrews *et al.*, 1995a). Based on studies of vWF modula-tors (discussed below) we speculated also that the tertiary structure of the inactive A1 loop is maintained by electrostatic interactions between anionic sequences flanking the 509-695 bond and cationic sequences within the loop (Figure 3.4).

Recombinant techniques have recently provided further evidence for such a model. Simple deletions of sequences within the vWF 509-695 loop have consistently resulted in a loss of GPIb-binding (Sugimoto *et al.*, 1993; Prior *et al.*, 1993). Matsushita and Sadler (1995) studied an extensive series of full-length vWF mutants, in which clusters of charged residues within the A1 domain were mutated to alanine. They concluded that two anionic flanking segments (Glu497-Arg511 and Arg687-Val698) and a cationic intraloop region (Met540-Arg578) do indeed cooperate to inhibit GPIb binding, and proposed that the GPIb binding site includes the segment Glu596-Lys599 and other predominantly cationic regions from within the loop. It is highly likely that the binding pocket for GPIb is formed by discontinuous residues in a complex three-dimensional structure (Figure 3.5).

vWF Binding to Heparin and Sulfatides

The interaction between vWF and the sulfated polysaccharide, heparin, is potentially important as a model for vWF binding to closely related proteoglycans in the extracellular matrix and in view of the widespread use of heparins in anticoagulant therapy. Heparin inhibits vWF binding to platelets, both *in vitro* and *in vivo,* and this may contribute to hep-arin-associated bleeding despite adequate monitoring with anticoagulant assays (Sobel *et al.*, 1991). This effect appears due to the presence of a heparin-binding site within the vWF A1 domain. A peptide corresponding to vWF residues Tyr565-Ala587 was found to bind heparin with a comparable affinity to native vWF and to compete with intact vWF for hep-arin binding in solution (Sobel *et al.*, 1992; Figure 3.5). In later kinetic studies, a core sequence, Lys569-Ile580, bound heparin with similar affinity to 565-587 *via* electrostatic interactions (Tyler-Cross *et al.*, 1993). The structure of the vWF 565-587 region is unknown, although it shows sequence homology with other heparin-binding domains (Sobel *et al.*, 1992). Heparin binding to the A1 domain blocks the binding of other ligands including GPIb (Fujimura *et al.*, 1987b) and botrocetin (Andrews *et al.*, 1995a), as do sev-eral other polyanionic compounds. The mechanism of this heparin inhibition remains un-clear, but could include direct competition for cationic binding sites within the A1 domain, steric hindrance or heparin-induced shifts in A1 domain conformation. Thus, when vWF

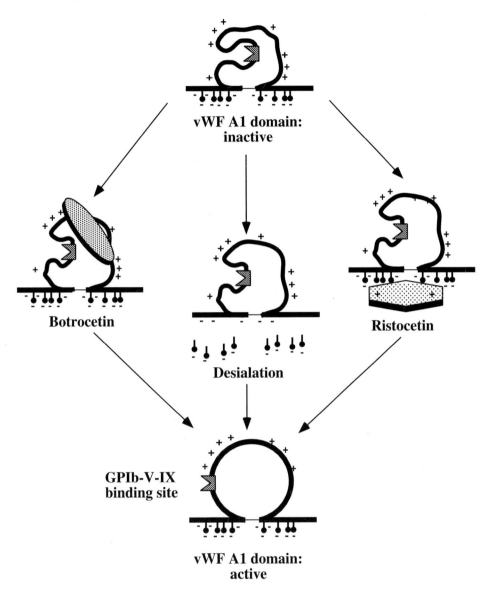

Figure 3.4 **An electrostatic model of vWF modulation**. The vWF A1 domain is shown schematically as a cationic disulfie-linked loop and anionic flanking sequences with sialated O-glycans (solid circles). The binding of modulators, botrocetin or ristocetin, and the loss of sialic acid residues may expose a GPIb-V-IX binding site by altering the conformation of the A1 domain.

associates *in vivo* with matrix and cell surfaces rich in sulfated proteoglycans, these heparin-like molecules may be important regulators of vWF-platelet binding.

vWF Binding to Fibrillar Collagens

Fibrillar collagens, particularly types I and III, form important structural components of the extracellular matrix and can directly bind and activate platelets *via* receptors such as the integrin $\alpha_2\beta_1$ (GPIa-IIa) and GPIV (reviewed by Kroll, 1994). The role of collagen as a

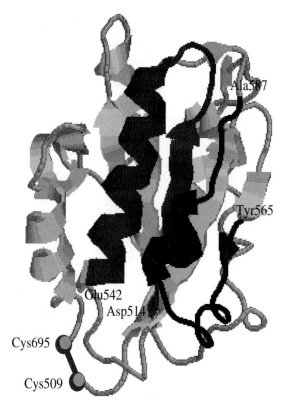

Figure 3.5   A theoretical structural model of vWF A1 domain. The sequence of the vWF A1 domain has been modeled in RasMac using coordinates kindly provided by J.E. Sadler and based on the crystal structures of homologous integrin I domains (Lee *et al.*, 1995; Qu and Leahy, 1995). The disulfide bond between Cys509-Cys695 is shown at lower left. The position of two potential ligand binding sites, the GPIb binding sequence Asp514-Glu542 (grey) and the heparin-binding sequence Tyr565-Ala587 (black) is also indicated.

platelet agonist has been studied mainly with collagen fibrils in suspension and will not be discussed here. Platelet adhesion to exposed collagen fibrils in the subendothelial matrix, on the other hand, is largely mediated by vWF (Aihara *et al.*, 1984) particularly under conditions of high shear forces (Houdijk *et al.*, 1985). vWF binds specifically to monomeric collagen, and coagulation Factor XIIIa can covalently cross-link vWF and collagen (Bockenstedt *et al.*, 1986). As for other binding interactions, the affinity of vWF for collagen (Santoro, 1983) and the ability of collagen-bound vWF to support platelet adhesion (Wu *et al.*, 1996) is directly related to multimer size. Defining a vWF binding site for fibrillar collagens has proved more difficult than for other ligands, with initial studies reporting that both the A1 and A3 domains contain recognition sites. Both purified vWF and a *S. aureus* V8 protease fragment of vWF, which includes all three A domains (Girma *et al.*, 1987), mediated platelet adhesion to fibrillar collagen in a perfusion chamber at high shear rates (Sakariassen *et al.*, 1986a). Subsequent studies of proteolytic vWF fragments (Pareti *et al.*, 1987) and reduced and alkylated cyanogen-bromide vWF peptides (Roth *et al.*, 1986) demonstrated that two distinct sites could interact with type I and III collagen. These sites, identified as 449-728 and 911-1114 (Pareti *et al.*, 1987) and more precisely as 542-622 and 948-998 (Roth *et al.*, 1986) lie within the A1 and A3 domain repeats, respectively (Titani *et al.*, 1986). In the case of vWF-collagen binding, functional studies with isolated peptides may not be an accurate guide to the interaction of two large and complex macromolecules.

In analyzing the relative contribution of the A1 and A3 domains to collagen binding, the use of recombinant vWF (rvWF) variants has been more informative. An rvWF fragment lacking the A1 domain (residues 478-716) bound collagen and Factor VIII with equal affinity to intact rvWF, but failed to bind heparin or platelets in the presence of ristocetin (Sixma *et al.*, 1991). Subsequently, an rvWF A3 domain polypeptide, but not an A1 domain fragment, was shown to inhibit the binding of multimeric vWF to immobilized collagen, whereas a hybrid A domain fragment (comprising residues 475-598 of the A1 domain and residues 1018-1114 of A3) inhibited vWF binding to both GPIb (as ristocetin-induced platelet agglutination) and collagen (Cruz *et al.*, 1995). These results are consistent with the previously described mapping of the heparin- and GPIb-binding sites to the A1 domain, but suggest that collagen binding is mediated solely by the A3 domain.

vWF Binding to Integrins

vWF binding to the activated platelet integrin, GPIIb-IIIa, provides one mechanism for platelet aggregation (reviewed by Peerschke, 1994) and may be an important part of thrombus formation at high shear rates (Weiss *et al.*, 1989). Integrin receptors, including platelet GPIIb-IIIa, bind to the sequence, Arg-Gly-Asp, in a divalent cation-dependent manner (Ginsberg *et al.*, 1993). An Arg-Gly-Asp-Ser (RGDS) sequence (residues 1744-1747) is present at the C-terminal end of the vWF subunit (Titani *et al.*, 1986). Monoclonal antibodies directed against the Arg-Gly-Asp sequence of vWF partially inhibited thrombus formation on rabbit subendothelium at high shear rates but had no effect at low shear rates (Weiss *et al.*, 1993). Arg-Ala-Asp-Ser- and Arg-Gly-Glu-Ser-containing vWF mutants competed with normal vWF for binding to collagen and to platelets in the presence of ristocetin, but did not compete for vWF binding to platelets activated by thrombin or ADP and did not support binding of cultured endothelial cells (Beacham *et al.*, 1992). This study provides evidence that GPIIb-IIIa and endothelial cell integrins bind vWF *via* the RGDS sequence. Platelets adhered to a similar vWF Arg-Gly-Gly-Ser mutant under static conditions, and to Type III collagen coated with this mutant under flow conditions, but did not spread or form large aggregates, suggesting that a GPIIb-IIIa-vWF interaction is necessary for these post-adhesion events (Lankhof *et al.*, 1995).

MODULATION OF vWF-GPIB BINDING

Despite extensive research on vWF, a key question remains to be answered: how is the inactive A1 domain of circulating vWF converted into the activated, GPIb-binding A1 domain of matrix-bound vWF? At a molecular level, this appears to be a complex process requiring the interaction of multiple regions of the A1 domain, inducing a specific shift in conformation. The study of the vWF-GPIb interaction *in vitro* is further complicated by the need to add modulators of vWF activity, which mimic the effect of vWF binding to subendothelial matrix. A diverse set of compounds, interventions and mutations share the ability to activate vWF-GPIb binding *in vitro*; the best characterized of these are described briefly below.

Ristocetin

Ristocetin is a glycoside antibiotic which induces vWF-dependent platelet agglutination (Howard and Firkin, 1971). Modulation of vWF by the cationic molecule ristocetin requires both electrostatic and specific hydrophobic interactions and appears to involve dimerization of ristocetin and cross-linking of vWF (Scott *et al.*, 1991). However, the interpretation of ristocetin-induced vWF-platelet binding is complicated by ristocetin's ability to bind nonspecifically to the platelet surface and to flocculate a variety of plasma proteins, including fibrinogen (Scott *et al.*, 1991). Several lines of evidence suggest that ristocetin interacts with anionic amino acid sequences flanking the Cys509-Cys695 bond of the vWF A1 domain, Cys474-Pro488 and Leu694-Pro708 (Girma *et al.*, 1990; Berndt *et al.*, 1992; Azuma *et al.*, 1993). The vWF sequences, Cys474-Pro488 and Leu694-Pro708 both contain polyproline repeats and anionic residues which form a specific recognition motif for ristocetin (Berndt *et al.*, 1992). Charge-to-alanine mutagenesis studies, however, suggest that additional residues within the loop (Glu626 and Asp520-Lys534) are involved specifically in ristocetin-induced modulation (Matsushita and Sadler, 1995); these residues may be important in the shape change that results from ristocetin binding to the flanking sequences.

Botrocetin

"Botrocetin" is the collective name for a group of proteins derived from the venom of the viper *Bothrops jararaca* (Read *et al.*, 1978; Andrews *et al.*, 1989a; Fujimura *et al.*, 1991), which also activate vWF, allowing it to bind to GPIb. Studies with anti-vWF monoclonal antibodies (Andrews *et al.*, 1989a; Girma *et al.*, 1990), a vWD variant (Howard *et al.*, 1984; Rabinowitz *et al.*, 1992) and selective mutagenesis (Matsushita and Sadler, 1995) indicate that botrocetin induces vWF activation *via* a different mechanism than ristocetin. Botrocetin binds directly to the A1 domain of vWF and, unlike ristocetin, does not appear to require an intact Cys509-Cys695 bond for modulation (Andrews *et al.*, 1989a; Sugimoto *et al.*, 1991). This suggests that botrocetin interacts with a different region of the A1 domain from ristocetin, possibly with residues within the disulfide loop. Synthetic peptides corresponding to three discontinuous sequences from within the A1 domain, Val539-Val553, Lys569-Gln583 and Arg629-Lys643, were reported to inhibit botrocetin binding to immobilized vWF (Sugimoto *et al.*, 1991). However, none of these peptides inhibited botrocetin-mediated vWF binding to platelets, and the three putative botrocetin-binding regions overlap with other identified binding sequences, namely the GPIbα-binding sequence Asp514-Glu542 (Berndt *et al.*, 1992), the heparin-binding sequence Tyr565-Ala587 (Sobel *et al.*, 1992) and a sulfatide-binding sequence Glu626-Val646 (Andrews *et al.*, 1995b). This may indicate that the same regions of the A1 domain can bind multiple ligands, including botrocetin.

Desialation

Removal of sialic acids with neuraminidase results in a form of vWF (asialo-vWF) with enhanced GPIb-binding affinity. Asialo-vWF agglutinates platelets independently of modulators (De Marco and Shapiro, 1981; Gralnick *et al.*, 1985). Asialo-vWF appears to agglutinate platelets by first binding to GPIb, then triggering activation of GPIIb-IIIa and further platelet-platelet interactions *via* GPIIb-IIIa binding to asialo-vWF and fibrinogen (De Marco *et al.*, 1985). This may explain the observation that asialo-vWF-induced platelet agglutination is more effective in the presence of fibrinogen and calcium ion (De Marco *et al.*, 1985). This modulation of the vWF A1 domain by desialation is consistent with an electrostatic model of activation (Figure 3.5), proposed by our group (Berndt *et al.*, 1992) and others (Matsushita and Sadler, 1995).

Shear Forces

Recent technical advances, in particular the development of the cone-plate aggregometer (Fukuyama *et al.*, 1989) have allowed continuous measurement of platelet agglutination generated by shear forces. This shear-induced agglutination involves vWF binding to GPIb (Chow *et al.*, 1992; Ikeda *et al.*, 1991; 1993). As in other systems, subsequent GPIIb-IIIa binding to vWF or fibrinogen reinforces the platelet-platelet interactions. The mechanism by which shear forces modulate vWF-GPIb binding is unknown. However, it may parallel the effect of high shear rates on matrix-bound vWF. In experimental models, platelet adhesion at high shear is exclusively vWF-dependent. This is thought to be a selection effect, whereby high shear disrupts most receptor-ligand interactions except the shear-resistant binding of vWF to GPIb (see above).

Mutations: Type 2B vWD

In most cases of vWD, GPIb-binding is deficient due to quantitative or qualitative changes in the vWF molecule (Ginsburg, 1992). There are also rare variants in which the affinity of vWF for GPIb is increased by mutations of vWF (Type 2B vWD) (Randi *et al.*, 1991). Type 2B vWD variants of vWF also mediate platelet agglutination in response to lower levels of shear stress than normal vWF and GPIb (Murata *et al.*, 1993). *In vivo*, the increased affinity of vWF-GPIb binding results in accelerated clearance of vWF and platelets from the circulation due to aggregate formation (Ruggeri *et al.*, 1982; Randi *et al.*, 1991). A considerable number of vWF mutations associated with type 2B vWD (Figure 3.6) have now been characterized (for references, see Ginsburg and Sadler, 1993; Sadler *et al.*, 1995; and their functional effects confirmed in recombinant vWF mutants (Hilbert *et al.*, 1995; Cooney and Ginsburg, 1996). All except five of the known Type 2B vWD mutations occur within a short segment of the A1 domain and mutations of Arg543, Arg545, Val553 and Arg578 make up over 90% of the known cases (Ginsburg and Sadler, 1993). This suggests that mutations in this region, particularly the loss of a charged Arg residue, activates the A1 domain. The clustering of Type 2B vWD mutations within a region that lies between proposed binding sequences for GPIb (Asp514-Glu542: Berndt *et al.*, 1992) and heparin (Tyr565-Ala587: Sobel *et al.*, 1992) raises the possibility that the relative spatial positions of these two binding sequences are important in binding. The recently-reported Type 2B vWD mutations at His505, Leu697, Ala698 (Ginsburg and Sadler, 1993) indicate that residues adjacent to the critical Cys509-Cys695 bond are also involved in maintaining native vWF in an inactive conformation.

vWF AND PLATELET ADHESION

vWF Binds to the Subendothelial Matrix

Platelet adhesion to exposed subendothelium is mediated by vWF bound to the extracellular matrix, a complex structure of macromolecules, including collagen and elastin fibers, proteoglycans and other glycoproteins, which supports and regulates the cells within it. Several early studies identified vWF as a component of the subendothelial matrix (Bloom *et al.*, 1973; Rand *et al.*, 1980) and showed that exogenous vWF can mediate platelet adhesion to matrix in a perfusion chamber (Sakariassen *et al.*, 1979; Bolhuis *et al.*, 1981). High molecular-weight vWF multimers released from Weibel-Palade bodies bound more avidly to a fibroblast-secreted matrix than smaller multimers (Sporn *et al.*, 1987) and vWF within human umbilical vein endothelial cell matrix consisted of extremely large multimers (Tannenbaum *et al.*, 1989). The selective incorporation of high molecular-weight vWF multimers into the matrix may maximize vWF adhesion to platelets and matrix macromolecules. Studies of platelet adhesion to matrix under shear (Baruch *et al.*, 1991) have suggested that matrix-bound vWF is more important than plasma vWF in mediating platelet adhesion.

The components of the subendothelial matrix which bind vWF *in vivo* have not been satisfactorily defined. Early studies focussed on the ability of vWF to interact *in vitro* with purified matrix components, particularly the fibrillar type I and III collagens and heparin (see below). Direct binding of vWF to fibrillar collagen has been demonstrated in both static (Santoro *et al.*, 1983; Bockenstedt *et al.*, 1986) and perfusion systems (Houdijk *et al.*, 1985) with preferential binding of high-molecular weight vWF multimers (Santoro *et al.*, 1983). These studies reinforced the concept that vWF interacted with fibrillar collagens

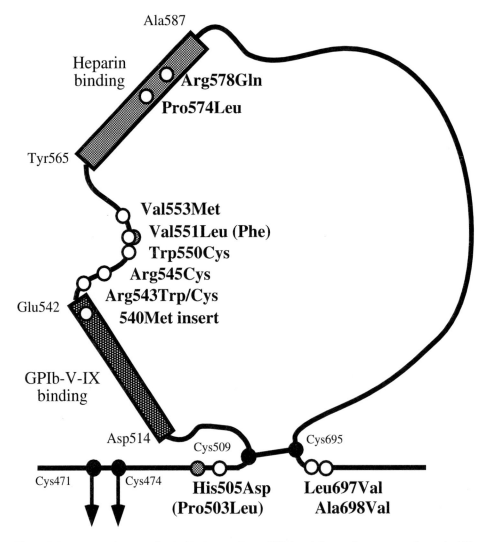

Figure 3.6 vWF mutations associated with the type 2B von Willebrand disease phenotype are clustered within the A1 domain of vWF.

within the matrix. However, two studies of vWF binding to intact matrices have provided evidence that fibrillar collagens are not the primary binding site for vWF. Firstly, both endogenous and exogenous vWF bound endothelial cell matrices from which fibrillar collagens had been removed by digestion with collagenase (Wagner *et al.*, 1984). Secondly, in perfusion experiments, anti-vWF monoclonal antibodies have been described which selectively inhibit vWF binding to collagen but not to subendothelial matrix, and *vice versa* (de Groot *et al.*, 1988).

The identity of the matrix vWF-binding component remains controversial. A 150-kDa vWF-binding protein from subendothelial matrix has been described and identified as type VI collagen (Rand *et al.*, 1991). Unlike the large fibrillar collagens, type VI collagen is resistant to collagenase digestion, consistent with the binding of vWF to collagenase-treated matrix. The molecule is composed of three chains, all of which have a short central collagenous triple-helical domain flanked by globular domains homologous to the A domains of vWF (Bonaldo *et al.*, 1989). The function of these A domains has not been characterized, but they may mediate type VI collagen adhesion to cells and other structures, in similar fashion to the vWF A1 and A3 domains (Chu *et al.*, 1990). This structural homology also raises the possibility of direct interactions between the A domains of vWF and type VI collagen, or of interactions with a third component which modulates their binding functions. The functional significance of vWF-type VI collagen binding is still under investigation. Platelets in whole blood adhered weakly to purified type VI collagen, but with optimal binding at low shear rates (300 s^{-1}) (Saelman *et al.*, 1994). This binding could be inhibited completely by an antibody against the collagen receptor, $\alpha_2\beta_1$. In another study, maximal platelet adhesion to type VI collagen occurred at low shear rates (100 s^{-1}), with poor adhesion at higher rates (1000 s^{-1}), and antibodies to GPIIb-IIIa inhibited platelet adhesion more effectively than anti-GPIb or aurintricarboxylic acid (an inhibitor of vWF-GPIb interaction) (Ross *et al.*, 1995). These studies suggest that platelet adhesion to type VI collagen may be mediated primarily by binding of the platelet integrins, GPIIb-IIIa and $\alpha_2\beta_1$, to Arg-Gly-Asp and other sequences within the collagenous regions of type VI collagen. Alternatively, platelet GPIIb-IIIa and GPIb could be recognising vWF bound to type VI collagen; blood from a severe vWD patient did show reduced platelet adhesion to type VI collagen (Ross *et al.*, 1995). A further question arising from these studies is the discrepancy in shear-rate dependence. Whereas several studies have shown that vWF-mediated platelet adhesion to matrix is maximal at high shear rates (see below), flow studies with purified type VI collagen show significant platelet adhesion only at low-shear rates (Saelman *et al.*, 1994; Ross *et al.*, 1995). This suggests that vWF ligands other than type VI collagen are necessary for platelet adhesion under shear.

vWF Mediates Platelet Adhesion to the Subendothelium

Thrombus formation on normal endothelium is prevented by a variety of secreted and membrane-bound inhibitors of platelet adhesion and activation (reviewed by Ware and Heistad, 1993). In contrast, exposure of the subendothelium triggers platelet adhesion and the formation of a platelet and fibrin thrombus. The role of vWF in this process has been investigated extensively over the last 15 years, largely through the use of annular (Baumgartner *et al.*, 1980a) or flat (Sakariassen *et al.*, 1983) perfusion chambers, as a means of modelling the interaction of platelets with matrix under flow conditions. A general consensus has emerged on the function of vWF in platelet-matrix interactions.

Firstly, vWF is the major mediator of platelet adhesion at high shear rates. Shear forces are generated by differences in velocity of adjacent blood elements as they approach the vessel wall and shear rates in the human body vary between 50 to 3000 s^{-1}, but could be significantly higher in a pathologically stenosed vessel (Slack *et al.*, 1993; Malek and Izumo, 1994). Anti-vWF antibodies inhibited platelet adhesion to subendothelium at shear rates (1300 s^{-1} and above) typical of the microvasculature, but had no effect at low shear rates (Baumgartner *et al.*, 1980b). Similar studies by other groups have confirmed that vWF is required for platelet adhesion to subendothelium at high shear rates (Weiss *et al.*, 1978; Stel *et al.*, 1985). At lower shear rates, more abundant matrix proteins such as fibronectin mediate platelet adhesion. However, fibronectin-mediated adhesion is readily reversed by increasing shear (Sixma and de Groot, 1991). Under *ex vivo* rather than *in vitro* conditions, vWF-mediated platelet adhesion can also be demonstrated at low shear rates (Badimon *et al.*, 1989) but the unique ability of vWF to bind platelets under high shear is likely to be more important physiologically.

Secondly, vWF-dependent platelet adhesion is primarily mediated *via* the platelet GPIb-IX-V complex. Patients with Bernard-Soulier syndrome, whose platelets lack functional GPIb-IX-V complex showed impaired platelet adhesion to subendothelium at high shear rates (Weiss *et al.*, 1978). An anti-GPIb antibody which blocked ristocetin-dependent platelet agglutination also inhibited platelet adhesion to subendothelium at shear rates of 500 and 1800 s^{-1}, but not at 300 s^{-1} (Sakariassen *et al.*, 1986b). Recombinant vWF fragments comprising the A1 domain have also been shown to inhibit shear-dependent platelet adhesion to extracellular matrix (Gralnick *et al.*, 1992; Dardik *et al.*, 1993). These studies suggest that GPIb is the major receptor for vWF-dependent platelet adhesion to matrix components.

In many of the experimental systems studied, however, inhibition of the vWF-GPIb interaction does not completely abolish platelet adhesion. This has led to the recognition that vWF is also binding to platelet GPIIb-IIIa, *via* the Arg1744-Gly1745-Asp1746 sequence. The binding of platelets to immobilized vWF under high shear was inhibited by both anti-GPIb and anti-GPIIb-IIIa monoclonal antibodies (Danton *et al.*, 1994). Monoclonal antibodies directed against GPIIb-IIIa (Sakariassen *et al.*, 1986b; Weiss *et al.*, 1989; Alevriadou *et al.*, 1993), antibodies against the vWF Arg1744-Asp1746 sequence (Weiss *et al.*, 1993) and GPIIb-IIIa-binding peptides (Weiss *et al.*, 1989) cause a partial inhibition of platelet binding to subendothelium at high shear rates. Thus, both vWF receptors may be required for maximal platelet binding. The relative contribution of GPIb-IX-V and GPIIb-IIIa may vary depending on the substrate. Platelet binding to collagen at 2600 s^{-1} could only be completely blocked by a combination of anti-vWF antibodies inhibiting vWF binding to GPIb, GPIIb-IIIa and collagen, suggesting that three functional regions of vWF may be involved (Fressinaud *et al.*, 1988). However, in a similar study of platelet adhesion to collagen type III under high shear, an Arg-Gly-Gly-Ser mutant vWF supported platelet binding and a mutant lacking the A1 domain did not (Lankhof *et al.*, 1995). In the case of platelet binding to fibrin, GPIIb-IIIa may be the primary receptor with vWF-GPIb acting as a secondary, shear-dependent pathway (Hantgan *et al.*, 1990). The overall evidence, however, favours GPIb as the primary receptor for matrix-associated vWF with GPIIb-IIIa binding occurring secondarily. This sequence of events is consistent with the observation that functional GPIb-IX-V complex is present on unactivated platelets whereas GPIIb-IIIa requires platelet activation before it can recognise ligand (see below).

A new insight into these sequential events has come from dynamic studies of platelet adhesion under flow (Savage *et al.*, 1996). This group noted that GPIb-dependent adhesion to immobilized vWF supported slow and continuous movement of platelets, even under extreme shear, followed eventually by arrest, whereas GPIIb-IIIa-dependent adhesion onto fibrinogen was irreversible but could occur only at low shear. They propose that the shear-resistant GPIb-vWF interaction captures and slows platelets on the substrate, allowing

GPIIb-IIIa to become activated and mediate permanent platelet adhesion. This conceptual model directly parallels leukocyte adhesion to endothelium (Carlos and Harlan, 1994), in which reversible, selectin-mediated cell rolling proceeds to irreversible integrin-mediated adhesion. Further studies may show that similar mechanisms have evolved to allow the selective capture of flowing platelets and leukocytes to a damaged vessel wall.

vWF-GPIb Binding Mediates Platelet Activation

Until recently, platelet GPIb-IX-V complex binding to vWF was viewed as a purely adhesive interaction with subsequent events required for platelet activation. Several reports now indicate that vWF binding to GPIb itself can trigger platelet activation. vWF-GPIb binding induced by ristocetin (Kroll *et al.*, 1991; Bertolino *et al.*, 1995), shear stress (Chow *et al.*, 1992; Ikeda *et al.*, 1993; Kroll *et al.*, 1993) and spontaneous binding of porcine vWF to human GPIb (Mazzucato *et al.*, 1996) caused a rise in intracellular calcium ion, with associated intracellular events such as phosphatidylinositol 4,5-bisphosphate degradation, protein kinase C activation, and thromboxane A_2 production (Kroll *et al.*, 1991; 1993). These events could be blocked by anti-GPIb monoclonal antibodies or aurintricarboxylic acid (Kroll *et al.*, 1991; Chow *et al.*, 1992), whereas anti-GPIIb-IIIa antibodies had no, or a partial effect (Bertolino *et al.*, 1995). These findings implicate vWF-GPIb interactions directly in the signalling mechanism. Details of this signalling pathway remain to be defined. Inhibition of cyclo-oxygenase metabolism with aspirin did not block shear-induced calcium influx (Chow *et al.*, 1992), but indomethacin inhibited the signaling response induced by ristocetin (Kroll *et al.*, 1991). The vWF-dependent protein kinase C activation induced by shear did not appear to require phosphatidylinositol metabolism (Kroll *et al.*, 1993). Activation of platelet tyrosine kinases (Razdan *et al.*, 1994) and protein tyrosine phosphorylation (Ozaki *et al.*, 1995) can be triggered by vWF binding to platelet GPIb.

The first link between vWF-GPIb binding and intracellular signalling pathways may have been established by recent collaborative work from this laboratory (Du *et al.*, 1994), identifying an intracellular 29-kDa protein associated with the GPIb-IX-V complex as the ζ-isoform of the 14-3-3 family of proteins (Du *et al.*, 1994). Experience from other biological systems suggests that 14-3-3 proteins modulate diverse cellular mitogenic and transformation signalling events (reviewed by Morrison, 1994). Evidence that vWF-GPIb binding can send signals to other surface glycoproteins came with the report that platelet GPIb binding to immobilized vWF modified the ligand binding ability of GPIIb-IIIa, such that it also bound to vWF (Savage *et al.*, 1992; 1996). This process of GPIIb-IIIa "activation" was inhibited by blocking protein kinase C activity, or by elevated intracellular cAMP levels, consistent with a GPIb-initiated signalling pathway. In addition, the cytoplasmic tail of GPIb interacts with the cytoskeleton, allowing vWF binding to trigger changes in cell morphology (Cunningham *et al.*, 1996).

CONCLUSIONS AND FUTURE DIRECTIONS

The multimeric glycoprotein vWF serves two major functions in normal haemostasis: binding and protecting Factor VIII from proteolysis and allowing platelets to adhere to exposed subendothelial matrix, under conditions of high shear severe enough to disrupt other receptor-ligand interactions. Deficient or abnormal vWF results in a bleeding disorder in both humans and animals, but the selective inhibition of vWF-dependent platelet adhesion could be a potent means of preventing arterial thrombosis in individuals at high risk due to stenosis or endothelial damage, such as that following coronary angioplasty. Detailed studies of vWF, outlined above, have localized binding sites for the platelet receptor GPIb-IX-V and ligands such as fibrillar collagens and heparin-like proteoglycans to the first and third A domains of vWF. Further work is needed to define the molecular mechanisms by which the A1 domain of matrix-bound vWF becomes capable of binding GPIb. *In vitro* modulators such as ristocetin, botrocetin and shear forces, appear to activate vWF by inducing a change in conformation of the A1 domain, exposing a cryptic GPIb-binding site. Understanding the structure and function of vWF A domains may allow the development of novel *in vivo* inhibitors of thrombosis and offer new insights into other adhesion receptors and matrix molecules belonging to the A-domain superfamily.

REFERENCES

Aihara, M., Cooper, H.A. and Wagner, R.H. (1984). Platelet-collagen interactions: increase in rate of adhesion of fixed washed platelets by Factor VIII-related antigen. *Blood*, **63**:495–501.

Alevriadou, B.R., Moake, J.L., Turner, N.A., Ruggeri, Z.M., Folie, B.J., Phillips, M.D. *et al.* (1993). Real-time analysis of shear-dependent thrombus formation and its blockade by inhibitors of von Willebrand Factor binding to platelets. *Blood*, **81**:1263–1276.

Andrews, R.K., Booth, W.J., Gorman, J.J., Castaldi, P.A. and Berndt, M.C. (1989a). Purification of botrocetin from *Bothrops jaracara* venom. Analysis of the botrocetin-mediated interaction between von Willebrand Factor and the human platelet membrane glycoprotein Ib-IX complex. *Biochemistry*, **28**:8317–8326.

Andrews, R.K., Gorman, J.J., Booth, W.J., Corino, G.L., Castaldi, P.A. and Berndt, M.C. (1989b). Cross-linking of a monomeric 39/34-kDa dispase fragment of von Willebrand Factor (Leu-480/Val-481-Gly-718) to the N-terminal region of the α-chain of membrane glycoprotein Ib on intact platelets with bis (sulfosuccinimidyl) suberate. *Biochemistry*, **28**:8326–8336.

Andrews, R.K., Bendall, L.J., Booth, W.J. and Berndt, M.C. (1995a). Inhibition of binding of von Willebrand Factor to the platelet glycoprotein Ib-IX complex, heparin and sulfatides by polyanionic compounds. The mechanism of modulation of the adhesive function of von Willebrand Factor. *Platelets*, **6**:252–258.

Andrews, R.K., Booth, W.J., Bendall, L.J. and Berndt, M.C. (1995b). The amino acid sequence glutamine-628 to valine-646 within the A1 repeat domain mediates binding of von Willebrand Factor to bovine brain sulfatides and equine tendon collagen. *Platelets*, **6**:245–251.

Azuma, H., Sugimoto, M., Ruggeri, Z.M. and Ware, J. (1993). A role for von Willebrand Factor proline residues 702–704 in ristocetin-mediated binding to platelet glycoprotein Ib. *Thrombosis and Haemostasis*, **69**:192–196.

Azzam, K., Garfinkel, L.I., Sollier, C.B.D., Thiam, M.C. and Drouet, L. (1995). Antithrombotic effect of a recombinant von Willebrand factor, VCL, on nitrogen laser-induced thrombus formation in guinea pig mesenteric arteries. *Thrombosis and Haemostasis*, **73**:318–323.

Badimon, L., Badimon, J.J., Turitto, V.T. and Fuster, V. (1989). Role of von Willebrand Factor in mediating platelet-vessel wall interaction at low shear rate: the importance of perfusion conditions. *Blood*, **73**:961–967.

Bahnak, B.R., Lavergne, J.-M., Ferreira, V., Kerbiriou, D. and Meyer, D. (1992). Comparison of the primary structure of the functional domains of human and porcine von Willebrand Factor that mediate platelet adhesion. *Biochemical and Biophysical Research Communications*, **182**:561–588.

Bahou, W.F., Ginsburg, D., Sikkink, R., Litwiller, R. and Fass, D.N. (1989). A monoclonal antibody to von Willebrand Factor (vWF) inhibits Factor VIII binding. Localization of its antigenic determinant to a nonadecapeptide at the amino terminus of the mature vWF polypeptide. *The Journal of Clinical Investigation*, **84**:56–61.

Bakhshi, M.R., Meyers, J., Howard, P., Soprano, D. and Kirby, E. (1992). Sequencing of the primary adhesion domain of bovine von Willebrand Factor. *Biochimica et Biophysica Acta*, **1132**:325–328.

Baruch, D., Denis, C., Marteaux, C., Schoevart, D., Coulombel, L. and Meyer, D. (1991). Role of von Willebrand Factor associated to extracellular matrices in platelet adhesion. *Blood*, **77**:519–527.

Baumgartner, H.R., Turitto, V. and Weiss, H.J. (1980a). Effect of shear rate on platelet interaction with subendothelium in citrated and native blood II. Relationships among platelet adhesion, thrombus dimensions, and fibrin formation. *Journal of Laboratory and Clinical Medicine*, **95**:208–221.

Baumgartner, H.R., Tschopp, T.B. and Meyer, D. (1980b). Shear rate dependent inhibition of platelet adhesion and aggregation on collagenous surfaces by antibodies to human Factor VIII/von Willebrand Factor. *British Journal of Haematology*, **44**:127–139.

Beacham, D.A., Wise, R.J., Turci, S.M. and Handin, R.I. (1992). Selective inactivation of the Arg-Gly Asp-Ser (RGDS) binding site in von Willebrand Factor by site-directed mutagenesis. *The Journal of Biological Chemistry*, **267**:3409–3415.

Berndt, M.C., Du, X. and Booth, W.J. (1988). Ristocetin-dependent reconstitution of binding of von Willebrand Factor to purified human platelet membrane glycoprotein Ib-IX complex. *Biochemistry*, **27**:633–640.

Berndt, M.C., Ward, C.M., Booth, W.J., Castaldi, P.A., Mazurov, A.V. and Andrews, R.K. (1992). Identification of aspartic acid 514 through glutamic acid 542 as a glycoprotein Ib-IX complex receptor recognition sequence in von Willebrand Factor. Mechanism of modulation of von Willebrand Factor by ristocetin and botrocetin. *Biochemistry*, **31**:11144–11151.

Bertolino, G., Noris, P., Spedini, P. and Balduini, C.L. (1995). Ristocetin-induced platelet agglutination stimulates GPIIb/IIIa-dependent calcium influx. *Thrombosis and Haemostasis*, **73**:689–692.

Bloom, A.L., Giddings, J.C. and Wilks, C.J. (1973). Factor VIII on the vascular intima: possible importance in haemostasis and thrombosis. *Nature*, **241**:217–219.

Bockenstedt, P., McDonagh, J. and Handin, R.I. (1986). Binding and covalent cross-linking of purified von Willebrand Factor to native monomeric collagen. *The Journal of Clinical Investigation*, **78**:551–556.

Bolhuis, P.A., Sakariassen, K.S., Sander, H.J., Bouma, B.N. and Sixma, J.J. (1981). Binding of factor VIII-von Willebrand factor to human arterial subendothelium precedes increased platelet adhesion and enhances platelet spreading. *Journal of Laboratory and Clinical Medicine*, **97**:568–576.

Bonaldo, P., Russo, V., Bucciotti, F., Bressan, G.M. and Colombatti, A. (1989). α1 chain of chick Type VI collagen. The complete cDNA sequence reveals a hybrid molecule made of one short collagen and three von Willebrand Factor type A-like domains. *The Journal of Biological Chemistry*, **264**:5575–5580.

Bonthron, D.T., Handin, R.I., Kaufman, R.J., Wasley, L.C., Orr, E.C., Mitsock, L.M. *et al.* (1986). Structure of pre-pro-von Willebrand factor and its expression in heterologous cells. *Nature*, **324**:270–273.

Booth, W.J., Furby, F.H., Berndt, M.C. and Castaldi, P.A. (1984). Factor VIII/von Willebrand Factor has potent lectin activity. *Biochemical and Biophysical Research Communications*, **118**:495–901.

Booth, W.J., Andrews, R.K., Castaldi, P.A. and Berndt, M.C. (1990). The interaction of von Willebrand Factor and the platelet glycoprotein Ib-IX complex. *Platelets*, **1**:169–176.

Carlos, T.M. and Harlan, J.M. (1994). Leukocyte-endothelial adhesion molecules. *Blood*, **84**:2068–2101.

Castaman, G. and Rodeghiero, F. (1995). Current management of von Willebrand's disease. *Drugs*, **50**:602–614.

Chow, T.W., Hellums, J.D., Moake, J.L. and Kroll, M.H. (1992). Shear-stress-induced von Willebrand Factor binding to platelet glycoprotein Ib initiates calcium influx associated with aggregation. *Blood*, **80**:113–120.

Chu, M.-L., Pan, T.-C., Conway, D., Saitta, B., Stokes, D., Kuo, H.-J. *et al.* (1990). The structure of Type VI collagen. *Annals of the New York Academy of Sciences, USA*, **580**:55–63.

Colombatti, A. and Bonaldo, P. (1991). The superfamily of proteins with von Willebrand Factor type A-like domains: one theme common to components of extracellular matrix, hemostasis, cellular adhesion, and defense mechanisms. *Blood*, **77**:2305–2315.

Colombatti, A., Bonaldo, P. and Doliana, R. (1993). Type A modules: interacting domains found in several non-fibrillar collagens and in other extracellular matrix proteins. *Matrix*, **13**:297–306.

Cooney, K.A., Lyons, S.E. and Ginsburg, D. (1992). Functional analysis of a type IIb von Willebrand disease missense mutation: increased binding of large von Willebrand Factor multimers to platelets. *Proceedings of the National Academy of Science, USA*, **89**:2869–2872.

Cooney, K.A., and Ginsburg, D. (1996). Comparative analysis of Type 2B von Willebrand disease mutations: implications for the mechanism of von Willebrand Factor binding to platelets. *Blood*, **87**:2322–2328.

Cruz, M.A., Yuan, H., Lee, J.R., Wise, R.J. and Handin, R.I. (1995). Interaction of the von Willebrand factor (vWF) with collagen. Localization of the primary collagen binding site by analysis of recombinant vWF A domain polypeptides. *The Journal of Biological Chemistry*, **270**:10822–10827.

Cunningham, J.G., Meyer, S.C. and Fox, J.E.B. (1996). The cytoplasmic domain of the α-subunit of glycoprotein (GP) Ib mediates attachment of the entire GPIb-IX complex to the cytoskeleton and regulates von Willebrand factor-induced changes in cell morphology. *The Journal of Biological Chemistry*, **271**:11581–7.

Danton, M.C., Zaleski, A., Nichols ,W.L. and Olson, J.D. (1994). Monoclonal antibodies to platelet glycoproteins Ib and IIb/IIIa inhibit adhesion of platelets to purified solid-phase von Willebrand Factor. *Journal of Laboratory and Clinical Medicine*, **124**:274–282.

Dardik, R., Ruggeri, Z.M., Savion, N., Gitel, S., Martinowitz, U., Chu, V. *et al.* (1993). Platelet aggregation on extracellular matrix: effect of a recombinant GPIb-binding fragment of von Willebrand Factor. *Thrombosis and Haemostasis*, **70**:522–526.

De Groot, P.G., Ottenhof-Rovers, M., van Mourik, J.A. and Sixma, J.J. (1988). Evidence that the primary binding site of von Willebrand Factor that mediates platelet adhesion on subendothelium is not collagen. *The Journal of Clinical Investigation*, **82**:65–73.

De Marco, L., Shapiro, S.S. (1981). Properties of human asialo-Factor VIII. A ristocetin-independent platelet-aggregating agent. *The Journal of Clinical Investigation*, **68**:321–328.

De Marco, L., Girolami, A., Russell, S. and Ruggeri, Z.M. (1985). Interaction of asialo von Willebrand Factor with glycoprotein Ib induces fibrinogen binding to the glycoprotein IIb/IIIa complex and mediates platelet aggregation. *The Journal of Clinical Investigation*, **75**:1198–1203.

Dent, J.A., Berkowitz, S.D., Ware, J., Kasper, C.K. and Ruggeri, Z.M. (1990). Identification of a cleavage site directing the immunochemical detection of molecular abnormalities in type IIA von Willebrand factor. *Proceedings of the National Academy of Science, USA*, **87**:6306.

Du, X., Harris, S.J., Tetaz, T.J., Ginsberg, M.H. and Berndt, M.C. (1994). Association of a phospholipase A_2 (14-3-3 protein) with the platelet glycoprotein Ib-IX complex. *The Journal of Biological Chemistry*, **269**:18287–18290.

Eikenboom, J.C.J., Rietsma, P.H. and Briet, E. (1995). The inheritance and molecular genetics of von Willebrand's disease. *Haemophilia*, **1**:77–90.

Foster, P.A., Fulcher, C.A., Marti, T., Titani, K. and Zimmerman, T.S. (1987). A major Factor VIII binding domain resides within the amino-terminal 272 amino acid residues of von Willebrand Factor. *The Journal of Biological Chemistry*, **262**:8443–8446.

Fressinaud, E., Baruch, D., Girma, J.-P., Sakariassen, K.S., Baumgartner, H.R. and Meyer, D. (1988). von Willebrand factor-mediated platelet adhesion to collagen involves platelet membrane glycoprotein IIb-IIIa as well as glycoprotein Ib. *Journal of Laboratory and Clinical Medicine*,**112**:58–67.

Fujimura, Y., Titani, K., Holland, L.Z., Russell, S.R., Roberts, J.R., Elder, J.H. *et al.* (1986). von Willebrand Factor. A reduced and alkylated 52/48-kda fragment beginning at amino acid residue 449 contains the domain interacting with platelet glycoprotein Ib. *The Journal of Biological Chemistry*, **261**:381–385.

Fujimura, Y., Holland, L.Z., Ruggeri, Z.M. and Zimmerman, T.S. (1987a). The von Willebrand Factor domain-mediating botrocetin-induced binding to glycoprotein Ib lies between Val_{449} and Lys_{728}. *Blood*, **70**:985–988.

Fujimura, Y., Titani, L., Holland, L.Z., Roberts, J.R., Kostel, P., Ruggeri, Z.M. *et al* (1987b). A heparin-binding domain of human von Willebrand Factor. Characterization and localization to a tryptic fragment extending from amino acid residue Val-449 to Lys-728. *The Journal of Biological Chemistry*, **262**:1734–1739.

Fujimura, Y., Titani, K., Usami, Y., Suzuki, M., Oyama, R., Matsui, T. *et al.* (1991). Isolation and chemical characterization of two structurally and functionally distinct forms of botrocetin, the platelet coagglutinin isolated from the venom of *Bothrops jararaca*. *Biochemistry*, **30**:1957–1964.

Fukuyama, M., Satai, K., Itagaki, I., Kawano, K., Murata, M., Kawai, Y. *et al.* (1989). Continuous measurement of shear-induced platelet aggregation. *Thrombosis Research*, **54**:253–260.

Furlan, M., Robles, R. and Lämmle, B. (1996). Partial purification and characterization of a protease from human plasma cleaving von Willebrand factor to fragments produced by in vivo proteolysis. *Blood*, **87**:4223–4234.

Ginsberg, M.H., Du, X., O'Toole,T.E., Loftus, J.C. and Plow, E.F. (1993). Platelet Integrins. *Thrombosis and Haemostasis*, **70**:87–93.

Ginsburg, D. (1992). Biology of inherited coagulopathies: von Willebrand Factor. *Haematology/Oncology Clinics of North America*, **6**:1011–1020.

Ginsburg, D., Bockenstedt, P.L., Allen, E.A., Fox, D.A., Foster, P.A., Ruggeri, Z.M. *et al.* (1992). Fine mapping of monoclonal antibody epitopes on human von Willebrand Factor using a recombinant peptide library. *Thrombosis and Haemostasis*, **67**:166–171.

Ginsburg, D., and Sadler, J.E. (1993). von Willebrand disease: a database of point mutations, insertions and deletions. *Thrombosis and Haemostasis*, **69**:177–184.

Girma, J.P., Chopek, M.W., Titani, K. and Davie, E.W. (1986). Limited proteolysis of human von Willebrand Factor by *Staphylococcus aureus* V-8 protease: isolation and partial characterization of a platelet-binding domain. *Biochemistry*, **25**:3156–3163.

Girma, J.P., Takahashi, Y., Yoshioka, A., Diaz, J. and Meyer, D. (1990). Ristocetin and botrocetin involve two distinct domains of von Willebrand factor for binding to platelet membrane glycoprotein Ib. *Thrombosis and Haemostasis*, **64**:326–332.

Girma, J.P., Fressinaud, E., Christophe, O., Roualt, C., Obert, B., Takahashi, Y. *et al.* (1992). Aurin tricarboxylic acid inhibits platelet adhesion to collagen by binding to the 509–695 disulphide loop of von Willebrand Factor and competing with glycoprotein Ib. *Thrombosis and Haemostasis*, **68**:707–713.

Goto, S., Salomon, D.R., Ikeda, Y. and Ruggeri, Z.M. (1995). Characterization of the unique mechanism mediating the shear-dependent binding of soluble von Willebrand factor to platelets. *The Journal of Biological Chemistry*, **270**:23352–23361.

Gralnick, H.R., Williams, S.B. and Morisato, D.K. (1981). Effect of the multimeric structure of the Factor VIII/von Willebrand Factor protein on binding to platelets. *Blood*, **58**:387–397.

Gralnick, H.R., Williams, S.B. and Coller, B.S. (1985). Asialo von Willebrand Factor interactions with platelets. Interdependence of glycoproteins Ib and IIb/IIIa for binding and aggregation. *The Journal of Clinical Investigation*,**75**:19–25.

Gralnick, H.R., Williams, S., McKeown, L., Kramer, W., Krutzsch, H., Gorecki, M. *et al.* (1992). A monomeric von Willebrand factor fragment, Leu-504-Lys-728, inhibits von Willebrand factor interaction with glycoprotein Ib-IX. *Proceedings of the National Academy of Science, USA*, **89**:7880–7884.

Hantgan, R.R., Hindriks, G., Taylor, R.G., Sixma, J.J. and de Groot, P.G. (1990). Glycoprotein Ib, von Willebrand Factor, and glycoprotein IIb:IIIa are all involved in platelet adhesion to fibrin in flowing whole blood. *Blood*, **76**:345–353.

Hilbert, L., Gaucher, C. and Mazurier, C. (1995). Effects of different amino-acid substitutions in the leucine 694-proline 708 segment of recombinant von Willebrand factor. *British Journal of Haematology*, **91**:983–990.

Houdijk, W.P.M., Sakariassen, K.S., Nievelstein, P.F.E.M. and Sixma, J.J. (1985). Role of Factor VIII-von Willebrand Factor and fibronectin in the interaction of platelets in flowing blood with monomeric and fibrillar human collagen Types I and III. *The Journal of Clinical Investigation*, **75**:531–540.

Howard, M.A., and Firkin, B.G. (1971). Ristocetin — a new tool in the investigation of platelet aggregation. *Thrombosis Diatheca Haematologia*, **26**:362–369.

Howard, M.A., Perkin, J., Salem, H.H. and Firkin, B.G. (1984). The agglutination of human platelets by botrocetin: evidence that botrocetin and ristocetin act at different sites on the factor VIII molecule and platelet membrane. *British Journal of Haematology*, **57**:25–35.

Ikeda, Y., Handa, M., Kawano, K., Kamata,T., Murata, M., Araki, Y. *et al.* (1991). The role of von Willebrand Factor and fibrinogen in platelet aggregation under varying shear stress. *The Journal of Clinical Investigation*,**87**:1234–1240.

Ikeda, Y., Handa, M., Kamata, T., Kawano, K., Kawai, Y.,Watanabe, K. *et al.* (1993). Transmembrane calcium influx associated with von Willebrand Factor binding to GP Ib in the initiation of shear-induced platelet aggregation. *Thrombosis and Haemostasis,* **69**:496–502.

Jahroudi, N., and Lynch, D.C. (1994). Endothelial-Cell-Specific Regulation of von Willebrand Factor Gene Expression. *Molecular and Cellular Biology*, **14**:999–1008.

Jakway, J.L. (1992). Acquired von Willebrand's disease in malignancy. *Seminars in Thrombosis and Haemostasis*, **18**:434–439.

Jorieux, S., Gaucher, C., Piétu, G., Chérel, G., Meyer, D. and Mazurier, C. (1994). Fine epitope mapping of monoclonal antibodies to the NH$_2$-terminal part of von Willebrand factor (vWF) by using recombinant and synthetic peptides: interest for the localization of the factor VIII binding domain. *British Journal of Haematology*, **87**:113–118.

Journet, A.M., Saffaripour, S. and Wagner, D.D. (1993). Requirement for both D domains of the propolypeptide in von Willebrand Factor multimerization and storage. *Thrombosis and Haemostasis*, **70**:1053–1057.

Kaufman, R.J. (1992). Biological regulation of Factor VIII activity. *Annual Review of Medicine*, **43**:325–339.

Koppelman, S.J., van Hoeij, M., Vink, T., Lankhof, H., Schiphorst, M.E.,Damas, C. *et al.* (1996). Requirements of von Willebrand factor to protect factor VIII from inactivation by activated protein C. *Blood*, **87**:2292–2300.

Kroll, M.H., Harris, T.S., Moake, J.L., Handin, R.I. and Schafer, A.I. (1991). von Willebrand Factor binding to platelet GPIb initiates signals for platelet activation. *The Journal of Clinical Investigation*, **88**:1568–1573.

Kroll, M.H., Hellums, J.D., Guo, Z., Durante, W., Razdan, K., Hroblich, J.K. *et al.* (1993). Protein kinase C is activated in platelets subjected to pathological shear stress. *The Journal of Biological Chemistry*, **268**:3520–3524.

Kroll, M.H. (1994). Mechanisms of platelet activation. In *Thrombosis and Hemorrhage*, edited by J. Loscalzo and A.I. Schafer, pp. 247–278. Boston:Blackwell Scientific Publications.

Lankhof, H., Wu, Y.-P., Vink, T., Schiphorst, M.E., Zerwes, H.-G., de Groot, P.G. *et al.* (1995). Role of the glycoprotein Ib-binding A1 repeat and the RGD sequence in platelet adhesion to human recombinant von Willebrand factor. *Blood*, **86**:1035–1042.

Larson, R.S., Corbi, A.L., Berman, L. and Springer, T. (1989). Primary structure of the Leukocyte Function-Associated molecule-1 α subunit: an integrin with an embedded domain defining a protein superfamily. *The Journal of Cell Biology,***108**:703–712.

Layet, S., Girma, J.-P., Obert, B., Peynaud-Debayle, E., Bihoreau, N. and Meyer, D. (1992). Evidence that a secondary binding and protecting site for Factor VIII on von Willebrand factor is highly unlikely. *Biochemical Journal*, **282**:129–137.

Lee, J., Rieu, P., Arnaout, M.A. and Liddington, R. (1995). Crystal structure of the A domain from the α subunit of integrin CR3 (CD11b/CD18). *Cell*, **80**:631–638.

Lind, S.E. (1994). Platelet morphology. In *Thrombosis and Hemorrhage*, edited by J. Loscalzo and A.I. Schafer, pp. 201–218. Boston:Blackwell Scientific Publications.

López, J.A. (1994). The platelet glycoprotein Ib-IX complex. *Blood Coagulation and Fibrinolysis*, **5**:97–119.

Lyons, S.E., and Ginsburg, D. (1994). Molecular and cellular biology of von Willebrand Factor. *Trends in Cardiovascular Medicine*, **4**:34–39.

Lyons, S.E., Cooney, K.A., Bockenstedt, P. and Ginsburg, D. (1994). Characterization of Leu777Pro and Ile865Thr Type IIA von WIllebrand disease mutations. *Blood*, **83**:1551–1557.

Malek, A.M. and Izumo, S. (1994). Molecular aspects of signal transduction of shear stress in the endothelial cell. *Journal of Hypertension*, **12**:989–999.

Mancuso, D.J., Tuley, E.A., Westfield, L.A., Worrall, N.K., Shelton-Inloes, B.B., Sorace, J.M. *et al.* (1989). Structure of the gene for human von Willebrand Factor. *The Journal of Biological Chemistry*, **264**:19514–19527.

Matsushita, T. and Sadler, J.E. (1995). Identification of amino acid residues essential for von Willebrand factor binding to platelet glycoprotein Ib. Charged-to-alanine scanning mutagenesis of the A1 domain of human von Willebrand factor. *The Journal of Biological Chemistry*, **270**:13406–13414.

Mazzucato, M., Marco, L.D., Pradella, P., Masotti, A. and Pareti, F.I. (1996). Porcine von Willebrand factor binding to human platelet GPIb induces transmembrane calcium influx. *Thrombosis and Haemostasis*, **75**:655–60.

Meyer, D. and Girma, J.-P. (1993). von Willebrand Factor: structure and function. *Thrombosis and Haemostasis*, **70**:99–104.

Mohri, H., Fujimura, Y., Shima, M., Yoshioka, A., Houghten, R.A., Ruggeri, Z.M. *et al.* (1988). Structure of the von Willebrand Factor domain interacting with glycoprotein Ib. *The Journal of Biological Chemistry*, **263**:17901–17904.

Morrison, D. (1994). 14-3-3: Modulators of Signaling Proteins? *Science*, **266**:56–57.

Murata, M., Russell, S.R., Ruggeri, Z.M. and Ware, J. (1993). Expression of the phenotypic abnormality of platelet-type von Willebrand disease in a recombinant glycoprotein Ibα fragment. *The Journal of Clinical Investigation*,**91**:2133–2137.

Ozaki, Y., Satoh, K., Yatomi, Y., Miura, S., Fujimura, Y. and Kume, S. (1995). Protein tyrosine phosphorylation in human platelets induced by interaction between glycoprotein Ib and von Willebrand factor. *Biochimica et Biophysica Acta*, **1243**:482–488.

Pareti, F.I., Fujimura, Y., Dent, J.A., Holland, L.Z., Zimmerman, T.S. and Ruggeri, Z.M. (1986). Isolation and characterization of a collagen binding domain in human von Willebrand Factor. *The Journal of Biological Chemistry*, **261**:15310–15315.

Pareti, F.I., Niiya, K., McPherson, J.M. and Ruggeri, Z.M. (1987). Isolation and characterization of two domains of human von Willebrand Factor that interact with fibrillar collagen Types I and III. *The Journal of Biological Chemistry*, **262**:13835–13841.

Peerschke, E.I.B. (1994). Platelet membranes and receptors. In *Thrombosis and Hemorrhage*, edited by J. Loscalzo and A.I. Schafer, pp. 219–246. Boston:Blackwell Scientific Publications.

Piétu,G., Ribba, A.-S., Chérel, G., Siguret, V., Obert, B., Roualt, C. *et al.* (1994). Epitope mapping of inhibitory monoclonal antibodies to human von Willebrand Factor by using recombinant cDNA libraries. *Thrombosis and Haemostasis*, **71**:788–92.

Prior, C.P., Chu, V., Cambou, B., Dent, J.A., Ebert, B., Gore, R. *et al.* (1993). Optimization of a recombinant von Willebrand Factor fragment as an antagonist of the platelet glycoprotein Ib receptor. *Bio/Technology*, **11**:709–713.

Qu, A., and Leahy, D.J. (1995). Crystal structure of the I-domain from the CD11a/CD18 (LFA-1, $\alpha_L\beta_2$). integrin. *Proceedings of the National Academy of Science, USA*, **92**:10277–10281.

Rabinowitz, I., Tuley, E.A., Mancuso, D.J., Randi, A.M., Firkin, B.G., Howard, M.A. *et al.* (1992). von Willebrand disease type B: A missense mutation selectively abolishes ristocetin-induced von Willebrand Factor binding to platelet glycoprotein Ib. *Proceedings of the National Academy of Science, USA*, **89**:9846–9849.

Rand, J.H., Sussman, I.I., Gordon, R.E., Chu, S.V. and Solomon, V. (1980). Localization of Factor-VIII-related antigen in human vascular subendothelium. *Blood*, **55**:752–756.

Rand, J.H., Patel, N.D., Schwartz, E., Zhou, S.-L. and Potter, B.J. (1991). 150-kD von Willebrand Factor binding protein extracted from human vascular subendothelium is Type VI collagen. *The Journal of Clinical Investigation*, **88**:253–259.

Randi, A.M., Rabinowitz, I., Mancuso, D.J., Mannucci, P.M. and Sadler, J.E. (1991). Molecular basis of von Willebrand disease Type IIB. Candidate mutations cluster in one disulfide loop between proposed platelet glycoprotein Ib binding sequences. *The Journal of Clinical Investigation*, **87**:1220–1226.

Randi, A.M., Jorieux, S., Tuley, E.A., Mazurier, C. and Sadler, J.E. (1992). Recombinant von Willebrand Factor Arg578>Gln. A Type IIB von Willebrand disease mutation affects binding to glycoprotein Ib but not to collagen or heparin. *The Journal of Biological Chemistry*, **267**:21187–21192.

Razdan, K., Hellums ,J.D. and Kroll, M.H. (1994). Shear-stress-induced von Willebrand factor binding to platelets causes the activation of tyrosine kinase(s). *Biochemical Journal*, **302**:681–686.

Read, M.S., Shermer R.W., and Brinkhous, K.M. (1978). Venom coagglutinin: An activator of platelet aggregation dependent on von Willebrand factor. *Proceedings of the National Academy of Science, USA*, **75**:4514–4518.

Ribba, A.-S., Voorberg,, J. Meyer, D., Pannekoek, H. and Pietu, G. (1992). Characterization of recombinant von Willebrand Factor corresponding to mutations in Type IIA and Type IIB von Willebrand disease. *The Journal of Biological Chemistry*, **267**:23209–23215.

Ribba, A.-S., Christophe, O., Derlon, A., Cherel, G., Siguret, V., Lavergne, J.M. *et al.* (1994). Discrepancy between IIA phenotype and IIB genotype in a patient with a variant of von Willebrand disease. *Blood*, **83**:833–841.

Ross, J.M., McIntyre, L.V., Moake, J.L. and Rand, J.H. (1995). Platelet adhesion and aggregation on human type VI collagen surfaces under physiological flow conditions. *Blood*, **85**:1826–1835.

Roth, G., Titani, K. Hoyer, L.W. and Hickey, M.J. (1986). Localization of binding sites within human von Willebrand Factor for monomeric Type III collagen. *Biochemistry*, **25**:8357–8361.

Ruggeri, Z.M., Lombardi, R., Gatti, L., Bader, R., Valsecchi C. and Zimmerman, T.S. (1982). Type IIB von Willebrand's disease: differential clearance of endogenous versus transfused large multimer von Willebrand factor. *Blood*, **60**:1453–1456.

Ruggeri, Z.M. and Ware, J. (1992). The structure and function of von Willebrand Factor. *Thrombosis and Haemostasis*, **67**:594–599.

Sadler, J.E. (1994). A revised classification of von Willebrand disease. *Thrombosis and Haemostasis*, **71**:520–525.

Sadler, J.E., Matsushita, T., Dong, Z., Tuley, E.A. and Westfield, L.A. (1995). Molecular mechanism and classification of von Willebrand disease. *Thrombosis and Haemostasis*, **74**:161–6.

Saelman, E.U.M., Nieuwenhuis, H.K., Hese, K.M., de Groot, P.G., Heijnen, H.F.G., Sage, E.H. *et al.* (1994). Platelet adhesion to collagen Types I through VIII under conditions of stasis and flow is mediated by GPIa/IIa ($\alpha_2\beta_1$-Integrin). *Blood*, **83**:1244–1250.

Sakariassen, K.S., Bolhuis, P.A. and Sixma, J.J. (1979). Human blood platelet adhesion to artery subendothelium is mediated by factor VIII-von Willebrand factor bound to the subendothelium. *Nature*, **279**:635–638.

Sakariassen, K.S., Aarts, P.A.M.M., de Groot, P.G., Houdijk, W.P.M. and Sixma, J.J. (1983). A perfusion chamber developed to investigate platelet interaction in flowing blood with human vessel wall cells, their extracellular matrix, and purified components. *Journal of Laboratory and Clinical Medicine*, **102**:522–535.

Sakariassen, K.S., Fressinaud, E., Girma, J.P., Baumgartner, H.R. and Meyer, D. (1986a). Mediation of platelet adhesion to fibrillar collagen in flowing blood by a proteolytic fragment of human von Willebrand Factor. *Blood*, **67**:1515–1518.

Sakariassen, K.J., Nievelstein, P.F.E.M., Coller, B.S. and Sixma, J.J. (1986b). The role of platelet membrane glycoproteins Ib and IIb-IIIa in platelet adherence to human artery subendothelium. *British Journal of Haematology*, **63**:681–691.

Santoro, S.A. (1983). Preferential binding of high molecular weight forms of von Willebrand Factor to fibrillar collagen. *Biochimica et Biophysical Acta*, **756**:123–126.

Savage, B., Shattil, S.J. and Ruggeri, Z.M. (1992). Modulation of platelet function through adhesion receptors. A dual role for Glycoprotein IIb-IIIa (Integrin $\alpha_{IIb}\beta_3$) mediated by fibrinogen and glycoprotein Ib-von Willebrand Factor. *The Journal of Biological Chemistry*, **267**:11300–11306.

Savage, B., Saldivar, E. and Ruggeri, Z.M. (1996). Initiation of platelet adhesion by arrest onto fibrinogen or translocation on von Willebrand factor. *Cell*, **84**:289–297.

Schneppenheim, R., Thomas, K.B. and Sutor, A.H. (1995). von Willebrand disease in childhood. *Seminars in Thrombosis and Hemostasis*, **21**:261–275.

Scott, J.P., Montgomery, R.R. and Retzinger, G.S. (1991). Dimeric ristocetin flocculates proteins, binds to platelets, and mediates von Willebrand factor-dependent agglutination of platelets. *The Journal of Biological Chemistry*, **266**:8149–8155.

Shelton-Inloes, B.B., Titani, K. and Sadler, J.E. (1986). cDNA sequences for human von Willebrand Factor reveal five types of repeated domains and five possible protein sequence polymorphisms. *Biochemistry*, **25**:3164–3171.

Sixma, J.J., Sakariassen, K.S., Stel, H.V., Houdijk, W.P.M., de Maur, D.W.I., Hamer, R.J. *et al.* (1984). Functional domains on von Willebrand Factor. Recognition of discrete tryptic fragments by monoclonal antibodies that inhibit interaction of von Willebrand Factor with platelets and with collagen. *The Journal of Clinical Investigation*, **74**:736–744.

Sixma, J.J., Schiphorst, M.E., Verwiej, C.L. and Pannekoek, H. (1991). Effect of deletion of the A1 domain of von Willebrand Factor on its binding to heparin, collagen and platelets in the presence of ristocetin. *European Journal of Biochemistry*, **196**:367–375.

Sixma, J.J. and de Groot, P.G. (1991). von Willebrand Factor and the blood vessel wall. *Mayo Clinic Proceedings*, **66**:628–633.

Slack, S.M., Cui, Y. and Turitto, V.T. (1993). The effects of flow on blood coagulation and thrombosis. *Thrombosis and Haemostasis*, **70**:129–134.

Slayter, H., Loscalzo, J., Bockenstedt, P. and Handin, R.I. (1985). Native conformation of human von Willebrand protein. Analysis by electron microscopy and quasi-elastic light scattering. *The Journal of Biological Chemistry*, **260**:8559–8563.

Sobel, M., McNeill, P.M., Carson, P.L., Kermode, J.C., Adelman, B., Conroy R. *et al.* (1991). Heparin inhibition of von Willebrand Factor-dependent platelet function in vitro and in vivo. *The Journal of Clinical Investigation*, **87**:1787–1793.

Sobel, M., Soler, D.F., Kermode, J.C. and Harris R.B. (1992). Localization and characterization of a heparin binding domain peptide of human von Willebrand Factor. *The Journal of Biological Chemistry*, **267**:8857–8862.

Sporn, J.L., Marder, V.J. and Wagner, D.D. (1987). von Willebrand Factor released from Weibel-Palade bodies binds more avidly to extracellular matrix than that secreted constitutively. *Blood*, **69**:1531–1534.

Stel, H.V., Sakariassen, K.S., de Groot, P.G., van Mourik, J.A. and Sixma J.J. (1985). von Willebrand Factor in the vessel wall mediates platelet adherence. *Blood*, **65**:85–90.

Sugimoto, M., Mohri, H., McClintock, R.A. and Ruggeri, Z.M. (1991). Identification of discontinuous von Willebrand Factor sequences involved in complex formation with botrocetin. *The Journal of Biological Chemistry*, **266**:18172–18178.

Sugimoto, M., Dent, J., McClintock, R., Ware, J. and Ruggeri, Z.M. (1993). Analysis of structure-function relationships in the platelet membrane glycoprotein Ib-binding domain of von Willebrand Factor by expression of deletion mutants. *The Journal of Biological Chemistry*, **268**:12185–12192.

Tannenbaum, S.H., Rick, M.E., Shafer, B. and Gralnick, H.R. (1989). Subendothelial matrix of cultured endothelial cells contains fully processed high molecular weight von Willebrand Factor. *Journal of Laboratory and Clinical Medicine*, **113**:372–378.

Titani, K., Kumar, S., Takio, K., Ericsson, L.H., Wade, R.D., Ashida, K. *et al.* (1986). Amino acid sequence of human von Willebrand Factor. *Biochemistry*, **25**:3171–3184.

Tsai, H.-M. (1996). Physiologic cleavage of von Willebrand Factor by a plasma protease is dependent on its conformation and requires calcium ion. *Blood*, **87**:4235–4244.

Tyler-Cross, R., Sobel, M., Marques, D., Soler, D.F. and Harris, R.B. (1993). Heparin-von Willebrand Factor binding as assessed by isothermal titration calorimetry and by affinity fractionation of heparins using synthetic peptides. *Archives of Biochemistry and Biophysiology*, **306**:528–533.

Vischer, U.M. and Wagner, D.D. (1994). von Willebrand Factor proteolytic processing and multimerization precede the formation of Weibel-Palade bodies. *Blood*, **83**:3536–3544.

Vlot, A.J., Koppelman, S.J., Meijers, J.C.M., Damas, C., van den Berg, H.M., Bouma, B.N. *et al.* (1996). Kinetics of factor VIII-von Willebrand factor association. *Blood*, **87**:1809–1816.

von Willebrand, E.A. (1926). Hereditär Pseudohemofili. *Finska Läkarsällskapetes Handl*, **67**:7–112.

Wagner, D.D., Urban-Pickering, M. and Marder, V.J. (1984). von Willebrand protein binds to extracellular matrices independently of collagen. *Proceedings of the National Academy of Science, USA*, **81**:471–475.

Wagner, D.D. (1990). Cell biology of von Willebrand factor. *Annual Reveiws of Cell Biology*, **6**:217.

Wagner, D.D. and Bonfanti, R. (1991). von Willebrand Factor and the endothelium. *Mayo Clinic Proceedings*, **66**:621–627.

Wagner, D.D. (1993). The Weibel-Palade body: the storage granule for von Willebrand Factor and P-selectin. *Thrombosis and Haemostasis*, **70**:105–110.

Ware, J., Dent, J.A., Azuma, H., Sugimoto, M., Kyrle, P.A., Yoshioka, A. *et al.* (1991). Identification of a point mutation in type IIB von Willebrand disease illustrating the regulation of von Willebrand Factor affinity for the platelet membrane glycoprotein Ib-IX receptor. *Proceedings of the National Academy of Science, USA*, **88**:2946–2950.

Ware, J.A. and Heistad, D.D. (1993). Platelet-endothelium interactions. *New England Journal of Medicine*, **328**:628–635.

Weiss, H.J., Turitto, V.T. and Baumgartner, H.R. (1978). Effect of shear rate on platelet interaction with subendothelium in citrated and native blood. I. Shear rate-dependent decrease of adhesion in von Willebrand's disease and the Bernard-Soulier syndrome. *Journal of Laboratory and Clinical Medicine*, **92**:750–764.

Weiss, H.J., Hawiger, J., Ruggeri, Z.M., Turitto, V.T., Thiagarajan, P. and Hoffmann, T. (1989). Fibrinogen-independent platelet adhesion and thrombus formation on subendothelium mediated by glycoprotein IIb-IIIa complex at high shear rate. *The Journal of Clinical Investigation*, **83**:288–297.

Weiss, H.J., Hoffmann, T., Yoshioka, A. and Ruggeri, Z.M. (1993). Evidence that the arg[1744] gly[1745] asp[1746] sequence in the GPIIb-IIIa-binding domain of von Willebrand factor is involved in platelet adhesion and thrombus formation on subendothelium. *Journal of Laboratory and Clinical Medicine*, **122**:324–332.

Wise, R.J., Dorner, A.J., Krane, M., Pittman, D.D. and Kaufman, R.J. (1991). The role of von Willebrand Factor multimers and propeptide cleavage in binding and stabilization of Factor VIII. *The Journal of Biological Chemistry*, **266**:21948–21955.

Wu, Y.-P., Bruegel, H.H.F.I., Lankhof, H., Wise, R.J., Handin, R.I., P.G. de Groot *et al.* (1996). Platelet adhesion to multimeric and dimeric von Willebrand factor and to collagen Type III preincubated with von Willebrand Factor. *Arteriosclerosis Thrombosis and Vascular Biology*, **16**:611–620.

4 A Unique Receptor for a Unique Function: The Glycoprotein Ib-IX-V Complex in Platelet Adhesion and Activation

José A. López*, David R. Smith and Jing-Fei Dong

*Division of Hematology/Oncology, Department of Internal Medicine,
and Department of Molecular and Human Genetics, Baylor College of Medicine
and Veterans Affairs Medical Center, Houston, TX 77030, USA
Tel: 713-794-7088, Fax: 713-794-7578*

In the most reductionist view, platelet function can be considered as having three components: stickiness, excitability, and the ability to secrete bioactive substances. Platelets have several adhesive functions: they adhere to the blood vessel wall, to other platelets, and under certain conditions, to other blood cells. Platelets are also among the most excitable of cells, responding not only to chemical agonists but also to physical forces such as shear. Finally, platelets are packed with a potent arsenal of agonists and growth factors, which they may secrete at the slightest provocation. All of these functions impinge on what is the primary physiological role of platelets: the arrest of blood loss. In some way, each function also involves proteins on the platelet plasma membrane; some of these proteins are involved in carrying out several of these functions. The subject of this chapter, the glycoprotein (GP) Ib-IX-V complex, is one such protein. Although its primary function is in adhesion of the platelets to the vessel wall through its binding of subendothelial von Willebrand factor (vWf), this protein complex also plays a role in the response of the platelet to the most potent of its agonists, thrombin. Finally, through mechanisms of signal transduction that are far from clear, the complex is involved in signaling events that lead to platelet aggregation and secretion.

STRUCTURE AND TOPOGRAPHY OF THE COMPLEX

The GP Ib-IX-V complex contains four polypeptides (GP Ibα, GP Ibβ, GP IX, and GP V) that are associated on the plasma membrane and whose synthesis is interrelated. Glycoprotein Ibα and GP Ibβ associate covalently to form the protein originally identified as GP Ib; GP IX and GP V associate with this complex noncovalently. Each of these polypeptides has the characteristics of a Type I transmembrane protein, their precursors containing a

---

Corresponding author: Veterans Affairs Medical Center, Hematology/Oncology (111H), 2002 Holcombe Boulevard, Houston, TX 77030. Tel: 713-794-7088, Fax: 713-794-7578, e-mail: josel @bcm.tmc.edu

hydrophobic N-terminal signal peptide for transport into the lumen of the endoplasmic reticulum and another hydrophobic domain that stops the transfer of the polypeptide across the membrane. The mature proteins are thus anchored on the plasma membrane with the N-terminus oriented toward the external environment and the carboxyl terminus residing in the interior of the cell. Each of the four polypeptides is synthesized from its own gene (Lopez *et al.*, 1987; Lopez *et al.*, 1988; Hickey *et al.*, 1989; Lanza *et al.*, 1993), although synthesis from these genes may be coordinated in a tissue-specific manner (see the section on genes) and their expression on the cell surface is coordinated at the level of cellular bio-synthesis. All of these components are required for the assembly of a fully functional com-plex and each is somehow dependent on other members of the complex for efficient surface expression (López *et al.*, 1992b; Li *et al.*, 1995). The polypeptides are also related at the structural level, each containing one or more tandem repeats of a phylogenetically wide-spread structural motif called the leucine-rich repeat (López, 1994).

Glycoprotein Ibα is the largest of the complex subunits with a molecular mass varying from 135–150 kDa, depending on which of four possible alleles specifying polymorphic size variants is expressed. These variants differ in the number of tandem repeats of a 13–amino acid sequence, containing either one, two, three, or four copies (López *et al.*, 1992a; Ishida *et al.*, 1995). Because the repeated sequence is within a highly glycosylated region of the polypeptide, each repeat is predicted to also add several sites for glycosyla-tion, which accounts for the approximately 5-kDa difference in molecular mass between the individual variant forms (López *et al.*, 1992a). Glycoprotein Ibβ has a mass of about 25 kDa, GP IX a mass of 19 kDa, and GP V a mass of 83 kDa.

Studies using monoclonal antibodies led to the estimate that there are approximately 25 000 copies of GP Ibα on the cell surface (Berndt *et al.*, 1985). The same study found approximately the same number of GP IX molecules on the cell surface and suggested that these components were synthesized in a coordinated fashion. Later, Du *et al.* (1987) demon-strated that GP IX forms a complex with GP Ib (consisting of a 1:1 complex of GP Ibα and GP Ibβ) and more recently Modderman *et al.* (1992), using mild detergent conditions to lyse platelets, demonstrated that GP V also associates with the complex, although the number of GP V molecules they found on the cell membrane was only about half (~ 11 000) of what had been reported for the other subunits. This stoichiometry of polypeptides led to the pro-posal that the complex exists on the platelet surface in a configuration such as the one depicted in Figure 4.1, with GP V sandwiched between two GP Ib-IX complexes (López, 1994). (The term GP Ib-IX complex is also appropriate here because a complex of GP Ibα, GP Ibβ, and GP IX is efficiently expressed on the plasma membrane of transfected hetero-logous cells and retains most of the functions of the full complex (López *et al.*, 1992b)). The arrangement of a supercomplex of the stoichiometry $(\alpha_2\beta_2\gamma_2\delta)_n$ is supported by a number of other lines of evidence. For instance, in one study Bernard-Soulier syndrome (BSS, the deficiency disorder of the complex) was reported to be transmitted in an autosomal-dominant manner (Miller *et al.*, 1992). The proband in this study was heterozygous for a mutant GP Ibα allele. Under most circumstances, expression of the Bernard-Soulier pheno-type requires two defective alleles, apparently because half of the normal complement of receptors is sufficient to carry out the functions of the complex. The most likely explanation for the autosomal-dominant variant is that the mutant polypeptide interferes with the function of its wild-type counterpart. This explanation has two important corollaries. The first is that the functional complex contains more than one copy of GP Ibα (and because of the 1:1 stoichiometry with GP Ibβ and GP IX, also of these polypeptides) and the sec-ond is that each of the associated polypeptides is necessary for the function of the receptor.

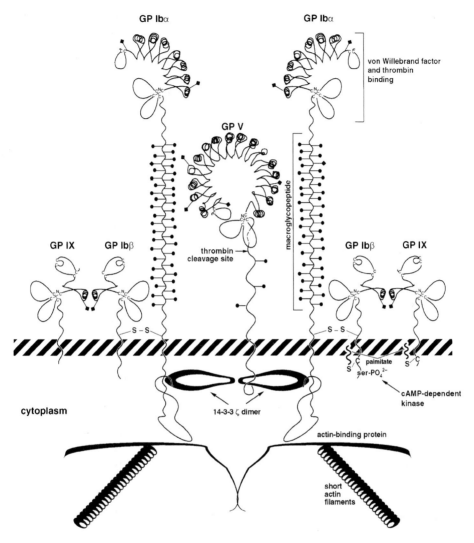

Figure 4.1   Schematic depiction of the GP Ib-IX-V complex on the platelet surface. Based on the stoichiometry determined for GP Ibα and GP V on the platelet surface and the demonstration of a direct association of these two polypeptides, the complex is depicted with one GP V polypeptide bridging two GP Ib-IX complexes. Other proteins associated with the complex are also depicted, including a 14-3-3ζ dimer, with each monomer bound to the C-terminus of one GP Ibα polypeptide (this arrangement has not been experimentally determined). The structures of the leucine-rich repeats are drawn based on the crystal structure of the porcine ribonuclease inhibitor (Kobe and Deisenhofer, 1993).

Yet another line of evidence for the complex containing more than one copy each of GP Ibα, GP Ibβ, and GP IX is suggested by the nature of high-affinity thrombin binding to the platelet. It has been known for some time that the platelet high-affinity receptor for thrombin is GP Ib, with the region containing the high-affinity site located in the 45 kDa N-terminus of GP Ibα that also contains the binding site for vWf, the receptor's major adhesive ligand. However, the number of high-affinity thrombin binding sites identified in radioligand binding studies is far less than the 25 000 GP Ibα molecules estimated from antibody binding studies; between 500 and 4000 sites have been predicted from Scatchard analysis of the binding data (Martin *et al.*, 1976; Harmon and Jamieson, 1986). One

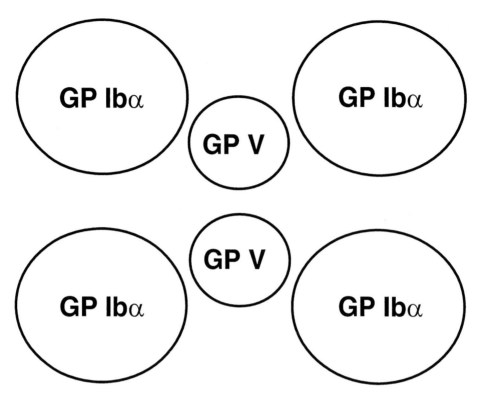

Figure 4.2 "Aerial view" of a potential arrangement of multiple complexes. In this model only the GP Ibα and GP V polypeptides are shown. An arrangement such as this one could account for the discrepancy between the number of high-affinity thrombin-binding sites on platelets and the number of GP Ibα polypeptides on the platelet surface. The molecular mass of this complex would be close to 900 000 daltons. See text for discussion.

arrangement that may accommodate these disparate facts is shown in Figure 4.2, where the complex contains four copies each of GP Ibα, GP Ibβ, and GP IX and two copies of GP V. This arrangement fits nicely with the molecular mass of 900 kDa that was predicted by Harmon and Jamieson (1985) for the high-affinity thrombin receptor, based on radiation inactivation studies. If one uses molecular masses of 135 kDa for GP Ibα, 25 kDa for GP Ibβ, 21 kDa for GP IX, and 83 kDa for GP V, the resulting molecular mass of the complex depicted in Figure 4.2 is almost precisely 900 kDa. Other functional consequences of a complex like this one are discussed in the sections on vWf and thrombin binding.

Other platelet proteins also appear to be constitutively associated with the GP Ib-IX-V complex. Ample evidence exists for a functional and structural relationship between the GP Ib-IX-V complex and the platelet Fc receptor, FcγRIIA (Moore *et al.*, 1978; Sullam *et al.*, 1998). Agents that bind FcγRIIA, such as aggregates of IgG, block vWf binding, while antibodies against the GP Ib-IX-V complex block the activation of platelets by antibody-coated streptococcus, a process known to involve FcγRIIA (Sullam *et al.*, 1998).

In addition to the proteins associated with the complex on its extracellular aspect, at least two proteins associate constitutively on the cytoplasmic side. Both actin-binding protein (ABP) and the signal-transducing protein 14-3-3ζ have been demonstrated to bind to the cytoplasmic domain of GP Ibα; ABP between Thr536 and Phe568 (Andrews and Fox, 1992) and 14-3-3ζ to the extreme C-terminus (Du *et al.*, 1994; Du *et al.*, 1996).

VON **WILLEBRAND FACTOR RECEPTOR**

Although the GP Ib-IX-V complex and vWf both are present in the blood, no meaning-
ful interaction between the two occurs under physiological blood flow conditions. Their
interaction requires that vWf be bound to a surface (Olson *et al.*, 1983), the presence of
a modulator such as ristocetin or botrocetin (Coller and Gralnick, 1977; Coller, 1978;
Andrews *et al.*, 1989a), or the existence of pathologically high shear stresses (Moake *et al.*,
1988).

The interaction between platelet GP Ib-IX-V and vWf on the blood vessel subendothelium
mediates the initial attachment of the platelets to regions of vessel injury (Sakariassen *et al.*,
1979). Studies using *in vitro* systems to model the interaction have demonstrated that this
mechanism of adhesion becomes increasingly important as fluid shear forces increase.
Savage *et al.* (1996) showed that the GP Ib-IX-V complex and the GP IIb-IIIa complex
cooperate in establishing the initial layer of platelets on vWf-coated surfaces. GP Ib-IX-V
mediates the initial interaction with vWf at shear rates above $600 \, \text{sec}^{-1}$. This interaction
slows the platelets and permits them to translocate along the surface at a much lower velocity than that of the flowing blood. The interaction between this ligand-receptor pair activates the platelet, enabling expression of the activated form of GP IIb-IIIa. This receptor
then forms a stable bond with vWf on the surface, fixing the position of the platelets and
allowing them to spread.

Despite intense study on this topic for many years, no precise region has been identified
as *the* vWf binding site. Clearly, the site in the GP Ib-IX-V complex responsible for vWf
binding resides in the 45 kDa N-terminal region of GP Ibα. Several techniques have been
employed to find that site, including studies with synthetic peptides (Vicente *et al.*, 1990;
Katagiri *et al.*, 1990), antibodies (Handa *et al.*, 1986), snake venom proteins (Ward *et al.*,
1996), site-directed mutagenesis (Murata *et al.*, 1991), and expression of recombinant
polypeptides (Cruz *et al.*, 1992). Perhaps the reason that one vWf-binding site has not been
found is that the interaction involves more than one site on GP Ibα. Several facts support
this conjecture. First, both GP Ibα and vWf are subject to activating mutations that allow
the ligand and the receptor to interact in the absence of modulators. Thus, mutations in
platelet-type von Willebrand disease (vWD) produce GP Ibα proteins that can bind spontaneously to vWf (Miller and Castella, 1982; Miller *et al.*, 1991; Russell and Roth, 1993;
Takahashi *et al.*, 1995) and mutations in type 2B vWD lead to vWf molecules that bind
platelets spontaneously (Sadler, 1991). Second, some reagents are able to selectively block
vWf binding to GP Ibα induced by one modulator and not by the others (Vicente *et al.*,
1990; Katagiri *et al.*, 1990; Ward *et al.*, 1996); several have been shown to bind to different
regions within GP Ibα (Ward *et al.*, 1996; Andrews *et al.*, 1996). Finally, both GP Ib-IX-V
and vWf have been implicated as being capable of activation by shear forces. Peterson
et al. (1987) demonstrated that washed platelets lacking granule vWf (from a patient with
severe vWD) failed to aggregate when sheared alone, but aggregated if vWf was added to
the suspension immediately after discontinuance of shear. In contrast, sheared vWf added
to these platelets did not induce their aggregation, implying that shear induces an active
conformation of GP Ib. In contrast, Siedlecki *et al.* (1996) recently suggested that shear
may induce the vWf-platelet interaction by affecting the conformation of vWf. In elegant
studies using atomic force microscopy, these investigators demonstrated that when shear
was applied to vWf bound to a hydrophobic surface, the molecule went from a globular
conformation to a more extended and open structure, presumably in the process exposing
binding sites for GP Ib. Thus, the question is still open as to which is more important in
terms of the shear response, the ligand or the receptor.

José A. López et al.

Glycoprotein Ibα

Botrocetin β-chain

Figure 4.3 Alignment of acidic sequences from GP Ibα and the botrocetin β chain. Asterisks indicate sites of sulfation on GP Ibα and sites of potential sulfation on botrocetin.

Almost every region of the GP Ibα N-terminus has been found to be important for binding vWf and certain regions appear to be relatively modulator specific. Studies using synthetic peptide identified two regions as important for ristocetin-induced vWf binding. Vicente and coworkers found that a peptide based on the sequence Ser251-Tyr279 blocked ristocetin-induced vWf binding, although it blocked botrocetin-induced vWf binding only partially at high concentrations (Vicente *et al.*, 1990). In contrast, another peptide corresponding to Gly271-Glu285 was much more effective at blocking botrocetin-induced binding. Katagiri and coworkers found that a peptide encompassing residues Asp235-Lys262 was the most potent in blocking ristocetin-induced binding (Katagiri *et al.*, 1990). Another study showed that site-directed mutations of the anionic region between Pro280 and Ala302 affect botrocetin-induced vWf binding without affecting ristocetin-induced binding (Murata *et al.*, 1991).

Other structural determinants are also important for protein binding. The requirement for disulfide structure was demonstrated by the studies of Cruz *et al.* (1992), who produced a recombinant bacterial polypeptide corresponding to the sequence Glu220-Leu318. This polypeptide blocked ristocetin-induced binding of vWf to platelets, but lost its activity when subjected to disulfide-bond reduction.

Tyrosine sulfation is another posttranslational modification vital to the normal function of the GP Ib-IX-V complex. Three tyrosine residues are found in the region spanned by residues Glu269-Glu287, in an environment that favors their sulfation. The three tyrosines, Tyr276, Tyr278, and Tyr279, were found to be completely sulfated in the recombinant complex (Dong *et al.*, 1994) and almost fully sulfated in the platelet polypeptide (Ward *et al.*, 1996). Sulfation is required for optimal binding of vWf induced by *both* ristocetin and botrocetin (Dong *et al.*, 1994; Marchese *et al.*, 1995; Ward *et al.*, 1996), although the effect on botrocetin-induced binding is greater than on ristocetin-induced binding, consistent with the mutagenesis studies that identified the anionic region as most important for botrocetin-induced binding (Murata *et al.*, 1991). Interestingly, recombinant cells expressing the GP Ib-IX complex are able to aggregate in the absence of modulators when shaken vigorously (presumably by a mechanism analogous to shear-induced platelet aggregation), and require tyrosine sulfation to aggregate (Dong *et al.*, 1995). With less vigorous shaking, the cells aggregate in the presence of ristocetin and can do so in the sulfate-depleted state. However, unlike the aggregates of cells with a fully sulfated complex, the aggregates of cells with the unsulfated complex are very unstable. This suggests the possibility that shear forces may induce a conformational change that is stabilized by the anionic sulfated region. One possibility for how this region may function is suggested by a comparison of the sequence of this region to a similar sequence in botrocetin, which binds within the A1 domain of vWf and activates binding to GP Ibα (Andrews *et al.*, 1989a). Botrocetin, from the venom of the South American pit viper, *Bothrops jararaca*, contains in its β-chain a sequence remarkably similar to the anionic region of GP Ibα (Figure 4.3) (Usami *et al.*, 1993; Andrews *et al.*, 1997). It has not been determined whether the tyrosines in this region of botrocetin are also sulfated. Botrocetin binding to the vWf A1 domain induces a conformational change that exposes another binding site within the same domain or within an adjacent A1 domain (vWf functions as a dimer) that then binds GP Ibα. The anionic sulfated sequence of GP Ibα may function similarly. Under normal circumstances, this region may be unavailable for interaction with vWf but could become available when the platelets are exposed to shear stresses (possibly also to ristocetin). Once exposed, this region could bind vWf and induce a botrocetin-like effect, allowing contact between the ligand and receptor at another site.

The GP Ib-IX-V complex was the first membrane receptor demonstrated to require sulfation for its activity, but it appears now that this modification may be used more widely in cell adhesion. The interaction between P-selectin and its ligand on blood neutrophils, PSGL-1 (P-selectin glycoprotein ligand 1), also requires sulfation of residues in an anionic region of PSGL-1 that closely resembles the sulfated region of GP Ibα (Pouyani and Seed 1995; Sako *et al.*, 1995; Wilkins *et al.*, 1995; De Luca *et al.*, 1995).

A candidate site on GP Ibα that may regulate exposure of the anionic region is the disulfide loop between Cys209 and Cys248. Two gain-of-function mutations have been identified in this region, which produce a receptor that associates with vWf in the absence of modulators or at lower concentrations of modulators than are necessary for the wild-type receptor (Miller *et al.*, 1991; Russell and Roth, 1993; Takahashi *et al.*, 1995). In both mutations (Gly233 → Val and Met239 → Val), the native amino acid is replaced by valine. Based on molecular modeling, Pincus and coworkers have postulated that both of these mutations render more energetically favorable a conformation that is less favorable in the wild-type protein (Pincus *et al.*, 1991). Presumably, the higher energy conformer is induced by shear or by ristocetin and favors the interaction with vWf. Whether this region regulates access to the anionic sulfated region or whether it itself forms part of the ligand-binding site has not been determined.

The leucine-rich repeats have also been shown to be important for vWf binding by mutations identified in BSS patients. Three BSS mutants affecting this region have been described that are expressed on the cell surface but are functionally defective. One (Leu57→Phe) affects the first leucine-rich repeat and expresses its phenotype as an autosomal dominant trait (Miller *et al.*, 1992). The others (Ala156→Val and deletion of Leu179) both are significantly farther carboxyl terminal in the domain and close to the anionic sulfated region (Ware *et al.*, 1993; Li *et al.*, 1996). Both of these mutations were transmitted as recessive traits.

How the leucine-rich repeat region functions in vWf binding is unclear, although there are two obvious possibilities. First, this region may simply have a structural role, perhaps to bring the sequences directly involved in the interaction with vWf into the appropriate spatial relationship. Alternatively, the leucine-rich repeats themselves may directly interact with vWf. Such a role in ligand binding was recently demonstrated for the leucine-rich repeats of porcine ribonuclease inhibitor, which is made up of 15 tandem repeats (Kobe and Deisenhofer, 1993). Each repeat contains a β-strand connected to an α-helix; tandem repetition of the repeats allows the formation of a partial donut-like structure (such as is depicted for GP V in Figure 4.1). The β strands form a parallel β-sheet surface on the inside of the donut, with the α-helices facing the outside. The ribonuclease inhibitor accommodates ribonuclease A within the center of the semicircle, contacting ribonuclease A primarily through the β-sheet and the loops (Kobe and Deisenhofer, 1995; Kobe and Deisenhofer, 1996). The interaction involves several of the leucine-rich repeats. A number of the amino acid residues that interact with ribonuclease are clustered within the fifteenth repeat, which is highly negatively charged and contains a potentially sulfated tyrosine. This negatively charged sequence is similar to that of GP Ibα, suggesting a means for how both the leucine-rich repeat and anionic sulfated regions of GP Ibα could be involved in binding vWf.

Other receptors with leucine-rich repeats use these sequences to directly interact with their ligands. The lutropin receptor, a seven-transmembrane-domain receptor for lutropin (LH) and human chorionic gonadotropin (hCG), contains leucine-rich repeats in its N-terminal extracytoplasmic domain. The first through sixth repeats (of fourteen) play a vital role in binding both LH and hCG, while the receptor is fully capable of binding the gonadotropins in the absence of leucine-rich repeats eight through 14 (Thomas *et al.*, 1996). Similarly, the nerve growth factor co-receptor TrkA employs one leucine-rich repeat to bind nerve growth factor (Windisch *et al.*, 1995a) as does the closely related TrkB to bind brain-derived neurotrophic factor, neurotrophin-3, and neurotrophin-4 (Windisch *et al.*, 1995b). As yet, no evidence has surfaced defining such requirements for the vWf-binding function of GP Ibα.

Thus, at least three regions of GP Ibα are important for binding vWf: the anionic tyrosine-sulfated sequence encompassed by Asp269 and Asp287, the disulfide loop formed by the pairing of Cys209 with Cys248 that contains the gain-of-function mutations of platelet-type vWD, and the leucine-rich repeats. Therefore, virtually the entire N-terminus of GP Ibα is involved in vWf binding (Figure 4.4).

Although the sequences that interact with vWf are entirely contained within the 300 N-terminal amino acid residues of GP Ibα, on the cell surface the appropriate arrangement of these domains appears to be necessary for optimal binding of the polyvalent ligand vWf. This requirement was demonstrated in a study of the role of the GP Ibα cytoplasmic domain in GP Ib-IX-V complex functions (Dong *et al.*, 1997a). Eight mutant receptors were studied, with truncations of GP Ibα of as few as six and as many as 92 amino acids (the latter mutant missing almost the entire cytoplasmic domain). Studies of receptor mobility on the plasma membrane showed that the wild-type complex was fixed on the cell surface, while even the smallest truncation greatly increased the mobility of the complex. An interesting correlate of increased mobility was a decrease in vWf binding of cells expressing the mutant complexes. Other investigators studying similar mutants did not find that cytoplasmic truncation decreased vWf binding (Cunningham *et al.*, 1996) but different modulators were employed in the two studies.

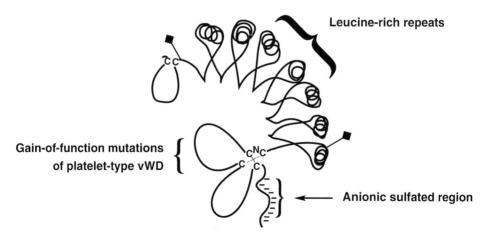

Figure 4.4 Schematic depiction of the regions within the GP Ibα N-terminus (amino acids His1–Thr300) involved in the interaction with vWf. Current evidence indicates that three regions are either directly or indirectly involved in vWf binding: the leucine-rich repeats, the disulfide loop between Cys209 and Cys248, and the anionic sulfated region between Asp269 and Asp287.

The binding of vWf is thus influenced by a number of structural elements of the GP Ib-IX-V complex, some in regions directly involved in the binding, and some that appear to influence the distribution or tethering of the receptor on the cell surface.

THROMBIN-BINDING

Two important sites of interaction for thrombin have been identified within the GP Ib-IX-V complex, GP Ibα and GP V. Glycoprotein Ibα was identified as a major binding site by crosslinking studies (Tollefsen and Majerus, 1976); by studies on platelets from patients with BSS, which respond poorly to low concentrations of the agonist (Jamieson and Okumura, 1978; Greco et al., 1996b); with the use of proteases that selectively cleave this polypeptide from the platelet surface (Harmon and Jamieson, 1988; Yamamoto et al., 1991; Greco et al., 1996a; Greco et al., 1996b); and with GP Ibα antibodies that block thrombin binding (Yamamoto et al., 1986; Mazurov et al., 1991; De Marco et al., 1994; Greco et al., 1996b). These studies provided a number of key findings: first, that GP Ibα contained the high-affinity site on thrombin for platelet activation, with 500–4000 high affinity sites identified per platelet; second, that binding of thrombin to this site was necessary for platelet activation at nanomolar thrombin concentrations (evidence from protease-treated and Bernard-Soulier platelets) (Tollefsen and Majerus, 1976; Harmon and Jamieson, 1986; Shuman and Majerus 1975; Martin et al., 1976; Workman et al., 1977); and third, that GP V was one of the only substrates for thrombin on the platelet surface (Phillips and Agin, 1977; Mosher et al., 1979) (although the rate and extent of GP V hydrolysis was difficult to correlate with the rate and extent of platelet activation (Berndt et al., 1986; McGowan et al., 1983)).

The small number of high-affinity binding sites was difficult to reconcile with the number of anti-GP Ibα monoclonal antibody binding sites on the platelet membrane (~25 000), suggesting that the relationship between high-affinity sites and polypeptides sub-units was a complicated one. Two early studies provided potential explanations for this dis-crepancy. Tollefson and Majerus (1976) provided evidence for an allosteric relationship between the high and moderate affinity sites. They demonstrated a negative cooperativity of receptor binding, indicating that once thrombin bound to one high-affinity site, nearby sites were converted to sites of lower affinity. This suggested a complicated receptor struc-ture on the platelet surface, but lack of structural information about the nature of the recep-tor limited further work. The radiation inactivation studies of Harmon and Jamieson (1985) led these investigators to suggest that the receptor could be composed of complexes of sev-eral polypeptides on the cell surface.

Interest in the role of the GP Ib-IX-V complex in thrombin-induced platelet aggregation waned considerably after the discovery of the seven-transmembrane-domain receptor that is proteolytically activated by thrombin and is fully capable of transmitting activation signals (Vu et al., 1991). However, this receptor alone (now called PAR-1 for protease-activated receptor-1) cannot account for all of thrombin's actions on platelets. For example, at throm-bin concentrations in the nanomolar range, thrombin activation absolutely requires GP Ib-IX-V, the most convincing evidence being provided by the studies of BSS platelets. In addition, Greco and coworkers have recently developed immunological reagents capable of selectively blocking thrombin binding to either PAR-1 or GP Ibα and found that thrombin binding to GP Ibα was necessary for the calcium fluxes observed in platelets at nanomolar concentrations of added thrombin (Greco et al., 1996a; Greco et al., 1996b).

Other evidence also suggests that PAR-1 may not be responsible for all of thrombin's actions on platelets. Cleavage of this receptor by thrombin exposes a new N-terminus, which functions as a "tethered ligand." Peptides based on the sequence of the new N-terminus mimic the effect of thrombin on the receptor (Connolly *et al.*, 1996), but do not faithfully reproduce the effect of thrombin on platelets (Kinlough-Rathbone *et al.*, 1993; Kinlough-Rathbone *et al.*, 1995).

Our recent studies on the synthesis and associations of the GP Ib-IX-V polypeptides has inadvertently provided insight as to why both binding and cleavage sites for thrombin should exist on this complex and provide further support for an allosteric model for high-affinity thrombin binding. We found that the recombinant GP Ib-IX complex (lacking GP V) had many of the features of the platelet complex, including its ability to bind vWf and its association with the cytoskeleton, but lacked high-affinity thrombin binding, even though the thrombin-binding site on GP Ibα was present and properly folded (evidence from conformation-specific monoclonal antibodies). However, when GP V was expressed in these cells, they then became capable of binding thrombin with high affinity (Dong *et al.*, 1997b). High-affinity thrombin binding by the GP Ib-IX complex therefore requires GP V, but this polypeptide does not itself constitute the thrombin-binding site, as cleavage of the GP Ibα N-terminus with the cobra venom protease mocarhagin, which specifically removes this region (Ward *et al.*, 1996), completely ablates high-affinity thrombin binding. Coupled with the aforementioned data on molecular mass derived from the radiation-inactivation studies of Harmon and Jamieson, we are now able to propose a model for thrombin binding to the complex (Figure 4.2). This model, while not proven, is consistent with much of the data so far obtained. Here, each of the more-or-less equivalent GP Ibα sites would provide a high-affinity binding site for thrombin. Upon thrombin binding, each of the remaining sites is converted to one of lower affinity, which fits well with the description of negative cooperativity proposed by Tollefsen and Majerus (1976). The molecular mass of this receptor is precisely that predicted from the radiation-inactivation studies. Presumably, signal transduction through this receptor requires proteolytic cleavage of GP V.

COMPLEX SYNTHESIS

Elucidation of the structures of the GP Ib-IX-V complex polypeptides by cDNA or gene cloning clearly showed that each was produced from its own gene. Nevertheless, it has been suggested that they arose from a common ancestor because of the similarity of their polypeptide and gene structures (López *et al.*, 1988; López *et al.*, 1992a; Hickey *et al.*, 1989). Interpretation of the combined deficiencies of the polypeptides on the surfaces of BSS platelets (Berndt *et al.*, 1989; Clemetson *et al.*, 1982) in light of the knowledge that the components were products of several genes suggested that their expression on the platelet surface is coordinately controlled. It was subsequently shown that a complex lacking GP V but containing the other three components could be efficiently expressed on the plasma membranes of transfected heterologous cells (López *et al.*, 1992b). Surface expression of GP Ibα in the absence of either GP Ibβ or GP IX was markedly less efficient than with the full polypeptide complement, although incomplete subcomplexes could still be expressed on the cell surface (López *et al.*, 1992b; López *et al.*, 1995; López *et al.*, 1994), particularly if their expression was amplified. Expression of GP Ibα alone does not appear to prevent it from traversing the cell secretion pathway and reaching the cell membrane, but absence of the other polypeptides does increase its susceptibility to proteolysis. In cells expressing only GP Ibα, a considerable quantity of a 43-kDa degradation product was found in the culture medium (Petersen and Handin, 1992; Murata *et al.*, 1991).

Studies of subcomplexes expressed in heterologous cells also have proved useful for determining associations between individual subunits of the complex. In our laboratory, we found that GP Ibβ is able to associate independently with both GP Ibα and GP IX, an interaction that increases surface expression of both of the latter polypeptides (López *et al.*, 1994). In these studies, we were unable to detect an association between GP Ibα and GP IX when these polypeptides were expressed in the absence of GP Ibβ. In contrast, Wu *et al.*, (1996), using an antibody-capture technique to study the complex from platelets, found a stronger interaction between GP IX and GP Ibα and virtually no interaction between GP IX and GP Ibβ. Together, these findings suggest that an association may exist in the mature complex between GP IX and GP Ibα but GP Ibβ is required to facilitate that association. Nevertheless, strong evidence also exists that GP IX and GP Ibβ remain associated on the mature complex based on studies with a unique monoclonal antibody, SZ1. SZ1 recognized the full GP Ib-IX complex purified from platelets but neither GP IX nor GP Ib (α and β) alone (Du *et al.*, 1987). Reconstitution of the complex also restored the epitope. In cultured cells, SZ1 binds only to those cells that express GP IX. We found that SZ1 appeared to recognize GP IX but definitely was able to stain a complex of this polypeptide with GP Ibβ, in the absence of GP Ibα (López *et al.*, 1995). Thus, it appears that in the complex GP Ibα and GP Ibβ associate covalently, with GP IX associating with both polypeptides but capable of associating individually only with GP Ibβ.

But what of GP V? Studies with GP V were made possible with the cloning of its gene (Hickey *et al.*, 1993; Lanza *et al.*, 1993). Conflicting data were obtained about its ability to augment GP Ib-IX complex surface expression (Li *et al.*, 1995; Meyer and Fox, 1995; Calverley *et al.*, 1995), but it seems clear that whatever effect it has is small. In contrast to the other polypeptides of the complex, GP V is expressed alone on the cell surface in significant quantities, although its expression is further augmented by the presence of the GP Ib-IX complex (Li *et al.*, 1995). Several lines of evidence indicate that GP V associates in the complex directly with GP Ibα (Li *et al.*, 1995). These data are the basis for the working model of the complex depicted in Figure 4.1.

The subcellular locations and timing of assembly are the subjects of current investigation (Dong and López, 1996). Thus far, these studies indicate that the complex is formed extremely rapidly after synthesis of the polypeptide chains and their translocation into the membrane of the endoplasmic reticulum. The full complex is transported across the trans-Golgi and onto the plasma membrane about 160 minutes after synthesis, having undergone a number of processing steps in transit, including *N*- and *O*-glycosylation, tyrosine sulfation, and fatty acylation. Partial complexes are markedly less stable, with a smaller percentage of the total reaching the later compartments and even less reaching the cell membrane. Most of the attrition is due to degradation through lysosomal pathways. Further studies of the complicated phenomenon of complex assembly and transport will lay the foundation for thorough investigation of the biosynthetic basis of BSS and the development of pharmacological strategies to modulate expression.

SIGNAL TRANSDUCTION AND CYTOSKELETAL ASSOCIATION

Shear- or ristocetin-induced vWf binding to the GP Ib-IX-V complex transmits signals into the platelet cytosol that result in phosphorylation of a number of proteins, ultimately resulting in platelet activation (Kroll *et al.*, 1991; Chow *et al.*, 1992; Ikeda *et al.*, 1993; Kroll *et al.*, 1993; Razdan *et al.*, 1994). Recently, clues to the topological requirements for signaling were provided by a study that examined the effect of snake venom proteins of the C-type lectin family on the interaction between vWf and the GP Ib-IX-V complex (Andrews *et al.*, 1996). Most of these proteins consist of disulfide-linked heterodimers of polypeptides of approximately 15 kDa that contain one GP Ibα binding site. An exception is the 50-kDa alboaggregin, from the venom of *Trimeresurus albolabris*, which apparently is a dimer of the basic heterodimer structure and is hence divalent. All of the venom proteins blocked the binding of vWf by binding to a site on GP Ibα, but only the divalent 50-kDa alboaggregin activated platelets, indicating that signal transduction through the complex requires crosslinking of two GP Ibα molecules. This requirement had been suggested previously by the observation that a monomeric dispase fragment of vWf containing the A1 domain will bind GP Ibα and block the binding of native vWf but does not itself activate platelets (Andrews *et al.*, 1989b).

The signaling functions of the GP Ib-IX-V complex are likely to involve its interactions with cytoskeletal and cytoplasmic proteins. A link with the cytoskeleton was first shown by Solum *et al.* (1984), who noted that GP Ib sedimented with the Triton-insoluble fraction of blood platelets, which constitutes the cytoskeletal fraction. The component to which it was bound was later identified as actin-binding protein (ABP), or non-muscle filamin (Fox 1985; Okita *et al.*, 1985). ABP functions as a dimer, with each monomer having one binding site for GP Ibα, lending further support for the arrangement depicted in Figure 4.1. The region recognized by ABP lies approximately in the middle of the ~ 110 amino acid cytoplasmic domain of GP Ibα within the sequence Thr536-Phe568 (Andrews and Fox, 1992). Further support for this region being important for vWf binding was provided by the finding of Cunningham *et al.* (1996) that recombinant complexes with GP Ibα truncations lacking the region no longer associate with the cytoskeleton.

The 14-3-3ζ isoform has also been demonstrated to associate with the cytoplasmic domain of the complex (Du *et al.*, 1994), an association that is completely prevented by truncation of as few as six amino acids from the GP Ibα cytoplasmic domain (Du *et al.*, 1996). Recent studies indicate that the 14-3-3 proteins recognize phosphoserine residues and may function in a manner similar to proteins that contain SH2 domains, which recognize phosphotyrosine residues. As with the SH2-containing proteins, recognition of the phosphorylated residue by 14-3-3 proteins is context-dependent, i.e., the sequence surrounding the phosphorylated serine determines recognition (Muslin *et al.*, 1997). The GP Ibβ cytoplasmic domain contains a serine that is phosphorylated by protein kinase A (Wyler *et al.*, 1986; Lopez *et al.*, 1988; Wardell *et al.*, 1989), whose binding to 14-3-3 may be regulated by phosphorylation.

The 14-3-3 proteins have been postulated to be involved in cellular signaling. They have been shown to activate Raf-like kinases, which in nucleated cells are part of an enzymatic cascade of kinases that include the mitogen-activated protein kinase (MAP kinase) and a series of related kinases (Fu *et al.*, 1994; Irie *et al.*, 1996). Because many of these components of this system exist in platelets and megakaryocytes, it is tempting to speculate that this pathway, or a related one, is involved in the signals generated after ligand binding to the GP Ib-IX-V complex. This candidate signaling pathway suggests a possible mechanism for the thrombocytopenia and macrothrombocytes observed as part of the clinical complex of BSS. These phenotypic features have been traditionally attributed to defective membrane-skeletal association. However, certain BSS variants with normal complex expression but defective vWf binding function still manifest very large platelets on the blood smear. A defective GP Ib-IX-V initiated signaling pathway could thus also possibly lead to abnormalities of platelet number and morphology.

GENES AND TRANSCRIPTION FACTORS

In addition to sharing structural similarity in their polypeptides, the components of the GP Ib-IX-V complex also share similarity of the genes that encode them. Although these polypeptides seem to have arisen from a common ancestral protein, likely by gene duplication, the genes are not clustered in one region of the human genome but intead are scattered among several chromosomes. This distribution indicates either that the duplications occurred rather early phylogenetically and may have been clustered in the genomes of ancestral organisms or that the duplication resulted from transpositions or chromosomal rearrangements. However they arose, these genes have several common features, some of which are shared by genes encoding other members of the leucine-rich motif family (Mikol *et al.*, 1990) (Figure 4.5). The coding region of each of the genes is devoid of introns, with the exception of the gene encoding GP Ibβ which contains an intron 10 bases into the open reading frame (Yagi *et al.*, 1994). This simplicity allows for easy genetic analysis for mutations because the entire genes can be readily amplified and sequenced.

Transcription of the GP Ib-IX-V complex polypeptides is limited to a very small number of tissues, with the only constitutive expression being in cells of the megakaryocytic lineage. Part of this tissue restriction is accounted for by the unusual structure of the promoter regions. None of the promoters contain TATA or CAAT boxes, consensus transcription factor binding sequences that are found in the promoter regions of a high percentage of all genes. The upstream region of the GP V gene does contain two potential TATA boxes, but primer-elongation studies indicate that these regions are not functional (Lanza *et al.*, 1993). This feature is shared with other genes whose transcription is restricted to megakaryocytes, including the genes encoding platelet factor 4 and GP IIb (Lemarchandel *et al.*, 1993; Hashimoto and Ware, 1995; Block and Poncz, 1995). Like the promoters for these other megakaryocyte-specific genes, the promoters of the genes encoding the GP Ib-IX-V polypeptides contain GATA and ETS binding sites, which are thought to control transcription (Yagi *et al.*, 1994; Bastian *et al.*, 1996; Hashimoto and Ware, 1995; Wenger *et al.*, 1989; Lanza *et al.*, 1993). GATA and ETS are not specific for megakaryocytes; it has been

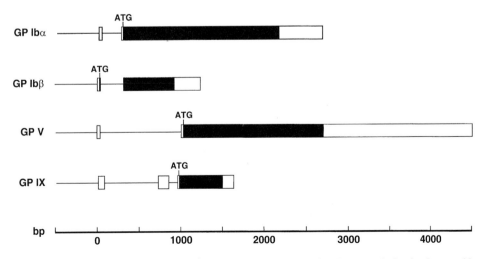

Figure 4.5   Genes encoding the polypeptides of the GP Ib-IX-V complex. Exons are depicted as boxes, with the open reading frames shown in black. The positions of the initiation codons are indicated (ATG).

suggested that the particular combinations and relative levels of members of the GATA and ETS families are what determine megakaryocyte specificity (Block and Poncz, 1995; Lemarchandel *et al.*, 1993). Mutations of the sites for either of these transcription factors in the promoter regions of the GP Ibα and GP IX genes have been shown to reduce or eliminate reporter gene expression in human erythroleukemia cells (Hashimoto and Ware, 1995; Bastian *et al.*, 1996).

The importance of the GATA sequence *in vivo* was highlighted by the discovery of a point mutation in the GATA site of the GP Ibβ promoter in a patient with BSS (Ludlow *et al.*, 1996). The causative nature of the mutation was demonstrated in cells transfected with a reporter construct in which the mutation greatly decreased transcription compared to that from the wild-type sequence.

SUMMARY

The GP Ib-IX-V complex is a unique receptor specialized for a unique and difficult task: to adhere platelets to a site of injury under conditions in which the flowing blood opposes the very tendency of the platelets to stick. Nature has evolved a solution for this problem in this very specialized receptor, which allows platelets to adhere better as the blood flow increases. This specialization has rendered the complex a one-of-a-kind receptor, with no similar receptors from which one might draw inferences as to structure-function relationships. Although much has been learned in the past ten years about how the receptor functions, it yields its secrets slowly. Still much remains to be learned about how the receptor senses shear stresses, how it binds its ligands and transduces signals into the cell, and how it might be involved in the normal physiology of the megakaryocyte.

References

Andrews, R.K., Booth, W.J., Gorman, J.J., Castaldi, P.A. and Berndt, M.C. (1989a). Purification of botrocetin from *Bothrops jararaca* venom. Analysis of the botrocetin-mediated interaction between von Willebrand factor and the human platelet membrane glycoprotein Ib-IX complex. *Biochemistry*, **28**:8317–8326.
Andrews, R.K., Gorman, J.J., Booth, W.J., Corino, G.L., Castaldi, P.A. and Berndt, M.C. (1989b). Cross-linking of a monomeric 39/34-kDa dispase fragment of von Willebrand Factor (Leu-480/Val-Gly-718) to the N-terminal region of the α-chain of membrane glycoprotein Ib on intact platelets with bis(sulfosuccinimidyl) suberate. *Biochemistry*, **28**:8326–8336.
Andrews, R.K. and Fox, J.E. (1992). Identification of a region in the cytoplasmic domain of the platelet membrane glycoprotein Ib-IX complex that binds to purified actin-binding protein. *J. Biol. Chem.*, **267**:18605–18611.
Andrews, R.K., Kroll, M.H., Ward, C.M., Rose, J.W., Scarborough, R.M., Smith, A.I. *et al.* (1996). Binding of a novel 50-kDa alboaggregin from *Trimeresurus albolabris* and related viper venom proteins to the platelet membrane glycoprotein Ib-IX-V complex. Effect on platelet aggregation and glycoprotein Ib-mediated platelet activation. *Biochemistry*, **35**:12629–12639.
Andrews, R.K., López, J.A. and Berndt, M.C. (1997). Molecular mechanisms of platelet adhesion and activation. *Int. J. Biochem. Cell Biol.*, **29**:91–105.
Bastian, L.S., Yagi, M., Chan, C. and Roth, G.J. (1996). Analysis of the megakaryocyte glycoprotein IX promoter identifies positive and negative regulatory domains and functional GATA and Ets sites. *J. Biol. Chem.*, **271**:18554–18560.
Berndt, M.C., Gregory, C., Kabral, A., Zola, H., Fournier, D. and Castaldi, P.A. (1985). Purification and preliminary characterization of the glycoprotein Ib complex in the human platelet membrane. *Eur. J. Biochem.*, **151**:637–649.
Berndt, M.C., Gregory, C., Dowden, G. and Castaldi, P.A. (1986). Thrombin interactions with platelet membrane proteins. *Ann. N. Y. Acad. Sci.*, **485**:374–386.
Berndt, M.C., Fournier, D.J. and Castaldi, P.A. (1989). Bernard-Soulier syndrome. *Baillière's Clin. Haematol.*, **2**:585–607.

Block, K.L. and Poncz, M. (1995). Platelet glycoprotein IIb gene expression as a model of megakaryocyte-specific expression. *Stem Cells*, **13**:135–145.

Calverley, D.C., Yagi, M., Stray, S.M. and Roth, G.J. (1995). Human platelet glycoprotein V: Its role in enhancing expression of the glycoprotein Ib receptor. *Blood*, **86**:1361–1367.

Chow, T.W., Hellums, J.D., Moake, J.L. and Kroll, M.H. (1992). Shear stress-induced von Willebrand factor binding to platelet glycoprotein Ib initiates calcium influx associated with aggregation. *Blood*, **80**:113–120.

Clemetson, K.J., McGregor, J.L., James, E., Dechavanne, M. and Lüscher, E.F. (1982). Characterization of the platelet membrane glycoprotein abnormalities in Bernard-Soulier syndrome and comparison with normal by surface-labeling techniques and high-resolution two-dimensional gel electrophoresis. *J. Clin. Invest.*, **70**:304–311.

Coller, B.S. and Gralnick, H.R. (1977). Studies on the mechanism of ristocetin-induced platelet agglutination. Effects of structural modification of ristocetin and vancomycin. *J. Clin. Invest.*, **60**:302–312.

Coller, B.S. (1978). The effects of ristocetin and von Willebrand factor on platelet electrophoretic mobility. *J. Clin. Invest.*, **61**:1168–1175.

Connolly, A.J., Ishihara, H., Kahn, M.L., Farese, R.V.J. and Coughlin, S.R. (1996). Role of the thrombin receptor in development and evidence for a second receptor. *Nature*, **381**:516–519.

Cruz, M.A., Petersen, E., Turci, S.M. and Handin, R.I. (1992). Functional analysis of a recombinant glycoprotein Ibα polypeptide which inhibits von Willebrand factor binding to the platelet glycoprotein Ib-IX complex and to collagen. *J. Biol. Chem.*, **267**:1303–1309.

Cunningham, J.G., Meyer, S.C. and Fox, J.E.B. (1996). The cytoplasmic domain of the α-subunit of glycoprotein (GP) Ib mediates attachment of the entire GP Ib-IX complex to the cytoskeleton and regulates von Willebrand factor-induced changes in cell morphology. *J. Biol. Chem.*, **271**:11581–11587.

De Luca, M., Dunlop, L.C., Andrews, R.K., Flannery, J.V., Ettling, R., Cumming, D.A. *et al.* (1995). A novel cobra venom metalloproteinase, mocarhagin, cleaves a 10-amino acid peptide from the mature N terminus of P-selectin glycoprotein ligand receptor, PSGL-1, and abolishes P-selectin binding. *J. Biol. Chem.*, **270**:26734–26737.

De Marco, L., Mazzucato, M., Masotti, A. and Ruggeri, Z.M. (1994). Localization and characterization of an α-thrombin-binding site on platelet glycoprotein Ibα. *J. Biol. Chem.*, **269**:6478–6484.

Dong, J.-F., Li, C.Q. and López, J.A. (1994). Tyrosine sulfation of the GP Ib-IX complex: Identification of sulfated residues and effect on ligand binding. *Biochemistry*, **33**:13946–13953.

Dong, J.-F., Hyun, W. and López, J.A. (1995). Aggregation of mammalian cells expressing the platelet glycoprotein (GP) Ib-IX complex and the requirement for tyrosine sulfation of GP Ibα. *Blood*, **86**:4175–4183.

Dong, J.-F., López, J.A. (1996). Complex formation and intracellular transport of the polypeptides of the platelet glycoprotein Ib-IX-V complex. *Blood*, **88**:624a.

Dong, J.-F., Li, C.Q., Sae-Tung, G., Hyun, W., Afshar-Kharghan, V. and López, J.A. (1997a). The cytoplasmic domain of glycoprotein (GP) Ibα constrains the lateral diffusion of the GP Ib-IX complex and modulates von Willebrand factor binding. *Biochemistry*, **36**:12421–12427.

Dong, J.-F., Sae-Tung, G. and López, J.A. (1997b). Role of glycoprotein V in the formation of the platelet high affinity thrombin-binding site. *Blood*, **89**:4355–4363.

Du, X., Beutler, L., Ruan, C., Castaldi, P.A. and Berndt, M.C. (1987). Glycoprotein Ib and glycoprotein IX are fully complexed in the intact platelet membrane. *Blood*, **69**:1524–1527.

Du, X., Harris, S.J., Tetaz, T.J., Ginsberg, M.H. and Berndt, M.C. (1994). Association of a phospholipase A$_2$ (14-3-3 protein) with the platelet glycoprotein Ib-IX complex. *J. Biol. Chem.*, **269**:18287–18290.

Du, X., Fox, J.E. and Pei, S. (1996). Identification of a binding sequence for the 14-3-3 protein within the cytoplasmic domain of the adhesion receptor, platelet glycoprotein Ibα. *J. Biol. Chem.*, **271**:7362–7367.

Fox, J.E. (1985). Linkage of a membrane skeleton to integral membrane glycoproteins in human platelets. Identification of one of the glycoproteins as glycoprotein Ib. *J. Clin. Invest.*, **76**:1673–1683.

Fu, H., Xia, K., Pallas, D.C., Cui, C., Conroy, K., Narsimhan, R.P. *et al.* (1994). Interaction of the protein kinase Raf-1 with 14-3-3 proteins. *Science*, **266**:126–129.

Greco, N.J., Jones, G.D., Tandon, N.N., Kornhauser, R., Jackson, B. and Jamieson, G.A. (1996a). Differentiation of the two forms of GPIb functioning as receptors for α-thrombin and von Willebrand factor: Ca^{2+} responses of protease-treated human platelets activated with α-thrombin and the tethered ligand peptide. *Biochemistry*, **35**:915–921.

Greco, N.J., Tandon, N.N., Jones, G.D., Kornhauser, R., Jackson, B., Yamamoto, N. *et al.* (1996b). Contribution of glycoprotein Ib and the seven transmembrane domain receptor to increases in platelet cytoplasmic [Ca^{2+}] induced by α-thrombin. *Biochemistry*, **35**:906–914.

Handa, M., Titani, K., Holland, L.Z., Roberts, J.R. and Ruggeri, Z.M. (1986). The von Willebrand factor-binding domain of platelet membrane glycoprotein Ib. Characterization by monoclonal antibodies and partial amino acid sequence analysis of proteolytic fragments. *J. Biol. Chem.*, **261**:12579–12585.

Harmon, J.T. and Jamieson, G.A. (1985). Thrombin binds to a high-affinity ~ 900 000-dalton site on human platelets. *Biochemistry*, **24**:58–64.

Harmon, J.T. and Jamieson, G.A. (1986). The glycocalicin portion of platelet glycoprotein Ib expresses both high and moderate affinity receptor sites for thrombin. A soluble radioreceptor assay for the interaction of thrombin with platelets. *J. Biol. Chem.*, **261**:13224–13229.

Harmon, J.T. and Jamieson, G.A. (1988). Platelet activation by thrombin in the absence of the high-affinity thrombin receptor. *Biochemistry*, **27**:2151–2157.

Hashimoto, Y. and Ware, J. (1995). Identification of essential GATA and Ets binding motifs within the promoter of the platelet glycoprotein Ibα gene. *J. Biol. Chem.*, **270**:24532–24539.

Hickey, M.J., Williams, S.A. and Roth, G.J. (1989). Human platelet glycoprotein IX: An adhesive prototype of leucine-rich glycoproteins with flank-center-flank structures. *Proc. Natl. Acad. Sci., USA*, **86**:6773–6777.

Hickey, M.J., Hagen, F.S., Yagi, M. and Roth, G.J. (1993). Human platelet glycoprotein V: Characterization of the polypeptide and the related Ib-V-IX receptor system of adhesive, leucine-rich glycoproteins. *Proc. Natl. Acad. Sci., USA*, **90**:8327–8331.

Ikeda, Y., Handa, M., Kamata, T., Kawano, K., Kawai, Y., Watanabe, K. *et al.* (1993). Transmembrane calcium influx associated with von Willebrand factor binding to GP Ib in the initiation of shear-induced platelet aggregation. *Thromb. Haemost.*, **69**:496–502.

Irie, K., Gotoh, Y., Yashar, B.M., Errede, B., Nishida, E. and Matsumoto, K. (1996). Stimulatory effects of yeast and mammalian 14–3–3 proteins on the Raf protein kinase. *Science*, **265**:1716–1719.

Ishida, F., Furihata, K., Ishida, K., Yan, J., Kitano, K., Kiyosawa, K. *et al.* (1995). The largest variant of platelet glycoprotein Ibα has four tandem repeats of 13 amino acids in the macroglycopeptide region and a genetic linkage with methionine 145. *Blood*, **86**:1357–1360.

Jamieson, G.A. and Okumura, T. (1978). Reduced thrombin binding and aggregation in Bernard-Soulier platelets. *J. Clin. Invest.*, **61**:861–864.

Katagiri, Y., Hayashi, Y., Yamamoto, K., Tanoue, K., Kosaki, G. and Yamazaki, H. (1990). Localization of von Willebrand factor and thrombin-interactive domains on human platelet glycoprotein Ib. *Thromb. Haemost.*, **63**:122–126.

Kinlough-Rathbone, R.L., Perry, D.W., Guccione, M.A., Rand, M.L. and Packham, M.A. (1993). Degranulation of human platelets by the thrombin receptor peptides SFLLRN: comparison with degranulation by thrombin. *Thromb. Haemost.*, **70**:1019–1023.

Kinlough-Rathbone, R.L., Perry, D.W. and Packham, M.A. (1995). Contrasting effects of thrombin and the thrombin receptor peptide, SFLLRN, on aggregation and release of ^{14}C-serotonin by human platelets pretreated with chymotrypsin or *Serratia marcescens* protease. *Thromb. Haemost.*, **73**:122–125.

Kobe, B. and Deisenhofer, J. (1993). Crystal structure of porcine ribonuclease inhibitor, a protein with leucine-rich repeats. *Nature*, **366**:751–756.

Kobe, B. and Deisenhofer, J. (1995). A structural basis of the interactions between leucine-rich repeats and protein ligands. *Nature*, **374**:183–186.

Kobe, B. and Deisenhofer, J. (1996). Mechanism of ribonuclease inhibition by ribonuclease inhibitor protein based on the crystal structure of its complex with ribonuclease A. *J. Mol. Biol.*, **264**:1028–1043.

Kroll, M.H., Harris, T.S., Moake, J.L., Handin, R.I. and Schafer, A.I. (1991). von Willebrand factor binding to platelet GpIb initiates signals for platelet activation. *J. Clin. Invest.*, **88**:1568–1573.

Kroll, M.H., Hellums, J.D., Guo, Z., Durante, W., Razdan, K., Hrbolich, J.K. *et al.* (1993). Protein kinase C is activated in platelets subjected to pathological shear stress. *J. Biol. Chem.*, **268**:3520–3524.

Lanza, F., Morales, M., de La Salle, C., Cazenave, J.-P., Clemetson, K.J., Shimomura, T. *et al.* (1993). Cloning and characterization of the gene encoding the human platelet glycoprotein V. A member of the leucine-rich glycoprotein family cleaved during thrombin-induced platelet activation. *J. Biol. Chem.*, **268**:20801–20807.

Lemarchandel, V., Ghysdael, J., Mignotte, V., Rahuel, C. and Roméo, P.-H. (1993). GATA and Ets cis-acting sequences mediate megakaryocyte-specific expression. *Mol. Cell. Biol.*, **13**:668–676.

Li, C., Martin, S. and Roth, G. (1996). The genetic defect in two well-studied cases of Bernard-Soulier syndrome: A point mutation in the fifth leucine-rich repeat of platelet glycoprotein Ibα. *Blood*, **86**:3805–3814.

Li, C.Q., Dong, J.-F., Lanza, F., Sanan, D.A., Sae-Tung, G. and López, J.A. (1995). Expression of platelet glycoprotein (GP) V in heterologous cells and evidence for its association with GP Ibα in forming a GP Ib-IX-V complex on the cell surface. *J. Biol. Chem.*, **270**:16302–16307.

López, J.A., Chung, D.W., Fujikawa, K., Hagen, F.S., Papayannopoulou, T. and Roth, G.J. (1987). Cloning of the α chain of human platelet glycoprotein Ib: A transmembrane protein with homology to leucine-rich α2-glycoprotein. *Proc. Natl. Acad. Sci., USA*, **84**:5615–5619.

López, J.A., Chung, D.W., Fujikawa, K., Hagen, F.S., Davie, E.W. and Roth, G.J. (1988). The α and β chains of human platelet glycoprotein Ib are both transmembrane proteins containing a leucine-rich amino acid sequence. *Proc. Natl. Acad. Sci., USA*, **85**:2135–2139.

López, J.A., Ludwig, E.H. and McCarthy, B.J. (1992a). Polymorphism of human glycoprotein Ibα results from a variable number of tandem repeats of a 13-amino acid sequence in the mucin-like macroglycopeptide region. Structure/function implications. *J. Biol. Chem.*, **267**:10055–10061.

López, J.A., Leung, B., Reynolds, C.C., Li, C.Q. and Fox, J.E. (1992b). Efficient plasma membrane expression of a functional platelet glycoprotein Ib-IX complex requires the presence of its three subunits. *J. Biol. Chem.*, **267**:12851–12859.

López, J.A. (1994). The platelet glycoprotein Ib-IX complex. *Blood Coagul. Fibrinolysis*, **5**:97–119.

López, J.A., Weisman, S., Sanan, D.A., Sih, T., Chambers, M. and Li, C.Q. (1994). Glycoprotein (GP) Ibβ is the critical subunit linking GP Ibα and GP IX in the GP Ib-IX complex. Analysis of partial complexes. *J. Biol. Chem.*, **269**:23716–23721.

López, J.A., Li, C.Q., Weisman, S. and Chambers, M. (1995). The GP Ib-IX "complex-specific" monoclonal antibody SZ1 binds to a conformation-sensitive epitope on GP IX: Implications for the target antigen of quinine/quinidine-dependent autoantibodies. *Blood*, **85**:1254–1258.

Ludlow, L.B., Schick, B.P., Budarf, M.L., Driscoll, D.A., Zackai, E.H., Cohen, A. *et al.* (1996). Identification of a mutation in a GATA binding site of the platelet glycoprotein Ibβ promoter resulting in the Bernard-Soulier syndrome. *J. Biol. Chem.*, **271**:22076–22080.

Marchese, P., Murata, M., Mazzucato, M., Pradella, P., De Marco, L., Ware, J. *et al.* (1995). Identification of three tyrosine residues of glycoprotein Ibα with distinct roles in von Willebrand factor and α-thrombin binding. *J. Biol. Chem.*, **270**:9571–9578.

Martin, B.M., Wasiewski, W.W., Fenton, J.W. II, and Detwiler, T.C. (1976). Equilibrium binding of thrombin to platelets. *Biochemistry*, **15**:4886–4893.

Mazurov, A.V., Vinogradov, D.V., Vlasik, T.N., Repin, V.S., Booth, W.J. and Berndt, M.C. (1991). Characterization of an antiglycoprotein Ib monoclonal antibody that specifically inhibits platelet-thrombin interaction. *Thromb. Res.*, **62**:673–684.

McGowan, E.B., Ding, A.-h. and Detwiler, T.C. (1983). Correlation of thrombin-induced glycoprotein V hydrolysis and platelet activation. *J. Biol. Chem.*, **258**:11243–11248.

Meyer, S.C. and Fox, J.E. (1995). Interaction of platelet glycoprotein V with glycoprotein Ib-IX regulates expression of the glycoprotein and binding of von Willebrand factor to glycoprotein Ib-IX in transfected cells. *J. Biol. Chem.*, **270**:14693–14699.

Mikol, D.D., Alexakos, M.J., Bayley, C.A., Lemons, R.S., Le Beau, M.M. and Stefansson, K. (1990). Structure and chromosomal localization of the gene for the oligodendrocyte-myelin glycoprotein. *J. Cell Biol.*, **111**:2673–2679.

Miller, J.L. and Castella, A. (1982). Platelet-type von Willebrand's disease: characterization of a new bleeding disorder. *Blood*, **60**:790–794.

Miller, J.L., Cunningham, D., Lyle, V.A. and Finch, C.N. (1991). Mutation in the gene encoding the α chain of platelet glycoprotein Ib in platelet-type von Willebrand disease. *Proc. Natl. Acad. Sci., USA*, **88**, 4761–4765.

Miller, J.L., Lyle, V.A. and Cunningham, D. (1992). Mutation of leucine-57 to phenylalanine in a platelet glycoprotein Ibα leucine tandem repeat occurring in patients with an autosomal dominant variant of Bernard-Soulier disease. *Blood*, **79**:439–446.

Moake, J.L., Turner, N.A., Stathopoulos, N.A., Nolasco, L. and Hellums, J.D. (1988). Shear-induced platelet aggregation can be mediated by vWF released from platelets, as well as by exogenous large or unusually large vWF multimers, requires adenosine diphosphate, and is resistant to aspirin. *Blood*, **71**:1366–1374.

Modderman, P.W., Admiraal, L.G., Sonnenberg, A. and von dem Borne, A.E.G.K. (1992). Glycoproteins V and Ib-IX form a noncovalent complex in the platelet membrane. *J. Biol. Chem.*, **267**:364–369.

Moore, A., Ross, G.D. and Nachman, R.L. (1978). Interaction of platelet membrane receptors with von Willebrand factor, ristocetin, and the Fc region of immunoglobulin G. *J. Clin. Invest.*, **62**:1053–1060.

Mosher, D.F., Vaheri, A., Choate, J.J. and Gahmberg, C.G. (1979). Action of thrombin on surface glycoproteins of human platelets. *Blood*, **53**:437–445.

Murata, M., Ware, J. and Ruggeri, Z.M. (1991). Site-directed mutagenesis of a soluble recombinant fragment of platelet glycoprotein Ibα demonstrating negatively charged residues involved in von Willebrand factor binding. *J. Biol. Chem.*, **266**:15474–15480.

Muslin, A.J., Tanner, J.W., Allen, P.M. and Shaw, A.S. (1997). Interaction of 14-3-3 with signaling proteins is mediated by the recognition of phosphoserine. *Cell*, **84**:889–897.

Okita, J.R., Pidard, D., Newman, P.J., Montgomery, R.R. and Kunicki, T.J. (1985). On the association of glycoprotein Ib and actin-binding protein in human platelets. *J. Cell Biol.*, **100**:317–321.

Olson, J.D., Moake, J.L., Collins, M.F. and Michael, B.S. (1983). Adhesion of human platelets to purified solid-phase von Willebrand factor: studies of normal and Bernard-Soulier platelets. *Thromb. Res.* **32**:115–122.

Peterson, D.M., Stathopoulos, N.A., Giorgio, T.D., Hellums, J.D. and Moake, J.L. (1987). Shear-induced platelet aggregation requires von Willebrand factor and platelet membrane glycoproteins Ib and IIb-IIIa. *Blood*, **69**:625–628.

Petersen, E. and Handin, R.I. (1992). Transient expression of recombinant glycoprotein Ibα polypeptides in COS cells that inhibit von Willebrand factor binding to the platelet glycoprotein Ib/IX complex. *Thromb. Haemost.*, **68**:203–207.

Phillips, D.R. and Agin, P.P. (1977). Platelet plasma membrane glycoproteins. Identification of a proteolytic substrate for thrombin. *Biochem. Biophys. Res. Commun.*, **75**:940–947.

Pincus, M.R., Dykes, D.C., Carty, R.P. and Miller, J.L. (1991). Conformational energy analysis of the substitution of Val for Gly 233 in a functional region of platelet GPIbα in platelet-type von Willebrand disease. *Biochim. Biophys. Acta*, **1097**:133–139.

Pouyani, T. and Seed, B. (1995). PSGL-1 recognition of P-selectin is controlled by a tyrosine sulfation consensus at the PSGL-1 amino terminus. *Cell*, **83**:333–343.

Razdan, K., Hellums, J.D. and Kroll, M.H. (1994). Shear-stress-induced von Willebrand factor binding to platelets causes the activation of tyrosine kinase(s). *Biochem. J.*, **302**:681–686.

Russell, S.D. and Roth, G.J. (1993). Pseudo-von Willebrand disease: a mutation in the platelet glycoprotein Ibα gene associated with a hyperactive surface receptor. *Blood*, **81**:1787–1791.

Sadler, J.E. (1991). von Willebrand factor. *J. Biol. Chem.*, **266**:22777–22780.

Sakariassen, K.S., Bolhuis, P.A. and Sixma, J.J. (1979). Human blood platelet adhesion to artery subendothelium is mediated by Factor VIII-von Willebrand factor bound to subendothelium. *Nature*, **279**:636–638.

Sako, D., Comess, K.M., Barone, K.M., Camphausen, R.T., Cumming, D.A. and Shaw, G.D. (1995). A sulfated peptide segment at the amino terminus of PSGL-1 is critical for P-selectin binding. *Cell*, **83**:323–331.

Savage, B., Saldivar, E. and Ruggeri, Z.M. (1996). Initiation of platelet adhesion by arrest onto fibrinogen or translocation on von Willebrand factor. *Cell*, **84**:289–297.

Shuman, M.A. and Majerus, P.W. (1975). The perturbation of thrombin binding to human platelets by anions. *J. Clin. Invest.*, **56**:945–950.

Siediecki, C.A., Lestini, B.J., Kottke-Marchant, K.K., Eppell, S.J., Wilson, D.L. and Marchant, R.E. (1996). Shear-dependent changes in the three-dimensional structure of human von Willebrand factor. *Blood*, **88**:2939–2950.

Solum, N.O. and Olsen, T.M. (1984). Glycoprotein Ib in the Triton-insoluble (cytoskeletal) fraction of blood platelets. *Biochim. Biophys. Acta*, **799**:209–220.

Sullam, P.M., Hyun, W.C., Szöllösi, J., Dong, J.-F., Foss, W.M. and López, J.A. Physical proximity and functional interplay of the glycoprotein Ib-IX-V complex and the Fc receptor FcγRIIA on the platelet plasma membrane. *J. Biol. Chem.*, **273**:5331–5336

Takahashi, H., Murata, M., Moriki, T., Anbo, H., Furukawa, T., Nikkuni, K. *et al.* (1995). Substitution of Val for Met at residue 239 of platelet glycoprotein Ibα in Japanese patients with platelet-type von Willebrand Disease. *Blood*, **85**:727–733.

Thomas, D., Rozell, T.G. and Segaloff, D.L. (1996). Mutational analyses of the extracellular domain of the full-length lutropin/choriogonadotropin receptor suggest leucine-rich repeats 1–6 are involved in hormone binding. *Mol. Endocrinol.*, **10**:760–768.

Tollefsen, D.M. and Majerus, P.W. (1976). Evidence for a single class of thrombin-binding sites of human platelets. *Biochemistry*, **15**:2144–2149.

Usami, Y., Fujimura, Y., Suzuki, M., Ozeki, Y., Nishio, K., Fukui, H. *et al.* (1993). Primary structure of two-chain botrocetin, a von Willebrand factor modulator purified from the venom of *Bothrops jararaca*. *Proc. Natl. Acad. Sci., USA*, **90**:928–932.

Vicente, V., Houghten, R.A. and Ruggeri, Z.M. (1990). Identification of a site in the α chain of platelet glycoprotein Ib that participates in von Willebrand factor binding. *J. Biol. Chem.*, **265**:274–280.

Vu, T.-K., Hung, D.T., Wheaton, V.I. and Coughlin, S.R. (1991). Molecular cloning of a functional thrombin receptor reveals a novel proteolytic mechanism of receptor activation. *Cell*, **64**:1057–1068.

Ward, C.M., Andrews, R.K., Smith, A.I. and Berndt, M.C. (1996). Mocarhagin, a novel cobra venom metalloproteinase, cleaves the platelet von Willebrand factor receptor glycoprotein Ibα. Identification of the sulfated tyrosine/anionic sequence Tyr-276-Glu-282 of glycoprotein Ibα as a binding site for von Willebrand factor and α-thrombin. *Biochemistry*, **35**:4929–4938.

Wardell, M.R., Reynolds, C.C., Berndt, M.C., Wallace, R.W. and Fox, J.E. (1989). Platelet glycoprotein Ibβ is phosphorylated on serine 166 by cyclic AMP-dependent protein kinase. *J. Biol. Chem.*, **264**:15656–15661.

Ware, J., Russell, S.R., Marchese, P., Murata, M., Mazzucato, M., De Marco, L. *et al.* (1993). Point mutation in a leucine-rich repeat of platelet glycoprotein Ibα resulting in the Bernard-Soulier syndrome. *J. Clin. Invest.*, **92**:1213–1220.

Wenger, R.H., Wicki, A.N., Kieffer, N., Adolph, S., Hameister, H. and Clemetson, K.J. (1989). The 5' flanking region and chromosomal localization of the gene encoding human platelet membrane glycoprotein Ibα. *Gene*, **85**:517–524.

Wilkins, P.P., Moore, K.L., McEver, R.P. and Cummings, R.D. (1995). Tyrosine sulfation of P-selectin glycoprotein ligand-1 is required for high affinity binding to P-selectin. *J. Biol. Chem.*, **270**:22677–22680.

Windisch, J.M., Marksteiner, R. and Schneider, R. (1995a). Nerve growth factor binding site on TrkA mapped to a single 24-amino acid leucine-rich motif. *J. Biol. Chem.*, **270**:28133–28138.

Windisch, J.M., Auer, B., Marksteiner, R., Lang, M.E. and Schneider, R. (1995b). Specific neurotrophin binding to leucine-rich motif peptides of TrkA and TrkB. *FEBS Lett.*, **374**:125–129.

Workman, E.F., Jr., White, G.C. II, and Lundblad, R.L. (1977). Structure-function relationships in the interaction of alpha-thrombin with blood platelets. *J. Biol. Chem.*, **252**:7118–7123.

Wu, G., Meloni, F.J. and Shapiro, S.S. (1996). Platelet glycoprotein (Gp) IX associates with GpIbα in the platelet membrane GpIb complex. *Blood*, **87**:2782–2787.

Wyler, B., Bienz, D., Clemetson, K.J. and Luscher, E.F. (1986). Glycoprotein Ibβ is the only phosphorylated major membrane glycoprotein in human platelets. *Biochem. J.*, **234**:373–379.

Yagi, M., Edelhoff, S., Disteche, C.M. and Roth, G.J. (1994). Structural characterization and chromosomal location of the gene encoding human platelet glycoprotein Ibβ. *J. Biol. Chem.*, **269**:17424–17427.

Yamamoto, K., Yamamoto, N., Kitagawa, H., Tanoue, K., Kosaki, G. and Yamazaki, H. (1986). Localization of a thrombin-binding site on human platelet membrane glycoprotein Ib determined by a monoclonal antibody. *Thromb. Haemost.*, **55**:162–167.

Yamamoto, N., Greco, N.J., Barnard, M.R., Tanoue, K., Yamazaki, H., Jamieson, G.A. *et al.* (1991). Glycoprotein Ib (GPIb)-dependent and GPIb-independent pathways of thrombin-induced platelet activation. *Blood*, **77**:1740–1748.

5 The Platelet Glycoprotein IIb-IIIa Complex (Integrin $\alpha_{IIb}\beta_3$)

Xiaoping Du

Department of Pharmacology, University of Illinois at Chicago, College of Medicine, 835 S. Wolcott Avenue, Chicago, IL 60612, USA

INTRODUCTION

To maintain normal circulation in the blood vessel while being able to form a hemastatic plug quickly at sites of vascular injury, the adhesiveness of platelets is dynamically regulated. Platelets in normal circulation are in a resting, non-adherent state. When vascular injury occurs, platelets are activated by contact with exposed subendothelial components such as collagen and collagen-bound von Willebrand factor or by soluble agonists such as thrombin and ADP. The activated platelets adhere and spread onto the subendothelial matrix, and aggregate to form a primary thrombus. Regulation of the adhesiveness of the platelets is achieved at the level of the ligand binding function of a major platelet adhesion receptor, the platelet glycoprotein IIb-IIIa complex (GPIIb-IIIa), also named integrin $\alpha_{IIb}\beta_3$ (Phillips *et al.*, 1991; Ginsberg *et al.*, 1995). Regulation of GPIIb-IIIa is a two-way process (Phillips *et al.*, 1991; Ginsberg *et al.*, 1995): on the one hand, the extracellular ligand-binding function of GPIIb-IIIa can be activated by intracellular signals of the platelet (inside-out signals) in response to agonists such as ADP and thrombin. Activated GPIIb-IIIa binds an abundant plasma protein, fibrinogen that forms bridges between platelets and thus mediates aggregation. Activated GPIIb-IIIa also binds ligands such as von Willebrand factor, fibronectin and vitronectin, which can mediate adhesion of platelets to the subendothelial matrix. On the other hand, ligand interaction with GPIIb-IIIa may be activated by activation of a ligand, and ligand binding to GPIIb-IIIa may trigger outside-in signals that result in a series of intracellular biochemical events leading to platelet responses such as spreading, secretion, the second wave of aggregation and release of pro-coagulant membrane vesicles. Nurden and Caen (Nurden *et al.*, 1974) first reported that a lack of GPIIb-IIIa is responsible for the dysfunction of platelets in a hereditary hemostatic disorder, Glanzmann's thrombasthenia. Since then, many aspects of the structure and function of GPIIb-IIIa and its role in hemostasis have been investigated and understanding has improved. This led to the development and clinical use of GPIIb-IIIa inhibitors for preventing and treating thrombosis (Coller *et al.*, 1995; Tcheng, 1996). Despite these achievements, mechanisms of GPIIb-IIIa regulation and signaling are still obscure. These will be the major focus of this chapter.

STRUCTURAL CHARACTERISTICS

GPIIb-IIIa is a Ca^{++}-dependent heterodimer complex of two type I membrane glycoproteins, GPIIb and GPIIIa (Plow *et al.*, 1989; Phillips *et al.*, 1991; Ginsberg *et al.*, 1993; Shattil,

1993; Ginsberg *et al.*, 1995). This nomenclature is based on the migration pattern of those two proteins on SDS-polyacrylamide gel electrophoresis (SDS-PAGE) and detection by carbohydrate-based labeling and staining techniques (Phillips *et al.*, 1977). DNA cloning revealed the primary structure of GPIIb and GPIIIa and their homology with several other adhesion receptors including the fibronectin receptor ($\alpha_2\beta_1$), vitronectin receptor ($\alpha_v\beta_3$), and the leukocyte complement receptor ($\alpha_L\beta_2$) (Fitzgerald *et al.*, 1987; Hynes, 1987; Poncz *et al.*, 1987; Rosa *et al.*, 1988; Zimrin *et al.*, 1988). These homologous adhesion receptors are named as "integrins" (Hynes, 1987). GPIIb is named as integrin α_{IIb} subunit, and GPIIIa as integrin β_3 subunit. The integrin family has since expanded to more than 20 members, each consisting of an α:β heterodimer complex (Hynes, 1992). The α_{IIb} subunit appears to be unique to platelets and megakaryocytes, and forms a complex only with β_3 subunit. β_3 is distributed in several cell types, and also forms a complex with the integrin α_v subunit (Zimrin *et al.*, 1988). The α_{IIb} consists of a heavy chain (GPIIbα) and a light chain (GPIIbβ) linked by a disulfide bond (Phillips *et al.*, 1977). The heavy and light chains have molecular masses of ~135 kDa and 25 kDa respectively as analyzed by SDS-PAGE, and are generated by proteolytic cleavage of a single polypeptide post-translationally (Poncz *et al.*, 1987; Loftus *et al.*, 1988). Disulfide bonds in α_{IIb} have been characterized by Calvete *et al.* (Calvete *et al.*, 1989; Calvete *et al.*, 1989). As deduced from its cDNA sequence (Poncz *et al.*, 1987), α_{IIb} heavy chain has 871 residues and light chain 137 residues. There is a single predicted transmembrane domain located in the light chain. The cytoplasmic domain of α_{IIb} has 26 residues and is heavily negatively charged. Features in α_{IIb}'s primary structure include four divalent cation binding motifs which are highly homologous to other integrin α subunits, and are important both to the binding of ligands and interaction with β_3 subunit (Poncz *et al.*, 1987; D'Souza *et al.*, 1990; D'Souza *et al.*, 1991). However, the I domain which constitutes ligand recognition regions in several other integrin α subunits is missing in α_{IIb} indicating a difference in ligand recognition from those integrins. Interestingly, a divalent cation binding motif similar to that found in I domain is present in β_3 and is important in ligand recognition (Loftus *et al.*, 1990). The β_3 subunit is a single chain glycosylated polypeptide with a molecular mass of ~95 kDa as analyzed by polyacrylamide gel electrophoresis under non-reduced conditions (Phillips *et al.*, 1977). Mature β_3 consists of 762 amino acid residues with a 692-residue extracellular domain, ~25 residue transmembrane domain and a 45 residue cytoplasmic domain (Fitzgerald *et al.*, 1987; Rosa *et al.*, 1988). In its extracellular domain, there is a protease-resistant cysteine rich region similar to that found in several growth factor receptors. There are several intra-molecular disulfide bonds, including a long range disulfide bond that links the cysteine-rich region to the N-terminal region (Calvete *et al.*, 1991). Disruption of these bonds by reducing agents such as β-mecaptoethanol results in unfolding of the molecule that may be responsible for the characteristic upwards shift on reduced SDS-PAGE. The cytoplasmic domain of β_3 contains two NXXY motifs which are similar to the NPXY motif identified in LDL receptor as an internalization signal (Chen *et al.*, 1990). The NPXY motif, when phosphorylated on the tyrosine residue, is the recognition site for the SH2 domain binding proteins, SHC and GRB2, which are linked to the mitogen-activated protein (MAP) kinase signaling pathway (Gustafson *et al.*, 1995; Law *et al.*, 1996). There are four calpain cleavage sites flanking the two NXXY motifs in the cytoplasmic domain of β_3 (Du *et al.*, 1995).

By rotary-shadowing electron microscopy (Carrell *et al.*, 1985; Weisel *et al.*, 1992), The $\alpha_{IIb}\beta_3$ complex appears to have an ~10 nm globular head domain which probably contains the complexed N-terminal regions of α_{IIb} and β_3, and two 18 nm long flexible tails which probably contain C-terminal portions of α_{IIb} (Weisel *et al.*, 1992) and β_3 (Du *et al.*, 1993) respectively. Removal of the C-terminal regions of both α_{IIb} and β_3 does not affect the formation of complex, suggesting that the N-terminal portions of both subunit is sufficient to form stable complex (Lam, 1992; Wippler *et al.*, 1994). This, however, does not exclude the possibility that C-terminal regions of both subunits also interact. In fact, synthetic peptides corresponding to cytoplasmic domains of both subunits have been implicated as interacting with each other in a divalent cation dependent manner (Haas *et al.*, 1996). α_{IIb} and β_3 complex formation is required for correct processing and expression of both subunits (O'Toole *et al.*, 1989).

LIGAND BINDING FUNCTION OF INTEGRIN $\alpha_{IIB}\beta_3$

Interaction of $\alpha_{IIb}\beta_3$ with Extracellular Ligands

The first identified ligand for $\alpha_{IIb}\beta_3$ was the abundant plasma coagulation factor, fibrinogen. The multivalent fibrinogen, by binding to $\alpha_{IIb}\beta_3$, cross-links platelets to form aggregates. Several other proteins, such as von Willebrand factor, vitronectin, fibronectin, and thrombospondin, can also bind to $\alpha_{IIb}\beta_3$. Most ligands of $\alpha_{IIb}\beta_3$ contain a common Arg-Gly-Asp (RGD) tripeptide sequence recognized by $\alpha_{IIb}\beta_3$ and several other integrins (Plow *et al.*, 1989). Fibrinogen contains four RGD sites, one at the N-terminal ($A\alpha^{95-97}$) and one at the C-terminal ($A\alpha^{572-574}$) ends of each of the two α-chains. Integrin $\alpha_{IIb}\beta_3$ also recognizes the fibrinogen γ-chain C-terminal 12 residues sequence, HHLGGAKQAGDV (γ400–411) (Kloczewiak *et al.*, 1984). Synthetic peptides containing the γ-chain C-terminal sequence cross-compete with RGD peptides indicating that they recognize the same binding site on $\alpha_{IIb}\beta_3$ (Lam *et al.*, 1987). The γ-chain C-terminal sequence is located at both ends of the fibrinogen molecule and is probably primarily responsible for the interaction with $\alpha_{IIb}\beta_3$ in the process of platelet aggregation. Fibrinogen lacking functional γ-chain C-terminal sequence fails to mediate platelet aggregation (Farrell *et al.*, 1992; Farrell *et al.*, 1994). However, the γ-chain C-terminal sequence may not be required for integrin interaction with fibrin during clot retraction (Rooney *et al.*, 1996). Rotary-shadowing imaging shows that the ligand binds to the N-terminal globular region of the $\alpha_{IIb}\beta_3$ complex (Weisel *et al.*, 1992). When both α_{IIb} and β_3 were truncated either by proteolytic cleavage or by site directed mutagenesis, the N-terminal domain of $\alpha_{IIb}\beta_3$ complex retains ligand binding function (Lam, 1992; Wippler *et al.*, 1994). However, dissociated α_{IIb} and β_3 lost ligand binding function suggesting that the $\alpha_{IIb}\beta_3$ complex is required for the formation of the ligand binding site (Shattil *et al.*, 1985).

Both α_{IIb} and β_3 are involved in ligand recognition. In β_3, several lines of evidence suggest that the region between residues 109–171 is likely to be a ligand binding site: (1) $\alpha_{IIb}\beta_3$-bound RGD peptide was cross-linked to a fragment containing residues 109–171, indicating its proximity to the bound RGD peptide (D'Souza *et al.*, 1988); (2) Point mutations at residues 119, 121, and 123 completely abrogate the ligand binding to the integrin (mutation at D119 to Y naturally occurred in a variant thrombasthenic patient) (Loftus *et al.*, 1990; Bajt *et al.*, 1994); (3) Antibodies that recognize this region inhibit ligand binding to $\alpha_{IIb}\beta_3$ (Andrieux *et al.*, 1991; Calvete *et al.*, 1991; D'Souza *et al.*, 1994); and (4) A peptide derived from this region was shown to directly bind to fibrinogen (D'Souza *et al.*, 1994). A second region of β_3 that may be involved in fibrinogen binding is located between residues 211–222. A synthetic peptide corresponding to this region of β_3 inhibited fibrinogen binding to $\alpha_{IIb}\beta_3$, and bound to fibrinogen (Charo *et al.*, 1991). A role for this region in ligand binding was supported by studies showing that point mutations within this region and antibodies against this region abolished ligand binding to $\alpha_{IIb}\beta_3$ (Charo *et al.*, 1991; Bajt *et al.*, 1992; Tozer *et al.*, 1996). However, the possibility that this region may have an alternative regulatory role was suggested by the observation that this peptide also bound to $\alpha_{IIb}\beta_3$ itself (Steiner *et al.*, 1993). When fibrinogen γ-chain peptide was used to probe the ligand binding site of $\alpha_{IIb}\beta_3$, it was cross-linked to a fragment of α_{IIb} (residues 294–314) (D'Souza *et al.*, 1990; D'Souza *et al.*, 1991), suggesting that this is also in proximity to the ligand binding sites. Further, synthetic peptides from this region as well as antibodies against this peptide inhibited fibrinogen binding and platelet aggregation, indicating that it is a ligand contact site (D'Souza *et al.*, 1991; Taylor *et al.*, 1992). It seems from these data that there are multiple ligand binding sites in $\alpha_{IIb}\beta_3$. However, considering the 1:1 stoichiometry of ligand binding to $\alpha_{IIb}\beta_3$ (Du *et al.*, 1991), it is likely that these different ligand contact sites in $\alpha_{IIb}\beta_3$ form a common ligand binding pocket.

The ligand binding to $\alpha_{IIb}\beta_3$ requires divalent cations. These are not only necessary for the formation of a functional $\alpha_{IIb}\beta_3$ complex, but probably also directly participate in the formation of the recognition site. The ligand contact region in β_3 (residues 109–171) is rich in oxygenated residues. Alignment of these residues showed similarity to the EF hand calcium binding region found in calmodulin (Loftus *et al.*, 1990). Mutations that substitute D119 to an alanine completely abolish the ligand binding function and also result in a conformational change in the integrin similar to the loss of cation binding (Loftus *et al.*, 1990). Further, divalent cation was shown to bind to a peptide corresponding to residues 118–131 (D'Souza *et al.*, 1994). Interestingly, this region contains the DxSxSx motif homologous to the conserved sequence found in the I domain of α subunits. I domain is an ~200 residue region in the α subunits of several integrins that is inserted possibly by exon shuffling, and is a ligand recognition site in these integrins (Corbi *et al.*, 1988; Michishita *et al.*, 1993; Kamata *et al.*, 1994; Kern *et al.*, 1994). I domain interacts with divalent cations and with the ligand (Michishita *et al.*, 1993). The DxSxSx motif is critical for both the ligand binding and divalent cation binding (Michishita *et al.*, 1993). Crystal structural analysis revealed that the bound divalent cation molecule coordinates the ligand binding surface (Lee *et al.*, 1995; Qu *et al.*, 1995). Similarity between the I domain cation binding region and ligand binding region of β_3 suggest a similar cation-dependent mechanism of ligand recognition.

Although integrin $\alpha_{IIb}\beta_3$ and several other integrins all recognize the RGD sequence which is present in several integrin ligands including fibrinogen, fibronectin, vitronectin, vWF and several snake venom proteins, different integrins have different preference for their specific ligands. For example, $\alpha_{IIb}\beta_3$ has a higher affinity for fibrinogen, while $\alpha_v\beta_3$ prefers vitronectin. This is presumably determined by the regions in the ligands flanking the RGD sequence and by the conformation of ligand binding pocket of these integrins. As $\alpha_{IIb}\beta_3$ shares a common β_3 subunit with $\alpha v\beta_3$, the integrin structure that determine the ligand specificity should be within the difference between α_v and α_{IIb}. Loftus *et al.* (Loftus *et al.*, 1996) found that an α_v/α_{IIb} chimera with the N-terminal portion of α_v substituted with a 334 residue fragment of α_{IIb} reconstituted the ligand specificity of $\alpha_{IIb}\beta_3$, indicating that the N-terminal region including four calcium binding repeats probably forms the structural basis for ligand specificity. In addition, a lack of I domain in α_{IIb} subunit may contribute to different ligand specificity of $\alpha_{IIb}\beta_3$ from the I-domain containing integrins.

Interaction of $\alpha_{IIb}\beta_3$ with Intracellular Proteins

Integrins mediate cell adhesion. This not only requires binding of integrins to extracellular ligands on the extracellular matrix or on the surface of other cells, but also requires integrins to be anchored onto the cell's framework, that is, the cytoskeleton. It is still not clear how $\alpha_{IIb}\beta_3$ or other integrins are anchored to the cytoskeleton. In many cells, integrins are colocalized with talin, α-actinin and several other proteins in a so-called focal adhesion complex which forms the cell contact site with extracellular matrix and the anchor point of intracellular stress fiber (actin filaments) (Burridge *et al.*, 1989; Clark *et al.*, 1995). In aggregated platelets, $\alpha_{IIb}\beta_3$ coprecipitates with triton X-100-insoluble actin-filaments and actin-associated cytoskeletal proteins (Phillips *et al.*, 1980; Jennings *et al.*, 1981). Several cytoskeletal proteins have been suggested to interact with integrins. These include talin (Horwitz *et al.*, 1986) and α-actinin (Otey *et al.*, 1990). Recent work by Knezevic *et al.* (Knezevic *et al.*, 1996) showed that purified $\alpha_{IIb}\beta_3$ bound directly to purified talin in a radio-labeled solid phase assay. Binding of $\alpha_{IIb}\beta_3$ to talin was inhibited by a monoclonal antibody directed against the cytoplasmic domain sequence of α_{IIb}, indicating the involvement of α_{IIb} cytoplasmic domain in the interaction. However, both the cytoplasmic domain of α_{IIb} and β_3 may be involved in the interaction with talin as the talin bound to the synthetic peptides corresponding to cytoplasmic domains of either α_{IIb} or β_3.

In addition to cytoskeletal proteins, several other intracellular proteins have also been described to interact with the cytoplasmic domains of either α_{IIb} or β_3. Calreticulin was shown to bind to the GFFKR motif in the membrane proximal region of α_{IIb} (Rojiani *et al.*, 1991). β_3-endonexin was shown to bind to the β_3 cytoplasmic domain (Shattil *et al.*, 1995). A novel intracellular protein kinase (integrin-linked kinase) interacts with the cytoplasmic domain of β_1 subunit (Hannigan *et al.*, 1996). pp125fak was shown to interact with the β_1 cytoplasmic domain peptide (Schaller *et al.*, 1995). In addition, $\alpha_{IIb}\beta_3$ was shown to coimmunoprecipitate with a membrane protein, integrin-associated protein (IAP) (Brown *et al.*, 1990). The functional importance of these interactions awaits future investigations. It is possible that some of these proteins may be involved in integrin signaling.

REGULATION OF LIGAND INTERACTION WITH INTEGRIN $\alpha_{IIB}\beta_3$

$\alpha_{IIb}\beta_3$ in Resting Platelets

$\alpha_{IIb}\beta_3$ in normal circulating platelets does not mediate platelet aggregation. This is probably due to a lack of high affinity binding of soluble ligands (fibrinogen, vWF etc.) to $\alpha_{IIb}\beta_3$ on the surface of resting platelets. High affinity binding of fibrinogen does not occur to $\alpha_{IIb}\beta_3$ in the isolated platelet membrane (Smyth *et al.*, 1993), even when $\alpha_{IIb}\beta_3$ is solubilized and purified (Du *et al.*, 1991), suggesting that inability to bind soluble macromolecular ligands is an intrinsic property of the $\alpha_{IIb}\beta_3$ molecule in resting platelets. Coller (Coller, 1986) demonstrated that an IgG monoclonal antibody that inhibits fibrinogen binding to $\alpha_{IIb}\beta_3$ was able to bind to resting platelets. However, cross-linked multimers of the antibody could not bind to platelets unless they are activated, suggesting that ligand binding sites in resting $\alpha_{IIb}\beta_3$ are not accessible to the recognition site in cross-linked antibodies. In fact, small soluble ligand mimetic peptides containing the RGD sequence do bind to $\alpha_{IIb}\beta_3$ in resting platelets, indicating that the RGD sequence in macromolecular ligands may be prevented from contact with $\alpha_{IIb}\beta_3$'s ligand binding site by other parts of the ligand molecule (Lam *et al.*, 1987; Du *et al.*, 1991). Monoclonal antibody AP7 contains an RGD-like recognition sequence, but can interact with $\alpha_{IIb}\beta_3$ without platelet activation. Exchange of the sequence flanking its RGD site with the sequence flanking the RGD-like recognition site of an activation-dependent ligand-mimicking antibody, PAC-1, resulted in activation-dependent binding of AP7 (Kunicki *et al.*, 1996). The access of the recognition sequence RGD to the binding sites in resting $\alpha_{IIb}\beta_3$ may be determined by the distance that the RGD sequence protrudes from the ligand surface. Beer *et al.* (Beer *et al.*, 1992) have shown that RGD peptide, protruding from a surface by the distance of 3 glycine residues, does not bind to platelets without platelet activation. However, an increase of the distance to 9 glycine residues results in its binding to resting platelets even in the presence of platelet activation inhibitor PGE1. Thus, it appears that ligand binding to $\alpha_{IIb}\beta_3$ may be regulated by controlling ligand accessibility to the binding sites in $\alpha_{IIb}\beta_3$.

Activation of Ligand Binding Function of $\alpha_{IIb}\beta_3$ (Inside-out Signaling)

When exposed to agonists, such as thrombin and ADP, platelets undergo changes including the hydrolysis of membrane phospholipids (generating secondary messengers such as thromboxane A_2 and phosphoinositides), phosphorylation and dephosphorylation of intracellular proteins, depolarization, Ca^{++} influx, and reorganization of the cytoskeleton. Morphologically, the platelets change from a discoid shape to irregular spheres with pseudopodia extending out from the surface. These changes in platelets precede the activation of the ligand-binding function of $\alpha_{IIb}\beta_3$ (for reviews, see (Ginsberg *et al.*, 1992; Shattil *et al.*, 1994)). $\alpha_{IIb}\beta_3$ activation requires active metabolism and is inhibited by reagents that elevate intracellular cAMP level such as PGE_1. $\alpha_{IIb}\beta_3$ activation induced by a specific agonist may be inhibited by inhibitors of a specific signaling pathway. For example, arachidonic acid-induced $\alpha_{IIb}\beta_3$ activation is inhibited by aspirin; thrombin-induced $\alpha_{IIb}\beta_3$ activation is inhibited by G-protein inhibitors such as pertussis toxin and GDPβS (Shattil *et al.*, 1987; Shattil *et al.*, 1992). Reagents that mimic or activate the action of intracellular signaling molecules, such as thromboxane analogues, Ca^{++} ionophores, GTPγS, and phorbol esters, activate $\alpha_{IIb}\beta_3$ (Shattil *et al.*, 1987; Shattil *et al.*, 1992). Thus, $\alpha_{IIb}\beta_3$ activation is a consequence of the intracellular signaling induced by platelet agonists.

$\alpha_{IIb}\beta_3$ activation is accompanied by a conformational change of the molecule. This is indicated by the resonance energy transfer of fluorescently labeled antibodies (Sims *et al.*, 1991), and by using monoclonal antibodies directed against the ligand binding site (Shattil *et al.*, 1985). Recombinant $\alpha_{IIb}\beta_3$ expressed in Chinese Hamster Ovary (CHO) cells is intrinsically unable to bind soluble fibrinogen and can not be activated by physiological platelet agonists such as thrombin and ADP. However, it can be activated by the binding of certain anti-$\alpha_{IIb}\beta_3$ monoclonal antibodies (O'Toole *et al.*, 1990). Furthermore, the platelet $\alpha_{IIb}\beta_3$ can be solubilized and purified in a resting state, and subsequently activated using monoclonal antibodies or the synthetic ligand mimetic peptides which induce conformational changes of $\alpha_{IIb}\beta_3$ (O'Toole *et al.*, 1990; Du *et al.*, 1991). Thus, activation of extracellular ligand binding function of $\alpha_{IIb}\beta_3$ results from intracellular signal-induced conformational change in the extracellular ligand binding sites of $\alpha_{IIb}\beta_3$. This is so-called inside-out signaling.

The extracellular ligand binding function of $\alpha_{IIb}\beta_3$ may be regulated by the cytoplasmic domains of $\alpha_{IIb}\beta_3$. This was first indicated by the finding that a single point mutation at the β_3 cytoplasmic domain (Ser_{752} to Pro) in a thrombasthenic patient resulted in abrogation of $\alpha_{IIb}\beta_3$ activation in platelets and a bleeding diathesis (Chen *et al.*, 1992). In the cytoplasmic domain of α_{IIb}, a GFFKR motif is highly conserved in different integrin α subunits. Mutations deleting this motif resulted in the expression of constitutively activated $\alpha_{IIb}\beta_3$, suggesting that this motif may be required to keep the integrin in a default resting conformation (O'Toole *et al.*, 1991; O'Toole *et al.*, 1994). In the membrane proximal region of the cytoplasmic domain of β_3, a negatively charged aspartic acid residue is spatially close to a positively charged lysine residue in the α_{IIb} GFFKR motif. Replacement of either one of these two charged residues resulted in activation of the ligand binding function of the integrin, indicating that a potential salt bridge between the GFFKR region of α_{IIb} and the membrane proximal region of β_3 may be involved in the maintenance of a resting state. In work to test whether the integrin cytoplasmic domains may be regulated by possible intracellular elements, Chen *et al.* (Chen *et al.*, 1994) found that over-expression of a fusion protein containing β_3 cytoplasmic domain could reverse the activated conformation of $\alpha_{IIb}\beta_3$. This suggests that the fusion protein containing β_3 cytoplasmic domain can compete with $\alpha_{IIb}\beta_3$ for an intracellular regulatory factor. Thus, it is likely that intracellular signals may regulate the ligand binding function by interacting with and modulating the cytoplasmic domain of the integrin.

Identification of the intracellular signaling pathway leading to integrin activation remains a major challenge to researchers. As $\alpha_{IIb}\beta_3$ activation is the common consequence of platelet activation induced by all platelet agonists via diversified pathways, it is likely that different signaling pathways should converge to a common regulator of the integrin. Searches for such a regulator of the integrin have led to many interesting findings. The cytoplasmic domain of β_3 can be phosphorylated probably by protein kinase C in activated platelets (Hillery *et al.*, 1991; van Willigen *et al.*, 1996). While Hillery *et al.* (Hillery *et al.*, 1991) reported that only a small percentage of $\alpha_{IIb}\beta_3$ was phosphorylated, Van Willigen *et al.* (van Willigen *et al.*, 1996) reported that the stoichiometry of phosphorylation was 80%. However, there has been no evidence that phosphorylated integrin is indeed in an activated state. Integrin activation is associated with actin polymerization and cytoskeleton reorganization. In fibroblasts, activation of the integrin-focal adhesion complex assembly and formation of stress fibres are dependent on the function of small GTP-binding proteins which activate phospatidylinositol-3 (PI-3) kinase. Inhibition of PI-3 kinase by a selective inhibitor, Wortmannin (Kovacsovics *et al.*, 1995), resulted in inhibition of high affinity binding of a ligand to $\alpha_{IIb}\beta_3$ and platelet aggregation in thrombin-stimulated platelets. It is still not clear how PI-3 kinase relay signals that activate the integrin. The effect of PI-3 kinase does not appear to be dependent on actin polymerization as inhibition of actin-polymerization by cytochalasin D does not affect the ligand binding to $\alpha_{IIb}\beta_3$ per se (Addo *et al.*, 1995), and Wortmannin did not affect actin polymerization during thrombin-induced platelet activation (Kovacsovics *et al.*, 1995).

Activation of the $\alpha_{IIb}\beta_3$ Binding Function of Fibrinogen

Although soluble fibrinogen does not bind to $\alpha_{IIb}\beta_3$ in resting platelets, it was observed (Coller, 1980; Savage *et al.*, 1995) that resting platelets adhere to a fibrinogen-coated surface. The adhesion occurs without adding platelet agonists, even in the presence of a potent platelet activation inhibitor, PGE$_1$ (Coller, 1980; Savage *et al.*, 1995). Such adhesion was not observed in platelets from a Glanzmann's thrombasthenia patient which lack $\alpha_{IIb}\beta_3$ (Coller, 1980), and was inhibited by anti-$\alpha_{IIb}\beta_3$ monoclonal antibodies (Savage *et al.*, 1995), suggesting that adhesion is mediated by $\alpha_{IIb}\beta_3$ interaction with fibrinogen. Recombinant $\alpha_{IIb}\beta_3$ expressed on the surface of CHO cells is in a resting state. However, the $\alpha_{IIb}\beta_3$-expressing CHO cells are also able to adhere to a fibrinogen-coated surface. As the $\alpha_{IIb}\beta_3$ expressed in CHO cells was not activatable by physiological platelet agonists, the possibility of prior $\alpha_{IIb}\beta_3$ activation was not likely (O'Toole *et al.*, 1990). The adhesion of PGE$_1$-treated platelets to immobilized fibrinogen requires fibrinogen γ-chain C-terminal sequence (Savage *et al.*, 1995), and is inhibited by RGD-containing peptides (Du *et al.*, unpublished data), suggesting that interaction of the immobilized fibrinogen with the RGD binding sites in resting $\alpha_{IIb}\beta_3$ is essential for the interaction.

The fact that immobilized fibrinogen, but not soluble fibrinogen, can bind to resting $\alpha_{IIb}\beta_3$ suggests that fibrinogen is changed following immobilization. This change in immobilized fibrinogen results in the exposure of the RGD or γ-chain C-terminal sequence sequences to the ligand binding site of resting $\alpha_{IIb}\beta_3$. Indeed, a monoclonal antibody, anti-RIBS1, reacted with immobilized fibrinogen but not soluble fibrinogen, indic-ating conformational changes in fibrinogen following immobilization (Zamarron *et al.*, 1991). Interestingly, anti-RIBS1 reacted with $\alpha_{IIb}\beta_3$-bound fibrinogen, suggesting that the receptor-bound fibrinogen undergoes a conformational change similar to that induced by immobilization (Zamarron *et al.*, 1991). Thus, the $\alpha_{IIb}\beta_3$-bound fibrinogen on one platelet is likely to interact with resting $\alpha_{IIb}\beta_3$ molecules on other pass-by platelets and mediate aggregation. Evidence in support of this was provided by Gawaz *et al.* (Gawaz *et al.*, 1991) using recombinant $\alpha_{IIb}\beta_3$ expressed in Chinese Hamster Ovary (CHO) cells. The wild type $\alpha_{IIb}\beta_3$ expressed in CHO cells is in a resting state and does not bind soluble fibrinogen unless activated by a monoclonal antibody. A mutant $\alpha_{IIb}\beta_3$ expressed in CHO cells is, however, constitutively active. Cells bearing the mutant $\alpha_{IIb}\beta_3$, but not cells bearing wild-type $\alpha_{IIb}\beta_3$, aggregate in the presence of fibrinogen without stimulation. However, when the cells bear-ing mutant $\alpha_{IIb}\beta_3$ were mixed with wild-type $\alpha_{IIb}\beta_3$, they coaggregate, suggesting that fib-rinogen bound to mutant $\alpha_{IIb}\beta_3$ interacts with resting wild-type $\alpha_{IIb}\beta_3$ and mediates the coaggregation of the two different types of cells. The ability of receptor-bound fibrinogen to interact with resting $\alpha_{IIb}\beta_3$ without prior platelet activation explains why platelet aggregates can be formed in arteries where blood flow rate is high and where soluble platelet agonists may be rapidly diluted.

Ligand-induced Activation of the Integrin $\alpha_{IIb}\beta_3$

Although $\alpha_{IIb}\beta_3$ is in a resting state when it is recognized by surface-coated fibrinogen, binding of a RGD-like sequence in fibrinogen to $\alpha_{IIb}\beta_3$ is likely to convert $\alpha_{IIb}\beta_3$ into an activated conformation. Evidence supporting this is provided in an experiment using syn-thetic RGD peptides. In this experiment (Du *et al.*, 1991), purified resting $\alpha_{IIb}\beta_3$ was allowed to bind to peptides containing RGD or HHLGGAKQAGDV sequences. After removing the bound peptides, $\alpha_{IIb}\beta_3$ remained in an active conformation and bound fibrino-gen with 1:1 stoichiometry and high affinity. The same treatment of $\alpha_{IIb}\beta_3$ also induced ex-posure of the epitopes for the anti-LIBS antibodies that preferentially recognize ligand-occupied $\alpha_{IIb}\beta_3$, indicating that the RGD-induced activation of $\alpha_{IIb}\beta_3$ is associated with ligand-induced conformational changes (Frelinger *et al.*, 1991). In intact platelets (Du *et al.*, 1991) and in purified platelet membranes (Smyth *et al.*, 1993), the RGD peptides revers-ibly converted $\alpha_{IIb}\beta_3$ to the activated state, which could be stabilized by fixation. After removing the RGD peptides, the fixed platelets bearing RGD-activated $\alpha_{IIb}\beta_3$ were able to aggregate in the presence of fibrinogen. The fact that ligand can induce activation of the integrin $\alpha_{IIb}\beta_3$ and $\alpha_{IIb}\beta_3$ can induce activation of fibrinogen indicates an induced-fit ligand binding mechanism involving initial recognition of an RGD-like sequence, recognition-induced conformational changes in the receptor and the ligand, and subsequent high affinity interaction. This explains why high affinity interaction between $\alpha_{IIb}\beta_3$ and fibrinogen may be initiated either by activation of the receptor or activation of the ligand (cf. Figure 5.1).

SIGNAL TRANSDUCTION INITIATED BY LIGAND BINDING TO INTEGRIN (OUTSIDE-IN SIGNALING)

Integrin $\alpha_{IIb}\beta_3$-Mediated Outside-in Signaling

Fibrinogen- and $\alpha_{IIb}\beta_3$-dependent platelet aggregation initiates secondary platelet responses such as the secretion of granule contents, second wave of platelet aggregation, the release

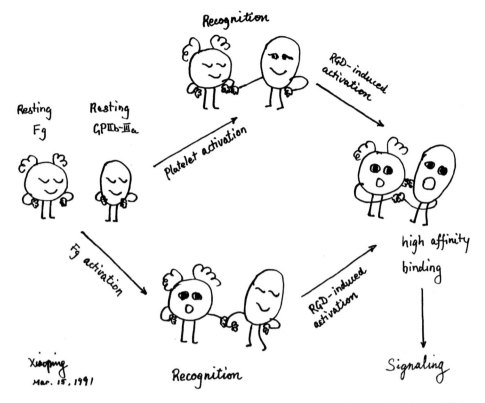

Figure 5.1 **Activation mechanisms of fibrinogen binding to integrin $\alpha_{IIb}\beta_3$.** The recognition between fibrinogen and $\alpha_{IIb}\beta_3$ may be initiated by the conformational change in $\alpha_{IIb}\beta_3$ that exposes its ligand binding sites, or by the conformational change in fibrinogen that make its recognition sequences accessible to resting $\alpha_{IIb}\beta_3$. Interaction between RGD-like sequences and $\alpha_{IIb}\beta_3$ induces further conformational changes in $\alpha_{IIb}\beta_3$ resulting in high affinity fibrinogen binding.

of procoagulant membrane vesicles and clot retraction. The interaction between surface-bound fibrinogen and $\alpha_{IIb}\beta_3$ in resting platelets also provokes platelet responses such as pseudopodia formation, spreading and aggregation (Savage *et al.*, 1995). Ligand-bound integrin moves selectively towards the open canalicular system and can be internalized (Coller *et al.*, 1991; Fox *et al.*, 1996; Wencel-Drake *et al.*, 1996). The platelet response to fibrinogen interaction with $\alpha_{IIb}\beta_3$ is associated with a series of intracellular biochemical changes characteristic of various intracellular signaling pathways. These include reorganization of cytoskeleton (Jennings *et al.*, 1981), elevation of intracellular Ca^{++} level (Powling *et al.*, 1985; Fujimoto *et al.*, 1991; Fujimoto *et al.*, 1991; Pelletier *et al.*, 1992) and pH (Banga *et al.*, 1986), hydrolysis of membrane lipids and phosphoinositide metabolism (Banga *et al.*, 1986; Sultan *et al.*, 1991), and activation of intracellular tyrosine kinases (Ferrell *et al.*, 1989; Golden *et al.*, 1990; Lipfert *et al.*, 1992; Haimovich *et al.*, 1993; Huang *et al.*, 1993) as well as serine/threonine protein kinases such as protein kinase C (Haimovich *et al.*, 1996). Thus, ligand binding to $\alpha_{IIb}\beta_3$ initiates transmembrane signal transduction. In many other cell types, integrin interaction with their extracellular ligands similarly results in the biochemical changes characteristic of various intracellular signaling pathways (Clark *et al.*, 1995; Schwartz *et al.*, 1995).

Integrin $\alpha_{IIb}\beta_3$-mediated outside-in signaling requires a functional cytoplasmic domain of $\alpha_{IIb}\beta_3$. Ligand binding to mutant $\alpha_{IIb}\beta_3$ that lacks the cytoplasmic domain of β_3 has significant reduced outside-in signals as detected by tyrosine-phosphorylation of intracellular proteins. Mutant β_3 with deleted cytoplasmic domain or disrupted NPXY sequence in the cytoplasmic domain, although able to bind ligand, failed to mediate cell spreading, and was unable to be incorporated into focal adhesion complexes or migrate on a fibrinogen-coated surface (Ylanne et al., 1993; Filardo et al., 1995; Ylanne et al., 1995). The point mutation of S^{752} of β_3 not only abolished inside-out signaling but also reduced $\alpha_{IIb}\beta_3$-mediated cell spreading on immobilized fibrinogen, focal adhesion, and fibrin clot retraction, suggesting that outside-in signaling was also impaired (Chen et al., 1994). It is thus possible that outside-in signals may be relayed by interaction of the cytoplasmic domain of $\alpha_{IIb}\beta_3$ with the intracellular molecules.

Regulation of Intracellular Calcium Level

Ligand interaction with $\alpha_{IIb}\beta_3$ induces changes in intracellular calcium levels. An early indication of this comes from the finding that a monoclonal antibody that blocks fibrinogen binding to $\alpha_{IIb}\beta_3$ inhibited Ca^{++} influx in activated platelets (Powling et al., 1985). Integrin-mediated increase in calcium level may involve IP-3-induced calcium mobilization or calcium influx through calcium channels. The function of $\alpha_{IIb}\beta_3$ or a $\alpha_{IIb}\beta_3$-associated protein as a calcium channel was suggested by Rybak et al. (Rybak et al., 1988) and Fujimoto et al. (Fujimoto et al., 1991) who showed that partially purified $\alpha_{IIb}\beta_3$ incorporated in liposomes or the planar phospholipid bilayer had Ca^{++} channel activity. Further, calcium channel activity was impaired in platelet membranes prepared from Glanzmann's thrombasthenic patients (Fujimoto et al., 1991). Prevention of fibrinogen binding to $\alpha_{IIb}\beta_3$ by monoclonal anti-$\alpha_{IIb}\beta_3$ antibodies or by RGDS peptides in thrombin-stimulated platelets inhibits the Ca^{++} influx into platelets and reduces the open probability of Ca^{++} channels in subsequently prepared membrane vesicle (Fujimoto et al., 1991). In cells expressing recombinant $\alpha_{IIb}\beta_3$, calcium level oscillations can be induced by surface coated fibrinogen or anti-$\alpha_{IIb}\beta_3$ antibody (Pelletier et al., 1992). A recent study suggests that the integrin-associated protein (IAP) may be involved in the integrin-mediated calcium influx (Schwartz et al., 1993). Intracellular calcium as an important second messenger regulates activities of many aspects of cellular functions and of numerous intracellular proteins including actin-myosin interaction, cytoskeleton organization, calmodulin activation, activation of phospholipases, protein kinases and proteases, as well as inhibition of adenylate cyclase (Clapham, 1995). These calcium-dependent intracellular processes are required for the integrin-mediated platelet responses such as spreading, secretion, and clot retraction (Sage et al., 1993).

$\alpha_{IIb}\beta_3$-mediated platelet aggregation is responsible for the activation of calcium dependent protease calpain, presumably by increasing intracellular calcium level (Fox et al., 1993). Calpain is colocalized with integrins in the focal adhesion complex (Beckerle et al., 1987), and catalyzes limited proteolysis of several important cytoskeletal and signaling molecules. Many identified calpain substrates are also colocalized with integrin in the focal adhesion complex and/or associated with integrin functions. These include protein tyrosine phosphatase IB (Frangioni et al., 1993), pp60src (Oda et al., 1993), pp125fak (Cooray et al., 1996), protein kinase C (Saido et al., 1991), talin (Fox et al., 1985), actin-binding protein (Fox et al., 1985), spectrin (Fox et al., 1987) and the cytoplasmic domain of the integrin β_3 subunit (Du et al., 1995). Cleavage of the β_3 subunit at sites flanking two functionally important NXXY motifs are likely to disrupt integrin interaction with the cytoskeleton and integrin signaling (Du et al., 1995). Thus, integrin-mediated calpain activation may be involved in regulating integrin interaction with the cytoskeleton and intracellular signaling pathways.

Activation of Protein Kinases

In the process of platelet aggregation, several intracellular proteins become phosphorylated on tyrosine residues. Tyrosine-phosphorylation of specific intracellular proteins was inhibited by blocking fibrinogen binding to $\alpha_{IIb}\beta_3$ and was absent in thrombasthenic platelets which lack $\alpha_{IIb}\beta_3$, indicating the activation of protein tyrosine kinases by ligand-$\alpha_{IIb}\beta_3$ interaction (Ferrell *et al.*, 1989; Golden *et al.*, 1990). The tyrosine-phosphorylation of intracellular proteins also occurs when resting platelets adhere to surface-coated fibrinogen (Haimovich *et al.*, 1993; Clark *et al.*, 1994). Some of these tyrosine-phosphorylated proteins were identified to be protein tyrosine kinases themselves. These include focal adhesion kinase (pp125fak) (Lipfert *et al.*, 1992) and pp72syk (Clark *et al.*, 1994). Integrin-mediated tyrosine phosphorylation of these protein tyrosine kinases is associated with activation of their enzymatic activities (Lipfert *et al.*, 1992; Clark *et al.*, 1994). Tyrosine-phosphorylation of pp72syk can be induced by fibrinogen binding to platelets without platelet activation by other agonists, and appeared to precede platelet aggregation process (Clark *et al.*, 1994). Tyrosine phosphorylation of pp125fak is dependent upon platelet aggregation suggesting that this is a late event (Shattil *et al.*, 1994; Haimovich *et al.*, 1996). Phosphorylation of pp125fak also requires co-stimulation of platelets by platelet agonists and is inhibited by protein kinase C inhibitors (Shattil *et al.*, 1994; Haimovich *et al.*, 1996). Phosphorylation and activation of these protein kinases can be inhibited by cytocalasin D which inhibits actin polymerization (Lipfert *et al.*, 1992; Clark *et al.*, 1994) suggesting that activation of protein tyrosine kinases is probably downstream the integrin-induced cytoskeleton reorganization.

Recently, Law *et al.* (Law *et al.*, 1996) found that in the process of integrin-mediated platelet aggregation, the cytoplasmic domain of β_3 becomes phosphorylated on its tyrosine residues. The only tyrosine residues in the cytoplasmic domain of β_3 are within the two NXXY motifs. Tyrosine phosphorylated peptides corresponding to the cytoplasmic domain of the integrin β_3 bind to two SH2 domain containing proteins GRG2 and SHC. Phosphorylation of tyrosine residues in both NXXY motifs is required for GRB2 binding, while phosphorylation of the C-terminal NXXY sequence is sufficient for SHC binding. Wary *et al.* (Wary *et al.*, 1996) reported that certain β_1 and β_3 integrins may interact with SHC and mediate SHC phosphorylation. This interaction, however, does not appear to be mediated by the cytoplasmic domains of either α or β subunits, but is specified by the transmembrane or extracellular domain of α subunits and may be mediated by caveolin. SHC-GRB2-Sos complex plays a central role in linking the tyrosine-phosphorylated receptors to the ras-raf-MEK-MAP kinase pathways. Ligand binding to integrins has been shown to activate MAP kinase pathway in nucleated cells (Zhu *et al.*, 1995; Renshaw *et al.*, 1996; Wary *et al.*, 1996), and this activation is mediated by SHC (Wary *et al.*, 1996). However, as cytocalasin D and Rho A inhibitor inhibited the integrin-mediated MAP kinase activation (Zhu *et al.*, 1995; Renshaw *et al.*, 1996), it appears that activation of MAP kinase pathway also requires assembly of integrin-cytoskeleton complex.

Regulation of Interaction Between $\alpha_{IIb}\beta_3$ and Cytoskeletal Proteins

In resting platelets, the majority of the $\alpha_{IIb}\beta_3$ population in a platelet lysate does not coprecipitate with Triton X-100-insoluble cytoskeleton after low speed ($\sim 10,000\,xg$) centrifugation (Fox *et al.*, 1993). A subpopulation of $\alpha_{IIb}\beta_3$ can coprecipitate with the short actin-filamental structure characteristic of a so-called membrane skeleton when centrifuged at $100\,000\,xg$ (Fox *et al.*, 1993). After platelet aggregation, which is dependent upon $\alpha_{IIb}\beta_3$ binding to fibrinogen, the majority of the $\alpha_{IIb}\beta_3$ population coprecipitates with the low speed-precipitable cytoskeletal protein (Jennings *et al.*, 1981). It is unclear whether ligand binding to $\alpha_{IIb}\beta_3$ directly regulates its interaction with intracellular cytoskeletal proteins or ligand binding initiates a signal that regulate the interaction between "membrane skeleton" actin filament and cytoskeleton. In either case, a ligand binding-initiated signal causes an intracellular change that results in the association of $\alpha_{IIb}\beta_3$ with the platelet cytoskeleton. $\alpha_{IIb}\beta_3$-mediated platelet aggregation also results in the incorporation of numerous intracellular cytoskeletal and signaling molecules into the integrin-cytoskeleton complex. These include talin, vinculin, spectrin, pp60src, pp125fak, pp62c-yes, pp21ras (Fox *et al.*, 1993) and phosphoinositide 3-kinase (Guinebault *et al.*, 1995). As both the translocation of protein tyrosine kinases to cytoskeleton and integrin-mediated activation of protein kinases are inhibited by cytochalasin D which inhibits actin polymerization, it is likely that formation of an integrin-cytoskeleton-signaling molecules complex may be a necessary and early event in the integrin-mediated signaling.

TRANSMEMBRANE SIGNALING MECHANISMS OF $\alpha_{IIB}\beta_3$

Conformational Changes as A Two-way Signaling Mechanism

The transmembrane signals mediated by integrin $\alpha_{IIb}\beta_3$ are bidirectional, and involve the extracellular ligand binding site and the cytoplasmic domain. As mentioned above, activation of the ligand binding function of $\alpha_{IIb}\beta_3$ by intracellular signals (inside-out signal) is associated with a conformational change in the extracellular ligand binding region of $\alpha_{IIb}\beta_3$. Conversely, the ligand binding also induces a conformational change in the integrin. The evidence of ligand-induced conformational change in $\alpha_{IIb}\beta_3$ was initially provided by Parise *et al.* (Parise *et al.*, 1987) who showed that incubation of $\alpha_{IIb}\beta_3$ with RGD peptides resulted in significant changes in its intrinsic fluorescence, sedimentation coefficient and sensitivity to proteases. Frelinger *et al.* (Frelinger *et al.*, 1988; Frelinger *et al.*, 1990; Frelinger *et al.*, 1991) and others (Kouns *et al.*, 1990) developed a series of monoclonal antibodies that had high affinity binding to ligand-occupied $\alpha_{IIb}\beta_3$ but not to unoccupied $\alpha_{IIb}\beta_3$, indicating that ligand binding to $\alpha_{IIb}\beta_3$ induced expression of new antibody binding sites (ligand-induced binding sites, LIBS). Ligand-induced conformation is different from the intracellular signaling induced conformational change as the latter does not express ligand-induced binding site (Frelinger *et al.*, 1991). Interestingly, the ligand binding induced-conformation of $\alpha_{IIb}\beta_3$ is an activated form with high affinity for fibrinogen (Du *et al.*, 1991). Thus, it appears that inside-out activation signals and ligand recognition synergetically transform $\alpha_{IIb}\beta_3$ into a conformation which has high affinity for fibrinogen.

Ligand binding-induced conformational change is not limited to the ligand binding region of $\alpha_{IIb}\beta_3$. This conformational change propagates through the entire 18 nm length of the $\alpha_{IIb}\beta_3$ molecule to the C-terminal domains of both α_{IIb} and β_3. The initial evidence suggesting a ligand-induced long range-propagated conformational change was obtained using a monoclonal antibody, anti-LIBS2, recognizing a ligand-induced binding site (Du *et al.*, 1993). We found that the binding site for anti-LIBS2 was located in the 89 residue section in the C-terminal region of $\alpha_{IIb}\beta_3$ immediately N-terminal to the transmembrane domain. Rotary shadowing electron microscopy indicates that anti-LIBS2 binds to a site close to the C-terminal end of β_3 tail. As the high affinity binding of the anti-LIBS2 is induced by RGD binding to the N-terminal globular region of $\alpha_{IIb}\beta_3$ complex, this suggests that ligand binding induced a conformational change that propagated to the C-terminal region of β_3.

Recently, Lam and coworkers (personal communications) developed a monoclonal antibody directed against the cytoplasmic domain of α_{IIb}. The affinity of this antibody for $\alpha_{IIb}\beta_3$ is also increased by ligand binding to $\alpha_{IIb}\beta_3$. This suggests that ligand-induced conformational change can be propagated to the cytoplasmic domain. It is possible that ligand-induced long range conformational changes may be a mechanism for transmission of outside-in signals that regulate the interaction between the integrin cytoplasmic domain and intracellular molecules.

Conversely, although the anti-LIBS antibodies (such as anti-LIBS2) interact with resting $\alpha_{IIb}\beta_3$ with much lower affinity than ligand-occupied $\alpha_{IIb}\beta_3$, these antibodies can induce ligand binding to $\alpha_{IIb}\beta_3$ (Kouns *et al.*, 1990; Frelinger *et al.*, 1991; Du *et al.*, 1993). Identification of the binding site for anti-LIBS2 in the C-terminal region of β_3 indicates that interaction of a molecule with the C-terminal region of $\alpha_{IIb}\beta_3$ can induce a propagated inside-out conformational change, resulting in activation of the ligand binding function (Du *et al.*, 1993). Interaction of the cytoplasmic domain of $\alpha_{IIb}\beta_3$ with intracellular molecules is probably responsible for regulating ligand binding function of the integrin (Chen *et al.*, 1992; Chen *et al.*, 1994; Chen *et al.*, 1994). Thus, it is likely that the inside-out propagated conformational changes are a mechanism for transmission of the intracellular signal to the extracellular ligand binding site of $\alpha_{IIb}\beta_3$.

Thus, studies on the conformational changes of $\alpha_{IIb}\beta_3$ indicate that there is a two-way propagated long range conformational change that is an intrinsic property of $\alpha_{IIb}\beta_3$ and may be a mechanism for the bidirectional cross-membrane signaling between the C-terminal cytoplasmic domain and the N-terminal ligand binding site of $\alpha_{IIb}\beta_3$.

The Role of $\alpha_{IIb}\beta_3$ Clustering in $\alpha_{IIb}\beta_3$ Signaling

Soluble monomeric peptide ligands of $\alpha_{IIb}\beta_3$, although able to initiate conformational changes of $\alpha_{IIb}\beta_3$, are not sufficient to induce significant platelet responses (Kouns *et al.*, 1991). $\alpha_{IIb}\beta_3$-mediated platelet responses appear to require multivalent ligands. Ligand-occupancy results in clustering of the integrin on the platelet surface (Isenberg *et al.*, 1987; Isenberg *et al.*, 1989). In fact, integrin mediated signaling can be initiated by clustering of integrin molecules with multivalent antibodies (Pelletier *et al.*, 1992). However, recent studies on the integrin $\alpha_5\beta_1$ showed the difference between the ligand-occupancy-induced signals and antibody cross-link induced integrin signals (Miyamoto *et al.*, 1995). When the integrin $\alpha_5\beta_1$ was bound to fibronectin-coated beads, actin, talin, vinculin α-actinin, tensin and focal adhesion kinase (pp125fak) were recruited to the cytoplasmic side of the focal adhesion sites. When integrins were clustered by non-inhibitory antibodies, however, only tensin and pp125fak were recruited. Adding RGD peptide to the antibody-clustered integrin resulted in an effect similar to using fibronectin or multivalent peptide-coated beads, suggesting that both the ligand occupancy-induced change in integrin and receptor clustering are required for the outside-in signals. It is possible that clustering of integrin may facilitate or amplify the ligand-induced conformational change, and initiate the formation of integrin-cytoskeleton-signaling molecule complex (Fox, 1993; Miyamoto *et al.*, 1995) which is critical to the outside-in signal transduction.

SUMMARY

Ligands containing RGD-like sequences interact with integrin $\alpha_{IIb}\beta_3$ at sites in the N-terminal globular region of the $\alpha_{IIb}\beta_3$ complex. Ligand binding to integrin $\alpha_{IIb}\beta_3$ may be activated either by an inside-out signal-induced conformational change in $\alpha_{IIb}\beta_3$ exposing ligand binding site, or by a conformational change in fibrinogen exposing RGD-like recognition sequences. Ligand binding to the integrin initiates outside-in signals, which involve activation of intracellular protein kinases, elevation of intracellular calcium level, phospholipid metabolism and cytoskeleton reorganization and result in platelet responses. Formation of integrin-cytoskeleton-signaling molecule complex appears to be an early event critical to the outside-in signaling. While it is not clear what the intracellular molecules are that regulate integrin and that are directly regulated by $\alpha_{IIb}\beta_3$, the cytoplasmic domain of the integrin plays a critical role in the interaction between integrin and intracellular molecules. Transmembrane signals between the extracellular ligand binding site and cytoplasmic domain of the integrin may be mediated by bidirectional propagated conformational changes and clustering of the integrin.

Acknowledgements

Xiaoping Du is an Established Investigator of the American Heart Association. This work is in part supported by grant HL52547 from NIH.

References

Addo, J.B., Bray, P.F., Grigoryev, D., Faraday, N. and Goldschmidt, C.P. (1995). Surface recruitment but not activation of integrin $\alpha_{IIb}\beta_3$ (GPIIb-IIIa) requires a functional actin cytoskeleton. *Arterioscler. Thromb. Vasc. Biol.*, **15**:1466–1473.

Andrieux, A., Rabiet, M.J., Chapel, A., Concord, E. and Marguerie, G. (1991). A highly conserved sequence of the Arg-Gly-Asp-binding domain of the integrin β_3 subunit is sensitive to stimulation. *J. Biol. Chem.*, **266**:14202–14207.

Bajt, M.L., Ginsberg, M.H., Frelinger, A.d., Berndt, M.C. and Loftus, J.C. (1992). A spontaneous mutation of integrin $\alpha_{IIb}\beta_3$ (platelet glycoprotein IIb-IIIa) helps define a ligand binding site. *J. Biol. Chem.*, **267**:3789–3794.

Bajt, M.L. and Loftus, J.C. (1994). Mutation of a ligand binding domain of β_3 integrin. Integral role of oxygenated residues in $\alpha_{IIb}\beta_3$ (GPIIb-IIIa) receptor function. *J. Biol. Chem.*, **269**:20913–20919.

Banga, H.S., Simons, E.R., Brass, L.F. and Rittenhouse, S.E. (1986). Activation of phospholipases A and C in human platelets exposed to epinephrine: role of glycoproteins IIb/IIIa and dual role of epinephrine. *Proc. Natl. Acad. Sci., USA*, **83**:9197–9201.

Beckerle, M.C., Burridge, K., DeMartino, G.N. and Croall, D.E. (1987). Colocalization of calcium-dependent protease II and one of its substrates at sites of cell adhesion. *Cell*, **51**:569–577.

Beer, J.H., Springer, K.T. and Coller, B.S. (1992). Immobilized Arg-Gly-Asp (RGD) peptides of varying lengths as structural probes of the platelet glycoprotein IIb/IIIa receptor. *Blood*, **79**:117–128.

Brown, E., Hooper, L., Ho, T. and Gresham, H. (1990). Integrin-associated protein: a 50-kD plasma membrane antigen physically and functionally associated with integrins. *J. Cell Biol.*, **111**:2785–2794.

Burridge, K. and Fath, K. (1989). Focal contacts: transmembrane links between the extracellular matrix and the cytoskeleton. *Bioessays*, **10**:104–108.

Calvete, J.J., Alvarez, M.V., Rivas, G., Hew, C.L., Henschen, A. and Gonzalez, R.J. (1989). Interchain and intrachain disulphide bonds in human platelet glycoprotein IIb. Localization of the epitopes for several monoclonal antibodies. *Biochem. J.*, **261**:551–560.

Calvete, J.J., Arias, J., Alvarez, M.V., Lopez, M.M., Henschen, A. and Gonzalez-Rodriguez, J. (1991). Further studies on the topography of the N-terminal region of human platelet glycoprotein IIIa: Localization of monoclonal antibody epitopes and the putative fibrinogen-binding sites. *Biochem. J.*, **274**:457–463.

Calvete, J.J., Henschen, A. and Gonzalez, R.J. (1989). Complete localization of the intrachain disulphide bonds and the N-glycosylation points in the α-subunit of human platelet glycoprotein IIb. *Biochem. J.*, **261**:561–568.

Calvete, J.J., Henschen, A. and Gonzalez, R.J. (1991). Assignment of disulphide bonds in human platelet GPIIIa. A disulphide pattern for the β-subunits of the integrin family. *Biochem. J.*, **274**:63–71.

Carrell, N.A., Fitzgerald, L.A., Steiner, B., Erickson, H.P. and Phillips, D.R. (1985). Structure of human platelet membrane glycoproteins IIb and IIIa as determined by electron microscopy. *J. Biol. Chem.*, **260**:1743–1749.

Charo, I.F., Nannizzi, L., Phillips, D.R., Hsu, M.A. and Scarborough, R.M. (1991). Inhibition of fibrinogen binding to GPIIb-IIIa by a GP IIIa peptide. *J. Biol. Chem.*, **266**:1414–1421.

Chen, W.J., Goldstein, J.L. and Brown, M.S. (1990). NPXY, a sequence often found in cytoplasmic tails, is required for coated pit-mediated internalization of the low density lipoprotein receptor. *J. Biol. Chem.*, **265**:3116–3123.

Chen, Y.P., Djaffar, I., Pidard, D., Steiner, B., Cieutat, A.M., Caen, J.P. and Rosa, J.P. (1992). Ser-752→Pro mutation in the cytoplasmic domain of integrin β_3 subunit and defective activation of platelet integrin $\alpha_{IIb}\beta_3$ (glycoprotein IIb-IIIa) in a variant of Glanzmann thrombasthenia. *Proc. Natl. Acad. Sci., USA*, **89**:10169–10173.

Chen, Y.P., O'Toole, T.E., Shipley, T., Forsyth, J., LaFlamme, S.E., Yamada, K.M., Shattil, S.J. and Ginsberg, M.H. (1994). "Inside-out" signal transduction inhibited by isolated integrin cytoplasmic domains. *J. Biol. Chem.*, **269**:18307–18310.

Chen, Y.P., O'Toole, T.E., Ylanne, J., Rosa, J.P. and Ginsberg, M.H. (1994). A point mutation in the integrin β_3 cytoplasmic domain (S752→P) impairs bidirectional signaling through $\alpha_{IIb}\beta_3$ (platelet glycoprotein IIb-IIIa). *Blood*, **84**:1857–1865.

Clapham, D.E. (1995). Calcium signaling. *Cell*, **80**:259–268.

Clark, E.A. and Brugge, J.S. (1995). Integrins and signal transduction pathways: the road taken. *Science*, **268**:233–239.

Clark, E.A., Shattil, S.J., Ginsberg, M.H., Bolen, J. and Brugge, J.S. (1994). Regulation of the protein tyrosine kinase pp72syk by platelet agonists and the integrin $\alpha_{IIb}\beta_3$. *J. Biol. Chem.*, **269**:28859–28864.

Coller, B.S. (1980). Interaction of normal, thrombasthenic, and Bernard-Soulier platelets with immobilized fibrinogen: Defective platelet-fibrinogen interaction in thrombasthenia. *Blood*, **55**:169–178.

Coller, B.S. (1986). Activation affects access to the platelet receptor for adhesive glycoproteins. *J. Cell Biol.*, **103**:451–456.

Coller, B.S., Anderson, K. and Weisman, H.F. (1995). New antiplatelet agents: Platelet GPIIb/IIIa antagonists. *Thrombosis and Haemostasis*, **74**:302–308.

Coller, B.S., Seligsohn, U., West, S.M., Scudder, L.E. and Norton, K.J. (1991). Platelet fibrinogen and vitronectin in Glanzmann thrombasthenia: evidence consistent with specific roles for glycoprotein IIb/IIIA and αv β_3 integrins in platelet protein trafficking. *Blood*, **78**:2603–2610.

Cooray, P., Yuan, Y., Schoenwaelder, S.M., Mitchell, C.A., Salem, H.H. and Jackson, S.P. (1996). Focal adhesion kinase (pp125FAK) cleavage and regulation by calpain. *Biochem. J.*, **318**:41–47.

Corbi, A.L., Kishimoto, T.K., Miller, L.J. and Springer, T.A. (1988). The human leukocyte adhesion glycoprotein Mac-1 (complement receptor type 3, CD11b) α-subunit. Cloning, primary structure, and relation to the integrins, von Willebrand factor and factor B. *J. Biol. Chem.*, **263**:12403–12411.

D'Souza, S.E., Ginsberg, M.H., Burke, T.A., Lam, S.C. and Plow, E.F. (1988). Localization of an Arg-Gly-Asp recognition site within an integrin adhesion receptor. *Science*, **242**:91–93.

D'Souza, S.E., Ginsberg, M.H., Burke, T.A. and Plow, E.F. (1990). The ligand binding site of the platelet integrin receptor GPIIb-IIIa is proximal to the second calcium binding domain of its α-subunit. *J. Biol. Chem.*, **265**:3440–3446.

D'Souza, S.E., Ginsberg, M.H., Matsueda, G.R. and Plow, E.F. (1991). A discrete sequence in a platelet integrin is involved in ligand recognition. *Nature*, **350**:66–68.

D'Souza, S.E., Haas, T.A., Piotrowicz, R.S., Byers, W.V., McGrath, D.E., Soule, H.R., Cierniewski, C., Plow, E.F. and Smith, J.W. (1994). Ligand and cation binding are dual functions of a discrete segment of the integrin β_3 subunit: cation displacement is involved in ligand binding. *Cell*, **79**:659–667.

Du, X., Gu, M., Weisel, J.W., Nagaswami, C., Bennett, J.S., Bowditch, R. and Ginsberg, M.H. (1993). Long range propagation of conformational changes in integrin $\alpha_{IIb}\beta_3$. *J. Biol. Chem.*, **268**:23087–23092.

Du, X., Plow, E.F., Frelinger, A.L.I., O'Toole, T.E., Loftus, J.C. and Ginsberg, M.H. (1991). Ligands "activate" integrin $\alpha_{IIb}\beta_3$ (platelet GPIIb-IIIa). *Cell*, **65**:409–416.

Du, X., Saido, T.C., Tsubuki, S., Indig, F.E., Williams, M.J. and Ginsberg, M.H. (1995). Calpain cleavage of the cytoplasmic domain of the integrin β_3 subunit. *J. Biol. Chem.*, **270**:26146–26151.

Farrell, D.H. and Thiagarajan, P. (1994). Binding of recombinant fibrinogen mutants to platelets. *J. Biol. Chem.*, **269**:226–231.

Farrell, D.H., Thiagarajan, P., Chung, D.W. and Davie, E.W. (1992). Role of fibrinogen α and gamma chain sites in platelet aggregation. *Proc. Natl. Acad. Sci., USA*, **89**:10729–10732.

Ferrell, J.J. and Martin, G.S. (1989). Tyrosine-specific protein phosphorylation is regulated by glycoprotein IIb-IIIa in platelets. *Proc. Natl. Acad. Sci., USA*, **86**:2234–2238.

Filardo, E.J., Brooks, P.C., Deming, S.L., Damsky, C. and Cheresh, D.A. (1995). Requirement of the NPXY motif in the integrin β3 subunit cytoplasmic tail for melanoma cell migration *in vitro* and *in vivo*. *J. Cell Biol.*, **130**:441–450.

Fitzgerald, L.A., Steiner, B., Rall, S.C.J., Lo, S.S. and Phillips, D.R. (1987). Protein sequence of endothelial glycoprotein IIIa derived from a cDNA clone. Identity with platelet glycoprotein IIIa and similarity to "integrin". *J. Biol. Chem.*, **262**:3936–3939.

Fox, J.E. (1993). Regulation of platelet function by the cytoskeleton. *Adv. Exp. Med. Biol.*, **344**:175–185.

Fox, J.E., Goll, D.E., Reynolds, C.C. and Phillips, D.R. (1985). Identification of two proteins (actin-binding protein and P235) that are hydrolyzed by endogenous Ca^{2+}-dependent protease during platelet aggregation. *J. Biol. Chem.*, **260**:1060–1066.

Fox, J.E., Lipfert, L., Clark, E.A., Reynolds, C.C., Austin, C.D. and Brugge, J.S. (1993). On the role of the platelet membrane skeleton in mediating signal transduction. Association of GP IIb-IIIa, pp60c-src, pp62c-yes, and the p21ras GTPase-activating protein with the membrane skeleton. *J. Biol. Chem.*, **268**:25973–25984.

Fox, J.E., Reynolds, C.C., Morrow, J.S. and Phillips, D.R. (1987). Spectrin is associated with membrane-bound actin filaments in platelets and is hydrolyzed by the Ca^{2+}-dependent protease during platelet activation. *Blood*, **69**:537–545.

Fox, J.E., Shattil, S.J., Kinlough, R.R., Richardson, M., Packham, M.A. and Sanan, D.A. (1996). The platelet cytoskeleton stabilizes the interaction between $\alpha_{IIb}\beta_3$ and its ligand and induces selective movements of ligand-occupied integrin. *J. Biol. Chem.*, **271**:7004–7011.

Fox, J.E., Taylor, R.G., Taffarel, M., Boyles, J.K. and Goll, D.E. (1993). Evidence that activation of platelet calpain is induced as a consequence of binding of adhesive ligand to the integrin, glycoprotein IIb-IIIa. *J. Cell Biol.*, **120**:1501–1507.

Frangioni, J.V., Oda, A., Smith, M., Salzman, E.W. and Neel, B.G. (1993). Calpain-catalyzed cleavage and subcellular relocation of protein phosphotyrosine phosphatase 1B (PTP-1B) in human platelets. *Embo J.*, **12**:4843–4856.

Frelinger, A.L.I., Cohen, I., Plow, E.F., Smith, M.A., Roberts, J., Lam, S.C. and Ginsberg, M.H. (1990). Selective inhibition of integrin function by antibodies specific for ligand-occupied receptor conformers. *J. Biol. Chem.*, **265**:6346–6352.

Frelinger, A.L.I., Du, X.P., Plow, E.F. and Ginsberg, M.H. (1991). Monoclonal antibodies to ligand-occupied conformers of integrin $\alpha_{IIb}\beta_3$ (glycoprotein IIb-IIIa) alter receptor affinity, specificity, and function. *J. Biol. Chem.*, **266**:17106–17111.

Frelinger, A.L.I., Lam, S.C., Plow, E.F., Smith, M.A., Loftus, J.C. and Ginsberg, M.H. (1988). Occupancy of an adhesive glycoprotein receptor modulates expression of an antigenic site involved in cell adhesion. *J. Biol. Chem.*, **263**:12397–12402.

Fujimoto, T., Fujimura, K. and Kuramoto, A. (1991). Electrophysiological evidence that glycoprotein IIb-IIIa complex is involved in calcium channel activation on human platelet plasma membrane. *J. Biol. Chem.*, **266**:16370–16375.

Fujimoto, T., Fujimura, K. and Kuramoto, A. (1991). Functional Ca^{2+} channel produced by purified platelet membrane glycoprotein IIB-IIIA complex incorporated into planar phospholipid bilayer. *Thromb. Haemost.*, **66**:598–603.

Gawaz, M.P., Loftus, J.C., Bajt, M.L., Frojmovic, M.M., Plow, E.F. and Ginsberg, M.H. (1991). Ligand bridging mediates integrin $\alpha_{IIb}\beta_3$ (platelet GPIIB-IIIA) dependent homotypic and heterotypic cell-cell interactions. *J. Clin. Invest.*, **88**:1128–1134.

Ginsberg, M.H., Du, X., O'Toole, T.E. and Loftus, J. C. (1995). Platelet Integrins. *Thrombosis and Haemostasis*, **74**:352–359.

Ginsberg, M.H., Du, X., O'Toole, T.E., Loftus, J.C. and Plow, E.F. (1993). Platelet integrins. *Thromb. Haemost.*, **70**:87–93.

Ginsberg, M.H., Du, X. and Plow, E.F. (1992). Inside-out integrin signalling. *Curr. Opin. Cell Biol.*, **4**:766–771.

Golden, A., Brugge, J.S. and Shattil, S.J. (1990). Role of platelet membrane glycoprotein IIb-IIIa in agonist-induced tyrosine phosphorylation of platelet proteins. *J. Cell Biol.*, **111**:3117–3127.

Guinebault, C., Payrastre, B., Racaud-Sultan, C., Mazarguil, H., Breton, M., Mauco, G., Plantavid, M. and Chap, H. (1995). Integrin-dependent translocation of phosphoinositide 3-kinase to the cytoskeleton of thrombin-activated platelets involves specific interactions of p85 α with actin filaments and focal adhesion kinase. *J. Cell Biol.*, **129**:831–842.

Gustafson, T.A., He, W., Craparo, A., Schaub, C.D. and O'Neill, T.J. (1995). Phosphotyrosine-dependent interaction of SHC and insulin receptor substrate 1 with the NPEY motif of the insulin receptor via a novel non-SH2 domain. *Mol. Cell Biol.*, **15**:2500–2508.

Haas, T.A. and Plow, E.F. (1996). The cytoplasmic domain of $\alpha_{IIb}\beta_3$. A ternary complex of the integrin α and β subunits and a divalent cation. *J. Biol. Chem.*, **271**:6017–6026.

Haimovich, B., Kaneshiki, N. and Ji, P. (1996). Protein kinase C regulates tyrosine phosphorylation of pp125FAK in platelets adherent to fibrinogen. *Blood*, **87**:152–161.

Haimovich, B., Lipfert, L., Brugge, J.S. and Shattil, S.J. (1993). Tyrosine phosphorylation and cytoskeletal reorganization in platelets are triggered by interaction of integrin receptors with their immobilized ligands. *J. Biol. Chem.*, **268**:15868–15877.

Hannigan, G.E., Leung, H.C., Fitz, G.L., Coppolino, M.G., Radeva, G., Filmus, J., Bell, J.C. and Dedhar, S. (1996). Regulation of cell adhesion and anchorage-dependent growth by a new β 1-integrin-linked protein kinase. *Nature*, **379**:91–96.

Hillery, C.A., Smyth, S.S. and Parise, L.V. (1991). Phosphorylation of human platelet glycoprotein IIIa (GPIIIa). Dissociation from fibrinogen receptor activation and phosphorylation of GPIIIa *in vitro*. *J. Biol. Chem.*, **266**:14663–14669.

Horwitz, A., Duggan, K., Buck, C., Beckerle, M.C. and Burridge, K. (1986). Interaction of plasma membrane fibronectin receptor with talin — a transmembrane linkage. *Nature*, **320**:531–533.

Huang, M.M., Lipfert, L., Cunningham, M., Brugge, J.S., Ginsberg, M.H. and Shattil, S.J. (1993). Adhesive ligand binding to integrin $\alpha_{IIb}\beta_3$ stimulates tyrosine phosphorylation of novel protein substrates before phosphorylation of pp125FAK. *J. Cell Biol.*, **122**:473–483.

Hynes, R.O. (1987). Integrins: A family of cell surface receptors. *Cell*, **48**:549–554.

Hynes, R.O. (1992). Integrins: versatility, modulation, and signaling in cell adhesion. *Cell*, **69**:11–25.

Isenberg, W.M., McEver, R.P., Phillips, D.R., Shuman, M.A. and Bainton, D.F. (1987). The platelet fibrinogen receptor: an immunogold-surface replica study of agonist-induced ligand binding and receptor clustering. *J. Cell Biol.*, **104**:1655–1663.

Isenberg, W.M., McEver, R.P., Phillips, D.R., Shuman, M.A. and Bainton, D.F. (1989). Immunogold-surface replica study of ADP-induced ligand binding and fibrinogen receptor clustering in human platelets. *Am. J. Anat.*, **185**:142–148.

Jennings, L.K., Fox, J.E., Edwards, H.H. and Phillips, D.R. (1981). Changes in the cytoskeletal structure of human platelets following thrombin activation. *J. Biol. Chem.*, **256**:6927–6932.

Kamata, T., Puzon, W. and Takada, Y. (1994). Identification of putative ligand binding sites within I domain of integrin $\alpha_2 \beta_1$ (VLA-2, CD49b/CD29). *J. Biol. Chem.*, **269**:9659–9663.

Kern, A., Briesewitz, R., Bank, I. and Marcantonio, E.E. (1994). The role of the I domain in ligand binding of the human integrin $\alpha_1 \beta_1$. *J. Biol. Chem.*, **269**:22811–22816.

Kloczewiak, M., Timmons, S., Lukas, T.J. and Hawiger, J. (1984). Platelet receptor recognition site on human fibrinogen. Synthesis and structure-function relationship of peptides corresponding to the carboxy-terminal segment of the γ-chain. *Biochemistry*, **23**:1767–1774.

Knezevic, I., Leisner, T.M. and Lam, S. (1996). Direct binding of the platelet integrin $\alpha_{IIb}\beta_3$ (GPIIb-IIIa) to talin. Evidence that interaction is mediated through the cytoplasmic domains of both α_{IIb} and β_3. *J. Biol. Chem.*, **271**:16416–16421.

Kouns, W.C., Fox, C.F., Lamoreaux, W.J., Coons, L.B. and Jennings, L.K. (1991). The effect of glycoprotein IIb-IIIa receptor occupancy on the cytoskeleton of resting and activated platelets. *J. Biol. Chem.*, **266**:13891–13900.

Kouns, W.C., Wall, C.D., White, M.M., Fox, C.F. and Jennings, L.K. (1990). A conformation-dependent epitope of human platelet glycoprotein IIIa. *J. Biol. Chem.*, **265**:20594–20601.

Kovacsovics, T.J., Bachelot, C., Toker, A., Vlahos, C.J., Duckworth, B., Cantley, L.C. and Hartwig, J.H. (1995). Phosphoinositide 3-kinase inhibition spares actin assembly in activating platelets but reverses platelet aggregation. *J. Biol. Chem.*, **270**:11358–11366.

Kunicki, T.J., Annis, D.S., Deng, Y.J., Loftus, J.C. and Shattil, S.J. (1996). A molecular basis for affinity modulation of Fab ligand binding to integrin $\alpha_{IIb}\beta_3$. *J. Biol. Chem.*, **271**:20315–20321.

Lam, S.C. (1992). Isolation and characterization of a chymotryptic fragment of platelet glycoprotein IIb-IIIa retaining Arg-Gly-Asp binding activity. *J. Biol. Chem.*, **267**:5649–5655.

Lam, S.C.-T., Plow, E.F., Smith, M.A., Andrieux, A., Ryckwaert, J.-J., Marguerie, G.A. and Ginsberg, M.H. (1987). Evidence that Arginyl-Glycyl-Aspartate peptides and fibrinogen gamma chain peptides share a common binding site on platelets. *J. Biol. Chem.*, **262**:947–950.

Law, D.A., Nannizzi, A.L. and Phillips, D.R. (1996). Outside-in integrin signal transduction. $\alpha_{IIb}\beta_3$-(GPIIb-IIIa) tyrosine phosphorylation induced by platelet aggregation. *J. Biol. Chem.*, **271**:10811–10815.

Lee, J.O., Bankston, L.A., Arnaout, M.A. and Liddington, R.C. (1995). Two conformations of the integrin A-domain (I-domain): a pathway for activation? *Structure*, **3**:1333–1340.

Lipfert, L., Haimovich, B., Schaller, M.D., Cobb, B.S., Parsons, J.T. and Brugge, J.S. (1992). Integrin-dependent phosphorylation and activation of the protein tyrosine kinase pp125FAK in platelets. *J Cell Biol* **119**:905–912.

Loftus, J.C., Halloran, C.E., Ginsberg, M.H., Feigen, L.P., Zablocki, J.A. and Smith, J.W. (1996). The amino-terminal one-third of α_{IIb} defines the ligand recognition specificity of integrin $\alpha_{IIb}\beta_3$. *J. Biol. Chem.*, **271**:2033–2039.

Loftus, J.C., O'Toole, T.E., Plow, E.F., Glass, A., Frelinger, A.d. and Ginsberg, M.H. (1990). A β_3 integrin mutation abolishes ligand binding and alters divalent cation-dependent conformation. *Science*, **249**:915–918.

Loftus, J.C., Plow, E.F., Jennings, L.K. and Ginsberg, M.H. (1988). Alternative proteolytic processing of platelet membrane glycoprotein IIb. *J. Biol. Chem.*, **263**:11025–11028.

Michishita, M., Videm, V. and Arnaout, M.A. (1993). A novel divalent cation-binding site in the A domain of the β_2 integrin CR3 (CD11b/CD18) is essential for ligand binding. *Cell*, **72**:857–867.

Miyamoto, S., Akiyama, S.K. and Yamada, K.M. (1995). Synergistic roles for receptor occupancy and aggregation in integrin transmembrane function. *Science*, **267**:883–885.

Miyamoto, S., Teramoto, H., Coso, O.A., Gutkind, J.S., Burbelo, P.D., Akiyama, S.K. and Yamada, K.M. (1995). Integrin function: molecular hierarchies of cytoskeletal and signaling molecules. *J. Cell Biol.*, **131**:791–805.

Nurden, A.T. and Caen, J.P. (1974). An abnormal platelet glycoprotein pattern in three cases of Glanzmann's thrombasthenia. *British Journal of Haematology*, **28**:253–260.

O'Toole, T.E., Katagiri, Y., Faull, R.J., Peter, K., Tamura, R., Quaranta, V., Loftus, J.C., Shattil, S.J. and Ginsberg, M.H. (1994). Integrin cytoplasmic domains mediate inside-out signal transduction. *J. Cell Biol.*, **124**:1047–1059.

O'Toole, T.E., Loftus, J.C., Du, X.P., Glass, A.A., Ruggeri, Z.M., Shattil, S.J., Plow, E.F. and Ginsberg, M.H. (1990). Affinity modulation of the $\alpha_{IIb}\beta_3$ integrin (platelet GPIIb-IIIa) is an intrinsic property of the receptor. *Cell Regul.*, **1**:883–893.

O'Toole, T.E., Loftus, J.C., Plow, E.F., Glass, A.A., Harper, J.R. and Ginsberg, M.H. (1989). Efficient surface expression of platelet GPIIb-IIIa requires both subunits. *Blood*, **74**:14–18.

O'Toole, T.E., Mandelman, D., Forsyth, J., Shattil, S.J., Plow, E.F. and Ginsberg, M.H. (1991). Modulation of the affinity of integrin $\alpha_{IIb}\beta_3$ (GPIIb-IIIa) by the cytoplasmic domain of α_{IIb}. *Science* **254**:845–847.

Oda, A., Druker, B.J., Ariyoshi, H., Smith, M. and Salzman, E.W. (1993). pp60src is an endogenous substrate for calpain in human blood platelets. *J. Biol. Chem.*, **268**:12603–12608.

Otey, C.A., Pavalko, F.M. and Burridge, K. (1990). An interaction between α-actinin and the β_1 integrin subunit *in vitro. J. Cell Biol.*, **111**:721–729.

Parise, L.V., Helgerson, S.L., Steiner, B., Nannizzi, L. and Phillips, D.R. (1987). Synthetic peptides derived from fibrinogen and fibronectin change the conformation of purified platelet glycoprotein IIb-IIIa. *J. Biol. Chem.*, **262**:12597–12602.

Pelletier, A.J., Bodary, S.C. and Levinson, A.D. (1992). Signal transduction by the platelet integrin $\alpha_{IIb}\beta_3$: induction of calcium oscillations required for protein-tyrosine phosphorylation and ligand-induced spreading of stably transfected cells. *Mol. Biol. Cell*, **3**:989–998.

Phillips, D.R. and Agin, P.P. (1977). Platelet plasma membrane glycoproteins. Evidence for the presence of nonequivalent disulfide bonds using nonreduced-reduced two-dimensional gel electrophoresis. *J. Biol. Chem.*, **252**:2121–2126.

Phillips, D.R., Charo, I.F. and Scarborough, R.M. (1991). GPIIb-IIIa: the responsive integrin. *Cell*, **65**:359–362.

Phillips, D.R., Jennings, L.K. and Edwards, H.H. (1980). Identification of membrane proteins mediating the interaction of human platelets. *J. Cell Biol.*, **86**:77–86.

Plow, E.F. and Ginsberg, M.H. (1989). Cellular adhesion: GPIIb-IIIa as a prototypic adhesion receptor. *Prog. Hemost. Thromb.*, **9**:117–156.

Poncz, M., Eisman, R., Heidenreich, R., Silver, S.M., Vilaire, G., Surrey, S., Schwartz, E. and Bennett, J.S. (1987). Structure of the platelet membrane glycoprotein IIb. Homology to the α subunits of the vitronectin and fibronectin membrane receptors. *J. Biol. Chem.*, **262**:8476–8482.

Powling, M.J. and Hardisty, R.M. (1985). Glycoprotein IIb-IIIa complex and Ca^{2+}-influx into stimulated platelets. *Blood*, **66**:731–734.

Qu, A. and Leahy, D.J. (1995). Crystal structure of the I-domain from the CD11a/CD18 (LFA-1, $\alpha_L\beta_2$) integrin. *Proc. Natl. Acad. Sci., USA*, **92**:10277–10281.

Renshaw, M.W., Toksoz, D. and Schwartz, M.A. (1996). Involvement of the small GTPase rho in integrin-mediated activation of mitogen-activated protein kinase. *J. Biol. Chem.*, **271**:21691–21694.

Rojiani, M.V., Finlay, B.B., Gray, V. and Dedhar, S. (1991). *in vitro* interaction of a polypeptide homologous to human Ro/SS-A antigen (calreticulin) with a highly conserved amino acid sequence in the cytoplasmic domain of integrin α-subunits. *Biochemistry*, **30**:9859–9866.

Rooney, M.M., Parise, L.V. and Lord, S.T. (1996). Dissecting clot retraction and platelet aggregation. Clot retraction does not require an intact fibrinogen gamma chain C terminus. *J. Biol. Chem.*, **271**:8553–8555.

Rosa, J.P., Bray, P.F., Gayet, O., Johnston, G.I., Cook, R.G., Jackson, K.W., Shuman, M.A. and McEver, R.P. (1988). Cloning of glycoprotein IIIa cDNA from human erythroleukemia cells and localization of the gene to chromosome 17. *Blood*, **72**:593–600.

Rybak, M.E., Renzulli, L.A., Bruns, M.J. and Cahaly, D.P. (1988). Platelet glycoproteins IIb and IIIa as a calcium channel in liposomes. *Blood*, **72**:714–720.

Sage, S.O., Sargeant, P., Heemskerk, J.W. and Mahaut, S.M. (1993). Calcium influx mechanisms and signal organisation in human platelets. *Adv. Exp. Med. Biol.*, **344**:69–82.

Saido, T.C., Mizuno, K. and Suzuki, K. (1991). Proteolysis of protein kinase C by calpain: effect of acidic phospholipids. *Biomed. Biochim. Acta*, **50**:485–489.

Savage, B., Bottini, E. and Ruggeri, Z.M. (1995). Interaction of integrin $\alpha_{IIb}\beta_3$ with multiple fibrinogen domains during platelet adhesion. *J. Biol. Chem.*, **270**:28812–28817.

Schaller, M.D., Otey, C.A., Hildebrand, J.D. and Parsons, J.T. (1995). Focal adhesion kinase and paxillin bind to peptides mimicking β integrin cytoplasmic domains. *J. Cell Biol.*, **130**:1181–1187.

Schwartz, M.A., Brown, E.J. and Fazeli, B. (1993). A 50-kDa integrin-associated protein is required for integrin-regulated calcium entry in endothelial cells. *J. Biol. Chem.*, **268**:19931–19934.

Schwartz, M.A., Schaller, M.D. and Ginsberg, M.H. (1995). Integrins: emerging paradigms of signal transduction. *Ann. Rev. Cell Dev. Biol.*, **11**:549–599.

Shattil, S.J. (1993). Regulation of platelet anchorage and signaling by integrin $\alpha_{IIb}\beta_3$. *Thromb. Haemost.*, **70**:224–228.

Shattil, S.J. and Brass, L.F. (1987). Induction of the fibrinogen receptor on human platelets by intracellular mediators. *J. Biol. Chem.*, **262**:992–1000.

Shattil, S.J., Brass, L.F., Bennett, J.S. and Pandhi, P. (1985). Biochemical and functional consequences of dissociation of the platelet membrane glycoprotein IIb-IIIa complex. *Blood*, **66**:92–98.

Shattil, S.J., Cunningham, M., Wiedmer, T., Zhao, J., Sims, P.J. and Brass, L.F. (1992). Regulation of glycoprotein IIb-IIIa receptor function studied with platelets permeabilized by the pore-forming complement proteins C5b-9. *J. Biol. Chem.*, **267**:18424–18431.

Shattil, S.J., Ginsberg, M.H. and Brugge, J.S. (1994). Adhesive signaling in platelets. *Curr. Opin. Cell Biol.*, **6**:695–704.

Shattil, S.J., Haimovich, B., Cunningham, M., Lipfert, L., Parsons, J.T., Ginsberg, M.H. and Brugge, J.S. (1994). Tyrosine phosphorylation of pp125FAK in platelets requires coordinated signaling through integrin and agonist receptors. *J. Biol. Chem.*, **269**:14738–14745.

Shattil, S.J., Hoxie, J.A., Cunningham, M. and Brass, L.F. (1985). Changes in the platelet membrane glycoprotein IIb.IIIa complex during platelet activation. *J. Biol. Chem.*, **260**:11107–11114.

Shattil, S.J., O'Toole, T., Eigenthaler, M., Thon, V., Williams, M., Babior, B.M. and Ginsberg, M.H. (1995). β_3-endonexin, a novel polypeptide that interacts specifically with the cytoplasmic tail of the integrin β_3 subunit. *J. Cell Biol.*, **131**:807–816.

Sims, P.J., Ginsberg, M.H., Plow, E.F. and Shattil, S.J. (1991). Effect of platelet activation on the conformation of the plasma membrane glycoprotein IIb-IIIa complex. *J. Biol. Chem.*, **266**:7345–7352.

Smyth, S.S. and Parise, L.V. (1993). Regulation of ligand binding to glycoprotein IIb-IIIa (integrin $\alpha_{IIb}\beta_3$) in isolated platelet membranes. *Biochem. J.*, **292**:749–758.

Steiner, B., Trzeciak, A., Pfenninger, G. and Kouns, W.C. (1993). Peptides derived from a sequence within β_3 integrin bind to platel $\alpha_{IIb}\beta_3$ (GPIIb-IIIa) and inhibit ligand binding. *J. Biol. Chem.*, **268**:6870–6873.

Sultan, C., Plantavid, M., Bachelot, C., Grondin, P., Breton, M., Mauco, G., Levy, T.S., Caen, J.P. and Chap, H. (1991). Involvement of platelet glycoprotein IIb-IIIa (α_{IIb}-β_3 integrin) in thrombin-induced synthesis of phosphatidylinositol 3,4-bisphosphate. *J. Biol. Chem.*, **266**:23554–23557.

Taylor, D.B. and Gartner, T.K. (1992). A peptide corresponding to GPIIbα 300–312, a presumptive fibrinogen gamma-chain binding site on the platelet integrin GPIIb/IIIa, inhibits the adhesion of platelets to at least four adhesive ligands. *J. Biol. Chem.*, **267**:11729–11733.

Tcheng, J.E. (1996). Glycoprotein IIb/IIIa receptor inhibitors: putting the EPIC, IMPACT II, RESTORE, and EPILOG trials into perspective. *Am. J. Cardiol.*, **78**:35–40.

Tozer, E.C., Liddington, R.C., Sutcliffe, M.J., Smeeton, A.H. and Loftus, J.C. (1996). Ligand binding to integrin $\alpha_{IIb}\beta_3$ is dependent on a MIDAS-like domain in the β_3 subunit. *J. Biol. Chem.*, **271**, 21978–21984.

van Willigen, G., Hers, I., Gorter, G. and Akkerman, J.W. (1996). Exposure of ligand-binding sites on platelet integrin α_{IIb}/β_3 by phosphorylation of the β_3 subunit. *Biochem. J.*, **314**:769–779.

Wary, K.K., Mainiero, F., Isakoff, S., Marcantonio, E.E. and Giancotti, F.G. (1996). The adaptor protein Shc couples a class of integrins to the control of cell cycle progression. *Cell*, **87**:733–743.

Weisel, J.W., Nagaswami, C., Vilaire, G. and Bennett, J.S. (1992). Examination of the platelet membrane glycoprotein IIb-IIIa complex and its interaction with fibrinogen and other ligands by electron microscopy. *J. Biol. Chem.*, **267**:16637–16643.

Wencel-Drake, J., Boudignon, P.C., Dieter, M.G., Criss, A.B. and Parise, L.V. (1996). Internalization of bound fibrinogen modulates platelet aggregation. *Blood*, **87**:602–612.

Wippler, J., Kouns, W.C., Schlaeger, E.J., Kuhn, H., Hadvary, P. and Steiner, B. (1994). The integrin α_{IIb}-β_3, platelet glycoprotein IIb-IIIa, can form a functionally active heterodimer complex without the cysteine-rich repeats of the β_3 subunit. *J. Biol. Chem.*, **269**:8754–8761.

Ylanne, J., Chen, Y., O'Toole, T.E., Loftus, J.C., Takada, Y. and Ginsberg, M.H. (1993). Distinct functions of integrin α and β subunit cytoplasmic domains in cell spreading and formation of focal adhesions. *J. Cell Biol.*, **122**:223–233.

Ylanne, J., Huuskonen, J., O'Toole, T.E., Ginsberg, M.H., Virtanen, I. and Gahmberg, C.G. (1995). Mutation of the cytoplasmic domain of the integrin β_3 subunit. Differential effects on cell spreading, recruitment to adhesion plaques, endocytosis, and phagocytosis. *J. Biol. Chem.*, **270**:9550–9557.

Zamarron, C., Ginsberg, M.H. and Plow, E.F. (1991). A receptor-induced binding site in fibrinogen elicited by its interaction with platelet membrane glycoprotein IIb-IIIa. *J. Biol. Chem.*, **266**:16193–16199.

Zhu, X. and Assoian, R.K. (1995). Integrin-dependent activation of MAP kinase: a link to shape-dependent cell proliferation. *Mol. Biol. Cell*, **6**:273–282.

Zimrin, A.B., Eisman, R., Vilaire, G., Schwartz, E., Bennett, J.S. and Poncz, M. (1988). Structure of platelet glycoprotein IIIa. A common subunit for two different membrane receptors. *J. Clin. Invest.*, **81**:1470–1475.

6 The Platelet Cytoskeleton

Joan E.B. Fox and Sylvie C. Meyer

Department of Molecular Cardiology, Joseph J. Jacobs Center for Thrombosis and Vascular Biology, Cleveland Clinic Foundation, 9500 Euclid Avenue, Cleveland, OH 44195, USA

INTRODUCTION

In the past, the major functions of the platelet cytoskeleton have been considered to be those of regulating the shape of the unstimulated platelet and inducing contractile events such as secretion of granules and retraction of clots by activated cells. However, in recent years, the realization that the cytoskeleton can interact with numerous signaling molecules has revolutionized our understanding of the importance of the platelet cytoskeleton in regulating platelet function. An important component of the platelet cytoskeleton in terms of regulating signal transduction is the membrane skeleton. The model that is emerging is that this structure lines the lipid bilayer, interacts with the cytoplasmic domain of transmembrane receptors, and recruits signaling molecules to a submembranous location (Fox, 1996a; Fox *et al.*, 1993a). As platelets are activated, the cytoskeleton reorganizes producing complexes of cytoskeletal proteins that recruit and activate additional signaling molecules. The cytoskeletal reorganizations that occur and the signaling molecules that are recruited vary depending on the conditions of activation.

Another major advance in recent years has been an increased understanding of the mechanisms regulating the recruitment of signaling molecules to the cytoskeleton. Much of the new information has come from work on other cells showing the importance of tyrosine phosphorylation in recruiting and activating SH2/SH3 containing molecules to specific locations (Bar-Sagi *et al.*, 1993; Pawson, 1995; van der Geer *et al.*, 1994). It is now clear that numerous proteins in the platelet cytoskeleton can be phosphorylated on tyrosine residues and that many of the signaling molecules recruited to the platelet cytoskeleton are SH2/SH3 containing molecules that in other cells are recruited to a submembranous location by interaction with phosphotyrosine residues (Clark and Brugge, 1993; Clark *et al.*, 1994b; Dash *et al.*, 1995a; Falet *et al.*, 1996; Fox, 1993; Fox *et al.*, 1993a; Fox, 1996a; Grondin *et al.*, 1991; Guinebault *et al.*, 1995; Horvath *et al.*, 1992; Li *et al.*, 1994; Lipfert *et al.*, 1992; Zhang *et al.*, 1992).

Finally, in recent years, work on other cells has provided a clear demonstration of the role of Rho family members in regulating distinct cytoskeletal reorganizations (Burridge and Chrzanowska-Wodnicka, 1996; Chrzanowska-Wodnicka and Burridge, 1992; Hall, 1994; Nobes and Hall, 1994; Nobes and Hall, 1995; Symons, 1996). Until recently, very little was known about the mechanisms regulating actin polymerization, focal contact formation, or cytoskeletal reorganizations in platelets. Because Rho family members and many of their activators and downstream effectors are present in platelets, it appears likely that the cytoskeletal reorganizations induced in platelets will be induced by mechanisms very similar to those in other cells. The present chapter will review the organization of the cytoskeleton, discuss the recruitment of signaling molecules to the cytoskeleton of activated and aggregating platelets, and describe ways in which the rapid reorganizations of the platelet cytoskeleton may be regulated by members of the Rho family.

The Cytoskeleton of Unstimulated Platelets

Organization of the cytoskeleton

The platelet cytoskeleton consists of a network of long actin filaments that are cross-linked by a variety of accessory proteins. This network is present throughout the platelet cyto-plasm but is concentrated at the cell periphery. Another component is a microtubule coil that is tightly coiled on itself just beneath the plasma membrane. Yet another component is a membrane skeleton that has been visualized by thin-section electron microscopy (Fox, 1988; Fox and Boyles, 1988b; Fox et al., 1988) and by high-resolution quick-freeze deep-etch approaches (Hartwig and DeSisto, 1991) and appears to consist of a lattice-work of short cross-linked actin filaments that coats the inner surface of the lipid bilayer. Connec-tions have been described between each of the components of the cytoskeleton (Boyles et al., 1985) and between the membrane skeleton and the cytoplasmic domain of the mem-brane receptors (Fox, 1996b; Fox, 1985c), in this way the cytoskeleton serves as a scaffold-ing, maintaining the shape of the cell and supporting the membrane. Moreover, by attaching to transmembrane receptors, the cytoskeleton can respond to signals from the outside of the cell.

Essentially nothing is known about the way in which the microtubule coil is maintained at the periphery of the cell, the identity of proteins that link it to other components of the cytoskeleton, or the identity of microtubule-associated proteins that regulate its organ-ization or function. In contrast, much more is known about the organization of the actin filament-based component of the cytoskeleton and the accessory proteins that maintain this organization. The actin filament-based components of the cytoskeleton have been studied in detail by lysing platelets with Triton X-100 containing buffers and sedimenting the detergent-insoluble material (Fox, 1996b; Fox et al., 1992; Phillips, 1980). Once platelets are lysed, the membrane skeleton readily breaks into fragments (Fox, 1985c; Fox et al., 1988; Hartwig and DeSisto, 1991), this is augmented by the shear forces of centrifugation. The network of cytoplasmic actin filaments can be sedimented from detergent-lysates at low g-forces but the membrane skeleton fragments remain in the detergent-soluble fraction and can subsequently be sedimented at higher g-forces. Fragmentation of the membrane skeleton has been useful because it has allowed the differential isolation of the cytoplasmic actin filaments and the membrane skeleton. Analysis of the composition of the low-speed detergent-insoluble material (Fox et al., 1988) has shown that the cytoplasmic actin fila-ments interact with proteins such as actin-binding protein (ABP) and α-actinin, two proteins known to cross-link actin filaments in *in vitro* binding assays. Another protein recovered with these filaments is tropomyosin, a protein known to bind to actin filaments and prevent their interaction with other proteins. Yet other proteins that may associate with cytoplasmic actin filaments are VASP (Reinhard et al., 1992) and caldesmon (Dingus et al., 1986; Pho et al., 1986).

Analysis of the high-speed detergent-insoluble fraction has identified actin and the cross-linking proteins spectrin, actin-binding protein (ABP), and dystrophin-related pro-tein (DRP) as components of the membrane skeleton (Earnest et al., 1995; Fox, 1985b; Fox and Boyles, 1988a; Fox et al., 1987). Additional cytoskeletal proteins in the membrane skeleton include talin, vinculin, and protein 4.1 (Fox *et al.*, 1993a), all of which are proteins present at the sites of interaction of transmembrane receptors with actin filaments in other cells. Signaling molecules that co-isolate with this structure include the tyrosine kinases pp60[c-src], pp62[c-yes], and pp72syk, the p21ras GTPase-activating protein (GAP), and the tyrosine phosphatase SHPTP1 (Ezumi *et al.*, 1995; Fox *et al.*, 1993a; Tohyama *et al.*, 1994). Membrane glycoproteins include the GP Ib-IX complex, the integrins $\alpha_{IIb}\beta_3$ (GP IIb-IIIa) and $\alpha_2\beta_1$ (GP Ia-IIa), and an unidentified glycoprotein of molecular weight ~ 250 kDa (Fox, 1985c; Fox *et al.*, 1993a). Because dystrophin-related protein is present in platelets (Earnest *et al.*, 1995), it is possible that the dystrophin-glycoprotein complex with which dystrophin-related protein interacts in other cells (Matsumura and Campbell, 1994; Matsumura *et al.*, 1992) provides another site of attachment for the membrane skeleton in platelets. Although this complex has not yet been described in platelets, syntrophin, one of its components, has been identified in the detergent-insoluble membrane skeleton fraction (Fox, unpublished observations). A schematic representation of the membrane skeleton showing the presence of some these molecules in a lattice work beneath the lipid bilayer is presented in Figure 6.1.

The best studied interaction of a membrane glycoprotein with the platelet membrane skeleton is that of the GP Ib-IX complex. Association of this glycoprotein complex with the membrane skeleton has been demonstrated by immunogold electron microscopy (Fox

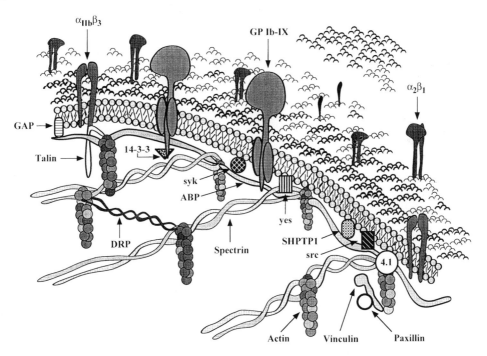

Figure 6.1 Schematic representation of the platelet membrane skeleton. This model of the platelet membrane skeleton shows some of the proteins that are recovered with the membrane skeleton from detergent-lysed platelets. Some of the interactions such as those between GP Ib-IX and actin-binding protein and GP Ib-IX and 14-3-3 have been clearly demonstrated and characterized in detail (Andrews and Fox, 1991; Cunningham *et al.*, 1996; Du *et al.*, 1996). However, the way in which many of the proteins are organized is unknown and the interactions shown in this model are based either on known interactions that occur *in vitro* (*e.g.*, paxillin interacts with vinculin) or are simply shown schematically to demonstrate the presence of the molecules. GAP, the p21[ras]GTPase-activating protein; Syk, pp72[syk]; Yes, pp62[c-yes]; Src, pp60[c-src].

et al., 1988; Hartwig and DeSisto, 1991). Furthermore, immunoprecipitation experiments have identified actin-binding protein as the component of the membrane skeleton with which GP Ib-IX interacts (Fox, 1985b). Subsequent experiments with purified proteins and peptides have identified a domain in the central region of GP Ibα as the site of interaction of actin-binding protein (Andrews and Fox, 1991; Andrews and Fox, 1992). More recent studies using transfected cells, have demonstrated that this region of GP Ibα mediates the interaction of the GP Ib-IX complex with actin-binding protein and have identified the region of actin-binding protein with which GP Ib-IX interacts (Cunningham *et al.*, 1996). The cytoplasmic domain of GP Ib-IX also interacts with 14-3-3, although the functional significance of the interaction of GP Ib-IX with this putative signaling molecule is not known (Du *et al.*, 1996; Du *et al.*, 1994).

Less is known about the way in which integrins interact with the membrane skeleton. Experiments utilizing cultured cells transfected with a variety of truncated forms of the α and β-subunits of integrins have shown that the β-subunit mediates the interaction of integrins with the cytoskeleton. Thus, the β-subunit is entirely responsible for incorporation of integrins into focal contacts (Ylänne *et al.*, 1993). In cells transfected with the $\alpha_{IIb}\beta_3$ complex, the β-subunit is responsible for the transmission of contractile forces to externally bound clots (Ylänne *et al.*, 1993). Immunofluorescence studies in cultured cells have shown that a number of cytoskeletal proteins including talin, α-actinin, vinculin, and tensin co-localize with integrins (Burridge and Chrzanowska-Wodnicka, 1996). Talin and α-actinin have also be shown to interact with the cytoplasmic domains of both the β_1 and β_3 subunits in vitro, (Horwitz *et al.*, 1986; Knezevic *et al.*, 1996; Otey *et al.*, 1993; Otey *et al.*, 1990) making them candidates for proteins that link the integrins to the cytoskeleton. It is entirely possible that other unidentified proteins mediate the interaction of the integrin with the membrane skeleton in unstimulated platelets. In search of new candidates for integrin-binding proteins, several groups have used the yeast two-hybrid system. In recent studies, a clone that encodes for an amino-acid sequence present in a previously identified protein known as skelemin was identified (Reddy *et al.*, 1996). Although it is not yet known whether skelemin is present in platelets, it is of interest that it is a cytoskeletal myosin-binding protein present in skeletal muscle cells (Price and Gomer, 1993). Further, it colocalizes with β_1 containing integrin in bovine endothelial cells (Reddy and Fox, unpublished observations). Thus, skelemin is another candidate for a protein that could mediate the interaction of integrins with the cytoskeleton under certain conditions.

Regulation of actin polymerization in unstimulated platelets

In the unstimulated platelet, only 30–40% of the total actin is polymerized into filaments (Fox, 1993). The rest of the actin is in a monomeric form and is prevented from polymerizing by two mechanisms. One is by proteins that cap the ends of actin filaments, preventing additional monomers from adding on. The other is by association of actin monomers with proteins that prevent the monomers from polymerizing. Actin filaments have an asymmetry; one end of a filament is referred to as the barbed end, the other as the pointed end. Actin monomers add on much more rapidly to the barbed than to the pointed ends. In activated platelets, polymerization of actin occurs at the barbed end of the pre-existing filaments (Fox and Phillips, 1981). One protein that has been suggested to bind to the barbed ends in unstimulated platelets is gelsolin (Hartwig *et al.*, 1989b). However, it has been reported that in unstimulated platelets, only about 1% of the gelsolin is associated with actin filaments (Nachmias *et al.*, 1996). Further, in gelsolin-deficient mice the actin filament content of unstimulated platelets is only slightly higher than normal, indicating that gelsolin is not the major capping protein in platelets (Witke *et al.*, 1995). These findings have led to the search for additional barbed-end capping proteins. One candidate is CAPZ, a Ca^{2+}-insensitive capping protein that was originally identified in Acanthamoeba and has subsequently been identified in every cell examined (Caldwell *et al.*, 1989; Hartmann *et al.*, 1989; Heiss and Cooper, 1991). This protein consists of two subunits of approximately 32 and 36 kDa. Its concentration in platelets is consistent with it being a major actin filament capping protein in unstimulated platelets. Two groups have found that significant amounts of CAPZ are recovered in association with the high-speed detergent-insoluble membrane skeleton fraction isolated from lysed platelets (Barkalow *et al.*, 1996; Nachmias *et al.*, 1996). Additional candidates for barbed-end capping proteins include the Ca^{2+}-dependent severing protein adseverin (otherwise known as scinderin) (Marcu *et al.*, 1996; Rodriguez del Castillo *et al.*, 1992), tensin (Lo *et al.*, 1994; Schafer and Cooper, 1995), and adducin (Kuhlamn *et al.*, 1996).

One protein that is known to bind to monomeric actin is thymosin β_4 (Safer *et al.*, 1990; Safer and Nachmias, 1994; Weber *et al.*, 1992). Purified thymosin β_4 forms a 1:1 complex with monomeric actin and inhibits its polymerization; much of the monomeric actin in permeabilized platelets is recovered in association with thymosin β_4. Another protein that sequesters monomeric actin is ASP-56 (Gieselmann and Mann, 1992), a protein that shows homology with a yeast protein that interacts with adenylate cyclase and regulates the cytoskeleton (Field *et al.*, 1990; Vojtek *et al.*, 1991). ASP-56 has been purified from pig platelets and shown to inhibit actin polymerization in *in vitro* assays. Elucidation of the function of this protein in intact platelets awaits further studies. Yet another protein that can bind to monomeric actin and has been considered a candidate for a protein that prevents monomeric actin from polymerizing is profilin (Carlsson *et al.*, 1977). However, it is now apparent that the concentration of profilin within platelets is considerably below that of monomeric actin (Lind *et al.*, 1987). Further, the binding constant of profilin for actin is relatively low (Pollard and Cooper, 1986). Profilin can also bind to phosphatidylinositol 4,5,bisphosphate (PtdIns(4,5)P2) with an affinity that is greater than that for actin (Goldschmidt-Clermont *et al.*, 1990; Lassing and Lindberg, 1985). It has been calculated that there is sufficient PtdIns(4,5)P2 present in platelets to bind all of the profilin, thus, it has been suggested that in the unstimulated platelet, most of the profilin may be associated with PtdIns(4,5)P2 in the plasma membrane rather than with actin monomers (Goldschmidt-Clermont *et al.*, 1990). The morphological demonstration that profilin has a submembranous location in unstimulated platelets is consistent with this possibility (Hartwig *et al.*, 1989a). There is evidence that under conditions that exist within the cell, profilin can accelerate rather than inhibit actin polymerization (Goldschmidt-Clermont *et al.*, 1992). Thus, as discussed below, profilin may function to initiate actin polymerization in activated platelets rather than to inhibit it in unstimulated cells.

Several other proteins have been described that can regulate the extent of polymerization of actin. For example, tropomodulin prevents the depolymerization of actin at the pointed ends of actin filaments (Weber *et al.*, 1994), proteins such as moesin (Arpin *et al.*, 1994) and 4.1 may play a role in regulating the length of actin filaments (Tsukita *et al.*, 1997; Elbaum *et al.*, 1984). Future studies will be needed to elucidate the functions of these and other accessory proteins.

The length of actin filaments can vary at different locations in the platelet. For example, long filaments are present throughout the cytoplasm while those in the membrane skeleton are thought to be only a few monomers in length (Fox *et al.*, 1988). It will be of particular interest to elucidate the mechanisms by which the polymerization of actin is differentially regulated at different locations in the cell.

Function of the cytoskeleton in unstimulated cells

A major function of the cytoskeleton in unstimulated platelets is presumably to provide a structural support maintaining the discoid shape of the unstimulated cells. However, it is becoming apparent that the cytoskeleton, particularly the membrane skeleton, may have several additional functions. Because the membrane skeleton coats the lipid bilayer and associates with membrane receptors, it appears that it would be ideally suited to bind signaling molecules, bringing them into an appropriate location to be activated and to act on membrane-associated substrates following transmembrane signaling. As discussed above, studies to test this possibility have revealed the co-isolation of a number of signaling molecules (e.g., pp60$^{c\text{-src}}$, pp62$^{c\text{-yes}}$, GAP, pp72^{syk}, and SHPTP1) with the detergent-insoluble membrane skeleton (Ezumi *et al.*, 1995; Fox *et al.*, 1993a; Tohyama *et al.*, 1994). Moreover, as platelets are activated, proteins co-isolating with the detergent-insoluble membrane skeleton become tyrosine phosphorylated (Fox *et al.*, 1993a) raising the possibility that membrane skeleton proteins may play a role in recruiting SH2-containing molecules to a submembranous location.

Another function of the membrane skeleton may be to stabilize the lipid bilayer and to regulate the expression of receptors in the membrane. Thus, in recent studies in which an ABP-deficient melanoma cell line was transfected with the cDNAs encoding ABP and each of the subunits of GP Ib-IX, expression of GP Ib-IX was increased in the membrane when actin-binding protein was present (Meyer *et al.*, 1997b). The use of cDNAs encoding truncated ABP that lacked the GP Ib-IX binding site indicated that it was the presence of ABP in the membrane skeleton rather than its interaction with the membrane glycoprotein that was important in maintaining the membrane so that the receptor could be expressed. The finding that a component of the membrane skeleton regulates properties of the plasma membrane is consistent with the existence of hereditary diseases in which mutations in cytoskeletal proteins lead to abnormalities in the content of membrane glycoproteins and changes in the stability of the plasma membrane. Often, the absence of one component of the membrane skeleton can lead to the loss of several proteins; common consequences of these diseases are decreased sialoglycoproteins on the cell surface, decreased survival time in the circulation, and activation of proteases within the cell. One such disease may be the Wiskott-Aldrich syndrome in which there are defects in a 62 kDa proline-rich protein known as WASP (Remold-O'Donnell *et al.*, 1996; Symons *et al.*, 1996). The finding that platelets from patients with this disease have abnormal cytoskeletons, decreased surface sialoglycoproteins, shortened survival time in the circulation, and increased activation of calpain are all consistent with WASP being an important component of the platelet membrane skeleton.

Another potential function of the membrane skeleton is to regulate the function of transmembrane receptors with which it associates. In recent studies, cultured cells were transfected with various truncated forms of GP Ib-IX or actin-binding protein (Cunningham *et al.*, 1996). The ability of GP Ib-IX to bind its ligand, von Willebrand factor, was the same whether the receptor was associated with the membrane skeleton or not. However, the subsequent signaling induced across GP Ib-IX was very different in cells expressing truncated GP Ib-IX. Another receptor that is associated with the membrane skeleton is $\alpha_{IIb}\beta_3$ (Fox *et al.*, 1993a). Only a subpopulation of this integrin is recovered with the membrane skeleton from lysates of unstimulated platelets. However, it is this subpopulation that is recruited first into the focal-contact like structures that form as the integrin binds ligand in activated platelets (Fox *et al.*, 1993a; Fox *et al.*, 1996). Thus, membrane skeleton-associated integrin may be selectively induced to bind ligand. Future studies will be needed to determine mechanisms by which association of transmembrane receptors with the membrane skeleton can regulate signaling across the receptors.

Yet another function of the membrane skeleton may be to provide spatial cues. A question of considerable interest is the way in which reorganizations of the cytoskeleton are localized to specific sites in the cell. For example, as platelets are activated, bundles of new actin filaments that induce the formation of filopodia are polymerized at only a few locations; the way in which these locations are specified is unknown. Presumably, the cell has mechanisms for maintaining functional domains. In yeast, genetic approaches have led to the discovery that Ras family members assemble at specific sites and serve as the markers for the assembly of complexes of signaling molecules that direct polymerization of actin at these sites (Chant and Stowers, 1995; Drubin and Nelson, 1996). One of these Ras family members is Cdc42Hs, a Rho protein that is recruited to the platelet membrane skeleton upon platelet activation (Dash *et al.*, 1995a) and has been shown to interact with WASP (Symons, 1996; Symons *et al.*, 1996). Thus, it is conceivable that in unstimulated platelets, the membrane skeleton is involved in defining functional domains at which cytoskeletal reorganizations can occur.

Organization of the Cytoskeleton in Activated Cells

As platelets are activated, they rapidly reorganize their cytoskeleton. The specific reorganizations that occur vary depending on the change in cell behavior that is required. For example, cells that are activated in suspension lose their discoid shape and extend a few long filopodia. Examination of their cytoskeletons reveals an increase in the total amount of filamentous actin from 30–40% to 60–70% of the total actin (Carlsson *et al.*, 1979; Fox *et al.*, 1984; Jennings *et al.*, 1981). The filamentous actin is present as a network at the periphery of the cell, as bundles in the filopodia, and as a contracting ring surrounding the granules which are forced towards the center of the activated platelet (Hartwig, 1992; Karlsson *et al.*, 1984; Lindberg and Markey, 1987; Nachmias and Yoshida, 1988; White, 1984). As discussed above, the proteins that bind to actin filaments in the unstimulated platelet include actin-binding protein, α-actinin, and tropomyosin. Immunofluorecence studies have revealed that these proteins are redistributed during platelet activation (Debus *et al.*, 1981).

If the platelet is adherent, the interaction of adhesion receptors with extracellular ligands induces different reorganizations of cytoskeletal proteins and the formation of complexes of cytoskeletal proteins and signaling molecules. These complexes not only serve to provide a transmembrane linkage between extracellular adhesive ligand and the structural scaffolding within the cell, but also allow the transmission of signals across the cell membrane. The composition and function of the signaling complexes have been studied best in the case of the integrin $\alpha_{IIb}\beta_3$. However, it is becoming apparent that that von Willebrand factor binding to GP Ib-IX also induces the formation of such complexes and leads to the generation of intracellular signals (Cunningham *et al.*, 1996; Jackson *et al.*, 1994). It is also apparent that with both GP Ib-IX (Meyer and Fox, unpublished observations) and $\alpha_{IIb}\beta_3$ (Clark *et al.*, 1994a; Clark *et al.*, 1994b; Clark *et al.*, 1994c; Fox *et al.*, 1993a; Huang *et al.*, 1993; Kouns *et al.*, 1991; Lipfert *et al.*, 1992), the cytoskeletal reorganizations and the signaling complexes that form can vary depending on whether the adhesive receptor binds to immobilized ligand or to soluble ligand, and perhaps also on whether the receptor is exposed to shear forces.

The activation-induced reorganizations of the cytoskeleton are dynamic. The platelet extends filopodia and subsequently retracts them, extending filopodia in new locations.

Shape changes and extension of membrane ruffles occur as actin filament networks contin-
uously form and disassemble in submembranous locations. Adhesive receptor interactions
are made and broken so that the cell can spread over a surface; presumably these changes
result from intracellular signals generated in the complexes of cytoskeletal proteins and
signaling molecules with which the receptor associates. The next sections will describe
some of the signals that are known to be involved in regulating these dynamic and varied
cytoskeletal reorganizations, discuss the way in which the signals may induce changes in
actin filament content and organization in platelets activated by agonists that act through
heterotrimeric G-protein coupled receptors, and describe the complexes of cytoskeletal
proteins and signaling molecules that form following signaling across the integrin $\alpha_{IIb}\beta_3$ in
aggregating platelets.

Signals regulating cytoskeletal changes

Recent advances in our understanding of potential mechanisms by which cytoskeletal reorganizations are induced in activated platelets have come from studies on the mechanisms
by which cytoskeletal reorganizations are induced in cultured cells. Much of this information has come from the study of the Ras superfamily of proteins. Members of this family
are GTP-binding proteins of Mr = 20–29 kDa (Boguski and McCormick, 1993; Valencia
et al., 1991). They are typically divided into classes on the basis of sequence homology; the
different classes have distinctly different functions. One class, known simply as Ras or
p21[ras], is responsible for activation of pathways involved in growth and differentiation.
Another class, that plays a critical role in regulating the cytoskeleton, is the Rho subfamily.
Members of this class include Rac, Cdc42Hs, and the Rho proteins (RhoA, RhoB, RhoC).
Studies in which constitutively active forms of Rho family members have been microin-
jected into cultured cells have shown that each family member has a dramatically different
effect (Hall, 1992; Hall, 1994; Nobes and Hall, 1994; Nobes and Hall, 1995; Ridley and
Hall, 1992). As shown in Figure 6.2, Rac induces the formation of lamellipodia, Cdc42
induces the formation of filopodia, and RhoA induces the formation of focal contacts.

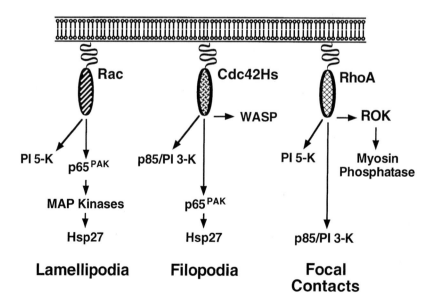

Figure 6.2   Downstream effectors of Rho family members.

Considerable information is available on the mechanisms by which Rho family members are activated and the actions of the different family members induced in cultured cells. Like other members of the Ras superfamily of proteins, Rho proteins are GTPases and are activated as they cycle between GTP-bound (active) and GDP-bound (inactive) states (Boguski and McCormick, 1993; Bokoch and Der, 1993; Bollag and McCormick, 1991). This cycle is regulated by exchange factors that allow GTP association. The best characterized GTP-exchange factor is a protein known as Sos1 that mediates the activation of Ras by growth factor receptors (Bonfini *et al.*, 1992; Egan *et al.*, 1993). Studies concerning the mechanism by which Sos1 does this have revealed the presence of adaptor proteins such as Grb2 (Lowenstein *et al.*, 1992) and Shc (Blaikie *et al.*, 1994; Kavanaugh and Williams, 1994; Pelicci *et al.*, 1992), which have no catalytic activity but have SH2 domains that interact with phosphotyrosine residues on the growth factor receptor and SH3 domains that interact with a proline-rich sequence in Sos1. In this way, Sos1 is linked to the growth factor receptor and localized to a submembranous region where it can activate membrane-bound Ras. It is thought that a variety of adaptors and exchange factors could be responsible for activating the different members of the Ras superfamily. For example, the exchange factor Vav (Bustelo *et al.*, 1992; Coppola *et al.*, 1991; Hart *et al.*, 1991; Katzav *et al.*, 1989) has been shown in transfected cells to induce cytoskeletal changes similar to those induced by Rac (Olson *et al.*, 1996), a finding that is consistent with the possibility that Vav is an upstream regulator of Rho family members.

Several downstream effectors for Rho family members have been identified. As indicated in Figure 6.2, one target for RhoA is PtdIns 3-kinase (p85/PI 3-K) (Rittenhouse, 1996; Zhang *et al.*, 1993; Zhang *et al.*, 1995b; Zhang *et al.*, 1996), which phosphorylates PtdIns(4,5)P2 at the 3-position of the inositol ring to form PtdIns(3,4,5)P3. Another candidate is PtdIns 4-phosphate 5-kinase (PI 5-K), which is activated when it interacts with RhoA (Chong *et al.*, 1994; Ren *et al.*, 1996). Yet another downstream target for RhoA is a family of ~ 160kDa serine/threonine protein kinases known as ROK (Ishizaki *et al.*, 1996; Leung *et al.*, 1995). In cultured cells, ROK is recruited to the membrane by active Rho and localizes with submembranous actin filaments. ROK in turn appears to phosphorylate myosin light chain (Amano *et al.*, 1996) and myosin phosphatase (Kimura *et al.*, 1996). Phosphorylation of myosin phosphatase inactivates this enzyme, thus, the net result is an increased phosphorylation of myosin light chain. This in turn leads to increased contractile forces exerted by actin-myosin interactions and, thus, to altered cell morphology (Amano *et al.*, 1997; Burridge and Chrzanowska-Wodnicka, 1996; Chrzanowska-Wodnicka and Burridge, 1996).

Downstream targets for Rac include PI 5-K (Hartwig *et al.*, 1995; Tolias *et al.*, 1995) and a family of serine/threonine kinases known as the p65[PAK] family (Manser *et al.*, 1995; Manser *et al.*, 1994; Martin *et al.*, 1995; Teo *et al.*, 1995). Members of the p65[PAK] family have been implicated in phosphorylating and activating a downstream cascade of kinases that includes a MAPK known as p38MAPK (Freshney *et al.*, 1994; Rouse *et al.*, 1994; Zhang *et al.*, 1995c). p38MAPK is an activator of MAPK-activated protein kinase-2, which phosphorylates the small heat shock protein, Hsp27. Downstream effectors for Cdc42Hs include p85/PI 3-K (Tolias *et al.*, 1995; Zheng *et al.*, 1994), p65[PAK], (Manser *et al.*, 1994; Martin *et al.*, 1995) and WASP (Symons *et al.*, 1996), all of which are associated with the cytoskeleton.

Each of the Rho family members mentioned above is present in platelets (Dash *et al.*, 1995a; Nemoto *et al.*, 1992; Polakis *et al.*, 1989), as are upstream regulators including the guanine nucleotide exchange factors Sos1 and Vav, and the adaptors Grb2 and Shc (Cichowski *et al.*, 1992; Cichowski *et al.*, 1996; Fox *et al.*, 1993a; Fox, 1996a; Katzav *et al.*,

1989; Law *et al.*, 1996; Meyer *et al.*, 1997a). Each of the downstream effectors shown in Figure 6.2 are also known to be present in platelets and, as discussed below, many have been identified in cytoskeletal fractions (Ishizaki *et al.*, 1996; Saklatvala *et al.*, 1996; Teo *et al.*, 1995; Zhang *et al.*, 1995b; Zhang *et al.*, 1996; Zhu *et al.*, 1994). The first indication that a Rho family member is activated in platelets came from the demonstration that addition of *botulinum* C3 ADP ribosyltransferase, an agent that specifically ADP ribosylates and inactivates RhoA, inhibits aggregation of thrombin-stimulated platelets (Morii *et al.*, 1992). More recently, evidence has has been provided that proteins involved in activating members of the Ras superfamily of proteins (e.g. $p21^{ras}GAP$, Vav, Grb2, Shc) or in mediating downstream effects (e.g. $p62^{PAK}$, Hsp27,) are phosphorylated during platelet activation.

Presumably, Rho family members in platelets are activated by the recruitment of adaptors and exchange proteins to phosphotyrosine residues in a submembranous location. Many proteins become phosphorylated by cytoplasmic tyrosine kinases in activated platelets (Clark *et al.*, 1994a; Ferrell and Martin, 1988; Golden and Brugge, 1989; Nakamura and Yamamura, 1989; Shattil and Brugge, 1991); immunofluorescence studies have revealed that these are concentrated in a submembranous location (Fox, unpublished observations). Further, many of them co-isolate with the membrane skeleton from platelets activated under non-aggregating conditions, or with the integrin-cytoskeletal complexes isolated from platelets activated under aggregating conditions (Fox *et al.*, 1993a). Thus, models have been proposed in which the tyrosine phosphorylation of membrane skeleton proteins, integrin-associated proteins, or the integrin cytoplasmic domain themselves (by membrane-skeleton associated tyrosine kinases) leads to the submembranous recruitment of upstream activators of Rho family members (Fox, 1996a; Law *et al.*, 1996). An area of active interest is the mechanism by which Rho family members regulate the actin polymerization and cytoskeletal reorganizations induced in activated platelets and the relative roles of these proteins and those of other signaling events such as elevated Ca^{2+}, serine phosphorylation, and phospholipid metabolism in inducing these changes.

Regulation of cytoskeletal reorganizations in non-aggregating platelets

Activation of platelets through heterotrimeric G-protein coupled receptors such as the thrombin receptor leads to elevated levels of cytoplasmic Ca^{2+}, activation of tyrosine kinases, activation of serine/threonine kinases, and metabolism of phosphoinositides. As discussed above, it is now apparent that Rho family members are also activated and involved in mediating the actions of some of these signals. Some of the signals involved in signaling through heterotrimeric G-protein coupled receptors are summarized in Figure 6.3 (see Brass *et al.*, 1993; Fox, 1996a; Rittenhouse, 1996 for reviews).

The importance of Ca^{2+} in regulating cytoskeletal reorganizations has been recognized for many years. One consequence of elevated Ca^{2+} levels is that calmodulin-dependent myosin light-chain kinase is activated. The resulting phosphorylation of myosin light chains allows the interaction of myosin with actin filaments (Fox and Phillips, 1982). In the past, it has been thought that the primary function of actin-myosin interactions was to induce the centralization of granules and to allow force to be generated so that fibrin bound to the platelet surface could be pulled inwards (Painter and Ginsberg, 1984; Pollard *et al.*, 1977; Stark *et al.*, 1991). As discussed below, it is now apparent that myosin-actin interactions may also be involved in the formation of stress fibers at focal contacts (Amano *et al.*, 1997; Burridge and Chrzanowska-Wodnicka, 1996; Chrzanowska-Wodnicka and Burridge, 1996). While activation of myosin light-chain kinase is the best characterized result of

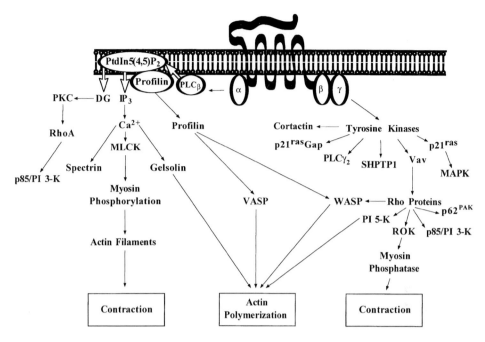

Figure 6.3 **Schematic representation of signal molecules potentially involved in induction of cytoskeletal reorganizations by heterotrimeric G-protein coupled receptors in platelets.** Activation of platelets through heterotrimeric G-protein coupled receptors leads to dissociation of the α-subunit of the G-protein from the βγ subunits. The α-subunit activates phospholipase Cβ while the βγ subunits activate other signaling molecules including tyrosine kinases. Ways in which activation of these pathways may induce cytoskeletal reorganizations are indicated.

Ca^{2+}-calmodulin interactions in activated platelets, there are other cytoskeletal proteins that interact with calmodulin and are therefore potential candidates for being regulated by elevated Ca^{2+} in platelets. One of these is spectrin (Glenney *et al.*, 1982; Wasenius *et al.*, 1989), a major component of the platelet membrane skeleton (Fox *et al.*, 1987; Hartwig and DeSisto, 1991). Future studies will be needed to determine whether Ca^{2+}-calmodulin interactions regulate the organization of the platelet membrane skeleton, the function of spectrin, or the function of other calmodulin-binding cytoskeletal proteins.

Another consequence of elevated Ca^{2+} in platelets is polymerization of actin. As discussed above, in unstimulated platelets, actin polymerization is prevented by proteins that sequester monomers and proteins that cap barbed ends of pre-existing actin filaments. In activated platelets at least some of the new barbed ends that are generated appear to be generated by a Ca^{2+}-dependent mechanism (Hartwig, 1992). A model has been suggested in which the Ca^{2+}-dependent protein, gelsolin, binds to pre-existing actin filaments, severs them, and is subsequently released from the new barbed ends by PtdIns(4,5)P2, a lipid that is generated during platelet activation (Hartwig *et al.*, 1995). Experiments in a permeabilized platelet system showed that constitutively active Rac stimulated the production of PtdIns(4,5)P2 and increased the number of barbed ends while constitutively active RhoA had no effect (Hartwig *et al.*, 1995). This suggested that the Rac-induced activation of PI 5-K (Figure 6.2) may be involved in the generation of new barbed filament ends. Because the affinity of actin monomers for barbed filament ends is greater than that for thymosin $β_4$, the generation of new barbed ends has been suggested to be sufficient for the dissociation of monomers from thymosin $β_4$ leading to the polymerization of actin (Cassimeris *et al.*, 1992; Weber *et al.*, 1992).

These results present a model by which new filament ends can be generated and actin filament polymerization induced by a Ca^{2+}-dependent mechanism in stimulated platelets. However, additional mechanisms for regulating actin polymerization must exist. For example, it has been suggested that some of the new filaments that form do so by a Ca^{2+}-independent mechanism (Hartwig, 1992). Moreover, the initial burst of actin polymerization in activated platelets is induced within 5 s of agonist addition (Fox and Phillips, 1981). It appears unlikely that a mechanism involving PtdIns(4,5)P2 is involved in inducing this early burst of actin filaments because the concentrations of this lipid decrease as phospholipase $C\beta$ is activated in activated platelets (Figure 6.3). It is only after 15–30 s that PtdIns(4,5)P2 concentrations reach levels higher than those present in the unstimulated cells (Agranoff *et al.*, 1983; Grondin *et al.*, 1991). Further, examination of shape changes in the platelets of gelsolin-deficient mice revealed that the formation of lamellipodia was decreased as these platelets were activated but that the extension of filopodia (which occurs as bundles of new actin filaments form) occurred normally (Witke *et al.*, 1995). Thus, mechanisms other than the proposed mechanism involing gelsolin and PtdIns(4,5)P2 must be involved in regulating the formation of actin filaments in filopodia.

Additional mechanisms for inducing actin polymerization could involve release of other capping proteins such as CAPZ or the regulation of candidate capping proteins such as moesin, adducin, or tensin. Alternatively, mechanisms may exist for releasing actin monomers from sequestering proteins, nucleating actin polymerization, or accelerating the addition of monomers onto filaments. Supporting the idea of multiple mechanisms of regulation of actin polymerization is the finding that microinjection of different Rho family members into cultured cells leads to very different cytoskeletal reorganizations (Hall, 1994; Nobes and Hall, 1995). Thus, in platelets, activation of PI 5-K by Rac may release gelsolin from barbed filament ends in specific locations while the Rac-induced activation of other downstream effectors or the Rho or Cdc42Hs-induced activation of their various effectors may be responsible for induction of actin polymerization and cytoskeletal reorganization in other locations.

As discussed below, there is evidence that RhoA is activated in platelets and that the Rho-induced activation of PI 3-K leads to stabilization of platelet aggregates (Kovacsovics *et al.*, 1995; Morii *et al.*, 1992; Zhang *et al.*, 1995a; Zhang *et al.*, 1995b). At this point, there is no direct evidence for Cdc42Hs activation in stimulated platelets, however, this Rho family member translocates to the membrane skeleton in activated platelets (Dash *et al.*, 1995b). One known downstream effector of Cdc42Hs is WASP, and as discussed above, WASP is probably a component of the platelet membrane skeleton, as demonstrated by unusual membrane properties of platelets from Wiscott-Aldrich syndrome patients (Remold-O'Donnell *et al.*, 1996). In other cells, there is evidence that WASP can induce actin polymerization (Symons *et al.*, 1996) and the observation that few filopodia are extended in platelets from Wiscott-Aldrich syndrome patients is consistent with the possibility that WASP is involved in regulating actin polymerization. It is of interest that WASP contains a proline-rich sequence that is assumed to interact with profilin (Symons *et al.*, 1996). As discussed above, profilin can accelerate the addition of actin monomers onto actin filaments and is thought to become available in activated platelets following release from PtdIns(4,5)P2 in the plasma membrane (Goldschmidt-Clermont *et al.*, 1990; Goldschmidt-Clermont *et al.*, 1992). Thus, a model has been proposed in which interactions between Cdc42Hs, WASP, and profilin provide a mechanism for recruiting profilin to specific submembranous locations and, thus, inducing actin polymerization at these sites (Symons *et al.*, 1996). Another WASP-related protein present in platelets is VASP (Halbrugge and Walter, 1989; Reinhard *et al.*, 1992). VASP also contains a proline-rich, profilin-binding sequence and is thought to regulate actin polymerization (Haffner *et al.*, 1995; Reinhard *et al.*, 1995a). As discussed below, VASP is typically present in focal contacts (Haffner *et al.*, 1995; Reinhard *et al.*, 1995a) so if it plays a role in the binding of profilin and the regulation of actin polymerization in platelets, it may do so following integrin engagement (Burridge and Chrzanowska-Wodnicka, 1996).

Additional mechanisms by which actin polymerization or cytoskeletal reorganizations may be regulated in activated platelets include the activation of tyrosine kinases and serine/threonine kinases. Proteins that are substrates for tyrosine kinases in activated, non-aggregating platelets include several proteins that co-isolate with the membrane skeleton from detergent-lysed platelets (Fox *et al.*, 1993a). Perhaps tyrosine phosphorylation leads to a reorganization of this structure or the phosphotyrosine residues in the membrane skeleton recruit SH2-containing signaling molecules to a submembranous location in platelets. One such SH2-containing signaling molecule is Vav which is tyrosine phosphorylated following thrombin receptor activation in platelets and is presumably involved in activation of Rho family members (Cichowski *et al.*, 1996). Tyrosine phosphorylation of other proteins may also be involved in the downstream effects of the Rho family members. Tyrosine phosphorylation of other proteins may be involved in regulating the organization of the cytoplasmic component of the cytoskeleton. One such protein is cortactin which contains an SH3 domain and a proline rich sequence so it may be involved in recruiting signaling molecules to the cytoplasmic filaments with which it associates in activated platelets (Fox *et al.*, 1993a).

Many different serine/threonine kinases appear to be activated in stimulated platelets and little is known about the identity or function of most of these (Ferrell and Martin, 1989a). Again, it appears likely that some of these kinases are involved in the induction of cytoskeletal reorganizations by both Rho family-dependent and -independent mechanisms. A serine/threonine kinase that is activated in platelets and that is a downstream effector of Rho family members is p62^{PAK} (Teo *et al.*, 1995), a member of the PAK family of proteins that is activated by Rac and Cdc42Hs. Others are members of the MAP kinase family which are downstream effectors of p21^{ras} (Nakashima *et al.*, 1994; Papkoff *et al.*, 1994) (Figure 6.3). A kinase that can be activated by a Rho-independent mechanism is protein kinase C (Figure 6.3). Proteins phosphorylated by serine/threonine kinases include talin (Beckerle and Yeh, 1990), ABP (Carroll and Gerrard, 1982), moesin (Nakamura *et al.*, 1995), and Hsp27 (Zhu *et al.*, 1994), all of which are components of the platelet cytoskeleton.

Clearly, considerable work needs to be done to elucidate the pathways leading to actin polymerization and cytoskeletal reorganizations in platelets following stimulation of heterotrimeric G-protein coupled receptors. It will be important to identify the role of the cytoskeleton in recruiting and regulating signaling molecules, and to understand the mechanisms by which specific cytoskeletal reorganizations occur in distinct locations in the cell.

Regulation of cytoskeletal reorganizations in aggregating platelets

The previous section discussed the cytoskeletal rearrangements induced by signaling through heterotrimeric G-protein coupled receptors under conditions in which signaling through adhesion receptors does not occur. Very different cytoskeletal reorganizations can be induced as different adhesion receptors engage their ligands. In platelets, it is becoming apparent that cytoskeletal reorganizations are induced as GP Ib-IX binds von Willebrand factor (Cunningham *et al.*, 1996; Jackson *et al.*, 1994; Meyer *et al.*, 1997a). However, as in other cell types, the best characterized adhesion-induced cytoskeletal reorganizations are those induced following integrin engagement. These cytoskeletal reorganizations lead to the formation of focal contacts, sites at which ligand-occupied integrins cluster and interact

Unstimulated Platelets **Aggregating Platelets**

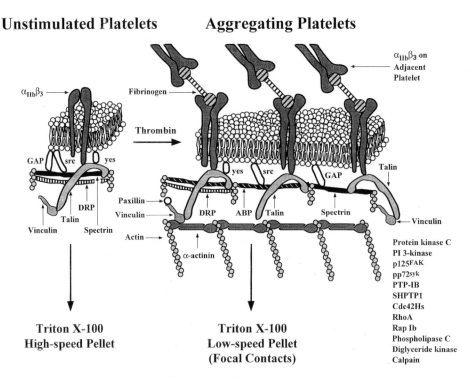

Figure 6.4 Schematic representation of the formation of focal contacts in platelets.

with complexes of cytoskeletal proteins, signaling molecules, and actin filaments (Burridge and Chrzanowska-Wodnicka, 1996). Focal contacts have a structural role, stabilizing the adhesive interaction. Thus, recent work has shown that in platelets in which the integrin $\alpha_{IIb}\beta_3$ has bound fibrinogen, cytochalasins have little effect on the initial binding of ligand to $\alpha_{IIb}\beta_3$, but markedly inhibit the stabilization of the interaction, as shown by the inhibition of aggregation and an inhibition of the conversion of ligand to an irreversibly bound form (Fox *et al.*, 1996). As in cultured cells, focal contacts in platelets allow a contractile force to be exerted on the ligand-occupied integrin, causing the ligand-occupied $\alpha_{IIb}\beta_3$ to be selectively pulled into the open canalicular system by the cytoskeleton (Fox *et al.*, 1996). Signaling molecules associated with focal contacts are presumably involved in the formation, stabilization, or breakdown of focal contacts, or in inducing the contractile force that is exerted on ligand-occupied integrin.

The composition of the integrin-cytoskeletal complexes that form in aggregating platelets has been studied in detail. As shown in Figure 6.4, cytoskeletal proteins include talin, vinculin, α-actinin, actin-binding protein, spectrin, protein 4.1, and dystrophin-related protein (Earnest *et al.*, 1995; Fox *et al.*, 1993a; Kouns *et al.*, 1991). Signaling molecules that have been recovered with the integrin-cytoskeletal complexes include protein kinase C, p85/PI 3-K, FAK, pp72[syk], PTP-1B, SHPTP1, Cdc42Hs, RhoA, Rap1B, phospholipase C, diglyceride kinase, paxillin, and calpain (Clark *et al.*, 1994b; Dash *et al.*, 1995b; Falet *et al.*, 1996; Fischer *et al.*, 1994; Fox, 1993; Fox *et al.*, 1993a; Fox *et al.*, 1995; Grondin *et al.*, 1991; Guinebault *et al.*, 1995; Horvath *et al.*, 1992; Li *et al.*, 1994; Lipfert *et al.*, 1992; Rittenhouse, 1996; Tohyama *et al.*, 1994; Torti *et al.*, 1994; Zhang *et al.*, 1992). Many of the cytoskeletal proteins that are recovered with the integrin-cytoskeletal complexes in platelets have been detected by immunofluorescence in the focal contacts of cultured cells. Similarly, many of the signaling molecules have been shown to co-localize with occupied integrin (Burridge and Chrzanowska-Wodnicka, 1996; Miyamoto *et al.*, 1995). Thus, it appears likely that mechanisms regulating the formation and functions of focal contacts in platelets may be similar to those regulating focal contacts in other cells.

It was originally thought that focal contacts in cultured cells were formed because ligand binding caused integrin clustering which in turn increased the affinity of the integrins for cytoskeletal proteins and induced the formation of integrin-cytoskeletal complexes. More recently, a model has been emerging in which the initial step in the formation of focal contacts involves the formation of complexes of integrin with signaling molecules and cytoskeletal proteins such as paxillin and vinculin (Burridge and Chrzanowska-Wodnicka, 1996; Miyamoto *et al.*, 1995; Nobes and Hall, 1995). Proteins within these complexes are tyrosine phosphorylated. Although the mechanisms involved in the formation of these focal complexes is not known, it is of interest that microinjection of constitutively active Cdc42Hs or Rac can induce similar complexes of proteins (Nobes and Hall, 1995). A second step is activation of RhoA which leads to actin polymerization and the formation of bundles of actin filaments known as stress fibers. The stress fibers interact with myosin and exert a contractile force that induces clustering of the integrins; this in turn leads to recruitment of additional signaling molecules, such as the tyrosine kinase FAK, and increased tyrosine phosphorylation of proteins within the resulting focal contacts (Amano *et al.*, 1997; Amano *et al.*, 1996; Burridge and Chrzanowska-Wodnicka, 1996; Chrzanowska-Wodnicka and Burridge, 1996; Kimura *et al.*, 1996). The signaling molecules recruited as a consequence of stress fiber formation are presumably involved in the turnover and function of focal contacts.

This model for mechanisms involved in the formation of focal contacts implicates RhoA as playing an important role in the stabilization of focal contacts and exertion of the contractile function. A number of mechanisms by which integrin engagement may induce activation of Rho have been proposed (Fox, 1996a; Law *et al.*, 1996; Parsons, 1996; Parsons *et al.*, 1994). Presumably tyrosine phosphorylation is involved and adaptors and exchange proteins such as Grb2, Shk, Sos1, and Vav are recruited to the sites of ligand-occupied integrin. Activation of PI 5-K, a downstream effector of Rac and RhoA (Figure 6.2), leads to the generation of PtdIns(4,5)P2 which induces actin polymerization. There is evidence that PtdIns(4,5)P2 can interact with cytoskeletal proteins inducing conformational changes that induce associations of the proteins with other molecules. For example, evidence has been presented that suggests that PtdIns(4,5)P2 may induce moesin to link actin filaments to the plasma membrane (Tsukita *et al.*, 1997) and induce vinculin to link talin to actin filaments (Gilmore and Burridge, 1996). Activation of p85/PI 3-K, an effector of Cdc42Hs and RhoA (Figure 6.2) has been shown to lead to activation of protein kinase C isoforms (Toker *et al.*, 1994) and may induce cytoskeletal reorganizations (Arcaro and Wymann, 1993; Kotani *et al.*, 1994; Wennstrom *et al.*, 1994; Wymann and Arcaro, 1994). Activation of ROK, a down-stream effector of RhoA (Figure 6.2), leads to phosphorylation and inactivation of myosin phosphatase (Kimura *et al.*, 1996), resulting in an increase in phosphorylation of myosin light chain. This results in increased actin-myosin interactions and exertion of a contractile force that stabilizes the focal contacts (Amano *et al.*, 1997; Burridge and Chrzanowska-Wodnicka, 1996; Chrzanowska-Wodnicka and Burridge, 1996).

Other proteins that may regulate focal contact formation are VASP and zyxin. As discussed above, VASP contains a proline-rich profilin binding site (Reinhard *et al.*, 1995a) and is present in focal contacts (Haffner *et al.*, 1995; Reinhard *et al.*, 1995a) so it has been suggested that by recruiting profilin to focal contacts it may regulate the polymerization of actin filaments at these sites (Burridge and Chrzanowska-Wodnicka, 1996). Because VASP interacts with zyxin (Reinhard *et al.*, 1995b), a model in which complexes of zyxin, VASP, and profilin regulate actin polymerization has been proposed (Reinhard *et al.*, 1995b; Burridge and Chrzanowska-Wodnicka, 1996). Yet another signaling molecule that is involved in the formation of focal contacts is calpain. This Ca^{2+}-dependent protease is present in focal contacts in cultured cells (Beckerle *et al.*, 1987). Recent experiments in which calpain was inhibited in cultured bovine aortic endothelial cells showed that focal complexes still form despite the absence of active calpain but the formation of stress fibers and focal contacts does not occur (Kulkarni *et al.*, 1999). Overexpression of calpain in these cells led to induction of an increased number of focal contacts (Kulkarni *et al.*, 1999). It is anticipated that these studies will lead to a rapid growth in our understanding of mechanisms directing specific cytoskeletal reorganizations in platelets, regulating specific locations at which cytoskeletal changes are induced, localizing and activating specific signaling molecules, and regulating the activation and function of Rho family members in platelets.

Evidence that focal contacts in platelets form by mechanisms similar to those proposed for cultured cells is accumulating. The finding that a subpopulation of $\alpha_{IIb}\beta_3$ co-isolates with complexes of cytoskeletal proteins from unstimulated platelets (Fox *et al.*, 1993a) suggests that structures comparable to focal complexes in cultured cells may already exist in unstimulated platelets. Moreover, because integrin in these complexes is selectively incorporated into focal contacts (Fox *et al.*, 1993a; Fox *et al.*, 1996), the association with these complexes appears to be important in allowing ligand binding or subsequent ligand-induced focal contact formation to occur. As ligand binds to $\alpha_{IIb}\beta_3$, the integrin is thought to dimerize and induce tyrosine phosphorylation of signaling molelules such as pp72[syk] (Clark *et al.*, 1994c). Upon platelet aggregation, cytoskeletal reorganizations (Fox *et al.*, 1993a; Kouns *et al.*, 1991) and tyrosine phosphorylation of additional proteins such as FAK, Grb2, Shk, pp72[syk], Vav, and the β-subunit of $\alpha_{IIb}\beta_3$ (Cichowski *et al.*, 1996; Clark *et al.*, 1994b; Ferrell and Martin, 1989b; Golden *et al.*, 1990; Law *et al.*, 1996; Lipfert *et al.*, 1992; Shattil *et al.*, 1994) occurs while other molecules including calpain (Fox *et al.*, 1993b; Fox *et al.*, 1995) and PI 3-K (Rittenhouse, 1996; Zhang *et al.*, 1992) are recruited to the integrin-cytoskeletal complexes where they are selectively active. Cytoskeletal reorganizations have been shown to be required for the activation and tyrosine phosphorylation of many of these signaling molecules (Clark *et al.*, 1994b; Fox *et al.*, 1993b; Fox *et al.*, 1995; Lipfert *et al.*, 1992). The consequence of these changes include stabilization of ligand binding, stabilization of platelet aggregates, and the exertion of a contractile force on ligand-occupied integrin (Fox *et al.*, 1996).

It appears likely that the initial ligand-induced signaling is comparable to the formation of focal complexes in cultured cells while subsequent cytoskeletal reorganizations in the aggregating platelets is comparable to the RhoA-induced formation of stress fibers. The finding that exposure of platelets to *botulinum* C3 ADP-ribosyl-transferase, which specifically inhibits RhoA, does not affect the initial aggregation of platelets but prevents the subsequent irreversible phase of aggregation is consistent with the aggregation-induced changes being induced by RhoA (Morii *et al.*, 1992).

As discussed above, two ways in which RhoA may be involved in the formation of focal contacts in other cells is through increased phosphorylation of myosin light chain (Amano *et al.*, 1997; Amano *et al.*, 1996; Burridge and Chrzanowska-Wodnicka, 1996; Chrzanowska-Wodnicka and Burridge, 1996; Kimura *et al.*, 1996) and PI 3-K activation (Arcaro and Wymann, 1993; Kotani *et al.*, 1994; Wennstrom *et al.*, 1994; Wymann and Arcaro, 1994). In platelets, little is known about the role of myosin in inducing the formation of focal contacts. However, it is known that p85/PI 3-K is recruited to the membrane skeleton and is activated in an $\alpha_{IIb}\beta_3$-dependent manner (Rittenhouse, 1996; Zhang *et al.*, 1992; Zhang *et al.*, 1996). Recent studies have shown that exposure of platelets to wortmannin, an inhibitor of this kinase, inhibits the ability of $\alpha_{IIb}\beta_3$ to bind ligand (Kovacsovics *et al.*, 1995). Wortmannin's effect was primarily on the stabilization of ligand interaction, not the initial interaction, indicating that p85/PI 3-K regulates focal contact formation.

Other proteins thought to be involved in the formation of focal contacts in cultured cells, such as VASP and zyxin, are present in platelets (Macalma *et al.*, 1996; Reinhard *et al.*, 1992), although their potential role in integrin-induced cytoskeletal reorganizations in platelets has not yet been investigated. More is known about the role of calpain in platelets. As in cultured cells, calpain associates with focal contacts where it is selectively activated (Fox *et al.*, 1993b; Fox *et al.*, 1995). Activation of calpain results in cleavage of the cytoplasmic domain of the β_3 integrin subunit (Du *et al.*, 1995) as well as cleavage of several components of the integrin-cytoskeletal complexes (including spectrin, talin, actin-binding protein, dystrophin-related protein, and protein kinase C) (Earnest *et al.*, 1995; Fox, 1985a; Fox *et al.*, 1985; Fox *et al.*, 1983; Fox *et al.*, 1987; Tapley and Murray, 1985). Since the work described above indicates an important role for calpain in inducing the formation of focal contacts (Kulkarni and Fox, unpublished observations), it is likely that the cleavage fragments of these proteins are important in mediating specific interactions required for the formation of stable focal contacts. It will be important to determine the precise mechanisms by which calpain exerts its effects and the contributions of the host of other signaling molecules that are present in the focal contacts.

References

Agranoff, B.W., Murthy, P. and Sequin, E.B. (1983). Thrombin-induced phosphodiesteratic cleavage of phosphatidylinositol bisphosphate in human platelets. *The Journal of Biological Chemistry*, **258**:2076–2078.

Amano, M., Chihara, K., Kimura, K., Fukata, Y., Nakamura, N., Matsuura, Y. and Kaibuchi, K. (1997). Formation of actin stress fibers and focal adhesions enhanced by Rho-kinase. *Science*, **275**:1308–1311.

Amano, M., Ito, M., Kimura, K., Fukata, Y., Chihara, K., Nakano, T., Matsuura, Y. and Kaibuchi, K. (1996). Phosphorylation and activation of myosin by Rho-associated kinase (Rho-kinase). *The Journal of Biological Chemistry*, **271**:20246–20249.

Andrews, R.K. and Fox, J.E.B. (1991). Interaction of purified actin-binding protein with the platelet membrane glycoprotein Ib-IX complex. *The Journal of Biological Chemistry*, **266**:7144–7147.

Andrews, R.K. and Fox, J.E.B. (1992). Identification of a region in the cytoplasmic domain of the platelet membrane glycoprotein Ib-IX complex that binds to purified actin-binding protein. *The Journal of Biological Chemistry*, **267**:18605–18611.

Arcaro, A. and Wymann, M. (1993). Wortmannin is a potent phosphatidylinositol 3-kinase inhibitor: The role of phosphatidylinositol 3,4,5-trisphosphate in neutrophil responses. *Biochemical Journal*, **296**:297–301.

Arpin, M., Algrain, M. and Louvard, D. (1994). Membrane-actin microfilament connections: An increasing diversity of players related to band 4.1. *Current Opinions in Cell Biology*, **6**:136–141.

Bar-Sagi, D., Rotin, D., Batzer, A. and Mandiyan, V. (1993). SH3 domains direct cellular localization of signaling molecules. *Cell*, **74**:83–91.

Barkalow, K., Witke, W., Kwiatkowski, D.J. and Hartwig, J.H. (1996). Coordinated regulation of platelet actin filament barbed ends by gelsolin and capping protein. *Journal of Cell Biology*, **134**:389–399.

Beckerle, M.C., Burridge, K., DeMartino, G.N. and Croall, D.E. (1987). Colocalization of calcium-dependent protease II and one of its substrates at sites of cell adhesion. *Cell*, **51**:569–577.

Beckerle, M.C. and Yeh, R.K. (1990). Talin: role at sites of cell-substratum adhesion. *Cell Motility Cytoskeleton*, **16**:7–13.

Blaikie, P., Immanuel, D., Wu, J., Li, N., Yajnik, V. and Margolis, B. (1994). A region in Shc distinct from SH2 domain can bind tyrosine-phosporylated growth factor receptors. *The Journal of Biological Chemistry*, **269**:32031–32034.

Boguski, M.S. and McCormick, F. (1993). Proteins regulating Ras and its relatives. *Nature*, **366**:643–654.

Bokoch, G.M. and Der, C.J. (1993). Emerging concepts in the Ras superfamily of GTP-binding proteins. *FASEB Journal*, **7**:750–759.

Bollag, G. and McCormick, F. (1991). Regulators and effectors of ras proteins. *Annual Review Cell Biology*, **7**:601–632.

Bonfini, L., Karlovich, C.A., Dasgupta, C. and Banerjee, U. (1992). The Son of sevenless gene product: A putative activator of Ras. *Science*, **255**:603–605.

Boyles, J., Fox, J.E.B., Phillips, D.R. and Stenberg, P.E. (1985). Organization of the cytoskeleton in resting, discoid platelets: preservation of actin filaments by a modified fixation that prevents osmium damage. *Journal of Cell Biology*, **101**:1463–1472.

Brass, L.F., Hoxie, J.A. and Manning, D.R. (1993). Signaling through G proteins and G protein-coupled receptors during platelet activation. *Thrombosis and Haemostasis*, **70**:217–223.

Burridge, K. and Chrzanowska-Wodnicka (1996). Focal adhesions, contractility, and signaling. *Annual Review of Cell and Developmental Biology*, **12**:463–519.

Bustelo, X.R., Ledbetter, J.A. and Barbacid, M. (1992). Product of vav proto-oncogene defines a new class of tyrosine protein kinase substrates. *Nature*, **356**:68–71.

Caldwell, J.E., Waddle, J.A., Cooper, J.A., Hollands, J.A., Casella, S.J. and Casella, J.F. (1989). cDNAs encoding the β subunit of Cap Z, the actin-capping protein of the Z line of muscle. *The Journal of Biological Chemistry*, **264**:12648–12652.

Carlsson, L., Markey, F., Blikstad, I., Persson, T. and Lindberg, U. (1979). Reorganization of actin in platelets stimulated by thrombin as measured by the DNase I inhibition assay. *Proceedings of the National Academy of Sciences, USA*, **76**:6376–6380.

Carlsson, L., Nyström, L.-E., Sundkvist, I., Markey, F. and Lindberg, U. (1977). Actin polymerizability is influenced by profilin, a low molecular weight protein in non-muscle cells. *Journal Molecular Biology*, **115**:465–483.

Carroll, R.C. and Gerrard, J.M. (1982). Phosphorylation of platelet actin-binding protein during platelet activation. *Blood*, **59**:466–471.

Cassimeris, L., Safer, D., Nachmias, V.T. and Zigmond, S.H. (1992). Thymosin β_4 sequesters the majority of G-actin in resting human polymorphonuclear leukocytes. *Journal of Cell Biology*, **119**:1261–1270.

Chant, J. and Stowers, L. (1995). GTPase cascades choreographing cellular behavior: Movement, morphogenesis, and more. *Cell*, **81**:1–4.

Chong, L.D., Traynor-Kaplan, A., Bokoch, G.M. and Schwartz, M.A. (1994). The small GTP-binding protein rho regulates a phosphatidylinositol 4-phosphate 5-kinase in mammalian cells. *Cell* **79**:507–513.

Chrzanowska-Wodnicka, M. and Burridge, K. (1992). Rho, rac and the actin cytoskeleton. *BioEssays*, **14**:777–779.

Chrzanowska-Wodnicka, M. and Burridge, K. (1996). Rho-stimulated contractility drives the formation of stress fibers and focal adhesions. *Journal of Cell Biology*, **133**:1403–1415.

Cichowski, K., Brugge, J.S. and Brass, L.F. (1996). Thrombin receptor activation and integrin engagement stimulate tyrosine phosphorylation of the proto-oncogene product, p95^{vav}, in platelets. *The Journal of Biological Chemistry*, **271**:7544–7550.

Cichowski, K., McCormick, F. and Brugge, J.S. (1992). p21^{ras}GAP association with Fyn, Lyn, and Yes in thrombin-activated platelets. *The Journal of Biological Chemistry*, **267**:5025–5028.

Clark, E.A. and Brugge, J.S. (1993). Redistribution of activated pp60^{c-src} to integrin-dependent cytoskeletal complexes in thrombin-stimulated platelets. *Molecular and Cellular Biology*, **13**:1863–1871.

Clark, E.A., Shattil, S.J. and Brugge, J.S. (1994a). Regulation of protein tyrosine kinases in platelets. *Trends in Biology Science*, **19**:464–469.

Clark, E.A., Shattil, S.J., Ginsberg, M.H., Bolen, J. and Brugge, J.S. (1994b). Regulation of the protein tyrosine kinase p72^{syk} by platelet agonists and the integrin $\alpha_{IIb}\beta_3$. *The Journal of Biological Chemistry*, **269**:28859–28864.

Clark, E.A., Trikha, M., Markland, F.S. and Brugge, J.S. (1994c). Structurally distinct disintegrins contortostatin and multisquamatin differentially regulate platelet tyrosine phosphorylation. *The Journal of Biological Chemistry*, **269**:21940–21943.

Coppola, S., Bryant, S., Koda, T., Conway, D. and Barbacid, M. (1991). Mechanism of activation of the vav proto-oncogene. *Cell Growth and Differentiation*, **2**:95–105.

Cunningham, J.G., Meyer, S.C. and Fox, J.E.B. (1996). The cytoplasmic domain of the α-subunit of glycoprotein Ib mediates attachment of the entire GP Ib-IX complex to the cytoskeleton and regulates von Willebrand factor-induced changes in cell morphology. *The Journal of Biological Chemistry*, **271**:11581–11587.

Dash, D., Aepfelbacher, M. and Siess, W. (1995a). The association of pp125^{FAK}, pp60^{Src}, CDC42Hs and Rap1B with the cytoskeleton of aggregated platelets is a reversible process regulated by calcium. *FEBS Letters*, **363**:231–234.

Dash, D., Aepfelbacher, M. and Siess, W. (1995b). Integrin $\alpha_{IIb}\beta_3$-mediated translocation of CDC42Hs to the cytoskeleton in stimulated human platelets. *The Journal of Biological Chemistry*, **270**:17321–17326.

Debus, E., Weber, K. and Osborn, M. (1981). The cytoskeleton of blood platelets viewed by immunofluorescence microscopy. *European Journal of Cell Biology*, **24**:45–52.

Dingus, J., Hwo, S. and Bryan, J. (1986). Identification by monoclonal antibodies and characterization of human platelet caldesmon. *Journal of Cell Biology*, **102**:1748–1757.

Drubin, D.G. and Nelson, W.J. (1996). Origins of cell polarity. *Cell* **84**:335–344.

Du, X., Fox, J.E.B. and Pei, S. (1996). Identification of a binding sequence for the 14-3-3 protein within the cytoplasmic domain of the adhesion receptor, platelet glycoprotein Ibα. *The Journal of Biological Chemistry*, **271**:7362–7367.

Du, X., Harris, S.J., Tetaz, T.J., Ginsberg, M.H. and Berndt, M.C. (1994). Association of a phospholipase A_2 (14-3-3 protein) with the platelet glycoprotein Ib-IX complex. *The Journal of Biological Chemistry*, **269**:18287–18290.

Du, X., Saido, T.C., Tsubuki, S., Indig, F.E., Williams, M.J. and Ginsberg, M.H. (1995). Calpain cleavage of the cytoplasmic domain of the integrin β_3 subunit. *The Journal of Biological Chemistry*, **270**:26146–26151.

Earnest, J.P., Santos, G.F., Zuerbig, S. and Fox, J.E.B. (1995). Dystrophin-related protein in the platelet membrane skeleton. Integrin-induced change in detergent-insolubility and cleavage by aggregating platelets. *The Journal of Biological Chemistry*, **270**:27259–27265.

Egan, S.E., Giddings, B.W., Brooks, M.W., Buday, L., Sizeland, A.M. and Weinberg, R.A. (1993). Association of Sos Ras exchange protein with Grb2 is implicated in tyrosine kinase signal transduction and transformation. *Nature*, **363**:45–51.

Elbaum, D., Mimms, L.T. and Branton, D. (1984). Modulation of actin polymerization by the spectrin-band 4.1 complex. *Biochemistry*, **23**:4813–4816.

Ezumi, Y., Takayama, H. and Okuma, M. (1995). Differential regulation of protein-tyrosine phosphatases by integrin $\alpha_{IIb}\beta_3$ through cytoskeletal reorganization and tyrosine phosphorylation in human platelets. *The Journal of Biological Chemistry*, **270**:11927–11934.

Falet, H., Ramos-Morales, F., Bachelot, C., Fischer, S. and Rendu, F. (1996). Association of the protein tyrosine phosphatase PTP1C with the protein tyrosine kinase c-Src in human platelets. *FEBS Letters*, **383**:165–169.

Ferrell, J.E., Jr. and Martin, G.S. (1988). Platelet tyrosine-specific protein phosphorylation is regulated by thrombin. *Molecular and Cellular Biology*, **8**:3603–3610.

Ferrell, F.E., Jr. and Martin, G.S. (1989a). Thrombin stimulates the activities of multiple previously unidentified protein kinases in platelets. *The Journal of Biological Chemistry*, **264**:20723–20729.

Ferrell, J.E., Jr. and Martin, G.S. (1989b). Tyrosine-specific protein phosphorylation is regulated by glycoprotein IIb-IIIa in platelets. *Proceedings of the National Academy of Sciences, USA*, **86**:2234–2238.

Field, J., Vojtek, A., Ballester, R., Bolger, G., Colicelli, J., Ferguson, K., Gerst, J., Kataoka, T., Michaeli, T., Powers, S., Riggs, M., Rodgers, L., Wieland, I., Wheland, B. and Wigler, M. (1990). Cloning and characterization of CAP, the S. cerevisiae gene encoding the 70 kd adenylyl cyclase-associated protein. *Cell*, **61**:319–327.

Fischer, T.H., Gatling, M.N., McCormick, F., Duffy, C.M. and White, G.C., II. (1994). Incorporation of Rap 1b into the platelet cytoskeleton is dependent on thrombin activation and extracellular calcium. *The Journal of Biological Chemistry*, **269**:17257–17261.

Fox, J.E.B. (1985a). Hydrolysis of cytoskeletal proteins by the Ca^{2+}-dependent protease during platelet activation. *Advances in Experimental Medicine and Biology*, **192**:201–213.

Fox, J.E.B. (1985b). Identification of actin-binding protein as the protein linking the membrane skeleton to glycoproteins on platelet plasma membranes. *The Journal of Biological Chemistry*, **260**:11970–11977.

Fox, J.E.B. (1985c). Linkage of a membrane skeleton to integral membrane glycoproteins in human platelets. Identification of one of the glycoproteins as glycoprotein Ib. *Journal of Clinical Investigation*, **76**:1673–1683.

Fox, J.E.B. (1988). Structure and function of the platelet membrane skeleton. Paper presented at the Cell Physiology of Blood, Chapter 11, Vol. 43, *Society of General Physiologists Series*, New York, 1988.

Fox, J.E.B. (1993). The Platelet Cytoskeleton. *Thrombosis and Haemostasis*, **70**:884–893.

Fox, J.E.B. (1996a). Platelet activation: New aspects. *Heamostasis*, **26** (Suppl. 4):102–131.

Fox, J.E.B. (1996b). Study of proteins associated with the platelet cytoskeleton. In *Platelets: A Practicle Approach*, eds. S.P. Watson and K.S. Authi, pp. 217–233. England: Oxford University Press.

Fox, J.E.B. and Boyles, J.K. (1988a). Characterization of the platelet membrane skeleton. In *Proceedings of the UCLA Conference on Signal Transduction in Cytoplasmic Organization and Cell Motility*, ed. C.F. Fox, pp. 313–324. New York: Alan R. Liss, Inc.

Fox, J.E.B. and Boyles, J.K. (1988b). The membrane skeleton — a distinct structure that regulates the function of cells. *BioEssays*, **8**:14–18.

Fox, J.E.B., Boyles, J.K., Berndt, M.C., Steffen, P.K. and Anderson, L.K. (1988). Identification of a membrane skeleton in platelets. *Journal of Cell Biology*, **106**:1525–1538.

Fox, J.E.B., Boyles, J.K., Reynolds, C.C. and Phillips, D.R. (1984). Actin filament content and organization in unstimulated platelets. *Journal of Cell Biology*, **98**:1985–1991.

Fox, J.E.B., Goll, D.E., Reynolds, C.C. and Phillips, D.R. (1985). Identification of two proteins (actin-binding protein and P235). that are hydrolyzed by endogenous Ca^{2+}-dependent protease during platelet aggregation. *The Journal of Biological Chemistry*, **260**:1060–1066.

Fox, J.E.B., Lipfert, L., Clark, E.A., Reynolds, C.C., Austin, C.D. and Brugge, J.S. (1993a). On the role of the platelet membrane skeleton in mediating signal transduction. Association of GP IIb-IIIa, pp60[c-src], pp62[yes], and the p21[ras] activating protein with the membrane skeleton. *The Journal of Biological Chemistry*, **268**:25973–25984.

Fox, J.E.B. and Phillips, D.R. (1981). Inhibition of actin polymerization in blood platelets by cytochalasins. *Nature*, **292**:650–652.

Fox, J.E.B. and Phillips, D.R. (1982). Role of phosphorylation in mediating the association of myosin with the cytoskeletal structures of human platelets. *The Journal of Biological Chemistry*, **257**:4120–4126.

Fox, J.E.B., Reynolds, C.C. and Boyles, J.K. (1992). Studying the platelet cytoskeleton in Triton X-100 lysates. *Methods in Enzymology*, **215**:42–58.

Fox, J.E.B., Reynolds, C.C., Morrow, J.S. and Phillips, D.R. (1987). Spectrin is associated with membrane-bound actin filaments in platelets and is hydrolyzed by the Ca^{2+}-dependent protease during platelet activation. *Blood*, **69**:537–545.

Fox, J.E.B., Reynolds, C.C. and Phillips, D.R. (1983). Calcium-dependent proteolysis occurs during platelet aggregation. *The Journal of Biological Chemistry*, **258**:9973–9981.

Fox, J.E.B., Shattil, S.J., Kinlough-Rathbone, R.L., Richardson, M., Packham, M.A. and Sanan, D.A. (1996). The platelet cytoskeleton stabilizes the interaction between $\alpha_{IIb}\beta_3$ and its ligand and induces selective movements of ligand-occupied integrin. *The Journal of Biological Chemistry*, **271**:7004–7011.

Fox, J.E.B., Taylor, R.G., Taffarel, M., Boyles, J.K. and Goll, D.E. (1993b). Evidence that activation of platelet calpain is induced as a consequence of binding of adhesive ligand to the integrin, glycoprotein IIb-IIIa. *Journal of Cell Biology*, **120**:1501–1507.

Fox, J.E.B., Zuerbig, S., Santos, G. and Saido, T.C. (1995). Role of calpain in regulating signaling across the integrin $\alpha_{IIb}\beta_3$. *Journal of Cell Biochemistry*, Supp: 19B, 136a.

Freshney, N.W., Rawlinson, L., Guesdon, F., Jones, E., Cowley, S., Hsuan, J. and Saklatvala, J. (1994). Interleukin-1 activtes a novel protein kinase cascade that results in the phosphorylation of Hsp27. *Cell*, **78**: 1039–1049.

Gieselmann, R. and Mann, K. (1992). ASP-56, a new actin sequestering protein from pig platelets with homology to CAP, an adenylate cyclase-associated protein from yeast. *FEBS*, **298**:149–153.

Gilmore, A.P. and Burridge, K. (1996). Regulation of vinculin binding to talin and actin by phospatidylinositol-4-5-bisphosphate. *Nature*, **381**:531–535.

Glenney, J.R., Jr., Glenney, P. and Weber, K. (1982). Erythroid spectrin, brain fodrin, and intestinal brush border proteins (TW-260/240) are related molecules containing a common calmodulin-binding subunit bound to a variant cell type-specific subunit. *Proceedings of the National Academy of Sciences, USA*, **79**:4002–4005.

Golden, A. and Brugge, J.S. (1989). Thrombin treatment induces rapid changes in tyrosine phosphorylation in platelets. *Proceedings of the National Academy of Sciences, USA*, **86**:901–905.

Golden, A., Brugge, J.S. and Shattil, S.J. (1990). Role of platelet membrane glycoprotein IIb-IIIa in agonist-induced tyrosine phosphorylation of platelet proteins. *Journal of Cell Biology*, **111**:3117–3127.

Goldschmidt-Clermont, P.J., Furman, M.I., Wachsstock, D., Safer, D., Nachmias, V.T. and Pollard, T.D. (1992). The control of actin nucleotide exchange by thymosin β_4 and profilin. A potential regulatory mechanism for actin polymerization in cells. *Molecular Biology of the Cell*, **3**:1015–1024.

Goldschmidt-Clermont, P.J., Machesky, L.M., Baldassare, J.J. and Pollard, T.D. (1990). The actin-binding protein profilin binds to PIP$_2$ and inhibits its hydrolysis by phospholipase C. *Science*, **247**:1575–1578.

Grondin, P., Plantavid, M., Sultan, C., Breton, M., Mauco, G. and Chap, H. (1991). Interaction of pp60^{c-src}, phospholipase C, inositol-lipid, and diacylglycerol kinases with the cytoskeletons of thrombin-stimulated platelets. *The Journal of Biological Chemistry*, **266**:15705–15709.

Guinebault, C., Payrastre, B., Racaud-Sultan, C., Mazarguil, H., Breton, M., Mauco, G., Plantavid, M. and Chap, H. (1995). Integrin-dependent translocation of phosphoinositide 3-kinase to the cytoskeleton of thrombin-activated platelets involves specific interactions of p85α with actin filaments and focal adhesion kinase. *Journal of Cell Biology*, **129**:831–842.

Haffner, C., Jarchau, T., Reinhard, M., Hoppe, J., Lohmann, S.M. and Walter, U. (1995). Molecular cloning, structural analysis and functional expression of the proline-rich focal adhesion and microfilament-associated protein VASP. *EMBO Journal*, **14**:19–27.

Halbrugge, M. and Walter, U. (1989). Purification of a vasodilator-regulated phosphoprotein from human platelets. *European Journal of Biochemistry* **185**:41–50.

Hall, A. (1992). Ras-related GTPases and the cytoskeleton. *Mol. Biol. Cell.* **3**:475–479.

Hall, A. (1994). Small GTP-binding proteins and the regulation of the actin cytoskeleton. *Annual Review of Cell Biology*, **10**:31–54.

Hart, M.J., Eva, A., Evans, T., Aaronson, S.A. and Cerione, R.A. (1991). Catalysis of guanine nucleotide exchange on the CDC42Hs protein by the *dbl* oncogene product. *Nature*, **354**:311–314.

Hartmann, H., Noegel, A.A., Eckerskorn, C., Rapp, S. and Schleicher, M. (1989). Ca^{2+}-independent F-actin capping proteins. *The Journal of Biological Chemistry*, **264**:12639–12647.

Hartwig, J.H. (1992). Mechanisms of actin rearrangements mediating platelet activation. *Journal of Cell Biology*, **118**:1421–1442.

Hartwig, J.H., Bokoch, G.M., Carpenter, C.L., Janmey, P.A., Taylor, L.A., Toker, A. and Stossel, T.P. (1995). Thrombin receptor ligation and activated Rac uncap actin filament barbed ends through phosphoinositide synthesis in permeabilized human platelets. *Cell*, **82**:643–653.

Hartwig, J.H., Chambers, K.A., Hopcia, K.L. and Kwiatkowski, D.J. (1989a). Association of profilin with filament-free regions of human leukocyte and platelet membranes and reversible membrane binding during platelet activation. *Journal of Cell Biology*, **109**:1571–1579.

Hartwig, J.H., Chambers, K.A. and Stossel, T.P. (1989b). Association of gelsolin with actin filaments and cell membranes of macrophages and platelets. *Journal of Cell Biology*, **108**:467–479.

Hartwig, J.H. and DeSisto, M. (1991). The cytoskeleton of the resting human blood platelet: structure of the membrane skeleton and its attachment to actin filaments. *Journal of Cell Biology*, **112**:407–425.

Heiss, S.G. and Cooper, J.A. (1991). Regulation of capZ, an actin capping protein of chicken muscle, by anionic phospholipids. *Biochemistry*, **30**:8753–8758.

Horvath, A.R., Muszbek, L. and Kellie, S. (1992). Translocation of pp60$^{c\text{-src}}$ to the cytoskeleton during platelet aggregation. *EMBO Journal*, **11**:855–861.

Horwitz, A., Duggan, K., Buck, C., Beckerle, M.C. and Burridge, K. (1986). Interaction of plasma membrane fibronectin receptor with talin — a transmembrane linkage. *Nature*, **320**:531–533.

Huang, M.-M., Lipfert, L., Cunningham, M., Brugge, J.S., Ginsberg, M.H. and Shattil, S.J. (1993). Adhesive ligand binding to integrin $\alpha_{IIb}\beta_3$ stimulates tyrosine phosphorylation of novel protein substrates before phosphorylation of pp125^{FAK}. *Journal of Cell Biology*, **122**:473–483.

Ishizaki, T., Maekawa, M., Fujisawa, K., Okawa, K., Iwamatsu, A., Fujita, A., Watanabe, N., Saito, Y., Kakizuka, A., Morii, N. and Narumiya, S. (1996). The small GTP-binding protein Rho binds to and activtes a 160 kDa Ser/Thr protein kinase homologous to myotonic dystrophy. *EMBO Journal*, **15**:1885–1893.

Jackson, S.P., Schoenwaelder, S.M., Yuan, Y., Rabinowitz, I., Salem, H.H. and Mitchell, C.A. (1994). Adhesion receptor activation of phosphatidylinositol 3-kinase: Von Willebrand factor stimulates the cytoskeletal association and activation of phosphatidylinositol 3-kinase and pp60$^{c\text{-src}}$ in human platelets. *The Journal of Biological Chemistry*, **269**:27093–27099.

Jennings, L.K., Fox, J.E.B., Edwards, H.H. and Phillips, D.R. (1981). Changes in the cytoskeletal structure of human platelets following thrombin activation. *The Journal of Biological Chemistry*, **256**:6927–6932.

Karlsson, R., Lassing, I., Höglund, A.-S. and Lindburg, U. (1984). The organization of microfilaments in spreading platelets: a comparison with fibroblasts and glial cells. *Journal of Cellular Physiology*, **121**:96–113.

Katzav, S., Martin-Zanca, D. and Barbacid, M. (1989). vav, a novel human oncogene derived from a locus ubiquitously expressed in hematopoietic cells. *EMBO Journal*, **8**:2283–2290.

Kavanaugh, W.M. and Williams, L.T. (1994). An alternative to SH2 domains for binding tyrosine-phosporylated proteins. *Science*, **266**:1862–1865.

Kimura, K., Ito, M., Amano, M., Chihara, K., Fukata, Y., Nakafuku, M., Chihara, K., Fukata, Y., Nakafuku, M., Yamamori, B., Feng, J., Nakano, T., Okawa, K., Iwamatsu, A. and Kaibuchi, K. (1996). Regulation of myosin phosphatase by Rho and Rho-associated kinase (Rho-kinase). *Science*, **273**:245–248.

Knezevic, I., Leisner, T.M. and Lam, S.C.-T. (1996). Direct binding of the platelet integrin $\alpha_{IIb}\beta_3$ (GPIIb-IIIa) to talin: Evidence that interaction is mediated through the cytoplasmic domains of both α_{IIb} and β_3. *The Journal of Biological Chemistry*, **271**:16416–16421.

Kotani, K., Yonezawa, K., Hara, K., Ueda, H., Kitamura, Y., Sakaue, H., Ando, A., Chavanieu, A., Calas, B., Grigorescu, F., Nishiyama, M., Waterfield, M. and Kasuga, M. (1994). Involvement of phosphoinositide 3-kinase in insulin- or IGF-1-induced membrane ruffling. *EMBO Journal*, **13**:2313–2321.

Kouns, W.C., Fox, C.F., Lamoreaux, W.J., Coons, L.B. and Jennings, L.K. (1991). The effect of glycoprotein IIb-IIIa receptor occupancy on the cytoskeleton of resting and activated platelets. *The Journal of Biological Chemistry*, **266**:13891–13900.

Kovacsovics, T.J., Bachelot, C., Toker, A., Vlahos, C.J., Duckworth, B., Cantley, L.C. and Hartwig, J.H. (1995). Phosphoinositide 3-kinase inhibition spares actin assembly in activating platelets, but reverses platelet aggregation. *The Journal of Biological Chemistry*, **270**:11358–11366.

Kuhlamn, P.A., Hughes, C.A., Bennett, V. and Fowler, V.M. (1996). A new function for adducin: Calcium/calmodulin-regulated capping of the barbed ends of actin filaments. *The Journal of Biological Chemistry*, **271**:7986–7991.

Kulkarni, S., Saido, T.C., Suzuki, K. and Fox, J.E.B. (1999). Calpain mediates integrin-induced signaling at a point upstream of Rho family members. *The Journal of Biological Chemistry* (In press).

Lassing, I. and Lindberg, U. (1985). Specific interaction between phosphatidylinositol 4,5-biphosphate and profilactin. *Nature*, **314**:472–474.

Law, D.A., Nannizzi-Alaimo, L. and Philips, D.R. (1996). Outside-in integrin signal transduction. $\alpha_{IIb}\beta_3$-(GP IIb-IIIa) tyrosine phosphorylation induced by platelet aggregation. *The Journal of Biological Chemistry*, **271**:10811–10815.

Leung, T., Manser, E., Tan, L. and Lim, L. (1995). A novel serine/threonine kinase binding the ras-related rhoA GTPase which translocates the kinase to peripheral membranes. *The Journal of Biological Chemistry*, **270**:29051–29054.

Li, R.Y., Gaits, F., Ragab, A., Ragab-Thomas, J.M.F. and Chap, H. (1994). Translocation of an SH2-containing protein tyrosine phosphate (SH-PTP1) to the cytoskeleton of thrombin-activated platelets. *FEBS Letters*, **343**:89–93.

Lind, S.E., Janmey, P.A., Chaponnier, C., Herbert, T.-J. and Stossel, T.P. (1987). Reversible binding of actin to gelsolin and profilin in human platelet extracts. *Journal of Cell Biology*, **105**:833–842.

Lindberg, U. and Markey, F. (1987). Platelet microfilaments and motility. In *Platelets in Biology and Pathology* III, ed. M.a. Gordon, Netherlands: Elsevier Science Publishers B.V.

Lipfert, L., Haimovich, B., Schaller, M.D., Cobb, B.S., Parsons, J.T. and Brugge, J.S. (1992). Integrin-dependent phosphorylation and activation of the protein tyrosine kinase pp125^{FAK} in platelets. *Journal of Cell Biology*, **119**:905–912.

Lo, S.H., Weisberg, E. and Chen, L.B. (1994). Tensin: A potential link between the cytoskeleton and signal transduction. *Bioessays*, **16**:817–823.

Lowenstein, E.J., Daly, R.J., Batzer, A.G., Li, W., Margolis, B., Lammers, R., Ullrich, A., Skolnik, E.Y., Bar-Sag, D. and Schlessinger, J. (1992). The SH2 and SH3 domain-containing protein GRB2 links receptor tyrosine kinases to ras signaling. *Cell*, **70**:431–442.

Macalma, T., Otte, J., Hensler, M.E., Bockholt, S.M., Louis, H.A., Kalff-Suske, M., Grzeschik, K.-H., von der Ahe, D. and Beckerle, M.C. (1996). Molecular characterization of human zyxin. *The Journal of Biological Chemistry*, **271**:31470–31478.

Manser, E., Chong, C., Zhao, Z.-S., Leung, T., Michael, G., Hall, C. and Lim, L. (1995). Molecular cloning of a new member of the p21-Cdc42/Rac-activated kinase (PAK) family. *The Journal of Biological Chemistry*, **270**:25070–25078.

Manser, E., Leung, T., Salihuddin, H., Zhao, Z.-S. and Lim, L. (1994). A brain serine/threonine protein kinase activated by Cdc42 and Rac1. *Nature*, **367**:40–46.

Marcu, M.G., Zhang, L., Nau-Staudt, K. and Trifaro, J.-M. (1996). Recombinant scinderin, an F-actin severing protein, increases calcium-induced release of serotonin from permeabilized platelets, an effect blocked by two scinderin-derived actin-binding peptides and phosphatidylinositol 4,5-bisphosphate. *Blood*, **87**:20–24.

Martin, G.A., Bollag, G., McCormick, F. and Abo, A. (1995). A novel serine kinase activated by rac1/CDC42Hs-dependent autophosphorylation is related to PAK65 and STE20. *EMBO Journal*, **14**:1970–1978.

Matsumura, K. and Campbell, K.P. (1994). Dystrophin-glycoprotein complex: Its role in the molecular pathogenesis of muscular dystrophies. *Muscle & Nerve* **17**:2–15.

Matsumura, K., Ervasti, J.M., Ohlendieck, K., Kahl, S.D. and Campbell, K.P. (1992). Association of dystrophin-related protein with dystrophin-associated proteins in *mdx* mouse muscle. *Nature*, **360**:588–593.

Meyer, S.C., Lowry, B.R., Cunningham, J.G., Ploplis, V.A. and Fox, J.E.B. (1997a). Role of the cytoplasmic domain of GP Ib-IX in regulating adhesion-induced cytoskeletal reorganizations. *Thrombosis and Haemostasis*, Submitted.

Meyer, S.C., Zuerbig, S., Cunningham, C.C., Hartwig, J.H., Bissell, T., Gardner, K. and Fox, J.E.B. (1997b). Identification of the region in actin-binding protein that binds to the cytoplasmic domain of glycoprotein Ibα. *The Journal of Biological Chemistry*, **272**:2914–2919.

Miyamoto, S., Teramoto, H., Coso, O.A., Gutkind, J.S., Burbelo, P.D., Akiyama, S.K. and Yamada, K.M. (1995). Integrin function: Molecular hierarchies of cytoskeletal and signaling molecules. *Journal of Cell Biology*, **131**:791–805.

Morii, N., Teru-uchi, T., Tominaga, T., Kumagai, N., Kozaki, S., Ushikubi, F. and Narumiya, S. (1992). A rho gene product in human blood platelets. II. Effects of the ADP-ribosylation by botulinum C3 ADP-ribosyltransferase on platelet aggregation. *The Journal of Biological Chemistry*, **267**:20921–20926.

Nachmias, V.T., Golla, R., Casella, J.F. and Barron-Casella, E. (1996). Cap Z, a calcium insensitive capping protein in resting and activated platelets. *FEBS Letters*, **378**:258–262.

Nachmias, V.T. and Yoshida, K.-I. (1988). The cytoskeleton of the blood platelet: a dynamic structure. *Advances in Cell Biology*, **2**:181–211.

Nakamura, F., Amieva, M.R. and Furthmay, H. (1995). Phosphorylation of threonine 558 in the carboxyl-terminal actin-binding domain of moesin by thrombin activation of human platelets. *The Journal of Biological Chemistry*, **270**:31377–31385.

Nakamura, S. and Yamamura, H. (1989). Thrombin and collagen induce rapid phosphorylation of a common set of cellular proteins on tyrosine in human platelets. *The Journal of Biological Chemistry*, **264**:7089–7091.

Nakashima, S., Yuji, C., Nakamura, M., Miyoshi, N., Kohno, M. and Nozawa, Y. (1994). Tyrosine phosphorylation and activation of mitogen-activated protein kinases by thrombin in human platelets: Possible involvement in late arachidonic acid release. *Biochemical and Biophysical Research Communications*, **198**:497–503.

Nemoto, Y., Namba, T., Teru-uchi, T., Ushikubi, F., Morii, N. and Narumiya, S. (1992). A rho gene product in human blood platelets. *The Journal of Biological Chemistry*, **267**:20916–20920.

Nobes, C. and Hall, A. (1994). Regulation and function of the Rho-subfamily of small GTPases. *Current Opinions in Genetic Development*, **4**:77–81.

Nobes, C.D. and Hall, A. (1995). Rho, Rac, and Cdc42 GTPase regulate the assembly of multimolecular focal complexes associated with actin stress fibers, lamellipodia, and filopodia. *Cell*, **81**:53–62.

Olson, M.F., Toksoz, D. and Hall, A. (1996). Small GTP-binding proteins and growth factor signaling pathways. Paper presented at the Keystone Symposia, Tamarron, Colorado, U.S.A., 1996.

Otey, C., Vasquez, G., Burridge, K. and Erickson, B. (1993). Mapping of the α-actinin binding site within the β_1 integrin cytoplasmic domain. *The Journal of Biological Chemistry*, **268**:21193–21197.

Otey, C.A., Pavalko, F.M. and Burridge, K. (1990). An interaction between α-actinin and the β_1 integrin subunit in vitro. *Journal of Cell Biology*, **111**:721–729.

Painter, R.G. and Ginsberg, M.H. (1984). Centripetal myosin redistribution in thrombin-stimulated platelets: relationship to platelet Factor 4 secretion. *Experimental Cell Research*, **155**:198–212.

Papkoff, J., Chen, R.-H., Blenis, J. and Forsman, J. (1994). p42 mitogen-activated protein kinase and p90 ribosomal S6 kinase are selectively phosphorylated and activated during thrombin-induced platelet activation and aggregation. *Molecular and Cellular Biology*, **14**:463–472.

Parsons, J.T. (1996). Integrin-mediated signalling: Regulation by protein tyrosine kinases and small GTP-binding proteins. *Current Opinions in Cell Biology*, **8**:146–152.

Parsons, J.T., Schaller, M.D., Hildebrand, J., Tzeng-Horng, L., Richardson, A. and Otey, C. (1994). Focal adhesion kinase: structure and signaling. *Journal of Cell Science*, **19** (Supplement):109–113.

Pawson, T. (1995). Protein modules and signaling networks. *Nature*, **373**:573–580.

Pelicci, G., Lanfrancone, L., Grignani, F., McGlade, J., Cavallo, F., Forni, G., Nicoletti, I., Grignani, F., Pawson, T. and Pelicci, P.G. (1992). A novel transforming protein (SHC) with an SH2 domain is implicated in mitogenic signal transduction. *Cell*, **70**:93–104.

Phillips, D.R. (1980). Membrane glycoproteins mediating platelet aggregation. In *The Regulation of Coagulation*, eds. K.G. Mann and J.F.B. Taylor, pp. 259–270. New York: Elsevier/North Holland Biomedical Press.

Pho, D.B., Desbruyeres, E., Der Terrossian, E. and Olomucki, A. (1986). Cytoskeletons of ADP- and thrombin-stimulated blood platelets. Presence of a caldesmon-like protein, α-actinin and gelsolin at different steps of the stimulation. *FEBS Letters*, **202**:117–121.

Polakis, P.G., Weber, R.F., Nevins, B., Didsbury, J.R., Evans, T. and Snyderman, R. (1989). Identification of the *ral* and *rac*1 gene products, low molecular mass GTP-binding proteins from human platelets. *The Journal of Biological Chemistry*, **264**:16383–16389.

Pollard, T.D. and Cooper, J.A. (1986). Actin and actin-binding proteins. A critical evaluation of mechanisms and functions. *Annals Review of Biochemistry*, **55**:987–1035.

Pollard, T.D., Fujiwara, K., Handin, R. and Weiss, G. (1977). Contractile proteins in platelet activation and contraction. *Annals N.Y. Academy of Science*, **283**:218–236.

Price, M.G. and Gomer, R.H. (1993). Skelemin, a cytoskeletal M-disc periphery protein, contains motifs of adhesion/recognition and intermediate filament proteins. *The Journal of Biological Chemistry*, **268**: 21800–21810.

Reddy, K.B., Gascard, P., Price, M.G., Schwartz Ebarb, J. M., Frank, E.G. and Fox, J.E.B. (1996). β-integrin cytoplasmic tails interact with the myosin-cytoskeleton linker protein skelemin. *Molecular Biology of the Cell*, **7**:385a.

Reinhard, M., Giehl, K., Abel, K., Haffner, C., Jarchau, T., Hoppe, V., Jockusch, B. M. and Walter, U. (1995a). The proline-rich focal adhesion and microfilament protein VASP is a ligand for profilins. *EMBO Journal*, **14**:1583–1589.

Reinhard, M., Halbrügge, M., Scheer, U., Wiegand, C., Jockusch, B.M. and Walter, U. (1992). The 46/50 kDa phosphoprotein VASP purified from human platelets is a novel protein associated with actin filaments and focal contacts. *EMBO Journal*, **11**:2063–2070.

Reinhard, M., Jouvenal, K., Tripier, D. and Walter, U. (1995b). Identification, purification, and characterization of a zyxin-related protein that binds the focal adhesion and microfilament protein VASP (vasodilator-stimulated phosphoprotein). *Proceedings of the National Academy of Sciences, USA*, **92**:7956–7960.

Remold-O'Donnell, E., Rosen, F.S. and Kenney, D.M. (1996). Defects in Wiskott-Aldrich Syndrome blood cells. *Blood*, **87**:2621–2631.

Ren, X.-D., Bokoch, G.M., Traynor-Kaplan, A., Jenkins, G.H., Anderson, R.A. and Schwartz, M.A. (1996). Physical association of the small GTPase Rho with a 68-kDa phospatidylinositol 4-phosphate 5-kinase in Swiss 3T3 cells. *Molecular Biology of Cell*, **7**:435–442.

Ridley, A.J. and Hall, A. (1992). The small GTP-binding protein rho regulates the assembly of focal adhesions and actin stress fibers in response to growth factors. *Cell*, **70**:389–399.

Rittenhouse, S.E. (1996). Phosphoinositide 3-kinase activation and platelet function. *Blood*, **88**:4401–4414.

Rodriguez del Castillo, A., Vitale, M.L., Tchakarov, L. and Trifaro, J.M. (1992). Human platelets contain scinderin, a Ca^{2+}-dependent actin filament-severing protein. *Thrombosis and Haemostasis*, **67**:248–251.

Rouse, J., Cohen, P., Trigon, S., Morange, M., Alonso-Llamazares, A., Zamanillo, D., Hunt, T. and Nebreda, A. (1994). A novel kinase cascade triggered by stress and heat shock that stimulates MAPKAP kinase-2 and phosphorylation of the small heat shock proteins. *Cell*, **78**:1027–1037.

Safer, D., Golla, R. and Nachmias, V.T. (1990). Isolation of a 5-kilodalton actin-sequestering peptide from human blood platelets. *Proceedings of the National Academy of Sciences, USA*, **87**:2536–2540.

Safer, D. and Nachmias, V.T. (1994). β-thymosins as actin binding peptides. *Bioessays*, **16**:473–479.

Saklatvala, J., Rawlinson, L., Waller, R.J., Sarsfield, S., Lee, J.C., Morton, L.F., Barnes, M.J. and Farndale, R.W. (1996). Role for p38 mitogen-activated protein kinase in platelet aggregation caused by collagen or a thromboxane analogue. *The Journal of Biological Chemistry*, **271**:6586–6589.

Schafer, D.A. and Cooper, J.A. (1995). Control of actin assembly at filament ends. *Annual Review of Cell Developmental Biology*, **11**:497–518.

Shattil, S., Haimovich, B., Cunningham, M., Lipfert, L., Parsons, J., Ginsberg, M. and Brugge, J. (1994). Tyrosine phosphorylation of pp125^{FAK} in platelets requires coordinated signaling through integrin and agonist receptors. *The Journal of Biological Chemistry*, **269**:14738–14745.

Shattil, S.J. and Brugge, J.S. (1991). Protein tyrosine phosphorylation and the adhesive functions of platelets. *Current Opinions in Cell Biology*, **3**:869–879.

Stark, F., Golla, R. and Nachmias, V.T. (1991). Formation and contraction of a microfilamentous shell in saponin-permeabilized platelets. *Journal of Cell Biology*, **112**:903–913.

Symons, M. (1996). Rho family GTPases: The cytoskeleton and beyond. *Trends in Biochemical Sciences*, **21**:178–181.

Symons, M., Derry, J.M.J., Kariak, B., Jiang, S., Lemahieu, V., McCormick, F., Francke, U. and Abo, A. (1996). Wiskott-Aldrich syndrome protein, a novel effector for the GTPase CDC42Hs, is implicated in actin polymerization. *Cell*, **84**:723–734.

Tapley, P.M. and Murray, A.W. (1985). Evidence that treatment of platelets with phorbol ester causes proteolytic activation of Ca^{2+}-activated, phospholipid-dependent protein kinase. *European Journal of Biochemistry*, **151**: 419–423.

Teo, M., Manser, E. and Lim, L. (1995). Identification and molecular cloning of a p21cdc$^{42/rac1}$-activated serine/threonine kinase that is rapidly activated by thrombin in platelets. *The Journal of Biological Chemistry*, **270**:26690–26697.

Tohyama, Y., Yanagi, S., Sada, K. and Yamamura, H. (1994). Translocation of p72^{syk} to the cytoskeleton in thrombin-stimulated platelets. *The Journal of Biological Chemistry*, **269**:32796–32799.

Toker, A., Meyer, M., Reddy, K., Falck, J., Aneja, R., Aneja, S., Parra, A., Burns, D., Ballas, L. and Cantley, L. (1994). Activation of protein kinase C family members by the novel polyphosphoinositides PtdIns-3,4-P$_2$ and PtdIns-3,4,5-P$_3$. *The Journal of Biological Chemistry*, **269**:32358–32367.

Tolias, K.F., Cantley, L.C. and Carpenter, C.L. (1995). Rho family GTPases bind to phosphoinositide kinases. *The Journal of Biological Chemistry*, **270**:17656–17659.

Torti, M., Ramaschi, G., Sinigaglia, F., Lapetina, E. and Balduini, C. (1994). Glycoprotein IIb-IIIa and the translocation of Rap2B to the platelet cytoskeleton. *Proceedings of the National Academy of Sciences, USA*, **91**:4239–4243.

Tsukita, S., Yonemura, S. and Tsukita, S. (1997). ERM proteins: head-to-tail regulation of actin-plasma membrane interaction. *Trends in Biochemical Science*, **22**:53–58.

Valencia, A., Chardin, P., Wittinghofer, A. and Sander, C. (1991). The Ras protein family: Evolutionary tree and role of conserved amino acids. *Biochemistry*, **30**:4637–4648.

van der Geer, P., Hunter, T. and Lindberg, R.A. (1994). Receptor protein-tyrosine kinases and their signal transduction pathways. *Annual Review of Cell Biology*, **10**:251–337.

Vojtek, A., Haarer, B., Field, J., Gerst, J., Pollard, T.D., Brown, S. and Wigler, M. (1991). Evidence for a functional link between profilin and CAP in the yeast S. cerevisiae. *Cell*, **66**:497–505.

Wasenius, V.-M., Saraste, M., Salvén, P., Erämaa, M., Holm, L. and Lehto, V.-P. (1989). Primary structure of the brain α-spectrin. *Journal of Cell Biology*, **108**:79–93.

Weber, A., Nachmias, V.T., Pennise, C.R., Pring, M. and Safer, D. (1992). Interaction of thymosin β_4 with muscle and platelet actin: Implications for actin sequestration in resting platelets. *Biochemistry*, **31**:6179–6185.

Weber, A., Pennise, C.R., Babcock, G.G. and Fowler, V.M. (1994). Tropomudulin caps that pointed ends of actin filaments. *Journal of Cell Biology*, **127**:1627–1635.

Wennstrom, S., Siegbahn, A., Yokote, K., Arvidsson, A.K., Heldin, C.H., Mori, S. and Claesson, W.L. (1994). Membrane ruffling and chemotaxis transduced by the PDGF β-receptor require the binding site for phosphatidylinositol 3' kinase. *Oncogene*, **9**:651–660.

White, J.G. (1984). Arrangements of actin filaments in the cytoskeleton of human platelets. *American Journal of Pathology*, **117**:207–217.

Witke, W., Sharpe, A.H., Hartwig, J.H., Azuma, T., Stossel, T.P. and Kwiatkowski, D.J. (1995). Hemostatic, inflammatory, and fibroblast responses are blunted in mice lacking gelsolin. *Cell*, **81**:41–51.

Wymann, M. and Arcaro, A. (1994). Platelet-derived growth factor-induced phosphatidylinositol 3-kinase activation mediates actin rearrangements in fibroblasts. *Biochemical Journal*, **298**:517–520.

Ylänne, J., Chen, Y., O'Toole, T.E., Loftus, J.C., Takada, Y. and Ginsberg, M.H. (1993). Distinct functions of integrin α and β subunit cytoplasmic domains in cell spreading and formation of focal adhesions. *Journal of Cell Biology*, **122**:223–233.

Zhang, J., Falck, J.R., Reddy, K.K., Abrams, C.S., Zhao, W. and Rittenhouse, S.E. (1995a). Phosphatidylinositol (3,4,5,)-trisphosphate stimulates phosphorylation of pleckstrin in human platelets. *The Journal of Biological Chemistry*, **270**:22807–22810.

Zhang, J., Fry, M.J., Waterfield, M.D., Jaken, S., Liao, L., Fox, J.E.B. and Rittenhouse, S.E. (1992). Activated phosphoinositide 3-kinase associates with membrane skeleton in thrombin-exposed platelets. *The Journal of Biological Chemistry*, **267**:4686–4692.

Zhang, J., King, W.G., Dillon, S., Hall, A., Feig, L. and Rittenhouse, S.E. (1993). Activation of platelet phosphatidylinositide 3-kinase requires the small GTP-binding protein Rho. *The Journal of Biological Chemistry*, **268**:22251–22254.

Zhang, J., Zhang, J., Benovic, J.L., Sugai, M., Wetzker, R., Gout, I. and Rittenhouse, S.E. (1995b). Sequestration of a G-protein $\beta\gamma$ subunit or ADP-ribosylation of Rho can inhibit thrombin-induced activation of platelet phosphoinositide 3-kinases. *The Journal of Biological Chemistry*, **270**:6589–6594.

Zhang, J., Zhang, J., Shattil, S.J., Cunningham, M.C. and Rittenhouse, S.E. (1996). Phosphoinositide 3-kinase γ and p85/phosphoinositide 3-kinase in platelets. *The Journal of Biological Chemistry*, **271**:6265–6272.

Zhang, S., Han, J., Sells, M.A., Chernoff, J., Knaus, U.G., Ulevitch, R.J. and Bokoch, G.M. (1995c). Rho family GTPases regulate p38 mitogen-activated protein kinase through the downstream mediator Pak1. *The Journal of Biological Chemistry*, **270**:23934–23936.

Zheng, Y., Baghrodia, S. and Cerione, R.A. (1994). Activation of phosphoinositide 3-kinase activity by Cdc42Hs binding to p85. *The Journal of Biological Chemistry*, **269**:18727–18730.

Zhu, Y., O'Neill, S., Saklatvala, J., Tassi, L. and Mendelsohn, M.E. (1994). Phosphorylated HSP27 associates with the activation-dependent cytoskeleton in human platelets. *Blood*, **84**:3715–3723.

7 The Role of the Phosphoinositide-Derived Second Messenger Molecules in Platelet Activation

Janine K. Campbell[#], Susan Brown[#], Adam D. Munday and Christina A. Mitchell*

Department of Medicine, Monash Medical School, Box Hill Hospital, Nelson Rd, Box Hill, Melbourne, VIC 3128, Australia

THE CLASSICAL PHOSPHOINOSITIDE SIGNALLING PATHWAY

The phosphoinositides and their derivatives are ubiquitous cellular phospholipids that are important components of several intracellular signalling pathways. The second messengers derived from the breakdown of phosphoinositides, mobilise intracellular calcium and activate protein kinase C. In addition, by regulating actin-binding proteins the polyphosphoinositides play a pivotal role in platelet actin rearrangement, an essential step for platelet aggregation.

Phosphoinositides represent 2–10% of the total phospholipid in the eukaryotic cell membrane, of which *myo*-phosphatidylinositol (PtdIns) comprises over 90%. Several distinct polyphosphoinositides can be generated by phosphorylation of PtdIns in either the 3, 4 and/or 5 position by the action of a series of phosphoinositide-specific kinases (Majerus, 1992; Berridge, 1993). The 3-position phosphoinositides have only been described in the last decade and although ubiquitously expressed are present in much smaller amounts than the classic phosphoinositides, phosphatidylinositol 4-phosphate (PtdIns (4)P) and phosphatidylinositol 4,5-bisphosphate (PtdIns (4,5)P$_2$). In addition the 3-position phosphoinositides probably represent messenger molecules in their own right, as they are not hydrolysed by phospholipase C (PLC), but are metabolised by series of phosphatase reactions (Lips and Majerus, 1989; Serunian *et al.*, 1989). These recently described phospholipids play diverse roles regulating secretion, mitogenesis, vesicular trafficking and platelet activation (reviewed Fry, 1994; Stephens *et al.*, 1993; Carpenter and Cantley, 1996b).

The classical pathway for phosphatidylinositol metabolism, follows stimulation by a variety of agonists leading to sequential phosphorylation of PtdIns by a PtdIns 4-kinase generating PtdIns (4)P and then by a PtdIns (4)P 5-kinase to form PtdIns(4,5)P$_2$. Following cellular stimulation this central phospholipid is hydrolysed by phospholipase C (PLC), generating at least two second messenger molecules: diacylglycerol, which activates protein kinase C (PKC) and inositol 1,4,5 trisphosphate (Ins(1,4,5)P$_3$) (Figure 7.1). The PLC-mediated cleavage of PtdIns(4,5)P$_2$ releases the water-soluble Ins(1,4,5)P$_3$ into the cytoplasm

---

Corresponding author: Department of Medicine, Box Hill Hospital, Nelson Rd, Box Hill, 3128, Melbourne, Australia.
[#]Equal first authors.

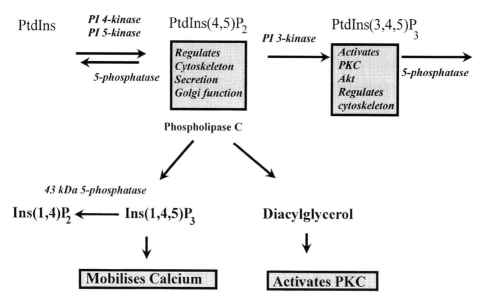

Figure 7.1   Pathway for the formation of the polyphosphoinositides and their second messengers. Phosphatidylinositol is sequentially phosphorylated by a PtdIns 4-kinase and PtdIns(4)P 5-kinase leading to the formation of PtdIns(4,5)P$_2$. Upon agonist stimulation phospholipase C hydrolyses PtdIns(4,5)P$_2$, generating Ins(1,4,5)P$_3$ which mobilises intracellular calcium and diacylglycerol, which activates protein kinase C. An alternative route for the metabolism of PtdIns(4,5)P$_2$ is mediated via PI 3-kinase, forming 3-position phosphoinositides. PtdIns(4,5)P$_2$ also serves to regulate cytoskeletal and Golgi function.

where its binds to and activates specific calcium channel receptors, principally located in the endoplasmic reticulum, resulting in the transient release of calcium (Berridge, 1993). In platelets the Ins(1,4,5)P$_3$ receptors are predominantly located in the dense tubular system.

Many signalling systems including those initiated by growth factors, neurotransmitters, and hormones stimulate phosphoinositide turnover. In platelets agonists such as thrombin, PAF and thomboxane α_2, adrenalin and serotonin stimulate the production of Ins(1,4,5)P$_3$ and thereby intracellular calcium release (reviewed Blockmans *et al.*, 1995; Heemskerk *et al.*, 1994a). The receptors for these agonists are all members of the seven membrane-spanning receptors that couple through G-proteins to PLC. PLC exists in multiple isoforms principally of three types: β, which are stimulated by heterotrimeric G proteins, γ which can be activated by both receptor and non-receptor tyrosine kinases and finally δ (reviewed Rhee, 1991; Lee *et al.*, 1995). Human platelets contain seven of the ten known PLC isoforms including γ_2, β_2, β_3, β_1, γ_1, δ_1, β_4. The significance of the PLC-mediated signalling pathway in platelet function is highlighted by the recent demonstration that a patient with an inherited bleeding disorder, had abnormal platelet secretion and aggregation caused by a deficiency in PLCβ_2 resulting in reduced Ins(1,4,5)P$_3$ generation and intracellular calcium mobilisation (Lee *et al.*, 1996).

PLC enzyme activity in human platelets may be regulated by several mechanisms which include tyrosine phosphorylation, translocation to the cytoskeleton and proteolytic cleavage by calpain. In many other cell types the γ isoform can be phosphorylated and activated by receptor tyrosine kinases (Todderud *et al.*, 1990; Lee *et al.*, 1995; Yang *et al.*, 1994). In platelets, thrombin, collagen and low affinity Fc receptor activation have been shown to stimulate a rapid transient tyrosine phosphorylation of PLCγ_2 (Tate and Rittenhouse, 1993; Blake *et al.*, 1994). Thrombin stimulation also causes an early translocation of PLCβ_{3b} to the cytoskeleton, which is dependent on integrin $\alpha_{IIb}\beta_3$ mediated platelet aggregation. A slower translocation of PLCβ_2 and γ_2 is also observed (Banno *et al.*, 1996). This may allow

PLC access to membrane substrates such as PtdIns(4,5)P_2 which plays an important role in regulating actin-binding and severing proteins and thereby platelet aggregation. Finally PLC activity in platelets may also be regulated via proteolytic cleavage of the enzyme by the calcium-dependent cysteine protease, calpain (Banno *et al.*, 1995). In intact thrombin or collagen-stimulated platelets, proteolytic cleavage of PLCβ_2 correlates with irreversible platelet aggregation.

Although a variety of different signalling systems contribute to the agonist-induced increase in intracellular calcium, a significant proportion of this transient rise is derived from the Ins(1,4,5)P_3-sensitive store. Recent investigations have demonstrated that the thrombin, serotonin and ADP-evoked rise in intracellular calcium is blocked by specific PLC inhibitors (Heemskerk *et al.*, 1994b). There is abundant evidence in intact and permeabilized platelets of the ability of Ins(1,4,5)P_3 to induce calcium release. Human platelets appear to contain relatively large calcium stores and are therefore less dependent on an influx of extracellular calcium to maintain oscillatory calcium waves. A family of Ins(1,4,5)P_3 receptors designated Type I, II or III have been well characterised in many tissues, including platelets (Taylor and Marshall, 1992; Mikoshiba, 1993). They all comprise a membrane spanning C-terminal domain of four subunits, which combine to form the Ins(1,4,5)P_3 channel. The N-terminal Ins(1,4,5)P_3 binding domain lies free in the cytoplasm at some distance from the channel opening. Ins(1,4,5)P_3 binding to the receptor opens the channel and thereby allows sequestered calcium to enter the cytosol. The platelet contains Ins(1,4,5)P_3 receptors predominantly of the Type 1b and Type II variety, located in the internal dense tubular membrane system (Quinton *et al.*, 1996a). In addition, Type II Ins(1,4,5)P_3 receptors have been reported in the platelet membrane (Buorguignon *et al.*, 1993). T cells also have plasma membrane Ins(1,4,5)P_3 receptors of the Type 2 class and it is postulated that they may serve to gate calcium influx at the cell surface, however, whether this occurs in platelets is unclear (Khan *et al.*, 1992).

Ins(1,4,5)P_3 receptor function may be regulated by calcium, ATP, phosphorylation and receptor inactivation (Taylor and Marshall, 1992; Furuichi and Mikoshiba, 1995). cAMP-dependent protein kinase phosphorylation of the receptor has divergent effects, depending on the receptor and cell type, varying from a ten-fold reduction in receptor sensitivity in rat brain, to a five fold increase in activity in liver microsomal preparations (Supattapone *et al.*, 1988; Burgess *et al.*, 1991). The platelet Type I Ins(1,4,5)P_3 receptor is phosphorylated by cAMP-dependent protein kinase, resulting in a decrease in the rate of calcium release from the store. As cAMP inhibits platelet activation, the mechanism of this inhibition may be via cAMP-dependent phosphorylation of the Ins(1,4,5)P_3 receptor, resulting in a decrease in the release of calcium from the store (Quinton *et al.*, 1996b; Cavallini *et al.*, 1996).

Finally, Ins(1,4,5)P_3 receptors are down regulated in response to chronic activation by Ins(1,4,5)P_3. The down regulation process targets the receptor for proteolytic degradation by a calcium-activated cysteine protease, most probably calpain (Wojcikiewicz and Oberdorf, 1996). Whether such calcium mediated proteolysis occurs in platelets is unknown, however calpain is likely to be activated by the high local concentration of calcium in the vicinity of the activated Ins(1,4,5)P_3 receptor.

INS(1,4,5)P_3 METABOLISM: THE INOSITOL POLYPHOSPHATE 5-PHOSPHATASES

There are two major pathways for the metabolism of Ins(1,4,5)P_3, which differ in relative importance depending on the cell type. Firstly, Ins(1,4,5)P_3 can be phosphorylated by an

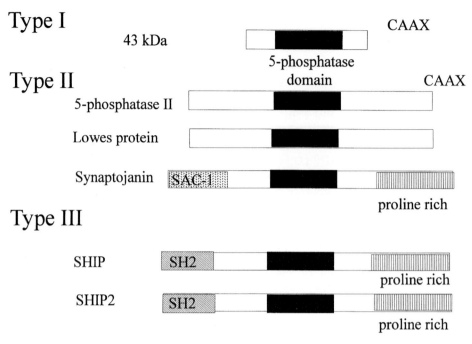

Figure 7.2   Structure of inositol polyphosphate 5-phosphatases.

Ins(1,4,5)P$_3$ 3-kinase to form inositol 1,3,4,5-tetrakisphosphate (Ins(1,3,4,5)P$_4$), which has been implicated in the regulation of Ca^{2+} homeostasis in certain cell types (Berridge, 1993). Secondly, Ins(1,4,5)P$_3$ is rapidly metabolised to a complex number of products within 10–20 seconds of its formation. The principle pathway for the first step in its metabolism is via dephosphorylation of the 5-position phosphate by the inositol polyphosphate 5-phosphatase (5-phosphatase) family. Many of these enzymes were initially identified in human platelets by their ability to hydrolyse the 5-position phosphate from the calcium mobilizing second messenger molecules, Ins(1,4,5)P$_3$ and Ins(1,3,4,5)P$_4$, forming Ins(1,4)P$_2$ and Ins(1,3,4)P$_3$ respectively. As the latter two inositol phosphates do not mobilize intracellular calcium, their formation by the actions of the 5-phosphatases represents a signal terminating reaction (Figure 7.2) (Berridge, 1993).

At least 9 mammalian and 4 yeast 5-phosphatases have been described with a widespread expression in many cell types (reviewed Majerus, 1996; Wolscholski and Parker, 1997). In the last two years there have been enormous advances in the identification of novel members of this family and the recognition of the role 5-phosphatases play in signal termination. The defining feature of the 5-phosphatases is the presence of a "5-phosphatase domain" a highly conserved 200–300 amino acid sequence (Laxminarayan *et al.*, 1994; Voss *et al.*, 1995, Jefferson and Majerus, 1995). Within this domain, two specific amino acid motifs, approximately 60 amino acids apart, are highly conserved in all 5-phosphatases. These two signature motifs comprise GDXN(F/Y)R and P(A/S)W(C/T)DRIL (Majerus, 1996). Recent studies suggest that these two regions play a critical role in both substrate binding and catalysis (Communi *et al.*, 1996; Jefferson and Majerus, 1996).

The 5-phosphatases were originally classified based upon the observation in human platelet cytosol that two isoforms eluted sequentially on anion exchange resins (Mitchell *et al*, 1989). The first eluting 5-phosphatase designated Type I, was subsequently purified as a 43 kDa enzyme and the second eluting enzyme as a 75 kDa polypeptide, designated Type II (Connolly *et al.*, 1985, Mitchell *et al.*, 1989). Several other members of the family have subsequently been identified and characterized. The 5-phosphatases are now variously designated Types I-IV on the basis of substrate specificity and the presence of other signalling domains (Majerus, 1996; Woscholski and Parker, 1997) (Figure 7.2).

The only Type I enzyme to be cloned and characterised is the 43 kDa enzyme which hydrolyses the water-soluble second messenger molecules Ins(1,4,5)P_3 and Ins(1,3,4,5)P_4 (Verjans *et al.*, 1994; Laxminarayan *et al.*, 1994). The 43 kDa 5-phosphatase (also called 5-phosphatase I, or the Type I 5-phosphatase) is considered to be the major cellular 5-phosphatase terminating the signalling function of Ins(1,4,5)P_3. Recent studies have demonstrated the 43 kDa 5-phosphatase plays a significant role in regulating Ins(1,4,5)P_3-mediated Ca^{2+} release. Erneux and colleagues have demonstrated that overexpression of the 43 kDa 5-phosphatase in Chinese hamster ovary cells (CHO-K1) abrogates ATP-induced Ca^{2+} oscillations (De Smedt *et al.*, 1997). The 43 kDa 5-phosphatase is predominantly membrane associated and this specific location is mediated via farnesylation of the enzyme's C-terminal CAAX motif (De Smedt *et al.*, 1996). Membrane localization of the 43 kDa 5-phosphatase is essential for regulation of Ins(1,4,5)P_3-induced calcium oscillations. In addition, we have shown underexpression of the 43 kDa 5-phosphatase in resting normal rat kidney cells (NRK) is associated with increases in Ins(1,4,5)P_3, Ins(1,3,4,5)P_4, basal Ca^{2+} concentration and a transformed phenotype (Speed *et al.*, 1996).

The Type II 5-phosphatases comprise a more diverse group of enzymes and include the 75 kDa 5-phosphatase (also called 5-phosphatase II, or Inpp5b), the protein encoded by the gene defective in Lowe's oculocerebrorenal syndrome (OCRL) and the neural synapse 5-phosphatase, synaptojanin. These enzymes all hydrolyse the 5-position phosphate of PtdIns(4,5)P_2 and PtdIns(3,4,5)P_3, in addition to Ins(1,4,5)P_3 and Ins(1,3,4,5)P_4 (Matzaris *et al.*, 1994; Jackson *et al.*, 1995; Zhang *et al.*, 1995; Majerus, 1996; McPherson *et al.*, 1996). From studies in our laboratory it appears that 5-phosphatase II is the major PtdIns(4,5)P_2 5-phosphatase in human platelets (Matzaris *et al.*, 1994). Although 5-phosphatase II was originally purified from platelet cytosol as a 75 kDa enzyme, recent cloning studies have revealed that the enzyme exists in many tissues as a 115 kDa polypeptide, which may be proteolytically cleaved to the smaller 75 kDa isoform. The 5-phosphatase II is also membrane associated, which is mediated by both a C-terminal CAAX motif and an N-terminal hydrophobic domain (Matzaris *et al.*, 1998). 5-phosphatase II is most homologous to the gene deficient in Lowe's oculocerebrorenal (OCRL) syndrome (Attree *et al.*, 1992). The latter is a rare X-linked disorder associated with renal-tubular abnormalities, growth and mental retardation, cataracts and renal bone disease (Lowe *et al.*, 1952). Lowe's protein is localised to the Golgi/lysosomes of affected cells and as the phenotype of the oculocerebrorenal syndrome implies is predominantly located in kidney, brain and lens of the eye and not in platelets (Olivos-Glander *et al.*, 1995; Zhang *et al.*, 1998). It appears this syndrome may be mediated by the accumulation of PtdIns(4,5)P_2 in these cells, due to decreased PtdIns(4,5)P_2 5-phosphatase activity (Zhang *et al.*, 1998). Surprisingly mice deficient in the OCRL gene, as a result of targeted gene disruption, do not develop the same phenotype as humans and appear normal (Janne *et al.*, 1998). In addition mice deficient in 5-phosphatase II appear apparently normal, apart from testicular degeneration after sexual maturation. The overlapping function between these highly related 5-phosphatases is demonstrated by the observation that no live mice were born which lack both OCRL and 5-phosphatase II.

The third member of the Type II 5-phosphatases is designated synaptojanin, as the enzyme was originally identified localised to nerve terminals, where synaptojanin may play a role in regulating synaptic vesicle recycling (McPherson *et al.*, 1996). Several synaptojanin isoforms have been identified and are all characterized by an N-terminal SAC1 domain and a C-terminal proline rich domain (Ramjaun *et al.*, 1996; Seet *et al.*, 1998; Khvotchev and Sudhof, 1998). The SAC1 domain is homologous to the N-terminus of the *Saccharomyces Cerevisiae* Sac1p protein, which plays a role in inositol metabolism in yeast. Synaptojanin has a much more widespread tissue and cellular distribution than originally described and forms complexes with various other signalling proteins including Grb2, amphiphysin I (a component of the synaptic vesicle compartment) and a novel protein family with Grb-2-related SH3 domains (Ringstad *et al.*, 1997). Three 5-phosphatases have been identified in *Saccharomyces cerevisiae*, which bear structural similiarity with synaptojanin, in that they comprise a SAC1 domain, a 5-phosphatase domain and a C-terminal proline rich domain (Srinivasan *et al.*, 1997). The yeast 5-phosphatases appear to hydrolyse $PtdIns(4,5)P_2$. Disruption of these individual genes results in a variable phenotype comprising abnormal vacuolar and plasma membrane morphology. A fourth yeast 5-phosphatase has been identified which lacks such N-and C-terminal domains.

The Type III 5-phosphatases mediate signalling events from the family of haematopoietic receptors. This recently identified 5-phosphatase has been designated SHIP, for SH2-containing inositol polyphosphate 5-phosphatase and is widely expressed in haematopoietic cells. The 145 kDa tyrosine phosphorylated 5-phosphatase associates with other signalling molecules such as Shc and Grb2 following cytokine stimulation (Damen *et al.*, 1996; Lioubin *et al.*, 1996). The 5-phosphatase comprises a N-terminal SH2 domain, a 5-phosphatase domain, two consensus sequences that in other proteins are targets for phosphotyrosine binding, and a proline rich domain. SHIP associates with Shc and Grb2. This novel 5-phosphatase has a unique substrate specificity, in that it hydrolyses only $PtdIns(3,4,5)P_3$ and $Ins(1,3,4,5)P_4$. More recently shorter versions of SHIP designated SIP (for signalling inositol phosphatase) have been identified and probably represent spliced variants of SHIP (Kavanaugh *et al.*, 1996). SHIP appears to act as a negative regulator of cytokine-mediated signalling in hematopoetic cells. Over expression of SHIP in hematopoetic cell lines inhibits both granulocyte-macrophage stimulating factor and interleukin-3 induced cell proliferation (Lioubin *et al.*, 1996). SHIP plays a role in negative regulation of B and mast cell activation mediated by FcγRIIB receptors (reviewed Coggeshall, 1998). Recent studies have shown SHIP recruitment to immune receptors regulates a proapoptotic signal initiated by the FcγRIIB receptor on B cells. Targeted disruption of SHIP in mice leads to a distinct phenotype comprising failure to thrive, and early death due to extensive consolidation of the lungs resulting from myeloid cell infiltration (Helgason *et al.*, 1998). In addition, the bone marrow of SHIP deficient mice shows increased numbers of granulocyte-macrophage progenitors, but no alteration in lymphoid or erythroid precursors. Collectively these studies reveal the critical role SHIP plays in controlling cytokine-mediated signalling in the haematopoietic system.

SHIP is also present in human platelet cytosol and upon thrombin stimulation the enzyme becomes tyrosine-phosphorylated and translocates to the actin cytoskeleton. Translocation of the 5-phosphatase occurs in an aggregation-dependent, integrin-mediated manner (Giuriato *et al.*, 1997). Although the tyrosine phosphorylation and translocation of SHIP has no effect on 5-phosphatase enzyme activity, it does correlate with the production of $PtdIns(3,4)P_2$. Therefore SHIP may serve to regulate two second messengers that are implicated in platelet signalling events, degradation of $PtdIns(3,4,5)P_3$, resulting in the formation of $PtdIns(3,4)P_2$.

Finally, a fourth class of 5-phosphatases has been described that only hydrolyse $PtdIns(3,4,5)P_3$, forming $PtdIns(3,4)P_2$. The enzyme was originally identified in human platelet cytosol, where it forms a complex with the p85 subunit of the PI 3-kinase (Jackson *et al.*, 1995). Platelet activation results in the disassociation of the kinase/phosphatase complex.

THE ROLE OF PHOSPHOINOSITIDE 3-KINASE IN PLATELET ACTIVATION

The phosphoinositide 3-kinase (PI 3-kinase) has been the focus of much recent interest and research (for reviews see Stephens *et al.*, 1993; Fry, 1994; Carpenter and Cantley, 1996a and 1996b; Vanhaesebroeck *et al.*, 1997). PI 3-kinase phosphorylates the D-3 position of the *myo*-inositol ring of the phosphoinositides: PtdIns, PtdIns(4)P and PtdIns(4,5)P$_2$ to produce phosphatidylinositol 3-phosphate (PtdIns(3)P), phosphatidylinositol 3,4-bisphosphate (PtdIns(3,4)P$_2$) and phosphatidylinositol 3,4,5-trisphosphate (PtdIns(3,4,5)P$_3$) respectively (Figure 7.3). Unlike the classic phosphoinositides such as PtdIns(4,5)P$_2$, these 3-position phosphoinositide products are not metabolised by PLC (Lips and Majerus, 1989; Serunian *et al.*, 1989). Therefore, they are unlikely to be intermediaries in the PLC- generated diacylglycerol and Ins(1,4,5)P$_3$ signalling pathway. Emerging evidence indicates the 3-position phosphoinositides represent second messenger molecules in their own right, with important functions in many cells, including platelets.

PI 3-kinase is involved in many signalling pathways including the regulation of mitogenic proliferation, vacuolar transport, prevention of apoptosis and cytoskeletal rearrangement. The role of PI 3-kinase is in many cases both cell type and stimulus specific, explaining the enzyme's diversity of signalling roles. Numerous studies have demonstrated the PI 3-kinase associates with agonist stimulated receptor and non-receptor protein tyrosine kinases, most notably the platelet derived growth factor (PDGF) receptor. PI 3-kinase mediates the transforming potential of oncogenes such as polyoma middle T-activated pp60[c-src] and pp60[v-src] (Whitman *et al.*, 1988), which is dependent on complex formation with the PI 3-kinase. Several studies indicate a role for the 3-position phosphoinositides in vesicle trafficking

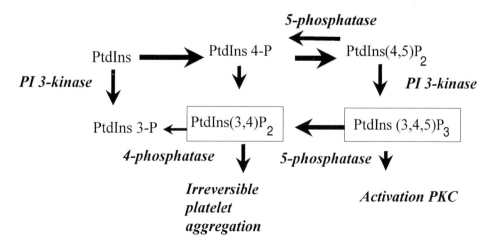

Figure 7.3   Formation of the 3-position phosphoinositides. Agonists stimulate the formation of PtdIns(3,4,5)P$_3$ and PtdIns(3,4)P$_2$ whilst levels of PtdIns 3-P remain unchanged. The possible pathways for the formation of these 3-position phosphoinositides are depicted. The metabolism of the 3-position phosphoinositides is by sequential dephosphorylation by 5 and then 4-phosphatases.

(reviewed De Camilli *et al.*, 1996; Shepherd *et al.*, 1996). In neural and haemopoietic cells, PI 3-kinase is necessary for the growth-factor dependent prevention of apoptosis (Yao and Cooper, 1995; Scheid *et al.*, 1995). Finally, although PI 3-kinase was originally identified in transformed cells, it is clearly present and active in anucleate cells such as platelets and plays a significant role in the signalling events that follow platelet activation.

The PI 3-kinases can be subdivided into three classes based upon substrate specificity (reviewed Vanhaesebroek *et al.*, 1997). Class I enzymes phosphorylate the 3-position of PtdIns, PtdIns(4)P and PtdIns(4,5)P$_2$, with the latter phosphoinositide probably representing the preferred *in vivo* substrate. These class I enzymes associate with adaptor subunits. The first isoform of the enzyme to be identified, was a heterodimer comprising a 85 kDa regulatory adaptor subunit and a 110 kDa catalytic subunit (Escobedo *et al.*, 1991). The p85 subunit contains multiple signalling domains: an N-terminal SH3 domain, proline rich sequences either side of a break point cluster (BCR) domain, and two C-terminal SH2 domains. SH2 domains bind phosphorylated tyrosine residues, SH3 domains bind proline rich motifs, whilst the binding partners for the BCR domain are unclear. The p85 inter-SH2 domain binds the p110 subunit. The p110 subunit represents the lipid kinase domain (Hiles *et al.*, 1992). This subunit also has serine kinase activity, which phosphorylates the associated p85 subunit (serine 608), resulting in significant inhibition of PI 3-kinase activity (Carpenter *et al.*, 1993a; Dhand *et al.*, 1994). Class I PI 3-kinases also include those stimulated by G-proteins which do not interact with SH2 adaptor subunits. The p110γ PI 3-kinase was first described in neutrophils (Stephens *et al.*, 1994) and subsequently cloned (Stoyanov *et al.*, 1995). This enzyme does not complex with the regulatory p85 subunit. Recently a regulatory p101 adaptor subunit has been identified, but the mechanisms by which it regulates the PI 3-kinase are as yet unclear.

The second class of PI 3-kinase enzymes only phosphorylate PtdIns and PtdIns(4)P and have a larger molecular mass of greater than 200 kDa. These enzymes contain a C2 domain, however they bind lipids in a calcium independent manner (reviewed Vanhaesebroek *et al.*, 1997).

The class III PI 3-kinases have a restricted substrate specificity in that they only phosphorylate PtdIns forming PtdIns(3)P. They include the yeast protein Vps34 which regulates trafficking of proteins from the Golgi to the vacuole, thereby implicating PtdIns(3)P in the control of the secretory pathway (Stack *et al.*, 1993). Recently, a mammalian PI 3-kinase Vps 34 homologue, which also phosphorylates only PtdIns has been identified and implicated in lysosomal processing (Volinia *et al.*, 1995). Members of all classes of PI 3-kinases, have been identified in platelets, suggesting multiple distinct roles in platelet activation (Banfic *et al.*, 1998).

One critical issue has been whether PI 3- kinase operates upstream or downstream of Ras. In fibroblasts, constituitively active PI 3-kinase induces *fos* transcription, an effect blocked by coexpression of dominant negative Ras (Hu *et al.*, 1995). In the same study, oocytes expressing constitutively active PI 3-kinase, demonstrated increased levels of GTP-bound Ras, thereby activating Raf-1, resulting in oocyte maturation. In addition, there is evidence that the class I PI 3-kinases interact with Ras in a GTP dependent manner. GTP-bound Ras binds to the catalytic subunit of PI 3-kinase, and enhances the production of 3-position phosphoinositides (Rodriguez-Viciana *et al.*, 1994 and 1996). It is probable that the interaction and relative importance of PI 3-kinase and Ras differs according to the stimulus, response and cell type.

The exact downstream targets of the 3-position phosphoinositides are currently being delineated. These phospholipids are implicated in the activation of a number of protein kinases, including protein kinase C (PKC), pp70^{S6k} and the serine-threonine protein kinase Akt. PtdIns(3,4,5)P$_3$ activates the calcium-independent, diacylglycerol-resistant protein kinase C ζ isoform (Nakanishi *et al.*, 1993). In addition PtdIns(3,4)P$_2$ and to a lesser extent PtdIns(3,4,5)P$_3$ activate other calcium-insensitive PKC isoforms, including δ, ε and η (Toker *et al.*, 1994). In this latter study, only minimal activation of PKC ζ was shown, possibly due to the higher constitutive kinase activity of the recombinant, as opposed to the purified PKC ζ. Platelet activating factor stimulation causes PKC δ to specifically associate with and activate PI 3-kinase in rabbit platelets (Ettinger *et al.*, 1996). Thus, there may be an association and mutual activation between PI 3-kinase and various PKC isoforms. The demonstration that PtdIns(3,4)P$_2$ and PtdIns(3,4,5)P$_2$ stimulate various PKC isoforms to a similar extent, questions the *in vivo* specificity of this activation (Palmer *et al.*, 1994).

In response to PDGF and insulin receptor stimulation, PI 3-kinase activates pp70^{S6k}, a serine-threonine kinase which phosphorylates ribosomal protein S6, thereby allowing G1 cell cycle transition (Chung *et al.*, 1994). The pp70^{S6k} is activated by multiple serine-threonine phosphorylation events, which involve both PI 3-kinase and the related kinase mTOR.

Growth factor activation of a cellular homologue of transforming *v-Akt*, the serine-threonine kinase Akt-1 (also known as protein kinase B), is mediated via PI 3-kinase (Burgering and Coffer, 1995). Akt comprises an N-terminal pleckstrin homology domain (PH) followed by a catalytic domain and a unique C-terminal domain. Three isoforms of Akt have been characterized and all are activated in response to insulin or growth factor stimulation of cells. This activation is abrogated following inhibition of PI 3-kinase (Hemmings, 1997). Akt is directly activated by binding of PtdIns(3,4,5)P$_3$ and or PtdIns(3,4)P$_2$ to its PH domain, recruiting Akt to the plasma membrane. This results in a conformational change in Akt, thereby allowing phosphorylation of Akt by upstream kinases (PDK1 and 2), on threonine 308 and serine 473, which significantly enhances Akt kinase activity (Marte and Downward, 1997). Membrane localisation and activity of PDK1 and perhaps PDK2, is also regulated by PtdIns(3,4,5)P$_3$. Activated Akt phosphorylates many substrates, resulting in a variety of biological effects including suppression of apoptosis, activation of GLUT-4 translocation, cell cycle regulation, differentiation and inactivation of glycogen synthase kinase 3 (Marte and Downward, 1997). Akt serves as a survival signal that protects cells from apoptosis. This protection may result from Akt-mediated phosphorylation of BAD. The association of overexpression of Akt in certain ovarian and pancreatic cancers may enable their prolonged survival. PtdIns(3,4)P$_2$ has also been implicated in Akt activation in platelets (Banfic *et al.*, 1998). In platelets Akt activation is inhibited by wortmannin and correlates with the delayed peak of PtdIns(3,4)P$_2$.

PATHWAY OF 3-PHOSPHORYLATED PHOSPHOINOSITIDE FORMATION IN PLATELETS

Platelets stimulated by thrombin or other thrombin receptor agonists, demonstrate a rapid increase in PtdIns(3,4,5)P$_3$ within the first 30 seconds. This is followed by a larger, delayed rise in PtdIns(3,4)P$_2$. The level of PtdIns(3)P appears to alter little on stimulation (Kucera and Rittenhouse, 1990; Sultan *et al.*, 1991). The temporal correlation of PtdIns(3,4,5)P$_3$ with early platelet activation events and PtdIns(3,4)P$_2$ with later events, suggest specific regulatory roles for these distinct 3-position phosphoinositides.

Two different pathways for the generation of PtdIns(3,4,5)P$_3$, have been proposed. These studies utilise ^{32}P incorporation into phosphoinositides in stimulated platelets, the last added phosphate group having the highest specific radioactivity. The initial report in thrombin- stimulated platelets exposed to ^{32}P for 60 minutes, suggested that PtdIns was the principal *in vivo* substrate for PI 3-kinase (Cunningham *et al.*, 1990). The proposed pathway being PtdIns $\rightarrow$ PtdIns(3)P $\rightarrow$ PtdIns(3,4)P$_2$ $\rightarrow$ PtdIns(3,4,5)P$_3$. However, later studies in thrombin-stimulated platelets exposed to ^{32}P for 10 minutes (Carter *et al.*, 1994) and in formyl peptide-stimulated neutrophils (Stephens *et al.*, 1991), suggest that the more likely pathway for the formation of PtdIns(3,4,5)P$_3$ is: PtdIns $\rightarrow$ PtdIns(4)P $\rightarrow$ PtdIns(4,5)P$_2$ $\rightarrow$ PtdIns(3,4,5)P$_3$.

Two main pathways for the generation of PtdIns(3,4)P$_2$ are proposed and may indeed each play a role in different platelet activation events. Firstly, given that in thrombin-stimulated platelets the rise in PtdIns (3,4,5)P$_3$ precedes the rise in PtdIns(3,4)P$_2$, it is possible that a 5-phosphatase hydrolyses PtdIns(3,4,5)P$_3$ to form PtdIns(3,4)P$_2$. In support of this contention, we have identified a platelet PtdIns(3,4,5)P$_3$ 5-phosphatase that complexes with the p85/p110 form of PI 3-kinase (Jackson *et al.*, 1995). Thus, PtdIns(3,4,5)P$_3$ formed by PI 3-kinase is subsequently converted to PtdIns(3,4)P$_2$ by the associated 5-phosphatase. Secondly, specific phosphorylation of PtdIns (3)P by a 4-kinase forming PtdIns(3,4)P$_2$ has been described (Yamamoto *et al.*, 1990). This putative pathway is supported by experiments in platelets activated by an antibody to the integrin $\alpha_{IIb}\beta_3$ receptor binding site that show the 4-position phosphate has a higher specific activity than the 3-position phosphate, indicating that the 4-position phosphate is added last (Banfic *et al.*, 1998). In these $\alpha_{IIb}\beta_3$ antibody stimulated platelets, an early rise in PtdIns(3)P is followed by a later peak of PtdIns(3,4)P$_2$. No increase in PtdIns(3,4,5)P$_3$ is observed. Furthermore, the later rise in PtdIns(3,4)P$_2$ may also be contributed to by an observed decrease in the activity of PtdIns(3,4)P$_2$ 4-phosphatase, due to thrombin-stimulated calpain-mediated cleavage of the 4-phosphatase (Norris *et al.*, 1997).

Therefore, several 3-phosphoinositide pathways appear to be triggered by platelet activation events, probably involving distinct classes of PI 3-kinases with differing substrate specificities. In early thrombin-triggered platelet activation, PtdIns(3,4,5)P$_3$ is formed and may represent the critical signalling species. Its metabolism by the 5-phosphatase possibly representing a signal terminating event producing "inactive" PtdIns(3,4)P$_2$. Whereas in later, particularly post-$\alpha_{IIb}\beta_3$ activation events, PtdIns(3,4)P$_2$ predominates and is probably the important signalling species. The later PtdIns(3,4)P$_2$ peak resulting from both the 5-phosphatase acting on early formed PtdIns(3,4,5)P$_3$ and sequential action of PI 3-kinase on PtdIns followed by PI 4-kinase acting on PtdIns(3)P.

PHOSPHOINOSITIDE 3-KINASE ACTIVATION IN PLATELETS

The PI 3-kinase is activated via a number of mechanisms: these include via G-proteins, by PKC, tyrosine phosphorylation, and by translocation of the enzyme to the cytoskeleton. Thrombin triggers the release of G-proteins which stimulate PI 3-kinase activity. In permeabilised platelets non-hydrolysable GTPγS results in more profound PI 3-kinase activation than that observed with thrombin alone (Kucera and Rittenhouse, 1990). Both Rho, a small GTP-binding protein, and heterotrimeric G proteins mediate this effect (Zhang *et al.*, 1995a). The $\beta\gamma$ subunits of dissociated heterotrimeric G-proteins have been shown to stimulate PI 3-kinase in platelets, an effect blocked by the addition of GDP-bound α subunits (Thomason *et al.*, 1994). In platelets, it has been shown that the p85/p110 form of PI 3-kinase is activated by Rho (Zhang *et al.*, 1993; Zhang *et al* 1995a) and p110γ activated by G-protein $\beta\gamma$ subunits (Stoyanov *et al.*, 1995; Zhang *et al.*, 1995a), both forms being initially activated by thrombin via GTP release.

Low molecular weight G-proteins such as the small GTP-binding, Ras-related protein Rho, are implicated in PI 3-kinase activation (Zhang *et al.*, 1993). ADP-ribosylation of Rho by *Clostridium botulinum* C3 ADP-ribosyltransferase prevents the GTPγS induced increase in PI 3-kinase activity and platelet aggregation. The inhibition of PI 3-kinase activity is overcome by the addition of recombinant Rho, but not by the related small GTP-binding protein Rac. Rho alone does not stimulate PI 3-kinase activity, it does so only in the presence of added GTPγS, indicating that it is GTP-bound Rho which stimulates the PI 3-kinase activity. Rho promotes the formation of stress fibres and focal adhesions in fibroblasts, an effect prevented by prior-ADP ribosylation of Rho (Ridley and Hall, 1992). A similar situation may occur in thrombin-stimulated platelets: GTP-Rho translocates to the cytoskeleton along with PI 3-kinase, directly or indirectly activating this lipid kinase and thereby regulating the formation of focal adhesions involving actin fibres and the $\alpha_{IIb}\beta_3$ integrin.

Several lines of evidence suggest protein kinase C (PKC) may also be responsible for PI 3-kinase activation. Firstly, phorbol esters which directly stimulate PKC result in an increase in 3-position phosphoinositides, however this increase is markedly less than that observed following thrombin or GTPγS treatment of platelets (Kucera and Rittenhouse, 1990; Yatomi *et al.*, 1994). Secondly, inhibition of PKC by the pseudosubstrate peptide RFARK inhibits thrombin-induced PtdIns(3,4)P$_2$ and PtdIns(3,4,5)P$_3$ accumulation by about 70 percent (King *et al.*, 1991). RFARK does not directly inhibit highly purified PI 3-kinase, therefore its affect is probably indirect via inhibition of PKC. Conversely, inhibition of serine-threonine phosphatases by okadaic acid, (which would partially simulate activation of PKC serine-threonine phosphorylation), promotes formation of these 3-position phosphoinositides. Therefore, PKC is most likely responsible for a significant proportion, but not all of the observed increase in these phospholipids, generated in response to thrombin and GTPγS. Unlike thrombin stimulation which activates both forms of PI 3-kinase, phorbol ester stimulation of PKC stimulates only the p85/p110 form, not p110γ (Zhang *et al.*, 1996) explaining the lower level of generated 3-phosphoinositides.

The activation of PI 3-kinase by tyrosine phosphorylation has been well described in multiple cell signalling pathways (reviewed in Fry, 1994). Stimulation of receptor and non-receptor tyrosine kinases, such as the PDGF receptor, results in phosphorylation of specific tyrosine residues in their cytoplasmic domains. PI 3-kinase binds to certain of these phosphotyrosine residues and is activated. Similarly binding of the SH2 domains of the p85 subunit, to synthetic phosphotyrosine peptides encompassing these regions, results in 2–3 fold increase in associated PI 3-kinase activity (Carpenter *et al.*, 1993b). Binding of both SH2 domains of p85 to phosphotyrosines, results in greater activation of PI 3-kinase activity (Rordorf-Nikolic *et al.*, 1995).

In platelets, the contribution of tyrosine phosphorylation to the activation of PI 3-kinase is less well defined than the role of G-proteins. Thrombin stimulation triggers sequential tyrosine phosphorylation of multiple platelet proteins (Golden and Brugge, 1989), much of which is dependent on $\alpha_{IIb}\beta_3$ ligand binding and platelet aggregation (reviewed in Clark *et al.*, 1994). Inhibition of protein kinase activity by inhibitors such as tyrophostin AG-213 or staurosporine, results in significant inhibition of platelet aggregation and 5-hydroxytryptamine release, which parallels the inhibition of PtdIns(3,4)P$_2$ production. However, the early small peak of PtdIns(3,4,5)P$_3$ is unaffected (Guinebault *et al.*, 1993; Yatomi *et al.*, 1992a and 1994). To what extent the decrease in PtdIns(3,4)P$_2$, is directly due to inhibition of tyrosine phosphorylation is unclear, as PKC (Yatomi *et al.*, 1994) and cellular events such as aggregation are also inhibited. In some studies (Guinebault *et al.*, 1993), but not others (Yatomi *et al.*, 1994) some p85, is recovered from antiphosphotyrosine immunoprecipitates from thrombin-stimulated, but not resting platelets. Staurosporine also decreases the amount of p85 found in antiphosphotyrosine immunoprecipitates after thrombin stimulation. Direct phosphorylation of the p85 subunit is clearly not the major mechanism of PI 3-kinase activation, as antiphosphotyrosine antibodies fail to detect phosphorylation of the p85 upon immunoblot analysis (Guinebault *et al.*, 1993; 1995; Yatomi *et al.*, 1994). It is therefore, possible that the p85 subunit of PI 3-kinase, associates with other tyrosine phosphorylated proteins following thrombin stimulation. The most abundant tyrosine kinase in platelets is pp60[c-src] which is activated initially following thrombin stimulation, and translocates to the cytoskeleton. In platelets only a small amount of PI 3-kinase has been shown to associate with pp60[c-src] following thrombin stimulation (Gutkind *et al.*, 1990).

PHOSPHOINOSITIDE 3-KINASE ACTIVATION BY OTHER PLATELET RECEPTORS

PI 3-kinase activity is also stimulated by binding of von Willebrand factor (vWF) to the glycoprotein Ib/IX receptor, an effect which is abolished in Bernard-Soulier platelets which lack this receptor (Jackson *et al.*, 1994). Bernard-Soulier platelets have normal increases in PtdIns$(3,4)$P$_2$ in response to thrombin. vWF results in translocation and activation of PI 3-kinase in a manner similar to thrombin. However, it is not inhibited by RGDS, indicating that vWF-stimulated increase in 3-position phosphoinositides occurs independently of thrombin-associated fibrinogen binding to $\alpha_{IIb}\beta_3$. Therefore, PI 3-kinase may function in the pathway resulting in platelet shape change and aggregation which follows binding of Ib/IX to vWF.

Other platelet receptors, independent of $\alpha_{IIb}\beta_3$ signalling, may be involved in transducing extracellular signals via the PI 3-kinase. Concanavalin A, which aggregates glycoprotein receptors is able to produce elevations of PtdIns$(3,4)$P$_2$ equivalent to those observed in thrombin-induced platelet aggregation, even in the absence of stirring and thus platelet aggregation, and also in thrombasthenic platelets lacking the $\alpha_{IIb}\beta_3$ receptor (Torti *et al.*, 1995). This suggests novel integrin and aggregation-independent pathways for the activation of PI 3-kinase, however, the *in vivo* correlates of this *in vitro* observation are unknown.

Stimulation of the platelet immunoglobulin Fc receptor, FcγRII, by an anti-CD9 monoclonal antibody also results in a delayed increase in PtdIns$(3,4)$P$_2$, which is dependent on tyrosine kinase activity (Yatomi *et al.*, 1993), again the functional significance of this activation is unclear. Upon activation and aggregation of the FcγRII receptor the cytoplasmic domain is tyrosine phosphorylated promoting binding of the p85 subunit of the PI 3-kinase by an adaptor molecule, probably p72[syk] (Chacko *et al.*, 1996).

TRANSLOCATION TO CYTOSKELETON

Recruitment of PI 3-kinase from the cytosol to the cytoskeletal fraction, moves the enzyme into closer proximity with its phospholipid substrates located in the membrane. In unstimulated platelets, PI 3-kinase, as assessed by enzyme activity and p85 in Western blots, is predominantly located in the Triton-soluble cytosol fraction (Zhang *et al.*, 1992). Upon thrombin stimulation, PI 3-kinase translocates to the Triton-insoluble cytoskeletal fraction (Grondin *et al.*, 1991). The amount of p85 in the cytosol decreases from 94 percent in resting platelets, to 53 percent in thrombin-stimulated platelets. The PI 3-kinase associated with cytoskeleton is activated, having a specific lipid kinase activity up to10-fold greater than the cytosolic enzyme (Zhang *et al.*, 1992). The greater specific activity of cytoskeletal PI 3-kinase is not dependent on thrombin activation. The cytosolic and cytoskeletal associated enzymes each have their own specific activity per p85 subunit, in resting and thrombin stimulated platelets. The total activity is increased with thrombin stimulation, because more of the PI 3-kinase is associated with the cytoskeleton.

We have observed that a PtdIns(3,4,5)P_3 5-phosphatase which complexes with the p85 subunit in the cytosol of resting platelets, dissociates upon thrombin stimulation when the p85/p110 form of PI 3-kinase is translocated to the cytoskeleton (Jackson *et al.*, 1995). Thus, in resting platelets any PtdIns(3,4,5)P_3 formed would be rapidly converted by the associated 5-phosphatase. Whereas upon platelet activation, dissociation of the 5-phosphatase and translocation of PI 3-kinase to the cytoskeleton, may allow PtdIns(3,4,5)P_3 to be formed and not immediately metabolised to PtdIns(3,4)P_2.

Translocation of p85 and PI 3-kinase activity correlates with the delayed production of PtdIns(3,4)P_2, rather than the early peak of PtdIns(3,4,5)P_3 (Guinebault *et al.*, 1995). Translocation of PI 3-kinase and production of PtdIns(3,4)P_2 is dependent on tyrosine phosphorylation, $\alpha_{IIb}\beta_3$ binding fibrinogen and platelet aggregation. However, translocation of PI 3-kinase has also been observed as an early event following thrombin stimulation. p72[syk], a src-related tyrosine kinase, associates with the p85 subunit of PI 3-kinase and they translocate to the cytoskeleton within 10 seconds of thrombin stimulation (Yanagi *et al.*, 1994). Perhaps a subset of PI 3-kinase translocates to the cytoskeleton in association with distinct signalling proteins, such as p72[syk], and is responsible for the early small peak of PtdIns(3,4,5)P_3. The majority of PI 3-kinase translocation is delayed and correlates with the late, larger rise of PtdIns(3,4)P_2.

ADP stimulation of platelets, results in reversible platelet aggregation. In ADP-stimulated platelets, the p85/p110 PI 3-kinase translocates reversibly to the cytoskeleton, in an amount equivalent to that seen upon thrombin stimulation, however, this is not associated with a rise in PtdIns(3,4)P_2 or PtdIns(3,4,5)P_3 (Gachet *et al.*, 1997). Unlike in thrombin-stimulated platelets, ADP stimulation fails to translocate Rho to the cytoskeletal fraction. This suggests that the enhanced production of 3-phosphoinositides in thrombin-stimulated platelets is dependent on the co-translocation of PI 3-kinase activators such as Rho and that such events are necessary to induce irreversible platelet aggregation.

The exact mechanism via which PI 3-kinase associates with the cytoskeletal fraction upon platelet activation is currently being elucidated. Association via one or more of the signalling domains of the regulatory p85 subunit with other signalling proteins, translocated to, or activated in the cytoskeleton is an attractive proposition. The interaction of the p85 subunit with the signalling domains of the other proteins results in activation of PI 3-kinase activity. This has been shown to occur via either binding of the SH2 domains to phosphotyrosine residues (Carpenter *et al.*, 1993b) or by binding of the proline rich domain (residues 84 to 99) to the SH3 domains of src-related kinases (Pleiman *et al.*, 1994).

Translocation of focal adhesion kinase (pp125[FAK]), p72[syk] and pp60[c-src] correlates with PI 3-kinase translocation (Grondin *et al.*, 1991; Zhang *et al.*, 1992; Yanagi *et al.*, 1994; Guinebault *et al.*, 1995). Rho, which stimulates the p85/p110 isoform, also translocates to the cytoskeleton after thrombin stimulation. (Zhang *et al.*, 1993). PI 3-kinase probably associates with several of these proteins. There is convincing evidence that PI 3-kinase associates with pp125[FAK]. pp125[FAK] immunoprecipitates from thrombin-stimulated, but not resting platelets, contain the p85 subunit of PI 3-kinase (Guinebault *et al.*, 1995). Thus, thrombin stimulation results in the association of the p85 subunit and pp125[FAK], which are both translocated to the cytoskeleton. Although pp125[FAK] is tyrosine phosphorylated by thrombin stimulation, this interaction does not seem to occur via the SH2 domain of p85 and a phosphorylated tyrosine residue of pp125^{FAK}. Rather the SH3 domain of p85 binds a proline-rich domain of pp125^{FAK}. This interaction, confirmed using a GST-fusion protein of the SH3 domain of p85 and a synthetic peptide of the proline-rich domain of pp125^{FAK}(Guinebault *et al.*, 1995), results in a 2.5 fold increase in PI 3-kinase activity.

The interaction between p85 and pp125^{FAK} seems to be direct, specific and in part explain the mechanism via which PI 3-kinase is translocated to the cytoskeleton and activated. p125^{FAK}, is a kinase involved in the interaction of integrins and the actin cytoskeleton. In platelets, the interaction between PI 3-kinase and pp125^{FAK}, may localise the generation of PtdIns(3,4)P$_2$ and PtdIns(3,4,5)P$_3$ to focal adhesion sites, potentially regulating the activation and interaction of $\alpha_{IIb}\beta_3$, with downstream signalling proteins, critical for platelet aggregation.

The p110γ form of PI 3-kinase also translocates to the cytoskeleton of thrombin-stimulated platelets (Zhang *et al.*, 1995a), however, the mechanism for this association is unclear. One potential interaction is via $\beta\gamma$ subunits which can bind p110γ and also the pleckstrin homology (PH) domains of other cytoskeletal proteins. Interestingly, the pleckstrin homology domains of β-adrenergic kinase (βARK) bind $\beta\gamma$ subunits and inhibit the PI 3-kinase activity of p110γ in response to thrombin. Platelets contain both βARK1 and βARK2, which possibly bind $\beta\gamma$ subunits to regulate the activation of p110γ PI 3-kinase activity. Potential roles of the differently regulated forms of PI 3-kinase in platelet signalling and function are discussed below.

FUNCTION OF THE PI 3-KINASE AND 3-PHOSPHOINOSITIDES IN PLATELETS

Pleckstrin phosphorylation and integrin $\alpha_{IIb}\beta_3$ activation have been identified as events mediated by PtdIns(3,4)P$_2$ and PtdIns(3,4,5)P$_3$ in platelets. Wortmannin is a fungal metabolite, which specifically inhibits PI 3-kinase at nanomolar concentrations (reviewed in Ui *et al.*, 1995). At higher concentrations, this inhibition may be less specific. Wortmannin has been used to determine the importance of the PI 3-kinase in platelet function. Interestingly, wortmannin administered to animals results in a bleeding diathesis, it is unknown whether this bleeding is the result of platelet dysfunction (Abbas *et al.*, 1989).

PLECKSTRIN PHOSPHORYLATION

Pleckstrin is the major phosphorylated protein in platelet cytosol following thrombin stimulation. Wortmannin inhibits thrombin and phorbol ester-stimulated phosphorylation of pleckstrin, as well as inhibiting platelet aggregation (Yatomi *et al.*, 1992b). Thrombin and GTPγS-induced pleckstrin phosphorylation are protected from the inhibitory effect of wortmannin by the addition of PtdIns(3,4,5)P$_3$ (Zhang *et al.*, 1995b). Pleckstrin phosphorylation occurs when PtdIns(3,4,5)P$_3$, but not PtdIns(4,5)P$_2$ or Ins(1,3,4,5)P$_4$, is added to permeabilised platelets. It is possible that this stimulatory effect on pleckstrin phosphorylation is due to PtdIns(3,4,5)P$_3$ itself, or to PtdIns(3,4)P$_2$ formed subsequent to the action of a platelet PtdIns(3,4,5)P$_3$ 5-phosphatase. However, the time course of pleckstrin phosphorylation following thrombin stimulation correlates better with the early production of PtdIns(3,4,5)P$_3$. The inhibitory effect of wortmannin on pleckstrin phosphorylation is likely due to inhibition of PI 3-kinase production of PtdIns(3,4,5)P$_3$, rather than inhibition of the intrinsic protein kinase activity of the p110 subunit, because it is overcome by the addition of PtdIns(3,4,5)P$_3$. The PtdIns(3,4,5)P$_3$-mediated phosphorylation of pleckstrin occurs on the same serine and threonine residues as by PKC. It is likely that PtdIns(3,4,5)P$_3$ acts to stimulate a platelet PKC isoform resulting in phosphorylation of pleckstrin.

The exact role pleckstrin plays in platelet function is unknown. Given its abundance in platelets and its striking phosphorylation upon thrombin stimulation, it is likely to perform an important role in the signalling events mediating platelet activation. PKC phosphorylation of pleckstrin can be stimulated by either by the classic PLC-diacylglycerol mechanism or PI 3-kinase generated PtdIns(3,4,5)P$_3$. Different PKC isoforms are probably involved. Whether these two different mechanisms reflect redundancy, or allow finer tuning of regulation in terms of duration or extent of pleckstrin phosphorylation is unknown.

INTEGRIN $\alpha_{IIb}\beta_3$ SIGNALLING

PI 3-kinase has been implicated in two main aspects of $\alpha_{IIb}\beta_3$ function: events following thrombin stimulation which result in integrin activation and events following fibrinogen binding to $\alpha_{IIb}\beta_3$ which lead to platelet aggregation. It has been demonstrated that the p85/p110 form of PI 3-kinase, rather than p110γ, is responsible for the thrombin-stimulated conformational activation of $\alpha_{IIb}\beta_3$ which results in fibrinogen binding (Zhang *et al.*, 1996). This conclusion was derived from studies utilising the differential sensitivity of PI 3-kinase isoforms to wortmannin. Wortmannin inhibits p85/p110 associated PI 3-kinase activity at lower concentrations than p110γ (Stephens, 1994). In phorbol ester-stimulated platelets, where direct PKC activation stimulates only the p85/110 isoform, wortmannin inhibition of $\alpha_{IIb}\beta_3$ activation correlates with the inhibition of the production of 3-position phosphoinositides (Zhang *et al.*, 1996). In contrast, in thrombin-treated platelets, low concentrations of wortmannin inhibit $\alpha_{IIb}\beta_3$ activation by 80 percent, even though PtdIns(3,4,5)P$_3$ production declines only 28 percent. In this case, thrombin stimulation of p110γ which remains relatively unaffected by the low concentration of wortmannin, stimulates 3-phosphoinositide production but not $\alpha_{IIb}\beta_3$ activation. In contrast, the thrombin stimulation of p85/110 is markedly inhibited by wortmannin, concomitant with inhibition of $\alpha_{IIB}\beta_3$ activation. Therefore, the p110/85 form of PI 3-kinase, not p110γ, regulates the "inside-out" regulation of platelet integrin function, which is necessary for the transduction of an activation signal to $\alpha_{IIb}\beta_3$. The reason for this functional difference between these two forms of PI 3-kinase is unknown.

PI 3-kinase signalling has also been implicated in regulating events that follow binding of fibrinogen to activated $\alpha_{IIb}\beta_3$. Inhibition of fibrinogen binding to $\alpha_{IIb}\beta_3$ by the RGDS tetrapeptide inhibits platelet aggregation. RGDS also results in a significant 40% decrease in the delayed rise of PtdIns(3,4)P$_2$ formation following thrombin stimulation (Sultan *et al.*, 1991), but not the earlier peak of PtdIns(3,4,5)P$_3$ (Sorisky *et al.*, 1992). Indeed, in thrombasthenic platelets, lacking the $\alpha_{IIb}\beta_3$ receptor, the increase in PtdIns(3,4)P$_2$ following thrombin stimulation is negligible (Sultan *et al.*, 1991). These results suggest that activated integrin $\alpha_{IIb}\beta_3$ binding of fibrinogen triggers at least part of the thrombin-induced increase in PtdIns(3,4)P$_2$, but not the earlier rise in PtdIns(3,4,5)P$_3$. However, because platelet aggregation was also inhibited, the observed decrease in PtdIns(3,4)P$_2$ may be a consequence of inhibition of aggregation, rather than inhibition of fibrinogen binding. In support of this contention, the thrombin-triggered increase in PtdIns(3,4)P$_2$ decreases in unstirred platelets which are unable to aggregate (Torti *et al.*, 1995). The delayed rise in PtdIns(3,4)P$_2$, unlike the early PtdIns(3,4,5)P3 peak, is calcium dependent (Sorisky *et al.*, 1992). RGDS inhibits the calcium increase in stimulated platelets, and this may be the mechanism by which the peptide inhibits the PtdIns(3,4)P$_2$ increase (Yamaguchi *et al.*, 1987). This effect may be mediated by calpain, a calcium-dependent protease, which either inactivates PtdIns(3,4)P$_2$ 4-phosphatase or increases PtdIns(3,4)P$_2$ production (Norris *et al.*, 1997; Banfic *et al.*, 1998). Calpeptin, an inhibitor of calpain, inhibits the delayed peak of PtdIns(3,4)P$_2$. Similarly fibrinogen binding to $\alpha_{IIB}\beta_3$, triggers tyrosine phosphorylation, which is implicated in the regulation of PI 3-kinase activity.

PI 3-kinase, therefore may be involved in regulating events that occur after fibrinogen binding to $\alpha_{IIb}\beta_3$, so-called post-occupancy events. Indeed, monoclonal antibodies which directly bind to and activate $\alpha_{IIb}\beta_3$, result in an increase in PtdIns(3,4)P$_2$ and platelet aggregation, wortmannin inhibits both of these effects (Kovacsovics *et al.*, 1995; Banfic *et al*, 1998). $\alpha_{IIb}\beta_3$ must be activated by an agonist such as thrombin, before it can bind soluble fibrinogen. However, it is able to adhere to an immobilized fibrinogen matrix without prior activation. Platelet adhesion and spreading on a fibrinogen matrix, results in a significant increase in PtdIns(3,4)P$_2$ which is dependent on ADP (Gironcel *et al.*, 1996). Again demonstrating that PI 3-kinase is activated by events downstream of fibrinogen binding to $\alpha_{IIb}\beta_3$.

The role of PI 3-kinase in platelet activation may be to maintain $\alpha_{IIB}\beta_3$ in its active conformation to ensure irreversible platelet aggregation (Kovacsovics *et al.*, 1995). In thrombin-stimulated, stirred platelets, aggregation is irreversible. However, wortmannin and LY294002, another PI 3-kinase inhibitor, influence platelet aggregation in a distinct manner. The initial slope of primary aggregation remains relatively unaffected, although at very high doses of wortmannin, this is also abolished. After primary aggregation, PI 3-kinase inhibitors, cause progressive platelet dissaggregation, in a dose-dependent manner. There is also a decrease in fibrinogen binding capacity with time, after thrombin stimulation in wortmannin-treated platelets. PI 3-kinase inhibition does not decrease the total number of surface $\alpha_{IIb}\beta_3$ receptors, but rather leads to a progressive decrease in those "active" receptors able to bind fibrinogen and mediate aggregation events. These results suggest that PI 3-kinase is less required for initial activation of $\alpha_{IIb}\beta_3$, binding of fibrinogen and primary aggregation but is necessary for maintenance of the activated conformation of $\alpha_{IIb}\beta_3$, which when impaired results in disaggregation.

ROLE OF PHOSPHOINOSITIDE 3-KINASE IN OTHER PLATELET FUNCTIONS

In basophils, wortmannin inhibits secretion of histamine (Yano *et al.*, 1993). A role for PI 3-kinase in secretion, is also suggested by inhibition of the insulin-regulatable glucose transporter (type 4) to the plasma membrane, both by wortmannin and dominant-negative p85 (Cheatham *et al.*, 1994). However in platelets a role in secretion has not been demonstrated. Wortmannin does not inhibit dense granule secretion, as measured by serotonin release, nor of α-granule release, as evidenced by normal P-selectin and total $\alpha_{IIb}\beta_3$ upregulation in response to thrombin (Kovacsovics *et al.*, 1995).

Cytoskeletal changes are dependent on PI 3-kinase in some cell types. Membrane ruffling and chemotaxis is prevented in fibroblasts with mutant PDGF receptors unable to bind and activate PI 3-kinase (Wennstrom *et al.*, 1994; Kundra *et al.*, 1994). In platelets however, there is no direct evidence of PI 3-kinase mediated regulation of cytoskeletal rearrangement. Wortmannin blocks thrombin-stimulated PtdIns(3,4,)P$_2$ and PtdIns(3,4,5)P$_3$ increase but has no effect on actin assembly, as measured by generation of filamentous actin and increase in actin barbed-end nucleation sites (Kovacsovics *et al.*, 1995). This is surprising given that addition of PtdIns(3,4)P$_2$ or PtdIns(3,4,5)P$_3$ to permeabilised resting platelets, is as effective as PtdIns(4,5)P$_2$ in uncapping F-actin, allowing addition of actin monomers to the barbed ends of F-actin (Hartwig *et al.*, 1995). The cellular concentrations of PtdIns(3,4)P$_2$ and PtdIns(3,4,5)P$_3$, even after thrombin-stimulation, are insignificant compared to the much more abundant PtdIns(4,5)P$_2$, which therefore may be the critical mediator of actin uncapping. Although PtdIns(3,4)P$_2$ and PtdIns(3,4,5)P$_3$ added to per-meabilised platelets uncap F-actin, they are unlikely to be important physiological actin regulators, given that complete inhibition of these 3-position phosphoinositides causes no disturbance of actin assembly. Platelet shape changes in response to thrombin are also unaffected by PI 3-kinase inhibition. Although PI 3-kinase is enriched in the F-actin rich fraction following thrombin stimulation (Guinebault *et al.*, 1995), the translocation of PI 3-kinase is not dependent on actin polymerisation. Translocation of p85 lags behind actin polymerisation, only becoming significant after 30 seconds of thrombin stimulation. Cyto-chalasin D which inhibits actin polymerisation, does not reduce the amount of p85 or PI3-kinase activity translocating to the cytoskeleton (Zhang *et al.*, 1992). Therefore, in platelets PI 3-kinase is neither a significant regulator of, nor directly regulated by cytoskeletal changes.

CONCLUSION

Although the platelet is a terminally differentiated cell it maintains many of the phosphoi-nositide-derived signalling pathways, utilised by both transformed and normal cells to mediate such diverse processes as cell division, secretion, vacuolar sorting, cytoskeletal rearrangement and contraction. It is clear phosphoinositides and the second messengers derived from their turnover, play an important role in many aspects of platelet physiology. The challenge for the future is to identify disease states and platelet abnormalities originat-ing from disordered phosphoinositide signalling.

References

Abbas, H.K., Mirocha, C.J. and Gunther, R. (1989). Mycotoxins produced by toxic Fusarium isolates obtained from agricultural and nonagricultural areas (Arctic) of Norway. *Mycopathologia*, **105**:143–151.

Attree, O., Olivos, I., Okabe, L.C., Bailey, D.L., Nelson, R.A., Lewis, R.R., McInnes, R.R. and Nussbaum, R.L. (1992). The Lowes oculocerebrorenal syndrome gene encodes a protein highly homologous to inositol polyphosphate 5-phosphatase. *Nature*, **358**:239–242.

Banfic, H., Downes, C.P. and Rittenhouse, S.E. (1998). Biphasic activation of PKB-α/Akt in platelets — evidence for stimulation both by phosphatidylinositol 3,4-bisphosphate, produced via a novel pathway, and by phos-phatidylinositol 3,4,5-trisphosphate. *The Journal of Biological Chemistry*, **273**:11630–11637.

Banno, Y., Nakashima, S., Hachiya, T. and Nozawa, Y. (1995). Endogenous cleavage of phospholipase β_3 by ag-onist induced activation of calpain in human platelets. *The Journal of Biological Chemistry*, **270**:4318–4324.

Banno Y., Nakashima, S, Ohzawa, M and Nozawa, Y (1996). Differential translocation of phospholipase C iso-enzyme to integrin-mediated cytoskeletal complexes in thrombin stimulated human platelets. *The Journal of Biological Chemistry*, **271**:14989–14994.

Berridge, M.J. (1993). Inositol trisphosphate and calcium signalling. *Nature*, **361**:315–325.

Blake R.A, Schive, G.L. and Watson, S.P (1994). Collagen stimulates tyrosine phosphorylation of phospholipase C γ_2 but not phospholipase γ_1 in human platelets. *FEBS Letters*, **353**:212–216.

Blockmans, D., Deckmyn, H. and Vermylen, J. (1995). Platelet Activation. *Blood Reviews*, **9**:143–156.

Bourguignon, L.Y.W., Brandt, N.R. and Zhang, S.H. (1993). The involvement of ankyrin in the regulation of inositol 1,4,5 trisphosphate receptor-mediated internal calcium release from calcium storage vesicles in mouse T-lymphoma cells. *The Journal of Biological Chemistry*, **268**:7290–7297.

Burgering, B.M. and Coffer, P.J. (1995). Protein kinase B (c-Akt) in phosphatidylinositol-3-OH kinase signal transduction. *Nature*, **376**:599–602.

Burgess, G.M., Bird, G.S., Obie, J.F. and Putney J.W., Jr. (1991). The mechanism for synergism between phospholipase C- and adenylylcyclase-linked hormones in liver. Cyclic AMP-dependent kinase augments inositol trisphosphate-mediated Ca^{2+} mobilization without increasing the cellular levels of inositol polyphosphates. *The Journal of Biological Chemistry*, **266**:4772–4781.

Carpenter, C.L., Suger, K.R., Duckworth, B.C., Hou, W.M., Schaffhausen, B. and Cantley, L.C. (1993a). A tightly associated serine threonine protein kinase regulates phosphoinositide 3-kinase. *Molecular Cellular Biology*, **13**:1657–1665.

Carpenter. C.L., Auger, K.R., Chanudhuri, M., Yoakim, M., Schaffhausen, B., Shoelson, S. and Cantley, L.C. (1993b). Phosphoinositide 3-kinase is activated by phosphopeptides that bind to the SH2 domains of the 85-kd subunit. *The Journal of Biological Chemistry*, **268**:9478–9483.

Carpenter, C.L. and Cantley, L.C. (1996a). Phosphoinositide kinases. *Current Opinion in Cell Biology*, **8**:153–158.

Carpenter, C.L. and Cantley, L.C. (1996b). Phosphoinositide 3-kinase and the regulation of cell growth. *Biochimica et Biophysica Acta*, **1288**:M11-M16.

Carter, A.N., Huang, R., Sorisky, A., Downes, C.P. and Rittenhouse, S.E. (1994). Phosphatidylinositol 3,4,5-trisphosphate is formed from phosphatidylinositol 4,5-bisphosphate in thrombin-stimulated platelets. *Biochemical Journal*, **301**:415–420.

Cavallini, L., Coassin, M., Boreau, A. and Alexandre, A (1996). Prostacyclin and sodium nitroprusside inhibit the activity of platelet inositol 1,4,5-trisphosphate receptor and promote its phosphorylation. *Journal of Biological Chemistry*, **271**:5545–5551.

Chacko, G.W., Brandy, J.T., Coggeshall, K.M. and Anderson, C.L. (1996). Phosphoinositide 3-kinase and p72[syk] noncovalently associate with the low affinity Fcr receptor on human platelets through an immunoreceptor tyrosine-based activation motif. *Journal of Biological Chemistry*, **271**:10775–10781.

Cheatham, B., Vlahos, C.J., Cheatham, L., Wang, L., Blenis, J. and Kahn, C.R. (1994). Phosphatidylinositol 3-kinase activation is required for insulin stimulation of pp70 S6 kinase DNA synthesis, and glucose transporter translocation. *Molecular Cellular Biology*, **14**:4902–4911.

Chung, J., Grammer, T.C., Lemon, K.P., Kazlauskas A. and Blenis, J. (1994). PDGF- and insulin-dependent pp70[S6k] activation mediated by phosphatidylinositol-3-OH kinase. *Nature*, **370**:71–75.

Clark, E.A., Shattil, S.J. and Brugge, J.S. (1994). Regulation of protein tyrosine kinases in platelets. *Trends in Biochemical Sciences*, **19**:464-469.

Coggeshall, K.M. (1998). Inhibitory signalling by B-cell Fcr RIIB. *Current Opinion in Immunology*, **10**:306–312.

Communi, D., Lecocq, R. and Emeux, C. (1996). Arginine 343 and 350 are two active site residues involved in substrate binding by human Type I D-myo-inositol 1,4,5-trisphosphate 5-phosphatase. *The Journal of Biological Chemistry*, **271**:11676–11683.

Connolly, T.M., Bross, T.E. and Majerus, P.W. (1985). Isolation of a phosphomonoesterase from human platelets that specifically hydrolyzes the 5-phosphate of inositol 1,4,5-trisphosphate. *The Journal of Biological Chemistry*, **260**:7868–7874.

Cunningham, T.W. Lips, D.L., Bansal, V.S., Caldwell, K.K., Mitchell, C.A. and Majerus, P.W. (1990). Pathway for the formation of D-3 phosphate containing inositol phospholipids in intact human platelets. *The Journal of Biological Chemistry*, **265**:21676–21683.

Damen, J.E., Liu, L., Rosten, P., Humphries, R.K., Jefferson, A.B., Majerus, P.W. and Krystal, G (1996). The 145-kDa protein induced to associate with Shc by multiple cytokines is an inositol tetraphosphate and phosphatidylinositol 3,4,5-trisphosphate 5-phosphatase. *Procedings of the National Academy of Sciences, USA*, **93**:1689–1693.

De Camilli, P., Emr, S.D., McPherson, PS. and Novick, P. (1996). Phosphoinositides as regulators in membrane traffic. *Science*, **271**:1533–1539.

De Smedt, F., Boom, A., Pesesse, X., Schiffmann, S.N. and Erneux, C. (1996). Post-translational modification of human brain Type I inositol-1,4,5 trisphosphate 5-phosphatase by. *The Journal of Biological Chemistry*, **271**:10419–10424.

De Smedt, F., Missian, L., Parys, J.B., Vanweyenbeg, V., De Smedt, H. and Erneux, C. (1997). Isoprenylated human brain Type I inositol 1,4,5-trisphosphate 5-phosphatase controls calcium oscillations induced by ATP in chinese hamster ovary cells. *Journal of Biological Chemistry*, **272**:367–17375.

Dhand, R., Hiles, I., Panayotou, G., Roche, S., Fry, M.J., Gout, I., Totty, N.F., Truong, O., Vicendo, P., Yonezawa, K., Kasuga, M., Courtneidge, S.A. and Waterfield, M.D. (1994). PI 3-kinase is a dual specificity enzyme: autoregulation by an intrinsic protein-serine kinase activity. *The EMBO Journal*, **13**:522–533.

Escobedo, J.A., Navankasattusas, S., Kavanaugh, W.M., Millay, D., Fried, V.A. and Williams, L.T. (1991). cDNA cloning of a novel 85 kDa protein that has SH2 domains and regulates binding of PI3-kinase to the PDGF β-receptor. *Cell*, **65**:75–82.

Ettinger, S.L., Lauener, R.W. and Duronio, V. (1996). Protein kinase C δ specifically associates with phosphatidylinositol 3-kinase following cytokine stimulation. *The Journal of Biological Chemistry*, **271**:14514–14518.

Fry, M.J. (1994). Structure, regulation and function of phosphoinositide 3-kinases. *Biochimica et Biophysica Acta*, **1226**:237–268.

Furuichi, T. and Mikoshiba, K. (1995). Inositol 1,4,5-trisphosphate receptor-mediated Ca^{2+} signaling in the brain. *Journal of Neurochemistry*, **64**:953–960.

Gachet, C., Payrastre, B., Guinebault, C., Trumel, C., Ohlmann, P., Mauco, G., Cazenave, J., Plantavid, M. and Chap, H. (1997). Reversible translocation of phophoinositide 3-kinase to the cytoskeleton of ADP-aggregated human platelets occurs independently of Rho A and without synthesis of phosphatidylinositol (3,4)-bisphosphate. *The Journal of Biological Chemistry*, **272**:4850–4854.

Gironcel, D., Racaud-Sultan, C., Payrastre, B., Haricot, M., Borchert, G., Kieffer, N., Breton, M. and Chap, H. (1996). $\alpha_{IIb}\beta_3$-integrin mediated adhesion of human platelets to a fibrinogen matrix triggers phospholipase C activation and phosphatidylinositol 3,4-bisphosphate accumulation. *FEBS Letters*, **389**:253–256.

Giuriato, S., Payrastre, B., Drayer, A., Plantavid, M., Woscholski, R., Parker, P., Erneux, C. and Chap, H. (1997). Tyrosine phosphorylation and relocation of SHIP are integrin-mediated in thrombin-stimulated human blood platelets. *The Journal of Biological Chemistry*, **272**:26857–26863.

Golden, A. and Brugge, J.C. (1989). Thrombin treatment induces rapid changes in tyrosine phosphorylation in platelets. *Proceedings of the National Academy of Science, USA*, **86**:901–905.

Grondin, P., Plantavid, M., Sultan, C., Breton, M., Mauco, G. and Chap, H. (1991). Interaction of pp60[c-src], phospholipase C, inositol-lipid, and diacylglycerol kinases with the cytoskeletons of thrombin-stimulated platelets. *The Journal of Biological Chemistry*, **266**:15705–15709.

Guinebault, C., Payrastre, B., Sultan, C., Mauco, G., Breton, M., Levy-Toledano, S., Plantavid, M. and Chap, H. (1993). Tyrosine kinases and phosphoinositide metabolism in thrombin-stimulated human platelets. *Biochemical Journal*, **292**:851–856.

Guinebault, C., Payrastre, B., Racaud-Sultan, C., Mazarguil, H., Breton, M., Mauco, G., Plantavid, M. and Chap, H. (1995). Integrin-dependent translocation of phosphoinositide 3-kinase to the cytoskeleton of thrombin-activated platelets involves specific interactions of p85α with actin filaments and focal adhesion kinase. *Journal of Cell Biology*, **129**:831–842.

Gutkind, J.S., Lacal, P.M. and Robbins, K.C. (1990). Thrombin-dependent association of phosphatidylinositol 3-kinase with pp60[c-src] and p59~ in human platelets. *Molecular Cellular Biology*, **10**:3806–3809.

Hartwig, J.H., Bokoch, G.M., Carpenter, C.L., Janmey, P.A., Taylor, L.A., Toker, A. and Stossel, T.P. (1995). Thrombin receptor ligation and activated Rac uncap actin filament barbed ends through phosphoinositide synthesis in permeabilized human platelets. *Cell*, **82**:643–653.

Heemskerk, J.W.M, Feijge, M.A.H., Sage, S.O. and Walter, U. (1994a). Indirect regulation of calcium entry by cAMP and cGMP-dependent protein kinases and phospholipase C in rat platelets. *European Journal of Biochemistry*, **223**:543–551.

Heemskerk, J.W.M. and Sage, S.O. (1994b). Calcium signalling in platelets and other cells. *Platelets*, **5**:295–316.

Helgason, C.D., Damen, J.E., Rosten, P., Grewal, R., Sorensen, P. Chappel, S.M., Borowski, A., Jirik, F., Krystal, G. and Humphries, R.K. (1998). Targeted disruption of SHIP leads to hematopoietic perturbations, lung pathology and a shortened life span. *Genes and Development*, **12**:1610–1620.

Hemmings, B.A. (1997). Akt signalling: Linking membrane events to life and death decisions. *Science*, **275**:628–630.

Hiles, I.D., Otsu, M., Volinia, S., Fry, M.J., Gout, I., Dhand, R., Panayotou, G., Ruiz-Larrea, F., Thompson, A., Totty, N.F., Hsuan, J.J., Courtneidge, S.A., Parker, P.J. and Waterfield, M.D. (1992). Phosphatidylinositol 3-kinase:structure and expression of the 110 kd catalytic subunit. *Cell*, **70**:419–429.

Hu, Q., Klippel, A., Muslin, A.J., Fantl, W.J. and Williams, L.T. (1995). Ras-dependent induction of cellular responses by constitutively active phosphatidylinositol- 3 kinase. *Science*, **268**:100–102.

Jackson, S.P., Schoenwaelder, S.M., Yuan, Y., Rabinowitz, I., Salem, H.H. and Mitchell, C.A. (1994). Adhesion receptor activation of phosphatidylinositol 3-kinase. *The Journal of Biological Chemistry*, **269**:27093–27099.

Jackson, S.P., Schoenwaelder, S.M., Matzaris, M., Brown, S. and Mitchell, C.A. (1995). Phosphatidylinositol 3,4,5-trisphosphate is a substrate for the 75 kDa inositol polyphosphate 5-phosphatase and a novel 5-phosphatase which forms a complex with the p85/p110 form of phosphoinositide 3-kinase. *The EMBO Journal*, **14**:4490–4500.

Janne, P.A., Suchy, S., Bernard, D., MacDonald, M., Crawley, J., Grinberg, A., Wynshaw-Boris, A., Hestphal, H. and Nussbaum, R.L. (1998). Functional overlap between murine *Inpp5b* and *Ocrl1* may explain why deficiency of the mouse ortholog for OCRL1 does not cause Lowe syndrome in mice. *Journal of Clinical Investigation*, **101**:2042–2053.

Jefferson, A.B. and Majerus, P.W. (1995). Properties of Type II inositol polyphosphate 5-phosphatase. *The Journal of Biological Chemistry*, **270**:9370–9377.

Jefferson and Majerus (1996). Mutation of the conserved domains of two inositol polyphosphate 5-phosphatases. *Biochemistry*, **35**:7890–7894.

Kavanaugh, W.M., Pot, D.A., Chin, S.M., Deuterreinhard, M., Jefferson, A.B., Norris, F.A., Masiarz, F.R., Cousens, L.S., Majerus, P.W. and Williams, L.T. (1996). Multiple forms of an inositol polyphosphate 5-phosphatase form signalling complexes with SHC and Grb2. *Current Biology*, **6**:438–445.

Khan, A.A., Steiner, J.P., Klein, M.G., Schneider, M.F. and Snyder, S.H. (1992). Inositol 1,4,5-trisphosphate receptor: localisation to plasma membrane of T cells and co-capping with the T cell receptor. *Science*, **257**:815–818.

Khvotchev, M. and Sudhof, T.C. (1998). Developmentally regulated alternative splicing in a novel synaptojanin. *The Journal of Biological Chemistry*, **273**:2306–2311.

King, W.G., Kucera, G.L., Sorisky, A., Zhang, J. and Rittenhouse, SE. (1991). Protein kinase C regulates the stimulated accumulation of 3-phosphorylated phosphoinositides in platelets. *Biochemical Journal*, **278**:475–480.

Kovacsovics, T.J., Bachelot, C., Toker, A., Vlahos, CJ., Duckworth, B., Cantley, L.C. and Hartwig, J.H. (1995). Phosphoinositide 3-kinase inhibition spares actin assembly in activating platelets but reverses platelet aggregation. *The Journal of Biological Chemistry*, **270**:11358–11366.

Kucera, G.L. and Rittenhouse, S.E. (1990). Human platelets form 3-phosphorylated phosphoinositides in response to α-thrombin, U46619, or GTPγS. *The Journal of Biological Chemistry*, **265**:5345–5348.

Kundra, V., Escobedo, J.A., Kazlauskas, A., Kim, H.K., Rhee, S.G., Williams, L.T. and Zetter, B.R. (1994). Regulation of chemotaxis by the platelet-derived growth factor receptor-β. *Nature*, **367**:474–476.

Laxminarayan, K.M., Chan, B.K., Tetaz, T. Bird, P.I. and Mitchell, C.A. (1994). Characterization of a cDNA encoding the 43-kDa membrane-associated inositol-polyphosphate 5-phosphatase *The Journal of Biological Chemistry*, **269**:17305–17310.

Lee, S.B and Rhee, S.G (1995). Significance of phosphatidylinositol 4,5-bisphosphate hydrolysis and regulation of phospholipase C isoenzymes. *Curr. Opin. Cell. Biol.*, **7**:183–189.

Lee, S.B, Rao, A.K, Lee, K.H, Yang, X., Bae, Y.S, and Rhee, S.G. (1996). Decreased expression of phospholipase C-β₂ isoenzyme in human platelets with impaired function. *Blood*, **88**:1684–1691.

Lioubin, M., Algate, P.A., Tsai, S., Carlberg, K., Aebersold, R. and Rohrschneider, L.R (1996). p150 (SHIP), a signal transduction molecule with inositol polyphosphate 5-phosphatase activity. *Genes and Development*, **10**:1084–1095.

Lips, D.L. and Majerus, P.W. (1989). Phosphatidylinositol 3-phosphate is present in normal and transformed fibroblasts and is resistant to hydrolysis by bovine brain phospholipase C II. *The Journal of Biological Chemistry*, **264**:8759–8763.

Lowe, C.V., Terrey, M. and MacLachan, E.A. (1952). Organic-aciduria, decreased renal ammonia production, hyropthalmos and mental retardation. *Am. J. Dis. Child*, **83**:164–168.

Majerus, P.W. (1992). Inositol phosphate biochemistry. *Annu. Rev. Biochem.*, **61**:225–250.

Majerus, P.W. (1996). Inositols do it all. *Genes and Development*, **10**:1051–52.

Marte, B.M. and Downward, J. (1997). PKB/Akt: connecting phosphoinositide 3-kinase to cell survival and beyond. *TIBS*, **22**:355–358.

Matzaris, M., Jackson, S.P., Laxminarayan, K.M., Speed, C.J. and Mitchell, C.A. (1994). Identification and characterization of the PIP₂ 5-phosphatase in human platelets. *The Journal of Biological Chemistry*, **269**:3397–3402.

Matzaris, M., O'Malley, C., Speed, C.J., Badger, A., Bird, P.I. and Mitchell, C.A. (1998). Distinct membrane and cytosolic forms of the 75 kDa 5-phosphatase. *The Journal of Biological Chemistry*, **273**:8256–8267.

McPherson, P.S., Garcia, E.P., Slepnev, V.I., David, C., Zhang, X.L., Grabs, D., Sossin, W.S., Bauerfeind, R., Nemoto, Y. and De Camilli, P. (1996). A presynaptic inositol-5-phosphatase. *Nature*, **379**:353–356.

Mikoshiba, K. (1993). Inositol 1,4,5-trisphosphate receptor. *Trends in Pharmacological Sciences*, **14**:86–89.

Mitchell, C.A., Connolly, T.M. and Majerus, P.W. (1989). Identification and isolation of a 75-kDa inositol polyphosphate-5-phosphatase from human platelets. *The Journal of Biological Chemistry*, **264**:8873–8877.

Nakanishi, H., Brewer, K.A. and Exton, J.H. (1993). Activation of the ζ of protein kinase C by phosphatidylinositol 3,4,5-trisphosphate. *The Journal of Biological Chemistry*, **268**:13–16.

Norris, F.A., Atkins, R.C. and Majerus, P.W. (1997). Inositol polyphosphate 4-phosphatase is inactivated by calpain-mediated proteolysis in stimulated human platelets. *The Journal of Biological Chemistry*, **272**:10987–10989.

Olivos-Glander, I.M., Janne, P. and Nussbaum, R.L. (1995). The oculocerebrorenal syndrome gene product is a 105-kD protein localized to the golgi complex. *Am. J. Genet.*, **57**:817–823.

Palmer, F.B.S., Theolis, R.T., Cook, H.W. and Byers, D.M. (1994). Purification of two immunologically related phosphatidylinositol-(4,5)-bisphosphate 5-phosphatases from bovine brain cytosol. *The Journal of Biological Chemistry*, **269**:3403–3410.

Pleiman, C.M., Hertz, W.M. and Cambier, J.C. (1994). Activation of phosphatidylinositol-3′ kinase by Src-family kinase SH3 binding to the p85 subunit. *Science*, **263**:1609–1612.

Quinton, T.M. and Dean, W.L. (1996a). Multiple inositol 1,4,5-trisphosphate receptor isoforms are present in platelets. *Biochemical and Biophysical Research and Communications*, **224**:740–746.

Quinton, T.M., Brown, K.D. and Dean, W.L. (1996b). Inositol 1,4,5-trisphosphate-mediated calcium release is regulated by differential phosphorylation. *Biochemistry*, **35**:6865–6871.

Ramjaun, A.R. and McPherson, P.S. (1996). Tissue specific alternative splicing generates two synaptojanin isoforms with differential membrane binding properties. *Journal of Biological Chemistry*, **271**:24856–24861.

Rhee, S.G. (1991). Inositol phospholipid specific phospholipase C: interaction of the γ1 isoform with tyrosine kinase. *Trends in Biochemical Sciences*, **16**:297–301.

Ridley, A.J. and Hall, A. (1992). The small GTP-binding protein rho regulates the assembly of focal adhesions and actin stress fibers. *Cell*, **70**:389–399.

Ringstad, N., Nemoto, Y. and De Camilli, P. (1997). The SH3p4/SH3p8/SH3p13 protein family: Binding partners for synaptojanin and dynamin via a Grb2-like Src homology domain. *Proceedings of the National Academy of Science, USA*. **94**:8569–8574.

Rodriguez-Viciana, P., Warne, P.H., Dhand, R., Vanhaesebroeck, B., Gout, I., Fry, M.J., Waterfield, M.D. and Downward, J. (1994). Phosphatidylinositol-3-OH kinase as a direct target of Ras. *Nature*, **370**:527–532.

Rodriguez-Viciana, P., Warne, P.H., Vanhaesebroeck, B., Waterfield, M.D. and Downward, J. (1996). Activation of phosphoinositide 3-kinase by interaction with Ras and by point mutation. *The EMBO Journal*, **15**:2442–2451.

Rordorf-Nikolic, T., VanHorn, D.J., Chen, D., White, M.F. and Backer, J.M. (1995). Regulation of phosphatidy-linositol 3′-kinase by tyrosyl phosphoprotein. *The Journal of Biological Chemistry*, **270**:3662–3666.

Scheid, M.P., Lauener, R.W. and Duronio. V. (1995). Role od phosphatidylinositol 3-OH-kinase activity in the inhibition of apoptosis in haemopoietic cells: phosphatidylinositol 3-OH-kibase inhibitors reveal a difference in signalling between interleukin-3 and granulocyte-macrophage colony stimulating factor *Biochemical Journal*, **312**:159–162.

Seet, L.F., Cho, S. Hessel, A. and Dumont, D.J. (1998). Molecular cloning of multiple isoforms of synaptojanin 2 and assignment of the gene to mouse chromosome $17\alpha_2$-3.1. *Biochemical Biophysical Research Communications*, **247**:116–122.

Serunian, L.A., Haber, M.T., Fukui, T., Kim, J.W., Rhee, S.G., Lowenstein, J.M. and Cantley, L.C. (1989). Polyphosphoinositides produced by phosphatidylinositol 3-kinase are poor substrates for phospholipases C from rat liver and bovine brain. *The Journal of Biological Chemistry*, **264**:17809–17815.

Shepherd, P.R., Reaves, B.J. and Davidson, H.W. (1996). Phosphoinositide 3-kinases and membrane traffic. *Trends in Cell Biology*, **6**:92–97.

Sorisky, A., King, W.G. and Rittenhouse, S.E. (1992). Accumulation of PtdIns(3,4)P_2 and PtdIns(3,4,5)P_3 in thrombin-stimulated platelets. Different sensitivities to Ca^{2+} or functional integrin. *Biochemical Journal*, **286**:581–584.

Speed, C.J., Little, P.J., Hayman, J.A. and Mitchell, C.A (1996). Underexpression of the 43kDa 5-phosphatase is associated with cellular transformation. *The EMBO Journal*, **15**:4852–4861.

Srinivasan, S., Seaman, M., Nemoto, Y., Daniell, L., Suchy, S.F., Emr, S., De Camilli, P. and Nussbaum, R. (1997). Disruption of three phosphatidylinositol-polyphosphate 5-phosphatase genes from Saccharomyces cerevisiae results in pleiotropic abnormalities of vacuole morphology, cell shape, and osmohomeostasis. *European Journal of Cell Biology*, **74**:350–360.

Stack, J.H., Herman, P.K., Schu, P.V. and Emr, S.D. (1993). A membrane-associated complex containing the Vps15 protein kinase and the Vps34 PI 3-kinase is essential for protein sorting to the yeast lysosome-like vacuole. *The EMBO Journal*, **12**:2195–2204.

Stephens, L.R., Hughes, K.T. and Irvine, R.F. (1991). Pathway of phosphatidylinositol (3,4,5)-trisphosphate synthesis in activated neutrophils. *Nature*, **351**:33–39.

Stephens, L.R., Jackson, T.R. and Hawkins, P.T. (1993). Agonist-stimulated synthesis of phosphatidylinositol(3,4,5)-trisphosphate: a new intracellular signalling system? *Biochimica et Biophysica Acta*, **1179**:27–75.

Stephens, L., Smrcka, A., Cooke, F.T., Jackson, T.R., Sternweis, L.C. and Hawkins, P.T. (1994). A novel phosphoinositide 3 kinase activity in myeloid-derived cells is activated by G protein β subunits. *Cell*, **77**:83–93.

Stoyanov, B., Volinia, S., Hanck, T., Rubio, I., Loubtchenkov, M., Malek, D., Stoyanova, S., Vanhaesebroeck, B., Dhand, R., Nurnberg, B., Gierschik, P., Seeedorf, K., Hsuan, J.J., Waterfield, M.D. and Wetzker, R. (1995). Cloning and characterization of a G protein-activated human phosphoinositide 3-kinase. *Science*, **269**:690–693.

Sultan, C., Plantavid, M., Bachelot, C., Grondin, P., Breton, M. Mauco, G., Levy-Toledano, S., Caen, J.P. and Chap, H. (1991). Involvement of platelet glycoprotein IIb-IIIa ($\alpha_{IIb}\beta_3$ integrin) in thrombin-induced synthesis of phosphatidylinositol 3,4-bisphosphate. *The Journal of Biological Chemistry*, **266**:23544–23557.

Supattapone, S., Worley, P.F., Baraban, J.M. and Snyder, S.H. (1988). Solubilisation and characterisation of an inositol trisphosphate receptor. *The Journal of Biological Chemistry*, **263**:1530–1534.

Tate, B.F and Rittenhouse S.E (1993). Thrombin activation of human platelets causes tyrosine phosphorylation of PLC-γ_2. *Biochimica et Biophysica Acta*, **1178**:281–285.

Taylor, C.W. and Marshall, C.B. (1992). Calcium and inositol 1,4,5-trisphosphate receptors: a complex relationship. *Trends in Biochemical Sciences*, **17**:403–407.

Thomason, P.A., James, S.R., Casey, P.J. and Downes, C.P. (1994). A G-protein β-subunit-responsive phosphoinositide 3-kinase activity in human platelet cytosol. *The Journal of Biological Chemistry*, **269**:16525–16528.

Todderud, G., Wahl, M.I., Rhee, S.G. and Carpenter, G. (1990). Stimulation of phospholipase Cγ_1 membrane association by epidermal growth factor. *Science*, **249**:296–298.

Toker, A., Meyer, M., Reddy, K.K., Falck, J.R., Aneja, R., Aneja, S., Parra, A., Burns, D.J., Ballas, L.M. and Cantley, L.C. (1994). Activation of protein kinase C family members by the novel polyphosphoinositides PtdIns-3,4-P_2 and PtdIns-3,4,5-P_3. *The Journal of Biological Chemistry*, **269**:32358–32367.

Torti, M., Ramaschi, G., Montsarrat, N., Sinigaglia, F., Balduini, C., Plantavid, M., Breton, M., Chap, H. and Mauco, G. (1995). Evidence for a glycoprotein IIb-IIIa- and aggregation-independent mechanism of phosphatidylinositol 3,4-bisphosphate synthesis in human platelets. *The Journal of Biological Chemistry*, **270**:13179–13185.

Ui, M., Okada, T., Hazeki, K. and Hazeki, O. (1995). Wortmannin as a unique probe for an intracellular signalling protein, phosphoinositide 3-kinase. *Trends in Biochemical Sciences*, **20**:303–307.

Vanhaesebroek, B., Leevers, S.J., Panayotou, G. and Waterfield, M.D. (1997). Phosphoinositide 3-kinases: a conserved family of signal transducers. *Trends in Biochemical Sciences*, **22**:267–272.

Verjans, B. De Smedt, F., Lecocq, R., Vaneyenberg, V., Moreau, C. and Erneux, C. (1994). Cloning and expression in *Escherichia coli* of a dog thyroid cDNA encoding a novel inositol 1,4,5-trisphosphate 5-phosphatase. *Biochemical Journal*, **300**:85–90.

Volinia, S., Dhand, R., Vanhaesebroeck, B., MacDougall, L.K., Stein, R., Zvelebil, M.J., Domin, J., Panaretou, C. and Waterfield, M.D. (1995). A human phosphatidylinositol 3-kinase complex related to the yeast Vps34p-Vps15p protein sorting system. *The EMBO Journal*, **14**:3339–3348.

Voss, H., Tamames, J., Teodoru, C., Valencia, A., Sensen, C., Wiemann, S., Schwagner, Zimmermann, C., Sander, C. and Ansorge, W. (1995). Nucleotide sequence and analysis of the centromeric region of yeast chromosome IX. *Yeast*, **11**:61–78.

Wennstrom, S., Siegbahn, A., Yokote, K., Arvidsson, A.K., Heldin, C.H., Mori, S. and Claesson, W.L. (1994). Membrane ruffling and chemotaxis transduced by the PDGF β-receptor require the binding site for phosphatidylinositol 3′ kinase. *Oncogene*, **9**:651–660.

Whitman, M. and Cantley, L.C. (1988). Phosphoinositide metabolism and the control of cell poliferation. *Biochimica et Biophysica Acta*, **948**:327–344.

Wojcikiewicz, R.J.H. and Oberdorf, J.A. (1996). Degradation of inositol 1,4, 5-trisphosphate receptors during cell stimulation is a specific process mediated by cysteine protease activity. *The Journal of Biological Chemistry*, **271**:16652–16655.

Woscholski, R. and Parker, P.J. (1997). Inositol lipid 5-phosphatases — traffic signals and signal traffic. *Trends in Biological Science*, **22**:427–431.

Yamaguchi, A., Yamamoto, N., Kitagawa, H., Tanoue, K. and Yamazaki, H. (1987). Ca^{2+} influx mediated through the GP IIb/IIIa complex during platelet activation. *FEBS Letters*, **225**:228–232.

Yamamoto, K., Graziani, A., Carpenter, C., Cantley, L.C. and Lapetina, E.G. (1990). A novel pathway for the formation of phosphatidylinositol 3,4-bisphosphate. *The Journal of Biological Chemistry*, **265**:22086–22089.

Yanagi, S., Sada, K., Tohyama, Y., Tsubokawa, M., Nagai, K., Yonezawa, K. and Yamamura, H. (1994). Translocation, activation and association of protein-tyrosine kinase p72′∼ with phosphatidylinositol 3-kinase are early events during platelet activation. *European Journal of Biochemistry*, **224**:329–333.

Yang, L.J., Rhee, S.G. and Williamson, J.R. (1994). Epidermal growth factor induced activation and translocation of PLCγ₁ to the cytoskeleton in rat hepatocytes. *The Journal of Biological Chemistry*, **269**:7156–7162.

Yano, H., Nakanishi, S., Kimura, K., Hanai, N., Saitoh, Y., Fukui, Y., Nonomura, Y. and Matsuda, Y. (1993). Inhibition of histamine secretion by wortmannin through the blockade of phosphatidylinositol 3-kinase in RBL-2H3 cells. *The Journal of Biological Chemistry*, **268**:25846–25856.

Yao, R. and Cooper, G.M. (1995). Requirement for phosphatidylinositol 3-kinase in the prevention of apoptosis by nerve growth factor. *Science*, **267**:2003–2006.

Yatomi, Y., Ozaki, Y. and Kume, S. (1992a). Synthesis of phosphatidylinositol 3,4-bisphosphate but not phosphatidylinositol 3,4,5-trisphosphate is closely correlated with protein-tyrosine phosphorylation in thrombin-activated human platelets. *Biochemical and Biophysical Research and Communications*, **186**:1480–1486.

Yatomi, Y., Hazeki, O., Kume, S. and Ui, M. (1992b). Suppression by wortmannin of platelet responses to stimuli due to inhibition of pleckstrin phosphorylation. *Biochemical Journal*, **285**:745–751.

Yatomi, Y., Ozaki, Y., Satoh, K. and Kume, S. (1993). Anti-CD9 monoclonal antibody elicits staurosporine inhibitable phosphatidylinositol 4,5-bisphosphate hydrolysis, phosphatidylinositol 3,4-bisphosphate synthesis, and protein-tyrosine phosphorylation in human platelets. *FEBS Letters*, **322**:285–290.

Yatomi, Y., Ozaki, Y., Satoh, K. and Kume, S. (1994). Synthesis of phosphatidylinositol 3,4-bisphosphate is regulated by protein-tyrosine phosphorylation but the p85 of subunit of phosphatidylinositol 3-kinase may not be a target for tyrosine kinases in thrombin-stimulated human platelets. *Biochimica et Biophysica Acta*, **1212**:337–344.

Zhang, J., Fry, M.J., Waterfield, M.D., Jaken, S., Liao, L., Fox, J.E.B. and Rittenhouse, S.E. (1992). Activated phosphoinositide 3-kinase associates with membrane skeleton in thrombin-exposed platelets. *The Journal of Biological Chemistry*, **267**:4686–4692.

Zhang, J., King, W.G., Dillon, S., Hall, A., Feig, L. and Rittenhouse, S.E. (1993). Activation of platelet phosphatidylinositide 3-kinase requires the small GTP-binding protein Rho. *The Journal of Biological Chemistry*, **268**:22251–22254.

Zhang, J., Zhang, J., Benovic, J.L., Sugai, M., Wetzker, R., Gout, I. and Rittenhouse, S.E. (1995a). Sequestration of a G-protein subunit βγ or ADP-ribosylation of rho can inhibit thrombin-induced activation of platelet phosphoinositide 3-kinases. *The Journal of Biological Chemistry*, **270**:6589–6594.

Zhang, J., Falck, J.R., Reddy, KK., Abrams, C.S., Zhao, W. and Rittenhouse S.E. (1995b). Phosphatidylinositol 3,4,5- trisphosphate stimulates phosphorylation of pleckstrin in human platelets. *The Journal of Biological Chemistry*, **270**:22807–22810.

Zhang, J., Zhang, J., Shattil, S.J., Cunningham, M.C. and Rittenhouse, S.E. (1996). Phosphoinositide 3-kinase γ and p85/phosphoinositide 3-kinase in platelets. *The Journal of Biological Chemistry*, **271**:6265–6272.

Zhang, X.L., Jefferson, A.B., Auethavekiat, V. and Majerus, P.W. (1995). The protein deficient in Lowe syndrome is a phosphatidylinositol 4,5-bisphosphate 5-phosphatase. *Procidings of the National Academy of Sciences, USA*. **92**:4853–4856.

Zhang, X.L, Hartz, P.A., Philip, E., Racusen, L.C. and Majerus, P.W. (1998). Cell lines from kidney proximal tubules of a patient with Lowe-syndrome lack OCRL inositol poylphosphate 5-phosphatase and accumulate phosphatylinositol 4,5-bisphosphate. *Journal of Biological Chemistry*, **273**:1574–1582.

8 The Biology of CD36

Rick F. Thorne, Douglas J. Dorahy, Robyn M. Wilkinson and Gordon F. Burns

Cancer Research Unit, Level 5, David Maddison Clinical Sciences Building, Royal Newcastle Hospital, Newcastle, NSW 2300, Australia

Independently identified over twenty years ago as a membrane glycoprotein of both platelets (Okumura and Jamieson, 1976; Podolsak, 1977) and milk fat globules (Kobylka and Carraway, 1973) that is highly resistant to proteolytic digestion, the primary function of CD36 remains something of an enigma. As will be discussed below, CD36 has been reported to bind a wide range of ligands, and reports that it is the thrombospondin membrane receptor (Asch *et al.*, 1987) or a primary receptor for platelet collagen adhesion (Tandon *et al.*, 1989a) have led to great controversy, with protagonists on both sides. The identification of a substantial proportion of Japanese blood donors that totally lack CD36 expression on platelets yet have no apparent haemostatic abnormalities (Yamamoto *et al.*, 1990) served to heighten this controversy in the literature (Tandon *et al.*, 1991; Kehrel *et al.*, 1991, 1993; Diaz-Ricart *et al.*, 1993; Daniel *et al.*, 1994). Notwithstanding subtle abnormalities that have been identified in *in vitro* studies of platelets from this population, the apparent healthy status of this phenotypic group probably suggests that any physiological role for CD36 as a receptor involved in platelet aggregation is minor. However recent evidence does indicate such individuals may be predisposed to development of the hereditary form of hypertrophic cardiomyopathy (Tanaka *et al.*, 1997). Additionally, there is accumulating evidence that CD36 is functionally involved in the development of atherosclerosis (Endemann *et al.*, 1993) and also in the clearance of apoptotic cells (Savill *et al.*, 1992), as well as functioning as a receptor for erythrocytes infected with the malaria parasite *Plasmodium falciparium* (Oquendo *et al.*, 1989). Therefore, further study of this intrinsically fascinating molecule is likely to prove rewarding and may lead to improvements in human health care.

Perhaps reflecting the controversy surrounding much of the literature on CD36, this molecule is still given a variety of titles and no uniform nomenclature has yet been established. Historically, the platelet glycoprotein (GP) was described as GPIIIb (McGregor *et al.*, 1989) or GPIV (Tandon *et al.*, 1989b) based on its mobility in SDS-polyacrylamide gels. Similarly the milk fat globule membrane (MFGM) form of the protein was named PAS IV because of its migration in SDS-PAGE and staining with periodic acid/Schiff (Greenwalt and Mather, 1985). While these titles do not ascribe a (possibly spurious) function to the molecule, they do imply a singular source which is clearly inappropriate for a widely distributed molecule. Equally, the appellations FAT, given to the rat homologue (Abumrad *et al.*, 1993), and class B scavenger receptor (Rigotti *et al.*, 1995) ascribe a single, specific function to a molecule which may have several functions. The more neutral term, GP88 has also been used (Barnwell *et al.*, 1989) but did not become popular, possibly because CD36 from different sources can exhibit different relative molecular weights.

Since the original cloning of the cDNA encoding CD36 was accomplished using an anti-CD36 monoclonal antibody (mAb) (Oquendo *et al.*, 1989), the CD (cluster designation antigen classification) would appear to be the most appropriate generic name for this molecule.

CELL AND TISSUE DISTRIBUTION

Strong CD36 expression is exhibited by some haemic cells (erythrocyte precursors, monocytes and macrophages, megakaryocytes and platelets), endothelial cells (reviewed in Greenwalt *et al.*, 1992), adipocytes (Jochen and Hayes, 1993; Abumrad *et al.*, 1993) and secretory mammary epithelial cells (Greenwalt and Mather, 1985; Greenwalt *et al.*, 1985). It is also highly expressed by melanoma cells (Oquendo *et al.*, 1989; Si and Hersey, 1994). CD36 expression has also been recorded on retinal pigment epithelial cells (Ryeom *et al.*, 1996b), cultured embryonic fibroblasts (Stomski *et al.*, 1992), and in skeletal and cardiac muscle cells (van Nieuwenhoven *et al.*, 1995).

Tissue Endothelial Cells

The general consensus is that CD36 is expressed by endothelial cells in most tissues except brain where it is weakly expressed, although discrepancies exist amongst those studies reporting these findings (Knowles *et al.*, 1984; Bordessoule *et al.*, 1993; Kuzu *et al.*, 1992; Swerlick *et al.*, 1992; Turner *et al.*, 1994). Most investigators find that CD36 expression is much more prominent on capillaries of the microvasculature than on larger vessels, but this is not universally agreed (Kuzu *et al.*, 1992). A confounding factor in immunohistological tissue surveys of CD36 is the variability of tissue staining that is found with different mAb, many of which appear conformation dependent. The reason for this is not clear, but it may reflect different post-translational changes to CD36 in various tissues or variations in the conformation of CD36, even within the same cell type. Such variations are well exemplified within the Proceedings of the Fifth International Workshop of White Cell Differentiation Antigens (see e.g. Silverstein *et al.*, 1995; Turner *et al.*, 1995; Thibert *et al.*, 1995; Dransfield *et al.*, 1995; Union *et al.*, 1995).

Failure to reach a consensus on the precise tissue distribution of CD36 is not trivial since, as pointed out by Greenwalt *et al.* (1995), this can provide valuable information regarding possible function. It may be that the distribution of CD36 varies between species, but it is more likely that the use of monospecific polyclonal antibodies (Greenwalt and Mather, 1985; Greenwalt *et al.*, 1995) or Northern blotting (Abumrad *et al.*, 1993) provides a more accurate reflection of the comparative levels of CD36 within different tissues.

Haemic Cells and Intracellular Localisation of CD36

Further clues about the possible functions of CD36 may be gleaned from its localisation within cells. CD36 is expressed on the surface of resting platelets at around 12000 copies per platelet (Aiken *et al.*, 1990). Interestingly, much of this material appears to be enriched within membrane microdomains known as DIGS (detergent insoluble glycosphinogolipid-enriched membrane domains) that may be related to caveolae (Dorahy *et al.*, 1996b). Caveolae are specialised plasma membrane structures involved in endocytosis and cell signalling, and CD36 is notably enriched within the caveolae isolated from murine endothelial cells (Lisanti *et al.*, 1994) and from transfected COS-7 cells (Thorne *et al.*, 1997). Haemic cells do not contain caveolin, the protein marker that identifies caveolae, but they do contain specialised lipid membrane microdomains (Fra *et al.*, 1994) and Dorahy *et al.* (1996b) found that platelet CD36 was also greatly enriched within this fraction.

However, much of the CD36 expressed by megakaryocytes and platelets is found within the cell rather than on the plasma membrane, being identified as lining the luminal side of the open canalicular system and around the inner face of α-granule membranes (Berger *et al*, 1993). In platelets this internal pool of CD36 was found to redistribute to the surface upon platelet activation (Michelson *et al.*, 1994; Berger *et al.*, 1993). On spread platelets, the majority of CD36 on the surface was located over the granulomere (an elevated structure formed by concentric circles of microtubules that centralise the granules) and was virtually absent from other regions of the surface except notably at the cell periphery, particularly at regions of cell-cell contact (Kieffer *et al.*, 1992).

An intracellular pool of CD36 has also been recorded in macrophages undergoing *in vitro* differentiation (Huh *et al.*, 1996). Analysis of monocytoid cell lines has provided a possible mechanism for this by showing that a 74 kDa precursor of CD36 is retained within a pre-medial Golgi compartment (Alessio *et al.*, 1996). CD36 has also been found to localise to intracellular compartments in non-haemic cells such as human embryonic fibroblasts (Stomski *et al.*, 1992) and in unpublished work we have established that CD36 within cultured melanoma cells and transfected COS cells is retained within a pre-medial Golgi compartment.

Melanoma Cells and Skin

CD36 is expressed by a number of melanoma cell lines (Panton *et al.*, 1987) and by the majority of primary and metastatic melanomas observed in tissue sections (Si and Hersey, 1994). Wong *et al.* (1989) observed that enrichment for high expression of CD36 in C32 melanoma cells enhanced cell growth both *in vitro* and *in vivo*, thereby suggesting that CD36 may define a malignant phenotype. However, this finding was not supported in an immunohistological examination of CD36 in melanoma (Si and Hersey, 1994). In our own laboratory studies (R.F. Thorne, unpublished data), transfection of CD36 into a CD36-negative melanoma cell line resulted in no differences in growth rates *in vitro* or in nude mice.

In normal human skin, the endothelium around sweat glands are particularly positive for CD36; in addition sebaceous glands are invariably positive for CD36 expression (Knowles *et al.*, 1984; Bordessoule *et al.*, 1993). Some keratin-positive cells in normal skin are also CD36 positive (Rouabhia *et al.*, 1994), but in inflamed tissue, and that surrounding melanomas a substantial proportion of keratinocytes display CD36 (Miklos and Janos, 1987; Si and Hersey, 1994).

Regulation of CD36 Expression

Expression of CD36 is tightly regulated during the differentiation of haemic cells, and is expressed at different stages of development of erythrocytes, monocytes and megakaryocytes (reviewed by Greenwalt *et al.*, 1992). Early erythroblasts express high levels of CD36 (Kieffer *et al.*, 1989) diminishing to very low levels on mature erythrocytes (van Schravendijk *et al.*, 1992), whereas expression of this molecule increased during the differentiation of monocytes into macrophages (Huh *et al.*, 1996) and with the development of mature megakaryocytes and platelets (Greenwalt *et al.*, 1992). In mammary epithelial cells, CD36 expression is also greatly increased during lactation where it becomes incorporated into the membranes of the released fat globules (Greenwalt and Mather, 1985).

Cellular expression of CD36 may also be induced by cell adhesion (Alessio *et al.*, 1991) or by binding to one of its putative ligands (Stomski *et al.*, 1992; Gani *et al.*, 1990; Union *et al.*, 1995). An instructive paper by Greenwalt *et al.* (1995) provided data to suggest that, in mice, CD36 expression by endothelial cells can be regulated by serum triglyceride levels, and that the distribution of the induced protein being expressed may relate to the use of fatty acids by the surrounding parenchymal cells.

Among cultured cells, CD36 expression can be modulated by cytokines and other soluble mediators. Noting that analysis of the 5′-flanking sequence of the CD36 gene contains possible sequences for transcriptional regulation by glucocorticoids and cytokines (Armesilla and Vega, 1994), Yesner *et al.* (1996) studied the expression of CD36 by cultured monocytes following their exposure to a variety of soluble mediators. It was found that interleukin-4, macrophage colony-stimulating factor and phorbol ester (PMA) all greatly increased CD36 expression, while lipopolysaccaride and dexamethasone caused decreased expression. Interestingly, this mediator-induced increase in expression was recorded in the CD36 mRNA and cytoplasmic protein levels, with little of the induced CD36 reaching the cell surface during the time course of these experiments, and presumably the newly synthesised CD36 was retained in a pre-medial Golgi compartment as was recorded for monocytoid cell lines also treated with PMA (Alessio *et al.*, 1996). In contrast, PMA treatment was reported to decrease CD36 expression on human dermal microvascular endothelial cells (Swerlick *et al.*, 1992). Whether the expressed CD36 was being sequestrated into an intracellular compartment was not measured in these studies, but treatment of the cells with interferon γ caused an upregulation of CD36 expression. Induction of CD36 expression by inteferon γ treatment has also been recorded on human umbilical vein endothelial (HUVEC) cells (Favaloro, 1993), although these cells do not normally express CD36 (Swerlick *et al.*, 1992; Huh *et al.*, 1995). However, it has been found that stimulated HUVEC, although themselves not expressing CD36, can induce CD36 mRNA and protein production by blood monocytes that adhere to them, perhaps through engagement of E-selectin displayed by the stimulated HUVEC (Huh *et al.*, 1995).

STRUCTURE OF CD36

Messenger RNA and Gene Structure

The mRNA coding for CD36 was first isolated from a human placental library (Oquendo *et al.*, 1989). The predicted sequence agreed with the partial amino acid sequence reported by Tandon *et al.* (1989b). Northern blot analysis revealed multiple CD36 mRNA transcripts; C32 melanoma cells displayed four transcripts of 7.6, 3.1, 2.4 and 2.0kb whereas human erythroleukemia (HEL) cells displayed only 7.6 and 2.9kb transcripts (Oquendo *et al.*, 1989). The cDNA for platelet CD36 has also been cloned (Wyler *et al.*, 1993). This study compared platelet mRNA species to placenta, monocytes, human umbilical vein (HUVEC) cells and HEL cells. Platelets expressed 1.9 and 1kb transcripts, monocytes four (4, 1.9, 1.5 1kb), while HUVEC and HEL cells expressed three transcripts (8.1, 4, 1.9kb). While sequences within the coding regions were in good agreement, the platelet cDNA contained an insertion of GTAA immediately following the termination codon. Noguchi *et al.* (1993) also observed that one CD36 cDNA clone from adriamycin resistant K562 cells exhibited the "platelet" GTAA insertion. The reasons behind the reported variations and inconsistencies in the size of CD36 mRNA transcripts may be purely technical or result from detection of pre-mRNA pools. However, as discussed below, alternate splicing of CD36 mRNA has been reported and may indicate an intricate regulatory mechanism for CD36 function.

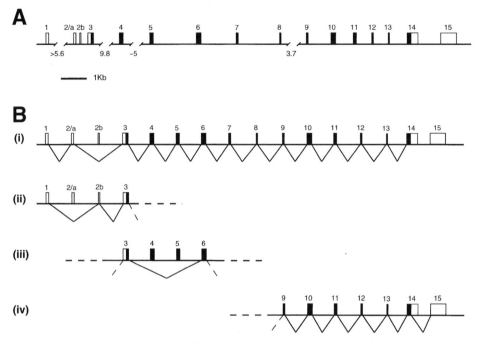

Figure 8.1 (A) **Structural organisation of the human CD36 gene**. Located on 7q11.2 (Fernandez-Ruiz *et al.*, 1993), the CD36 gene is comprised of 16 exons (Taylor *et al.*, 1993; Armesilla and Vega, 1994; Tang *et al.*, 1994). The coding sequences are denoted by filled boxes. (B) Patterns of Messenger RNA Splicing of CD36. Splicing for the archetypal human placental transcript of CD36 is shown (i) (Oquendo *et al.*, 1989; Armesilla and Vega, 1994). Variations are known to occur in the 5′-untranslated region (ii) (Taylor *et al.*, 1993), coding region (iii) (Tang *et al.*, 1994) and 3′-untranslated region (iv) (Armesilla and Vega, 1994) (see text for more information).

The structural organisation of the CD36 gene has been examined, and it was found that CD36 was encoded by 15 exons (Armesilla and Vega, 1994) (Figure 8.1A). The 5′-UTR consists of three exons with the transcription initiation codon residing in exon 3. The 3′-UTR region involves two exons which may encode two message species. The classical CD36 transcript (Oquendo *et al.*, 1989) appears to terminate with exon 14 whereas the alternative transcript is derived from splicing of exon 15 to an acceptor site within exon 14 (just outside the termination codon) (Armesilla and Vega, 1994) (Figure 8.1B).

The predicted amino- and carboxyterminus transmembrane domains of CD36, together with their respective cytoplasmic termini, are each contained within single exons, viz 3 and 14 (Armesilla and Vega, 1994). The intervening extracellular domain is encoded by 11 exons, where alternative splicing also has been documented: partial CD36 cDNA clones isolated from HEL cells and placenta lacked the 309 bases corresponding to exons 4 and 5 (Figure 8.1B) (Tang *et al.*, 1994).

Nucleotide sequence analysis of the 5′-flanking region of the CD36 gene from HEL cells revealed at least three untranslated region (UTR) exons denoted 2a, 2b and 3 as well as up-stream *Alu* tandem repeat regions (Taylor *et al.*, 1993). Exons 2a and 2b did not appear in the same mRNA when placental (2a) and HEL cell (2b) sequences were compared (see Figure 8.1B). Exon 2b containing message was also observed in platelets. Both exons contained two translation initiation codons with small ORF but only those in 2b conformed to the Kozak consensus. This raised the possibility that 2b containing mRNA may represent a negative regulatory feature of CD36 translation, but when transfected into COS-1 cells, mature CD36 was produced (Taylor *et al.*, 1993).

The proximal promoter region identifed by Armesilla and Vega (1994) contains a TATA box as well as several cis-regulatory motifs. Further analysis of this region has revealed that nucleotides –158/–90 (with respect to the transcription initiation site) are essential for optimal transcription (Armesilla *et al.*, 1996). Within this region exists a binding motif (ACCACA) for the polyomavirus enhancer-binding protein/core binding factor (PEBP2/ CBF) class of transcription factors, and mutation of this region resulted in perturbation of CD36 promoter activity. Another study by Konig *et al.* (1995) has also isolated CD36 cDNA clones as target of the Oct-2 transcription factor.

Protein Structure

The predicted CD36 polypeptide contains 471 amino acid residues with a calculated molecular mass of 53 kDa (Oquendo *et al.*, 1989). Hydrophobic regions exist adjacent to the amino and carboxyterminal regions and also near the centre of the molecule. The short aminoterminal hydrophobic region resembles an uncleaved signal peptide (Oquendo *et al.*, 1989) with only the initiator methionine being absent in the mature protein (Tandon *et al.*, 1989b). Whether this region is capable of anchoring the aminoterminus of CD36 in the membrane is uncertain and will be discussed below. The carboxyterminal hydrophobic region of 27 residues is characteristic of a membrane spanning domain and is followed by a short sequence of six residues that represents a carboxyterminal intracellular domain. The carboxyterminal half of the molecule is proline rich and contains six of the 10 cysteine molecules; the other four cysteines are in proximal pairs adjacent to the two terminal hydrophobic regions. The sequence predicts 10 potential sites for N-linked glycosylation, 8 of these towards the aminoterminal half of the protein. The Chou-Fasman algorithm (Figure 8.2) predicts a secondary structure consisting mainly of β strands and loops, as expected from the high proline content. There is compelling co-incidence of hydrophobic and flexible loops and turns in the aminoterminal two thirds of the sequence, with a frequency of around 10 residues, with intervening β strands. However the carboxyterminal

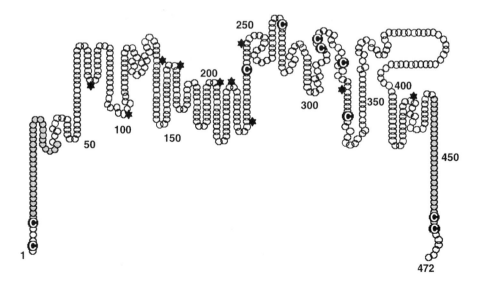

Figure 8.2   Schematic of the secondary structure of CD36 modelled using the Chou Fasman predictive algorithm (Chou and Fasman, 1978). The positions of cysteine residues (C), putative N-linked glycosylation sites (star) and transmembrane domains (shaded residues) are shown. Further details of the predictions are given in the text.

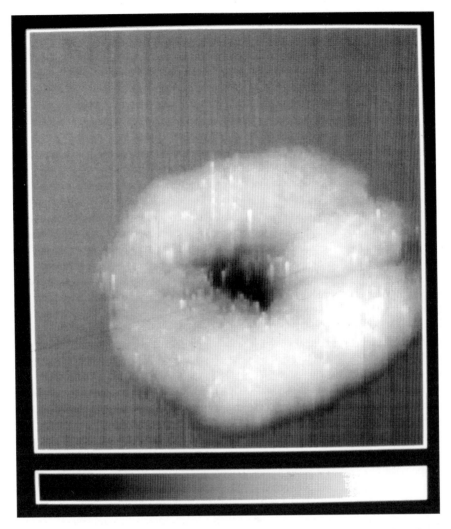

Figure 8.3 Scanning tunnelling microscopy of purified CD36. Based upon considerations of size and an apparent axis of symmetry, the captured image illustrated here and also seen in other samples of immunopurified platelet CD36 suggests that CD36 dimers may exist in a head to head or tail to head configuration. The authors thank Guy Cotterill, Department of Physics, The University of Newcastle for performing the Scanning Tunnelling Microscopy on the purified CD36.

third consists largely of unordered loops and turns (J. Leach and E. Minasian, personal communication).

Oligomer formation

In the original cloning paper, Oquendo *et al.* (1989) noted that some of the CD36 immuno-precipitated from lysates of transfected COS cells migrated as a dimer in SDS-polyacrylamide gels. Recently Coburn *et al.* cited unpublished observations that the rat homologue of CD36 exists as a dimer and possibly also as heterodimeric complexes in rat adipocyte membranes (cited in Ibrahimi *et al.*, 1996). By the use of native and 2D PAGE, gel filtration, and scanning tunnelling microscopy (Figure 8.3) we have shown homo-oligomerisation of CD36 that occurs through intermolecular cysteine bonding (Thorne *et al.*, 1997). Additionally, others have reported a 113 kDa form of CD36 on platelets following molecular cross-linking, these results suggesting either a distinct conformer of CD36 or a hetero-complex with an unidentifed molecule (Rhinehart-Jones and Greenwalt, 1996). The physiological role of CD36 oligomerisation is not fully understood, but it has been reported that thrombospondin, a CD36 ligand, could induce dimerisation of the receptor (Daviet *et al.*, 1997).

Post-translational modifications

CD36 displays different molecular weights depending on the source from which it is isolated, with reported values ranging from 78 kDa to 91 kDa. These differences in apparent molecu-lar mass probably represent cell type specific glycosylation, since treatment of CD36 puri-fied from bovine endothelial cells and bovine MFGMs with endoglycosidases to remove carbohydrate residues gave rise to core proteins of identical molecular weight (Greenwalt *et al.*, 1992). CD36 has been reported to be modified by the attachment of carbohydrate residues through both N-glycosidic (Nakata *et al.*, 1993) and O-linked glycosidic (Tandon *et al.*, 1989b) bonds. In platelets these modifications have been shown to account for 26% of the molecular mass of CD36 with N-linked glycosidic linkages comprising approxi-mately two-thirds of the total glycosylation (Tandon *et al.*, 1989b).

The most thorough investigation of CD36 glycosylation has involved the examination of the N-linked oligosaccharides of bovine MFGM CD36, where six N-linked sugar chains per molecule were observed (Nakata *et al.*, 1993). This glycosylation was found to be exceedingly diverse and to involve a number of complex type sugar chains, including high mannose type, hybrid-type, and bi-, tri-and tetraantennary structures. Furthermore most of the hybrid-type N-linked sugar chains include a novel hybrid-type oligosaccharide featur-ing a $Man\alpha_1 \rightarrow 2Man\alpha_2 \rightarrow 3Man$ group, the synthesis of which cannot be explained by any currently described biosynthetic pathway (Nakata *et al.*, 1993).

CD36 is also extensively modified by acylation. It has been shown to be palmitoylated in human platelets (Dorahy *et al.*, 1996b), and the rat homologue of CD36 is palmitoylated in rat adipocytes (Jochen and Hayes, 1993). The association of palmitate with CD36 was shown to be typical of covalent linkage via thioester or oxyester bonds (Jochen and Hayes, 1993), and it has been shown that cysteines at either end of the molecule can be modified in this way (Tao *et al.*, 1996). The function of such a modification to a transmembrane glyco-protein such as CD36 is not clear, but Jochen and Hayes (1993) suggest that it may indicate involvement of CD36 in membrane trafficking and provide support for this hypothesis by showing that a drug (cerulenin) that inhibits this event in adipocytes (Jochen *et al.*, 1995) also inhibited the palmitoylation of CD36. Dynamic modification with palmitate could also serve to facilitate conformational changes in the CD36 molecule and perhaps also modu-late protein-protein associations.

CD36 has also been reported to undergo modification by phosphorylation on threonine residue 92, this being located on an ectodomain (Asch *et al.*, 1993). Upon platelet activa-tion, it appears that released phosphatases may dephosphorylate CD36 expressed on the cell surface, with the possible consequence of this event being modulation of CD36 bind-ing function (discussed below).

Membrane orientation and conformation

The unusual structure of CD36 has lead to uncertainty about its topography within cell membranes. Two major models have been proposed. The first, expounded by Greenwalt *et al.* (1992) (Figure 8.4A) proposes that CD36 is anchored in the membrane by the two

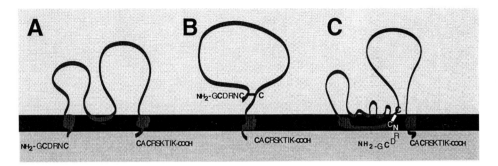

Figure 8.4 The membrane conformation of CD36. CD36 has been proposed to possess two transmembrane regions, as well as a central membrane-associated region (A) (Greenwalt *et al.*, 1992). An alternative model proposed by Asch *et al.* (1993) predicted a single transmembrane domain at the carboxyterminus, and a disulphide loop between cysteine residues 7 and 243. A unifying model is proposed in the text (C).

terminal hydrophobic regions, giving rise to two short intracellular domains each of less that eight residues. This configuration, if correct, is an unusual one for eukaryotic proteins, although LIMPII, another member of the CD36 family (see below) has also been proposed to share this membrane orientation (Vega *et al.*, 1991b).

Asch *et al.* (1993) propose an entirely different model. Work from this group, together with computer modelling, predicted that CD36 exhibited a single transmembrane domain at the carboxyterminus, and a disulphide loop between cysteine residues 7 and 243 defined an extracellular domain (Figure 8.4B). Apparent experimental support for this model was provided by Pearce *et al.* (1994) who reported that a CD36 mutant truncated just before the predicted carboxyterminal transmembrane segment was secreted into the culture medium in transfected cells. The characteristic protease resistance of CD36 makes difficult further analysis of the orientation of CD36 by traditional biochemical methods.

Our own unpublished work with CD36 enables us to propose a third, perhaps unifying, concept for the orientation of CD36. This is illustrated in Figure 8.4C and is based upon the following findings and considerations. (1) In mutational analysis we have found that deletions of cysteine 7 and mutations of cysteine 243 both result in an expressed protein that has undergone a conformational change based upon the binding of conformation-dependent antibodies. Thus, cysteines 7 and 243 are possibly bridged in CD36. (2) In transfected melanoma cells, carboxyterminal cytoplasmic domain deletion mutants of CD36 are not incorporated into the plasma membrane and are not detected as a secretion product. (3) The aminoterminal region of CD36 can be acylated (see above) which could facilitate its association with cell membranes. Further, the precedent of caveolin, which shares an intracellular localisation with CD36 in some cells (Lisanti *et al.*, 1994), and is proposed to form a transmembrane loop configuration with both the amino- and carboxyterminal domains exposed to the cytosol without the protein spanning the membrane (Monier *et al.*, 1995) suggests that this may be a feature of proteins contained within certain membrane microdomains. This notion is perhaps substantiated by the report of Abumrad *et al.* (1993) suggesting that the rat homologue of CD36 embeds in the membrane in multiple sites via stretches of hydrophobic amino acids too short to span the membrane.

We do not propose that this is the only configuration for CD36. Rather it is suggested that the conformation of CD36 is varied, differing in different cell types and in different locations within the same cell as reflected by the different staining patterns observed with individual monoclonal antibodies. One trigger able to drive such changes would be provided by dynamic acylation; other mechanisms could include protein-protein interactions and intracellular association between CD36 and its various ligands.

CD36 GENE FAMILY

Sequencing reports have now identified several proteins which share homology with CD36 (Figure 8.5). Historically, the first of these homologues to be identified was the lysosomal integral membrane protein (LIMP II) (Vega *et al.*, 1991b). Using PCR amplification with primers based on complementary amino acid segments between CD36 and LIMP II, Calvo and Vega (1993) identified a novel CD36 and LIMP II analogous protein, termed CLA-1. CLA-1 is the human homologue (81% identity) of hamster SR-B1. SR-B1, another member of the CD36 family, was so named because of its role as a scavenger receptor (type B1) (Acton *et al.*, 1994). Also identified was a rat adipocyte protein termed FAT (85% identical to human CD36) which has been implicated in fatty acid binding and transport (Abumrad *et al.*, 1993). CD36 gene family members have also been described for non-mammalian cells; the most thoroughly described being the *Drosophila* gene coding for an epithelial membrane protein, *emp* (Hart and Wilcox, 1993), and another *Drosophila* homologue termed croquemort (Franc *et al.*, 1996).

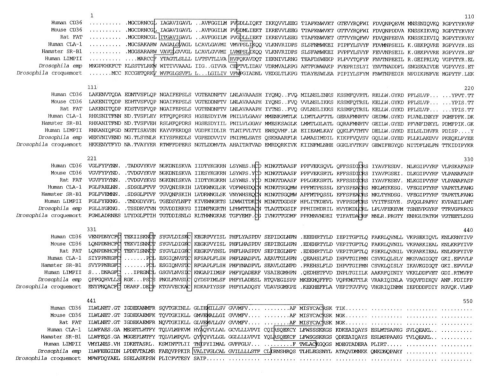

Figure 8.5 The CD36 gene family. Aligned amino sequences from CD36 (human form, Oquendo *et al.*, 1989; murine form, Endemann *et al.*, 1993), FAT (Abumrad *et al.*, 1993), CLA-1 (Calvo and Vega, 1993), SR-B1 (Acton *et al.*, 1994), LIMP-II (Calvo and Vega, 1995), *emp* (Hart and Wilcox, 1993) and croquemort (Franc *et al.*, 1996) are shown. Gaps have been introduced to optimise the alignments. The putative transmembrane regions and conserved cysteines are denoted by open boxes. The alignments were constructed using the PileUP program from the GCG Package (Madison, Wisconsin) and transmembrane segments predicted using the GES algorithm in the TopPred II program (Claros and von Heijne, 1994).

CD36 family members share several common structural features, most notably hydrophobic regions at the amino- and carboxyterminal consistent with transmembrane domains (Figure 8.5). The predicted amino-cytoplasmic tails for all gene family members are short, whereas the carboxy-cytoplasmic residues vary considerably (from six up to 45 residues) and some evidence suggests these residues may be critical for specific functions (Vega *et al.*, 1991a; Calvo *et al.*, 1995). In addition, several conserved cysteine residues were observed (Figure 8.5), as well as multiple N-linked glycosylation and PKC phosphorylation motifs (not shown).

The chromosomal location of the CD36 gene has been mapped to the q11.2 band on human chromosome 7 (Fernandez-Ruiz *et al.*, 1993). However, CD36 family members do not appear to constitute a gene cluster since CD36, LIMPII and CLA-1 were found on chromosomes 7, 4 and 12 respectively (Calvo *et al.*, 1995), although there is evidence that these molecules do share some functional properties (discussed below).

FUNCTIONS OF CD36

Nak[a] Phenotypes: A Model for the Study of CD36 Function

Specific antibodies to a platelet isoantigen, designated Nak[a], were first identified from an acute myelogenous leukemia patient refractory to HLA-matched platelet transfusions (Ikeda *et al.*, 1989). The incidence of Nak[a] negative individuals displays a racial bias and is highest in Asian populations with 3–11% and 2.3% occuring in Japan and Thailand respectively (Ikeda *et al.*, 1989; Urwijitaroon *et al.*, 1995). Nak[a] negative status is also elevated in African Americans (2.4%), whereas no negative white American donors were detected (Curtis and Aster, 1996).

Precipitation analysis with anti-Nak[a] antibodies of platelets from Nak[a]-positive donors revealed a 91 kDa antigen, which was subsequently confirmed to be CD36 (Tomiyama *et al.*, 1990). Intriguingly, Nak[a] negative platelets revealed a total absence of detectable CD36 (Yamamoto *et al.*, 1990). This phenotype has been further delineated by the occurrence of individuals whose monocytes also lack CD36 (type I deficiency) or whose monocytes express CD36 (type II) (Yamamoto *et al.*, 1994). Type II accounts for ~ 3% of the Japanese population but the occurrence of Type I has not yet been determined (Kashiwagi *et al.*, 1996).

The molecular basis of these defects has been investigated. In one Type I subject, deletions were found in two different 5′ products amplified from both platelet and monocyte CD36 cDNA resulting in premature translation stop codons (Kashiwagi *et al.*, 1994). Transcript levels were greatly reduced in both platelets and monocytes, and the likely mechanism suggested was that the deletions may be responsible for message instability or altered splicing. Yet another Type I individual displayed one of the same deletions described above, as well as an additional single nucleotide insertion encoding premature termination of translation (Kashiwagi *et al.*, 1996). This individual was heterozygous for both defects, but the single insertion was only found in monocyte and not platelet cDNA suggesting other defects might account for the deficiency in platelets.

Type II deficiency has been associated with predominantly a ^{478}C → T substitution defining a proline to serine change (Kashiwaga *et al.*, 1993). The monocytes of these Type II subjects were heterozygous for wildtype C478 and mutant T478 forms, while platelet transcripts contain only T478 (Kashiwagi *et al.*, 1995a). To account for the discrepant expression between platelets and monocytes, it was suggested that individuals with the T478 mutation may have a "platelet-specific silent allele" (Kashiwagi *et al.*, 1995b). Transfection of the T478 form revealed defective maturation in the protein, suggesting the deficiency arises from post-translation defects. Type I individuals may also display the $^{478}C \rightarrow T$ substitution, but in this form both platelet and monocyte transcripts are of the mutant T478 type (Kashiwagi *et al.*, 1995a).

While the Nak^a negative phenotype apparently exerts no overt symptoms of altered haemostasis, it should be noted that blood donors are selected on the basis of good health and therefore this observation cannot necessarily be extrapolated to long-term clinical outcomes (Tandon *et al.*, 1991). From the molecular characterisation of the Nak^a negative phenotype to date, it appears likely that CD36 deficiency may consist of a variety of mutations, the individual nature of which may result in as yet unknown consequences. Notably though, complications have been reported in the field of transfusion medicine; i.e.: refractoriness of HLA-matched platelet transfusions (Ikeda *et al.*, 1989) and thrombotic thrombocytopenic purpura syndrome (Tandon *et al.*, 1994; Bierling *et al.*, 1995). Additionally, as discussed below, the Nak^a negative phenotype appears associated with hereditary hypertrophic cardiomyopathy (Tanaka *et al.*, 1997).

Signalling

Agonist challenge of platelets is accompanied by a rapid and sustained increase in the tyrosine phosphorylation of multiple cellular proteins (Ferrell and Martin, 1988). Paradoxically, none of the major receptors involved in platelet activation events has been shown to exhibit intrinsic tyrosine kinase activity, perhaps suggesting the association or juxtapositioning of cytosolic tyrosine kinase(s) with these receptors.

Experimentation by Huang *et al.* (1991) showed that three members of the Src family, p60^{fyn}, p53/56^{lyn}, and p62^{c-Yes} were physically associated with CD36 in platelets and CD36 expressing cell lines. The association of p56^{lck} and the T-cell surface antigens CD4 and CD8 provides a prototypical model for the interaction of the Src-like tyrosine kinases with surface receptors. The specific interaction is dependent upon a short consensus sequence within the amino terminus of the p56^{lck} molecule and complementary regions within the cytoplasmic carboxy tails of the T cell antigens (Rudd, 1990). However, the CD36 associated Src-kinases lack homology to p56^{lck} in their equivalent domain, although CD36 does have homology with CD4 and CD8 in its short cytoplasmic tail. These findings suggest that the association with CD36 may occur through a unique set of consensus sequences present in the associated kinases acting via a similar mechanism to the T cell antigens, or alternatively through an intervening molecule which might provide the means of association (Shattil and Brugge, 1991). Which proposal is correct is not currently known.

A number of studies using antibodies to CD36 have shown that platelet activation can be induced by binding of intact anti-CD36 IgG whereas F(ab′)$_2$ fragments of the same antibodies failed to promote such activation (Masurov *et al.*, 1992; Aiken *et al.*, 1990; Ockenhouse *et al.*, 1989a). Collectively these data suggested an Fc receptor-mediated signal but the requisite for intact antibody against CD36 to induce activation has been contradicted in other studies examining the antibody-induced oxidative burst in monocytes (Trezzini *et al.*, 1990; Schuepp *et al.*, 1991). However, it has never been rigorously tested whether the clustering of CD36 by crosslinking of F(ab′)$_2$ fragments would be sufficient to induce platelet activation or if an additional signal involving the Fc receptor is required to trigger the activation response. Notably Fc receptors have recently been shown to be tightly associated with cell-type specific Src-like tyrosine kinases (e.g.: Ghazizadeh *et al.*, 1994) as well as p72^{syk} (Agarwal *et al.*, 1993), and they are also in physical, topographical association with both CD36 and the integrin GPIIb-IIIa (Mazurov *et al.*, 1992; Berndt *et al.*, 1993). It has also been shown that CD36 is in physical proximity to GPIIb-IIIa on the surface of resting plate-lets (Dorahy *et al.*, 1996a), and as is the case with most platelet agonists (Shattil and Brugge, 1991; Ferrell and Martin, 1989), platelet aggregation by antibodies to CD36 is absolutely dependent upon the GPIIb-IIIa receptor (Aiken *et al.*, 1990; Mazurov *et al.*, 1992).

The significance of the role of CD36 in platelet signalling and the activation must be tempered with the more recent findings from donors with the Nak[a] negative phenotype, where the absense of CD36 is not essential or even required for a normal aggregation response to most agonists (Kehrel *et al.*, 1993; Daniel *et al.*, 1994). As has been seen in the case of many studies with "knockout" mice, while these studies do not suggest that CD36 does not have a genuine role in platelet activation, they do indicate that, if such is the case, compensatory mechanisms exist to bypass this role.

Receptor Function and Ligands for CD36

Receptor for the extracellular matrix ligands, collagen and thrombospondin

A study by Tandon *et al.* (1989a) suggested that CD36 met four of seven defined criteria for a primary collagen receptor of platelets (Santoro, 1988). This work is thoroughly discussed in an excellent review of CD36 by Greenwalt *et al.* (1992). However, two aspects of the collagen binding role of CD36 are worthy of further comment.

Firstly, in the original study it was shown that rabbit anti-CD36 antibodies were able to antagonise the responses of platelets to collagen (Tandon *et al.*, 1989a), but a later report by the same group found that none of nine mAb tested had any inhibitory effect on platelet adhesion to collagen (Tandon *et al.*, 1995). This shortcoming has been addressed where mAb (raised against CD36 prepared using non-denaturing methods) were able to inhibit platelet adhesion to collagen under Mg^{2+} free conditions (Matsuno *et al.*, 1996).

Secondly, one important criterion not addressed by Tandon and colleagues (1989a) con-cerns the requirement that the induction of the putative receptor in non-adhesive cells should confer adhesive function (Santoro, 1988). Our own observations with CD36 transfected MV3 melanoma cells showed no difference in the rate or degree of adhesion to collagen under cation-independent or dependent conditions (R.F. Thorne and G.F. Burns, unpub-lished data). However, such studies are obviously complicated by the presence of multiple collagen receptors and the potential regulation of CD36 collagen binding (see below). This may also be a consideration in studies of platelets, although differences between Nak[a] negat-ive platelets and those from normal donors have been reported. Nak[a] negative platelets dis-played reduced adhesion during the earliest phase of attachment to collagen under both static (in Mg^{2+} free buffer; Tandon *et al.*, 1991) and flow conditions in whole blood (Diaz-Ricart *et al.*, 1993). However, McKeown and colleagues (1994) were unable to confirm these studies since they observed Nak[a] negative platelets were no different to normal con-trols in adhesion to collagen type I in the presence and absense of Mg^{2+}. It can also be noted that the methods employed by Tandon *et al.* (1991) were, by design, aimed toward low strength adhesive interactions, which may not be easily evaluable in other systems.

While most studies have concentrated on collagen type I, others have looked at other collagen types. Kehrel *et al.* (1993) found platelets from CD36 deficient donors aggregated normally in response to collagen types I and III, but were comparably deficient to normal platelets in activation and aggregation responses to collagen type V. Interestingly, collagen type V is a secretion product of endothelial and smooth muscle cells and is increased in athero-sclerotic plaques. This could indicate that platelet responses to this collagen type may be involved in plaque development (Kehrel *et al.*, 1993), and there is considerable evidence implicating CD36 in the etiology of this disease (see below).

Thrombospondin-1 (TSP), the major glycoprotein component of platelet α-granules, is so named because it is released following the platelet "response to thrombin" (Lawler *et al.*, 1977). Numerous investigators have implicated CD36 as a receptor for TSP, evidence primarly being based upon the blocking effects of antibodies raised against CD36 (Asch *et al.*, 1987; Silverstein *et al.*, 1989; McGregor *et al.*, 1989). However other studies utilising anti-CD36 antibodies appeared to contradict these findings (Roberts *et al.*, 1987; Aiken *et al.*, 1990). The molecular cloning of CD36 and the subsequent finding that CD36 transfected COS cells were unable to bind TSP (Oquendo *et al.*, 1989) exacerbated this doubt. However, a later molecular study was able to support the role of CD36 as a TSP receptor: the transfection of CD36 into CD36 negative Bowes melanoma cells conferred TSP binding, and anti-sense CD36 transfection of CD36 positive C32 melanoma cells ablated TSP binding (Silverstein *et al.*, 1992). With hindsight, some of the conflicting data can probably be ascribed to technical reasons, or to the variety of receptors now known to interact with thrombospondin (reviewed in Bornstein, 1995; Gao *et al.*, 1996; Dorahy *et al.*, 1997) and to also factors affecting the intricate regulation of CD36-TSP binding (see below).

A number of approaches have been used to define the region of TSP to which CD36 binds. Studies using TSP proteolytic fragments (Catimel *et al.*, 1992) and synthetic TSP peptides (Asch *et al.*, 1992) have indicated that a conserved peptide motif (CS)VTCG within the Type I (properdin homology) repeats of TSP could bind CD36. Additionally, an anti-TSP mAb that blocked the CD36-TSP interaction recognised an SVTCG synthetic peptide (Li *et al.*, 1993).

Other groups have also investigated what region in CD36 is responsible for TSP binding. Utilising synthetic peptides, Leung *et al.* (1992) identified two sequences encompassing CD36 amino acid residues 139–155 and 93–110. The 139–155 peptide could bind TSP but unexpectedly, this peptide augmented TSP binding to CD36. The aa93–110 peptide alone did not bind TSP, but when used in combination with aa139–155 could bind TSP with high affinity. This led these authors to suggest a two step mechanism of CD36 binding to TSP: initial TSP binding via aa139–155 in CD36, resulting in a conformational change in TSP which unmasks a cryptic high affinity aa93–110 ligand site. A separate study using GST/CD36 constructs found that a fusion protein encompassing the residues 93–120 of CD36 (inclusive of aa93–110) could bind TSP, but the role of aa139–155 was largely not supported (Pearce *et al.*, 1995).

Another approach by Asch *et al.* (1993) using random domain library screening identified a series of CD36 framents that bound to intact TSP as well as a CSVTCG peptide. Functional clones exhibited a common area of overlap (CD36 amino acids 87–99) which overlaps the high affinity aa93–110 ligand site (Leung *et al.*, 1992) and the aa93–120 fusion protein (Pearce *et al.*, 1995) described above. This region of CD36 is also reported to be involved in the binding of collagens and contains a PKC phosphorylation consensus site (Asch *et al.*, 1993). It was demonstrated that the ligand specificity of CD36 was controlled by the phosphorylation status of threonine 92, an extracellular residue within the binding domain. Phosphorylation of purified platelet CD36 or 87–99 peptide resulted in maximal collagen binding and, in a reciprocal fashion, reduced TSP binding. These data may help to explain the apparently conflicting findings concerning CD36 and TSP interactions described previously; for example in this study, CD36 transfected into COS cells appeared to be constitutively phosphorylated, perhaps accounting for the inability of CD36 transfected COS to bind TSP as observed by Oquendo *et al.* (1989).

In summary, the weight of evidence strongly suggests that CD36 can bind TSP and it appears to associate with TSP in certain situations (Kieffer *et al.*, 1992; Gani *et al.*, 1990). It has been questioned as to whether such interactions apply to any physiologically relevant situation (Roberts *et al.*, 1987; Kehrel *et al.*, 1991), although significantly it has been observed recently that CD36 sensitises endothelial cells to the *in vitro* inhibitory effects of TSP-1, suggesting that CD36 may play an essential role in the process of angiogenesis (Dawson *et al.*, 1997). Furthermore, and perhaps indicating a more universal role for CD36-TSP interactions, another member of the CD36 gene family LIMP II can also bind TSP-1 (Crombie and Silverstein, 1998).

Malarial parasite receptor

Erythrocytes infected with the human malarial parasite *Plasmodium falciparum* (IRBC) can form rosettes with uninfected cells, and the trophozoite and schizont forms of IRBC bind the fine capillaries of certain tissues leading to the clinical outcomes associated with malaria (Ockenhouse *et al.*, 1991). Platelets (Ockenhouse *et al.*, 1989a) and monocytes (Barnwell *et al.*, 1985) also bind IRBC, and since phagocytic monocytes are found in areas of sequestered IRBC (see Ockenhouse *et al.*, 1989a) this may suggest their involvement in malarial etiology, although this is still undefined. To identify those receptors mediating rosetting and cytoadherence of IRBC, *in vitro* binding assays of infected erythrocytes have been developed (Ockenhouse *et al.*, 1991). Several reports utilising such assays have observed blocking with anti-CD36 antibodies, and together with studies demonstrating a positive correlation with the expression of the OKM5 antigen (CD36), have implicated CD36 as a receptor for IRBC (Panton *et al.*, 1987; Ockenhouse and Chulay, 1988; Handunnetti *et al.*, 1992). In support of this, IRBC did not bind to Nak^a negative platelets and monocytes (Tandon *et al.*, 1991), nor did IRBC from a CD36 deficient patient rosette with their own uninfected erythrocytes (van Schravendijk *et al.*, 1991).

More direct evidence of the involvement of CD36 was obtained by Oquendo *et al.* (1989) who observed that CD36-transfected COS cells supported increased adhesion of *Plasmodium falciparum* infected erythrocytes. Furthermore, purifed CD36 immobilised to a solid phase could also support the adherence of IRBC (Barnwell *et al.*, 1989; Ockenhouse *et al.*, 1989b) and soluble CD36 could inhibit IRBC cell binding to endothelial cells, melanoma cells and monocytes (Ockenhouse *et al.*, 1989ab) thereby suggesting CD36 was indeed a primary receptor. Asch *et al.* (1993) were able to map the IRBC binding domains to two CD36 peptides, residues 8–21 and 97–110, which are distinct from p87–99 peptide that supported maximal thrombospondin binding (see above). The identity of the complementary ligand(s) for CD36 on IRBC has also been investigated and the reader is referred to the following articles for further information (Greenwalt *et al.*, 1992; Crandell *et al.*, 1994; Baruch *et al.*, 1995; Gardner *et al.*, 1996).

Scavenger Receptor and Lipid Transport Functions

Low density lipoproteins and atherogenesis

The cytopathology of atherosclerotic plaques is characterised by foam cells, which are macrophage burdened with lipid. Macrophage express few "classic" low density lipoprotein (LDL) receptors, and do not accumulate LDL, but do uptake oxidised (Ox) or acetylated (Ac) LDL via so called "scavenger" receptors (Steinberg, 1997). Scavenger receptors are implicated in the etiology of atherosclerosis, and Endemann *et al.* (1993) isolated the murine form of CD36 when screening to identify such macrophage OxLDL receptors. Transfection studies have confirmed this role for human CD36 (Nicholson *et al.*, 1995). Platelets (Endemann *et al.*, 1993) and monocyte-derived macrophage (Nozaki *et al.*, 1995) also can bind OxLDL, and this was inhibitable at least partially by an anti-CD36 mAb. Monocytes differentiated *in vitro* with M-CSF and treated with OxLDL displayed enhanced foam cell formation, which was also inhibited with anti-CD36 antibodies (Huh *et al.*, 1996). Consistent with this role for CD36 is the report that the macrophages derived from CD36-deficient subjects have ~ 40% reduced OxLDL binding and accumulation of cholesteryl ester compared to normal controls (Nozaki *et al.*, 1995), which may have implications for foam cell development *in vivo*. Most recently, insights have been gained into the regulation of CD36 during foam cell development. The exposure to native and modified LDL increased the expression of CD36 both *in vitro* and *in vivo* (Han *et al.*, 1997). A mechanism is suggested by the findings of Nagy *et al.* (1998), in showing linkage of CD36-mediated uptake of OxLDL and the actions of a nuclear receptor, PPARγ. The uptake of OxLDL resulted in signalling and activation by PPARγ, which in turn resulted in transcriptional upregulation of CD36, futher enhancing OxLDL uptake in a metacrine fashion.

Another member of the CD36 family, SR-B1 can bind native LDL as well as high density lipoproteins (HDL) (Acton *et al.*, 1994, 1996), suggesting that this receptor may have a more general role in lipid metabolism, however recently human CD36 has also been shown to be a receptor for both native LDL and HDL, perhaps also inditing CD36 in a similar role (Calvo *et al.*, 1998).

Transport of long chain fatty acids

CD36 has also been implicated in the binding and uptake of long-chain fatty acids (LCFA) (Abumrad *et al.*, 1993). The CD36 rat homologue, FAT (discussed above), could be specifically labelled with derivatives of LCFA. These compounds also caused a strong inhibition of fatty acid trafficking in adipocytes, suggesting a role for this protein in the transport process (Harmon and Abumrad, 1993). A myocardial LCFA transporter homologous to CD36 has also been isolated (Tanaka and Kawamura, 1995), and chemical perturbation of this receptor in rats caused cardiac hypertrophy (Kusaka *et al.*, 1995). Heart LCFA utilisation deficiencies were identified in human subjects with the Nak[a] negative phenotype, and this correlated with the increased occurrence of hereditary hypertrophic cardiomyopathy (Tanaka *et al.*, 1997). Perhaps significantly, CD36 has been shown to be in physical association with fatty-acid-binding protein (FABP) in bovine milk fat globule membranes (Spitsberg *et al.*, 1995). Clearly, the capacity of CD36 to bind LCFA and its proposed role in fatty acid transport has great implications for a primary function for CD36, particularly within the diverse range of tissues where it is abundantly expressed.

Receptor for Anionic Phospholipids

It has been suggested that scavenger receptors for modified low density lipoproteins (LDL) may in some cases function as receptors for anionic phospholipids. Rigotti *et al.* (1995) examined CD36 and SR-B1 as candidate anionic phospholipid receptors, since both are (class B) scavenger receptors. SR-B1 transfectants specifically bound phosphatidylserine incorporated liposomes in a saturable and high affinity fashion. The binding of AcLDL to CD36 transfected cells was also inhibited by both phosphatidylserine and phosphatidylinositol indicating an anionic phospholipid binding role. As noted above, CD36 is a receptor for Ox- and AcLDL, and it was shown that delipidation of the lipoprotein particle abolished binding to CD36, suggesting that CD36 either recognises a lipid moiety or that the lipid portion of the lipoprotein is essential for recognition by CD36 (Nicholson *et al.*, 1995). It is also interesting to note that the binding of anionic phospholipids and phagocytosis of receptor outer membrane segments by rat retinal pigment epithelial cells (Ryeom *et al.*, 1996ab) are both apparently mediated by CD36. Since elevated levels of phosphatidylserine and phoshatidylinositol are observed in the external membrane leaflet of *P. falciparum* infected red cells (Joshi and Gupta, 1988), binding of CD36 to anionic phospholipids may account for this function, as well as CD36-mediated phagocytosis of apoptotic cells (discussed below). However, the relationship between anionic phospholipid binding by scavenger receptors and these functions has yet to be rigorously validated.

Phagocytosis of Apoptotic Cells

The recognition and elimination of cells undergoing apoptosis or programmed cell death is an essential function in the maintenance of homeostasis of the tissue environment (Arends and Wyllie, 1992). Apoptotic cells are likely to undergo specific surface changes permitting the identification and ingestion by phagocytic cells (Fadok *et al.*, 1992a). The clearance of apoptotic cells or bodies may be performed by phagocytic macrophage or even by neighbouring tissue cells, as has been demonstrated for epithelial cells and hepatocytes (Arends and Wyllie, 1992).

By the use of an antagonist peptide (RGDS) and specific mAbs, the macrophage recognition of apoptotic cells (both neutrophils and lymphocytes) was shown to involve the vitronectin receptor (the integrin $\alpha_v\beta_3$) (Savill *et al.*, 1990). Other studies have also revealed that the phagocytosis of apoptotic cells by murine peritoneal macrophage was inhibited by phosphatidylserine (Fadok *et al.*, 1992ab). Instructively, the phagocytosis of apoptotic cells by murine bone marrow-derived macrophage was inhibited by an $\alpha_v\beta_3$ antagonist peptide (RGDS) but not by phosphatidylserine (Fadok *et al.*, 1992a). Therefore collectively this evidence suggested the existence of two independent and apparently mutually exclusive mechanisms able to mediate phagocytosis (Fadok *et al.*, 1992a).

Since CD36 is a receptor for anionic phospholipids, this raises the possibility that CD36 or related proteins may indeed function as the phosphoserine receptor eluded to in the experiments by Fadok *et al.* (1992ab). However, another study by Savill *et al.* (1992) suggested that CD36 was a cooperative receptor in the $\alpha_v\beta_3$-mediated phagocytic mechanism. Antibodies to either CD36 or $\alpha_v\beta_3$ inhibited phagocytosis of aged neutrophils by macrophages, and combination of these mAbs at suboptimal concentrations produced synergystic inhibition. Recognition of the neutrophils by macrophage was potentiated by purifed platelet thrombospondin, which was subsequently shown to bind to both the effector and target cells. These results suggested that TSP may act as a "molecular bridge" between neutrophil and macrophage, where it is bound by both CD36 and $\alpha_v\beta_3$, and this ligation leads to the phagocytic response. Note that both CD36 and β_3-integrins are spatially associated, at least in platelets (Dorahy *et al.*, 1996a), which lends some support for this cooperative receptor hypothesis.

While early evidence for the involvement of CD36 in phagocytosis relied exclusively on antibody inhibition studies, Ren *et al.* (1995) were able to demonstrate that transfection of CD36 into CD36 negative Bowes melanoma cells or COS-7 monkey kidney cells resulted in the dramatically enhanced phagocytosis of apoptotic cells. Antibody inhibition studies indicated that mAb to CD36, $\alpha_v\beta_3$ and TSP all inhibited phagocytosis whereas phospho-L-serine did not, confirming earlier results where CD36-mediated phagocytosis appeared to involve $\alpha_v\beta_3$ integrin and TSP. Other molecular studies have indicated that the extracellular amino acids 155–183 define a domain on CD36 essential for phagocytosis (Daviet *et al.*, 1995; Puente Navazo *et al.*, 1996b). Notably, this region of CD36 was also implicated in the OxLDL binding by CD36 (Puente Navazo *et al.*, 1996a).

Interestingly, the SR-B1 receptor shares the ability to phagocytose apoptotic cells (Fukasawa *et al.*, 1996), as does a newly described *Drosophila* protein termed croquemort or "catcher of cell death", which appears to be involved in recognition of apoptotic cells during larval development (Franc *et al.*, 1996).

References

Abumrad, N.A., El-Maghrabi, M.R., Amri, E.Z., Lopez, E. and Grimaldi, P.A. (1993). Cloning of a rat adipocyte membrane protein implicated in binding or transport of long-chain fatty acids that is induced in pre-adipocyte differentiation. Homology with human CD36. *Journal of Biological Chemistry*, **268**:17665–17668.

Acton, S., Rigotti, A., Landschulz, K.T., Xu, S., Hobbs, H.H. and Krieger, M. (1996). Identification of scavenger receptor SR-BI as a high density lipoprotein receptor. *Science*, **271**:518–520.

Acton, S.L., Scherer, P.E., Lodish, H.F. and Krieger, M. (1994). Expression cloning of SR-BI, a CD36 related class B scavenger receptor. *Journal of Biological Chemistry*, **269**:21003–21008.

Agarwal, A., Salem, P. and Robbins, K.C. (1993). Involvement of p72[syk], a protein-tyrosine kinase, in Fcg receptor signaling. *Journal of Biological Chemistry*, **268**:15900–15905.

Aiken, M.L., Ginsberg, M.H., Byers-Ward, V. and Plow, E.F. (1990). Effects of OKM5, a monoclonal antibody to glycoprotein IV, on platelet aggregation and thrombospondin surface expression. *Blood*, **76**:2501–2509.

Alessio, M., De Monte, L., Scirea, A., Gruarin, P., Tandon, N.N. and Sitia, R. (1996). Synthesis, processing, and intracellular transport of CD36 during monocytic differentiation. *Journal of Biological Chemistry*, **271**:1770–1775.

Alessio, M., Ghigo, D., Garbarino, G., Geuna, M. and Malavasi, F. (1991). Analysis of the human CD36 leucocyte differentiation antigen by means of the monoclonal antibody NL07. *Cellular Immunology*, **137**:487–500.

Arends, M.J. and Wyllie, A.H. (1992). Apoptosis: mechanisms and roles in pathology. *International Reviews of Experimental Pathololology*, **32**:223–254.

Armesilla, A.L., Calvo, D. and Vega, M.A. (1996). Structural and functional characterisation of the human CD36 gene promoter. Identification of a proximal PEBP2/CBF site. *Journal of Biological Chemistry*, **271**:7781–7787.

Armesilla, A.L. and Vega, M.A. (1994). Structural organization of the gene for human CD36 glycoprotein. *Journal of Biological Chemistry*, **269**:18985–18990.

Asch, A.S., Barnwell, J., Silverstein, R.L. and Nachman, R.L. (1987). Isolation of the thrombospondin membrane receptor. *Journal of Clinical Investigation*, **79**:1054–1061.

Asch, A.S., Liu, I., Bricetti, F.M., Barnwell, J.W., Kwakye-Berko, F., Dokun, A., Goldberger, J. and Pernambuco, M. (1993). Analysis of CD36 binding domains: ligand specificity controlled by dephosphorylation of an ectodomain. *Science*, **262**:1436–1440.

Asch, A.S., Silbiger, S., Heimer, E. and Nachman, R.L. (1992). Thrombospondin Sequence Motif (CSVTCG) is responsible for CD36 Binding. *Biochemical and Biophysical Research Communications*, **182**:1208–1217.

Barnwell, J.W., Asch, A.S., Nachman, R.L., Yamaya, M., Aikawa, M. and Ingravallo, P. (1989). A human 88-kDa membrane glycoprotein (CD36) functions *in vitro* as a receptor for a cytoadherence ligand on *Plasmodium falciparum*-infected erythrocytes. *Journal of Clinical Investigation*, **84**:765–772.

Barnwell, J.W., Ockenhouse, C.F. and Knowles, D.M. (1985). Monoclonal antibody OKM5 inhibits *in vitro* binding of *Plasmodium falciparum*-infected erythrocytes to monocytes, endothelial and C32 melanoma cells. *Journal of Immunology*, **135**:3494–3497.

Baruch, D.I., Pasloske, B.L., Singh, H.B., Bi, X., Ma, X.C., Feldman, M. *et al.* (1995). Cloning the *P. falciparum* gene encoding PfEMP1, a malarial variant antigen and adherence receptor on the surface of parasitized human erythrocytes. *Cell*, **82**:77–87.

Berger, G., Caen, J.P., Berndt, M.C. and Cramer, E.M. (1993). Ultrastructural demonstration of CD36 in the α-granule membrane of human platelets and megakaryocytes. *Blood*, **82**:3034–3044.

Berndt, M.C., Mazurov, A.V., Vinogradov, D.V., Burns, G.F. and Chesterman, C.N. (1993). Topographical association of the platelet Fc-receptor with the glycoprotein IIb-IIIa complex. *Platelets*, **4**:190–196.

Bierling, P., Godeau, B., Fromont, P., Bettaieb, A., Debili, N., el-Kassar, N. *et al.* (1995). Posttransfusion purpura-like syndrome associated with CD36 (Nak[a]). *Transfusion*, **35**:777–782.

Bordessoule, D., Jones, M., Gatter, K.C. and Mason, D.Y. (1993). Immunohistological patterns of myeloid antigens: tissue distribution of CD13, CD14, CD16, CD31, CD36, CD65, CD66 and CD67. *British Journal of Haematology*, **83**:370–383.

Bornstein, P. (1995). Diversity of function is inherent in matricellular proteins: an appraisal of thrombospondin 1. *Journal of Cell Biology*, **130**:503–506.

Calvo, D., Dopazo, J. and Vega, M.A. (1995). The CD36, CLA-1 (CD36L1), and LIMPII (CD36L2) gene family: cellular distribution, chromosomal location, and genetic evolution. *Genomics*, **25**:100–106.

Calvo, D., Gomez-Coronado, D., Suarez, Y., Lasuncion, M.A. and Vega, M.A. (1998). Human CD36 is a high affinity receptor for native lipoproteins HDL, LDL, and VLDL. *Journal of Lipid Research*, **39**:777–788.

Calvo, D. and Vega, M.A. (1993). Identification, primary structure, and distribution of CLA-1, a novel member of the CD36/LIMPII gene family. *Journal of Biological Chemistry*, **268**:18929–18935.

Catimel, B., Leung, L., Ghissasi, H.E., Mercier, N. and McGregor, J. (1992). Human platelet glycoprotein IIIb binds thrombospondin fragments bearing the C-terminal region, and/or the type I (CSVTCG motif), but not to the N-terminal heparin-binding region. *Biochemical Journal*, **284**:231–236.

Chou, P.Y. and Fasman, G.D. (1978). Predictions of protein conformation. *Annual Reviews in Biochemistry*, **47**:251–276.

Claros, M.G. and von Heijne, G. (1994). TopPred II: an improved software for membrane protein structure predictions. *Computer Applications in Biosciences*, **10**: 685–686.

Crandall, I., Land, K.M. and Sherman, I.W. (1994). Plasmodium falciparum: pfalhesin and CD36 form an adhesin/receptor pair that is responsible for the pH-dependent portion of cytoadherence/sequestration. *Experimental Parasitology*, **78**:203–209.

Crombie, R. and Silverstein, R.J. (1998). Lysosomal integral membrane protein II binds thrombospondin-1. Structure-function homology with the cell adhesion molecule CD36 defines a conserved recognition motif. *Journal of Biological Chemistry*, **273**:4855–4863.

Curtis, B.R. and Aster, R.H. (1996). Incidence of the Nak^a-negative platelet phenotype in African Americans is similar to that of Asians. *Transfusion*, **36**:331–334.

Daniel, J.L., Dangelmaier, C., Strouse, R. and Smith, J.B. (1994). Collagen induces normal signal transduction in platelets deficient in CD36 (platelet glycoprotein IV). *Thrombosis and Haemostasis*, **71**:353–356.

Daviet, L., Buckland, R., Puente Navazo, M.D. and McGregor, J.L. (1995). Identification of an immunodominant functional domain on human CD36 antigen using human-mouse chimaeric proteins and homologue-replacement mutagenesis. *Biochemical Journal*, **305**:221–224.

Daviet, L., Malvoisin, E., Wild, T.F. and McGregor, J.L. (1997). Thrombospondin induces dimerization of membrane-bound, but not soluble CD36. *Thrombosis and Haemostasis*, **78**:897–901.

Dawson, D.W., Pearce, S.F., Zhong, R., Silverstein, R.L., Frazier, W.A. and Bouck, N.P. (1997). CD36 mediates the *In vitro* inhibitory effects of thrombospondin-1 on endothelial cells. *Journal of Cell Biology*, **138**:707–717.

Diaz-Ricart, M., Tandon, N.N., Carretero, M., Ordinas, A., Bastida, E. and Jamieson, G.A. (1993). Platelets lacking functional CD36 (Glycoprotein IV) show reduced adhesion to collagen in flowing whole blood. *Blood*, **82**:491–496.

Dorahy, D.J., Berndt, M.C., Shafren, D.R. and Burns, G.F. (1996a). CD36 is spatially associated with glycoprotein IIb-IIIa ($\alpha_{IIb}\beta_3$) on the surface of resting platelets. *Biochemical and Biophysical Research Communications*, **218**:575–581.

Dorahy, D.J., Lincz, L.S.R., Meldrum, C.J. and Burns, G.F. (1996b). Biochemical isolation of a membrane microdomain from resting platelets highly enriched in the plasma membrane glycoprotein CD36. *Biochemical Journal*, **319**:67–72.

Dorahy, D.J., Thorne, R.F., Fecondo, J.V. and Burns, G.F. (1997). Stimulation of platelet activation and aggregation by a carboxyterminal peptide from thrombospondin binding to the integrin-associated protein receptor. *Journal of Biological Chemistry*, **272**: 1323–1330.

Dransfield, I., Haslett, C. and Savill, J. (1995). CD36 mAb bind to human macrophages and inhibit ingestion of senescent neutrophils undergoing apoptosis. In *Leukocyte Typing V: White cell differentiation antigens* edited by S.F. Schlossman *et al.*, pp 1280–1282. New York: Oxford University Press.

Endemann, G., Stanton, L.W., Madden, K.S., Bryant, C.M., White, R.T. and Protter, A.A. (1993). CD36 is a receptor for oxidised low density lipoprotein. *Journal of Biological Chemistry*, **268**:11811–11816.

Fadok, V.A., Savill, J.S., Haslett, C., Bratton, D.L., Doherty, D.E., Campbell, B.A. *et al.* (1992a). Different populations of macrophages use either the vitronectin receptor or the phosphatidylserine receptor to recognise and remove apoptotic cells. *Journal of Immunology*, **149**:4029–4035.

Fadok, V.A., Voelker, D.R., Campbell, P.A., Cohen, J.J., Bratton, D.L. and Henson, P.M. (1992b). Exposure of phosphatidylserine on the surface of apoptotic lymphocytes triggers specific recognition and removal by macrophages. *Journal of Immunology*, **148**:2207–2216.

Favaloro, E.J. (1993). Differential expression of surface antigens on activated endothelium. *Immunology and Cell Biology*, **71**:571–581.

Fernandez-Ruiz, E., Armesilla, A.L., Sanchez-Madrid, F. and Vega, M.A. (1993). Gene encoding the collagen type I and thrombospondin receptor CD36 is located on chromosome 7q11.2. *Genomics*, **17**:759–761.

Ferrell, J.E., Jr. and Martin, G.S. (1988). Platelet tyrosine-specific protein phosphorylation is regulated by thrombin. *Molecular and Cell Biology*, **8**:3603–3610.

Ferrell, J.E. and Martin, G.S. (1989). Tyrosine-specific protein phosphorylation is regulated by glycoprotein IIb-IIIa in platelets. *Proceedings of the National Academy of Science of the United States of America*, **86**:2234–2238.

Fra, A.M., Williamson, E., Simons, K. and Parton, R.G. (1994). Detergent-insoluble glycolipid microdomains in lymphocytes in the absence of caveolae. *Journal of Biological Chemistry*, **269**:30745–30748.

Franc, N.C., Dimarcq, J-L., Lagueux, M., Hoffmann, J. and Ezekowitz, R.A.B. (1996). Croquemort, a novel *Drosophila* hemocyte/macrophage receptor that recognises apoptotic cells. *Immunity,* **4**:431–443.

Fukasawa, M., Adachi, H., Hirota, K., Tsujimoto, M., Arai, H. and Inoue, K. (1996). SRB1, a class B scavenger receptor, recognizes both negatively charged liposomes and apoptotic cells. *Experimental Cell Research,* **222**:246–250.

Gani, J.S., Stomski, F., De Nichilo, M., Bates, R.C. and Burns, G.F. (1990). The role of GP88 as a receptor for the attachment of melanoma cells to thrombospondin. In *Biological agents in the treatment of cancer,* edited by P. Hersey, pp. 221–231. Sydney: N.S.W. Goverment Printers.

Gao, A-G., Lindberg, F.P., Finn, M.B., Blystone, S.D., Brown, E.J. and Frazier, W.A. (1996). Integrin-associated protein is a receptor for the C-terminal domain of thrombospondin. *Journal of Biological Chemistry,* **271**:21–24.

Gardner, J.P., Pinches, R.A., Roberts, D.J. and Newbold, C.I. (1996). Variant antigens and endothelial receptor adhesion in *Plasmodium falciparum. Proceedings of the National Academy of Science of the United States of America,* **93**:3503–3508.

Ghazizadeh, S., Bolen, J.B. and Fleit, H.B. (1994). Physical and functional association of Src-related protein tyrosine kinases with FcγRII in monocytic THP-1 cells. *Journal of Biological Chemistry,* **269**:8878–8884.

Greenwalt, D.E., Johnson, V.G. and Mather, I.H. (1985). Specific antibodies to PAS IV, a glycoprotein of bovine milk-fat-globule membrane, bind to a similar protein in cardiac endothelial cells and epithelial cells of lung bronchioles. *Biochemical Journal,* **228**:233–240.

Greenwalt, D.E., Lipsky, R.H., Ockenhouse, C.F., Ikeda, H., Tandon, N.N. and Jamieson, G.A. (1992). Membrane glycoprotein CD36: A review of its roles in adherence, signal transduction, and transfusion medicine. *Blood,* **80**:1105–1115.

Greenwalt, D.E. and Mather, I.H. (1985). Characterisation of an apically derived epithelial membrane glycoprotein from bovine milk, which is expressed in capillary endothelia in diverse tissues. *Journal of Cell Biology,* **100**:397–408.

Greenwalt, D.E., Scheck, S.H. and Rhinehart-Jones, T. (1995). Heart CD36 expression is increased in murine models of diabetes and in mice fed a high fat diet. *Journal of Clinical Investigation,* **96**:1382–1388.

Han, J., Hajjar, D.P., Febbraio, M. and Nicholson A.C. (1997). Native and modified low density lipoproteins increase the functional expression of the macrophage class B scavenger receptor, CD36. *Journal of Biological Chemistry,* **272**:21654–21659.

Handunnetti, S.M., van Schravendijk, M.R., Hasler, T., Barnwell, J.W., Greenwalt, D.E. and Howard, R.J. (1992). Involvement of CD36 on erythrocytes as a rosetting receptor for *Plasmodium falciparum-* infected erythrocytes. *Blood,* **80**:2097–2104.

Harmon, C.M. and Abumrad, N.A. (1993). Binding of sulfosuccinimidyl fatty acids to adipocyte membrane proteins: isolation and amino-terminal sequence of an 88-kD protein implicated in transport of long chain fatty acids. *Journal of Membrane Biology,* **133**:43–49.

Hart, K. and Wilcox, M. (1993). A *Drosophila* gene encoding an epithelial membrane protein with homology to CD36/LIMP II. *Journal of Molecular Biology,* **234**:249–253.

Huang, M.M., Bolen, J.B., Barnwell, J.W., Shattil, S.J. and Brugge, J.S. (1991). Membrane glycoprotein IV (CD36) is physically associated with the Fyn, Lyn, and Yes protein tyrosine kinases in human platelets. *Proceedings of the National Academy of Science of the United States of America,* **88**:7844–7849.

Huh, H.Y., Lo, S.K., Yesner, L.M. and Silverstein, R.L. (1995). CD36 induction on human monocytes upon adhesion to tumor necrosis factor-activated endothelial cells. *Journal of Biological Chemistry,* **270**:6267–6271.

Huh, H.Y., Pearce, S.F., Yesner, L.M., Schindler, J.L. and Silverstein, R.L. (1996). Regulated expression of CD36 during monocyte-to-macrophage differentiation: potential role of CD36 in foam cell formation. *Blood,* **87**:2020–2028.

Ibrahimi, A., Sfeir, Z., Magharaie, H., Amri, E-Z. and Grimaldi, P. (1996). Expression of the CD36 homolog (FAT) in fibroblasts cells: effects on fatty acid transport. *Proceedings of the National Academy of Science of the United States of America,* **93**:2646–2651.

Ikeda, H., Mitani, T., Ohnuma, M., Haga, S., Ohtsuka, S., Kato, T. *et al.* (1989). A new platelet specific antigen, Nakᵃ, involved in the refractoriness of HLA-matched platelet transfusion. *Vox Sang,* **57**:213–217.

Jochen, A. and Hays, J. (1993). Purification of the major substrate for palmitoylation in rat adipocytes: N-terminal homology with CD36 and evidence for cell surface acylation. *Journal of Lipid Research,* **34**:1783–1792.

Jochen, A.L., Hays, J. and Mick, G. (1995). Inhibitory effects of cerulenin on protein palmitoylation and insulin internalization in rat adipocytes. *Biochimica Biophysica Acta,* **1259**:65–72.

Joshi, P. and Gupta, C.M. (1988). Abnormal membrane phospholipid organisation in *Plasmodium falciparum* infected-human erythrocytes. *British Journal of Haematology,* **68**:255–259.

Kashiwagi, H., Honda, S., Tomiyama, Y., Mizutani, H., Take, H., Honda, Y. *et al.* (1993). A novel polymorphism in glycoprotein IV (replacement of proline-90 by serine) predominates in subjects with platelet GPIV deficiency. *Thrombosis and Haemostasis,* **69**:481–484.

Kashiwagi, H., Tomiyama, Y., Honda, S., Kosugi, S., Shiraga, M., Nagao, N. *et al.* (1995a). Molecular basis of CD36 deficiency. Evidence that a 478C→T substitution (proline90→serine) in CD36 cDNA accounts for CD36 deficiency. *Journal of Clinical Investigation,* **95**:1040–1046.

Kashiwagi, H., Tomiyama, Y., Kosugi, S., Shiraga, M., Lipsky, R.H., Kanayama, Y. *et al.* (1994). Identification of molecular defects in a subject with type I CD36 deficiency. *Blood,* **83**:3545–3552.

Kashiwagi, H., Tomiyama, Y., Kosugi, S., Shiraga, M., Lipsky, R.H., Nagao, N. *et al.* (1995b). Family studies of type II CD36 deficient subjects: linkage of a CD36 allele to a platelet-specific mRNA expression defect(s) causing type II CD36 deficiency. *Thrombosis and Haemostasis,* **74**:758–763.

Kashiwagi, H., Tomiyama, Y., Nozaki, S., Honda, S., Kosugi, S., Shiraga, M. *et al.* (1996). A single nucleotide insertion in codon 317 of the CD36 gene leads to CD36 deficiency. *Arteriosclerosis, Thrombosis and Vascular Biology,* **16**:1026–1032.

Kehrel, B., Kronenberg, A., Rauterberg, J., Niesing, Bresch, D., Niehues, U. *et al.* (1993). Platelets deficient in glycoprotein IIIb aggregate normally to collagens type I and III but not to collagen type V. *Blood,* **82**:3364–3370.

Kehrel, B., Kronenberg, A., Schippert, B., Niesing-Bresch, D., Niehues, U., Tschope, D. *et al.* (1991). Thrombospondin binds normally to glycoprotein IIIb deficient platelets. *Biochemical and Biophysical Research Communications,* **179**:985–991.

Kieffer, N., Bettaieb, A., Legrand, C., Coulombel, L., Vainchenker, W., Edelman, L. *et al.* (1989). Developmentally regulated expression of a 78kDa erythroblast membrane glycoprotein immunologically related to the platelet thrombospondin receptor. *Biochemical Journal,* **262**:835–842.

Kieffer, N., Guichard, J. and Breton-Gorius, J. (1992). Dynamic redistribution of major platelet surface receptors after contact-induced platelet activation and spreading. An immunoelectron microscopy study. *American Journal of Pathology,* **140**:57–73.

Knowles, D.M., Tolidjian, B., Marboe, C., Agati, V.D., Grimes, M. and Chass, L. (1984). Monoclonal anti-human monocyte antibodies OKM1 and OKM5 possess distinctive distributions including reactivity with vascular endothelium. *Journal of Immunology,* **132**:2170–2174.

Kobylka, D. and Carraway, K.L. (1973). Proteolytic digestion of proteins of the milk fat globule membrane. *Biochimica Biophysica Acta,* **307**:133–140.

Konig, H., Pfisterer, P., Corcoran, L.M. and Wirth, T. (1995). Identification of CD36 as the first gene dependent on the B-cell differentiation factor Oct-2. *Genes and Development,* **9**:1598–1607.

Kusaka, Y., Tanaka, T., Okamoto, F., Terasaki, F., Matsunaga, Y., Miyazaki, H. and Kawamura, K. (1995). Effect of sulfo-N-succinimidyl palmitate on the rat heart: myocardial long-chain fatty acid uptake and cardiac hypertrophy. *Journal of Molecular and cellular Cardiology,* **27**:1605–1612.

Kuzu, I., Bicknell, R., Harris, A.L., Jones, M., Gatter, K.C. and Mason, D.Y. (1992). Heterogeneity of vascular endothelial cells with relevance to diagnosis of vascular tumours. *Journal of Clinical Pathology,* **45**:143–148.

Lawler, J.W., Chao, F.C. and Fang, P.-H. (1977). Observation of a high molecular weight platelet protein released by thrombin. *Thromb. Haemostasis.* **37**:355–357.

Leung, L.L.K., Li, W-X., McGregor, J.L., Albrecht, G. and Howard, R.J. (1992). CD36 peptides enhance or inhibit CD36-thrombospondin binding: A two step process of ligand receptor interaction. *Journal of Biological Chemistry,* **267**:18244–18250.

Li, W-X., Howard, R.J. and Leung, L.L.K. (1993). Identification of SVTCG in thrombospondin as the conformation-dependent, high affinity binding site for its receptor, CD36. *Journal of Biological Chemistry,* **268**:16179–16184.

Lisanti, M.P., Scherer, P.E., Vidugiriene, J., Tang, Z., Hermanowski-Vosatka, A., Tu, Y. *et al.* (1994). Characterization of caveolin-rich domains isolated from an endothelial-rich source: implications for human disease. *Journal of Cell Biology,* **126**:111–126.

Matsuno, K., Diaz-Ricart, M., Montgomery, R.R., Aster, R.H., Jamieson, G.A. and Tandon, N.N. (1996). Inhibition of platelet adhesion to collagen by monoclonal anti-CD36 antibodies. *British Journal of Haematology,* **92**:960–967.

Mazurov, A.V., Vinogradov, D.V., Vlasik, T.N., Burns, G.F. and Berndt, M.C. (1992). Heterogeneity of platelet Fc-receptor-dependent response to activating monoclonal antibodies. *Platelets,* **3**:181–188.

McGregor, J.L., Catimel, B., Parmentier, S., Clezardin, P., Dechavanne, M. and Leung, L-L.K. (1989). Rapid purification and partial characterization of human platelet GPIIIb: Interaction with thrombospondin and its role in platelet aggregation. *Journal of Biological Chemistry,* **264**:501–506.

McKeown, L., Vail, M., Williams, S., Kramer, W., Hansmann, K. and Gralnick, H. (1994). Platelet adhesion to collagen in individuals lacking glycoprotein IV. *Blood,* **83**:2866–2871.

Michelson, A.D., Wencel-Drake, J.D., Kestin, A.S. and Barnard, M.R. (1994). Platelet activation results in a redistribution of glycoprotein IV (CD36). *Arteriosclerosis and Thrombosis,* **14**:1193–1201.

Miklos, S., Jr. and Janos, H. (1987). Expression of OKM5 antigen on human keratinocytes in positive intracutaneous tests for delayed-type hyersensitivity. *Dermatologica,* **175**: 121–125.

Monier, S, Parton, R.G., Vogel, F., Behlke, J., Henske, A. and Kurzchalia, T.V. (1995). VIP21-Caveolin, a membrane protein constituent of the caveolar coat, oligomerizes *in vivo* and *in vitro. Molecular Biology of the Cell,* **6**:911–927.

Nagy, L., Tontonoz, P., Alvarez, J.G.A., Chen, H. and Evans, R.M. (1998). Oxidised LDL regulates macrophage gene expression through ligand activation of PPARγ. *Cell,* **93**:229–240.

Nakata, N., Furukawa, K., Greenwalt, D.E., Sato, T. and Kobata, A. (1993). Structural study of the sugar chains of CD36 purified from bovine mammary epithelial cells: occurrence of novel hybrid-type sugar chains containing the Neu5Ac $\alpha_2 \rightarrow$ 6GalNAc $\beta_1 \rightarrow$ 4GlcNAc and the Man $\alpha_1 \rightarrow$ 2Man $\alpha_1 \rightarrow$ 3Man $\alpha_1 \rightarrow$ 6Man groups. *Biochemistry,* **32**:4369–4383.

Nicholson, A.C., Frieda, S., Pearce, A. and Silverstein, R.L. (1995). Oxidized LDL binds to CD36 on human monocyte-derived macrophages and transfected cell lines. Evidence implicating the lipid moiety of the lipoprotein as the binding site. *Arteriosclerosis, Thrombosis and Vascular Biology*, **15**:269–275.

Noguchi, K., Naito, M., Tezuka, K., Ishii, S., Seimiya, H., Sugimoto, Y. and Amann, E. (1993). cDNA expression cloning of a 85-kDa protein overexpressed in adriamycin-resistant cells. *Biochemical and Biophysical Research Communications*, **192**:88–95.

Nozaki, S., Kashiwagi, H., Yamashita, S., Nakagawa, T., Kostner, B., Tomiyama, Y. *et al.* (1995). Reduced uptake of oxidized low density lipoproteins in monocyte-derived macrophages from CD36-deficient subjects. *Journal of Clinical Investigation*, **96**:1859–1865.

Ockenhouse, C.F. and Chulay, J.D. (1988). Plasmodium falciparum sequestration: OKM5 antigen (CD36) mediates cytoadherence of parasitized erythrocytes to a myelomonocytic cell line. *Journal of Infectious Diseases*, **157**:584–588.

Ockenhouse, C.F., Ho, M., Tandon, N.N., Van Seventer, G.A, Shaw, S., White, N.J. *et al.* (1991). Molecular basis of sequestration in severe and uncomplicated *Plasmodium falciparum* malaria: differential adhesion of infected erythrocytes to CD36 and ICAM-1. *Journal of Infectious Diseases*, **164**:163–169.

Ockenhouse, C.F., Magowan, C. and Chulay, J.D. (1989a). Activation of monocytes and platelets by monoclonal antibodies or malaria-infected erythrocytes binding to the CD36 surface receptor *in vitro*. *Journal of Clinical Investigation*, **84**:468–475.

Ockenhouse, C.F., Tandon, N.N., Magowan, C., Jamieson, G.A. and Chulay, J.D. (1989b). Identification of a platelet membrane glycoprotein as *Falciparum* malaria sequestration receptor. *Science*, **243**:1469–1471.

Okumura, T. and Jamieson, G.A. (1976). Platelet glycocalicin: Orientation of glycoproteins of the human platelet surface. *Journal of Biological Chemistry*, **251**:5944–5949.

Oquendo, P., Hundt, E., Lawler, J. and Seed, B. (1989). CD36 directly mediates cytoadherence of *Plasmodium falciparum* parisitized erythrocytes. *Cell*, **58**:95–101.

Panton, L.J., Leech, J.H., Miller, L.H. and Howard, R.J. (1987). Cytoadherence of *Plasmodium falciparum*-infected erythrocytes to human melanoma cell lines correlates with surface OKM5 antigen. *Infection and Immunity*, **55**:2754–2758.

Pearce, S.F.A., Wu, J. and Silverstein, R.L. (1994). A carboxy terminal truncation mutant of CD36 is secreted and binds thrombospondin: Evidence for a single transmembrane domain. *Blood*, **84**:384–389.

Pearce, S.F.A., Wu, J. and Silverstein, R.L. (1995). Recombinant GST/CD36 fusion proteins define a thrombospondin binding domain. Evidence for a single calcium-dependent binding site on CD36. *Journal of Biological Chemistry*, **270**:2981–2986.

Podolsak, B. (1977). Effect of thrombin, chymotrypsin and aggregate gammaglobulins on the human platelet membrane. *Thrombosis and Haemostasis*, **37**:396–406.

Puente Navazo, M.D., Daviet, L., Ninio, E. and McGregor, J.L. (1996a). Identification on human CD36 of a domain (155–183) implicated in binding oxidized low-density lipoproteins (Ox-LDL). *Arteriosclerosis, Thrombosis and Vascular Biology*, **16**:1033–1039.

Puente Navazo, M.D., Daviet, L., Savill, J., Ren, Y., Leung, L.K. and McGregor, J.L. (1996b). Identification of a domain (155–183) on CD36 implicated in the phagocytosis of apoptotic neutrophils. *Journal of Biological Chemistry*, **271**:15381–15385.

Ren, Y., Silverstein, R.L., Allen, J. and Savill, J. (1995). CD36 gene transfer confers capacity for phagocytosis of cells undergoing apoptosis. *Journal of Experimental Medicine*, **181**:1857–1862.

Rhinehart-Jones, T. and Greenwalt, D.E. (1996). A detergent-sensitive 113-kDa conformer/complex of CD36 exists on the platelet surface. *Archives of Biochemistry and Biophysics*, **326**:115–118.

Rigotti, A., Acton, S.L. and Krieger, M. (1995). The class B Scavenger Receptors SR-B1 and CD36 are receptors for anionic phospholipids. *Journal of Biological Chemistry*, **270**:16221–16224.

Roberts, D.D., Sherwood, J.A. and Ginsberg, V. (1987). Platelet thrombospondin mediates attachment and spreading of human melanoma cells. *Journal of Cell Biology*, **104**:131–139.

Rouabhia, M., Jobin, N., Doucet, R., Bergeron, J. and Auger, F.A. (1994). CD36[+]-dendritic epidermal cells: a putative actor in the cutaneous immune system. *Cell Transplantation*, **3**:529–536.

Rudd, C.E. (1990). CD4, CD8 and the TCR-CD3 complex: a novel class of protein tyrosine kinase receptor. *Immunology Today*, **11**:400–406.

Ryeom, S.W., Silverstein, R.L., Scotto, A. and Sparrow, J.R. (1996a). Binding of anionic phospholipids to retinal pigment epithelium may be mediated by the scavenger receptor CD36. *Journal of Biological Chemistry*, **34**:20536–20539.

Ryeom, S., Sparrow, J. and Silverstein, R.L. (1996b). CD36 functions as a receptor on retinal pigment epithelium for binding and uptake of photoreceptor outer segments. *Journal of Cell Science*, **109**:387–395.

Santoro, S.A. (1988). Molecular basis of platelet adhesion to collagen. In *Platelet membrane receptors*, edited by G.A. Jamieson, p. 291. New York: Liss.

Savill, J., Dransfield, I., Hogg, N. and Haslett, C. (1990). Vitronectin receptor-mediated phagocytosis of cells undergoing apoptosis. *Nature*. **343**:170–173.

Savill, J., Nancy, H., Ren, Y. and Haslett, C. (1992). Thrombospondin cooperates with CD36 and the vitronectin receptor in macrophage recognition of neutrophils undergoing apoptosis. *Journal of Clinical Investigation*, **90**:1513–1522.

Schuepp, B.J., Pfister, H., Clemetson, K.J., Silverstein, R.L. and Jungi, T.W. (1991). CD36-mediated signal trans-duction in human monocytes by anti-CD36 antibodies but not by anti-thrombospondin antibodies recogniz-ing cell membrane bound thrombospondin. *Biochemical and Biophysical Research Communications*, **175**:236–270.

Shattil, S.J. and Brugge, J.S. (1991). Protein tyrosine phosphorylation and the adhesive functions of platelets. *Current Opinions in Cell Biology*, **3**:869–879.

Si, Z. and Hersey, P. (1994). Immunohistological examination of the relationship between metastatic potential and expression of adhesion molecules and "selectins" on melanoma cells. *Pathology*, **26**:6–15.

Silverstein, R.L., Asch, A.S. and Nachman, R.L. (1989). Glycoprotein IV mediates thrombospondin-dependent platelet-monocyte and platelet-U937 cell adhesion. *Journal of Clinical Investigation*, **84**:546–552.

Silverstein, R.L., Baird, M., Lo, S.K. and Yesner, L.M. (1992). Sense and antisense cDNA transfection of CD36 (Glycoprotein IV) in melanoma cells. Role of CD36 as a thrombospondin receptor. *Journal of Biological Chemistry*, **267**:16607–16612.

Silverstein, R.L., La Salla, J. and Pearce, S.F. (1995). CD36 cluster workshop report. In *Leukocyte Typing V: White cell differentiation antigens*, edited by S.F. Schlossman *et al.*, pp. 1269–1271. New York: Oxford University Press.

Spitsberg, V.L., Matitashvili, E. and Gorewit, R.C. (1995). Association and coexpression of fatty-acid-binding pro-tein and glycoprotein CD36 in the bovine mammary gland. *European Journal of Biochemistry*, **230**:872–878.

Steinberg, D. (1997). Low density lipoprotein oxidation and its pathobiological significance. *Journal of Biolo-gical Chemistry*, **272**:20963–20966.

Stomski, F.C., Gani, J.S., Bates, R.C. and Burns, G.F. (1992). Adhesion to thrombospondin by human embryonic fibroblasts is mediated by multiple receptors and includes a role for glycoprotein 88 (CD36). *Experimental Cell Research*, **198**: 85–92.

Swerlick, R.A., Lee, K.H., Wick, T.M. and Lawley, T.J. (1992). Human dermal microvascular endothelial but not human umbilical vein endothelial cells express CD36 *in vivo* and *in vitro*. *Journal of Immunology*, **148**:78–83.

Tanaka, T. and Kawamura, K. (1995). Isolation of myocardial membrane long-chain fatty acid-binding protein: homology with a rat membrane protein implicated in the binding or transport of long-chain fatty acids. *Journal of Molecular and Cellular Cardiology*, **27**:1613–1622.

Tanaka, T., Sohmiya, K. and Kawamura, K. (1997). Is CD36 deficiency an etiology of hereditary hypertrophic cardiomyopathy? *Journal of Molecular and Cellular Cardiology*, **29**:121–127.

Tandon, N.N., Kralisz, U. and Jamieson, G.A. (1989a). Identification of glycoprotein IV (CD36) as a primary receptor for platelet-collagen adhesion. *Journal of Biological Chemistry*, **264**:7576–7583.

Tandon, N.N., Lipsky, R.H., Burgess, W.H. and Jamieson, G.A. (1989b). Isolation and characterization of platelet glycoprotein IV (CD36). *Journal of Biological Chemistry*, **264**:7570–7575.

Tandon, N.N., Matsuno, K., Ockenhouse, C.F. and Jamieson, G.A. (1995). Effect of CD36 mAb on collagen-induced platelet aggregation, adhesion to immobilised collagen, and the adherence of malaria-infected red blood cells to C32 melanoma cells. In *Leukocyte Typing V: White cell differentiation antigens*, edited by S.F. Schlossman *et al.*, pp. 287–1289. New York: Oxford University Press.

Tandon, N.N., Ockenhouse, C.F., Greco, N.J. and Jamieson, G.A. (1991). Adhesive functions of platelets lacking glycoprotein IV (CD36). *Blood*, **78**:2809–2813.

Tandon, N.N., Rock, G. and Jamieson, G.A. (1994). Anti-CD36 antibodies in thrombotic thrombocytopenic pur-pura. *British Journal of Haematology*, **88**:816–825.

Tang, Y., Taylor, K.T., Sobieski, D.A., Medved, E.S. and Lipsky, R.H. (1994). Identification of a human CD36 isoform produced by exon skipping. Conservation of exon organization and pre-mRNA splicing patterns with a CD36 gene family member, CLA-1. *Journal of Biological Chemistry*, **269**:6011–6015.

Tao, N., Wagner, S.J., and Lublin, D.M. (1996). CD36 is palmitoylated on both N- and C-terminal cytoplasmic tails. *Journal of Biological Chemistry*, **271**:22315–22320.

Taylor, K.T., Tang, Y., Sobieski, D.A. and Lipsky, R.H. (1993). Characterization of two alternatively spliced 5′-untranslated exons of the human CD36 gene in different cell types. *Gene*, **133**:205–212.

Thibert, V., Cristofari, M., Romagne, O. and Legrand, C. (1995). Characterisation of platelet panel CD36 mAb and a study of their effects on platelet activation. In *Leukocyte Typing V: White cell differentiation antigens*, edited by S.F. Schlossman *et al.*, pp. 1274–1278. New York: Oxford University Press.

Thorne, R.F., Meldrum, C.J., Harris, S.J., Dorahy, D.J., Shafren, D.R., Berndt, *et al.* (1997). CD36 forms covalently associated dimers and multimers in platelets and transfected COS-7 cells. *Biochemical and Biophysical Research Communications*, **240**:812–818.

Tomiyama, Y., Take, H., Ikeda, H., Mitani, T., Furubayashi, T., Mizutani, H., *et al.* (1990). Identification of the platelet specific alloantigen, Nak[a], on platelet membrane glycoprotein IV. *Blood*, **75**:684–687.

Trezzini, C., Jungi, T.W., Spycher, M.O., Maly, F.E. and Rao, P. (1990). Human monocyte CD36 and CD16 are signalling molecules. Evidence from studies using antibody-induced chemiluminescence as a tool to probe signal transduction. *Immunology*, **71**:29–37.

Turner, G., Craig, A., Pinches, R., Newbold, C. and Berendt, A. (1995). Characterisation of platelet panel CD36 mAb and their effects on Plasmodium falciparum-infected erythrocyte binding *in vitro*. In *Leukocyte Typing V: White cell differentiation antigens.*, edited by S.F. Schlossman *et al.*, pp. 1271–1274. New York: Oxford University Press.

Turner, G.D, Morrison, H., Jones, M., Davis, T.M., Looareesuwan, S., Buley, I.D., *et al.* (1994). An immunohisto-chemical study of the pathology of fatal malaria. Evidence for widespread endothelial activation and a potential role for intercellular adhesion molecule-1 in cerebral sequestration. *American Journal of Pathology*, **145**:1057–1069.

Union, A., De Baetselier, P. and De Smet, W. (1995). Upregulation of CD36 and CDw17 antigens on *in vitro* generated human foam cells. In *Leukocyte Typing V: White cell differentiation antigens*, edited by S.F. Schlossman *et al.*, pp. 1282–1285. New York: Oxford University Press.

Urwijitaroon, Y., Barusrux, S., Romphruk, A. and Puapairoj, C. (1995). Frequency of human platelet antigens among blood donors in northeastern Thailand. *Transfusion*, **35**:868–870.

van Nieuwenhoven, F.A., Verstijnen, C.P., Abumrad, N.A., Willemsen, P.H., Van Eys, G.J., Van der Vusse, G.J., *et al.* (1995). Putative membrane fatty acid translocase and cytoplasmic fatty acid-binding protein are co-expressed in rat heart and skeletal muscles. *Biochemical and Biophysical Research Communications*, **207**:747–752.

van Schravendijk, M-R., Handunnetti, S.M., Barnwell, J.W. and Howard, R.J. (1992). Normal human erythro-cytes express CD36, an adhesion molecule of monocytes, platelets, and endothelial cells. *Blood*, **80**:2105–2114.

van Schravendijk, M-R., Rock, E.P., Marsh, K., Ito, Y., Aikawa, M., Neequaye, J. *et al.* (1991). Characterisation and localisation of *Plasmodium falciparum* surface antigens on infected erythrocytes from West Africa. *Blood*, **78**:226–236.

Vega, M.A., Rodriguez, F., Segui, B., Cales, C., Alcalde, J. and Sandoval, I.V. (1991a). Targeting of lysosomal integral membrane protein LIMP II. The tyrosine-lacking carboxyl cytoplasmic tail of LIMP II is sufficient for direct targeting to lysosomes. *Journal of Biological Chemistry*, **266**:16269–16272.

Vega, M.A., Segui-Real, B., Garcia, J.A., Cales, C., Rodriguez, F., Vanderkerckhove, J. and Sandoval, I.V. (1991b). Cloning, sequencing and expression of a cDNA encoding rat LIMP II, a novel 74-kDa lysosomal membrane protein related to the surface adhesion protein CD36. *Journal of Biological Chemistry*, **266**:16818–16824.

Wyler, B., Daviet, L., Bortkiewicz, H., Bordet, J.-C. and J.L. McGregor (1993). Cloning of the cDNA encoding human platelet CD36: Comparison to PCR amplified fragments of monocyte, endothelial and HEL cells. *Thrombosis and Haemostasis*, **70**:500–505.

Wong, J.E.L., Asch, A.S., Silverstein, R.L. and Nachman, R.L. (1989). Glycoprotein IV expression in C32 human melanoma cell culture is a marker for a more malignant phenotype. (Abstract) *Blood*, **74** Suppl.1:638.

Yamamoto, N., Akamatsu, N., Sakuraba, H., Yamazaki, H. and Tanoue, K. (1994). Platelet glycoprotein IV (CD36) deficiency is associated with the absence (type I) or the presence (type II) of glycoprotein IV on monocytes. *Blood*, **83**:392–397.

Yamamoto, N., Ikeda, H., Tandon, N.N., Herman, J., Tomiyama, Y., Mitani, T. *et al.* (1990). A platelet membrane glycoprotein (GP) deficiency in healthy blood donors: Nak^{a}platelets lack detectable GPIV (CD36). *Blood*, **76**:1698–1704.

Yesner, L.M., Huh, H.Y., Pearce, S.F. and Silverstein, R.L. (1996). Regulation of monocyte CD36 and throm-bospondin-1 expression by soluble mediators. *Arteriosclerosis, Thrombosis, and Vascular Biology.* **16**:1019–1025.

9 CD9 Structure and Function

Lisa K. Jennings, Joseph T. Crossno, Jr., and Melanie M. White

Department of Medicine and Biochemistry, University of Tennessee; 956 Court Avenue, Coleman Building Room H335, Memphis, TN 38163, Tel: 901-448-5067, Fax: 901-448-7181

CD9 has been described in the literature for almost two decades yet its structure was only identified within the last few years. CD9 belongs to the tetraspanin family of proteins that is also referred to as the Transmembrane 4 (TM4) Superfamily. This chapter describes the functional characterization of CD9 in platelets as well as in other cell types. Anti-CD9 mAbs activate platelets via crosslinking of CD9 and the FcγRII receptor. Evidence suggests that CD9 may be in complex with the adhesive protein receptors, the integrins, and function in mechanisms associated with cell spreading and movement.

THE IDENTIFICATION OF CD9

The p24/CD9 antigen is an integral transmembrane protein of apparent molecular weight of 24 kilodaltons (kDa) and is expressed on the surface of hematopoietic and nonhematopoietic cells. This protein, earlier termed p24/BA-2, was first reported by Kersey *et al.* (1981), who described a monoclonal antibody (mAb) raised against a pre-B cell line, NALM-6 (Hurwitz *et al.*, 1979). Immunofluorescent binding assays demonstrated that the p24/BA-2 protein was detectable on the surface of bone marrow lymphohemopoietic precursors and most non-T, non-B acute lymphocytic leukemia (ALL) cells. It was not detectable on mature peripheral blood T and B lymphocytes. Some epithelial carcinomas and neuroectodermal malignancies also expressed the p24/BA-2 protein.

Subsequently, other anti-p24 antibodies were later described to recognize cell surface antigens in the 24 kilodalton (kDa) molecular weight range. The J2 mAb recognized a cell surface protein on mitogen or alloantigen stimulated T cells of apparent molecular weight 25 to 28 kDa, designated glycoprotein 26 (GP 26). In normal unstimulated peripheral blood, GP 26 was not expressed on mature lymphocytes, monocytes, granulocytes, or erythrocytes but was present on platelets (Hercend *et al.*, 1981). Jones *et al.* (1982) generated and characterized the DU-ALL-1 mAb raised against common acute lymphoblastic leukemia (cALL) cells. It reacted with a 24 kDa molecular weight protein on cALL cells and platelets, but did not bind significantly to normal bone marrow cells, peripheral blood lymphocytes (PBL), monocytes, granulocytes, or mitogen-activated lymphocytes. Using an indirect immunoperoxidase technique on frozen tissue sections, the DU-ALL-1 mAb labeled a variety of nonhematopoietic tissues including kidney distal tubules and glomeruli, epithelial cells of mammary ducts, pulmonary alveoli, intestine, cardiac and smooth muscle cells, as well as blood vessel endothelium (Jones *et al.*, 1982). The CALL1 mAb generated against a human schwannoma tumor immunoprecipitated a 26 kDa protein from cALL cells (Deng *et al.*, 1983). The p26/CALL1 antigen was found to be expressed on platelets and CALL cell lines as well as megakaryocytes and lymphoblastic cells from a patient with chronic myelocytic leukemia (CML) in blast crisis (Deng *et al.*, 1983). In 1983, Komada *et al.* characterized the SJ-9A4 mAb generated against cALL cell lines, NALM-1 and NALM-16. SJ-9A4 reacted with a p24 antigen on cALL cells, B-cell chronic lymphocytic leukemia cells, platelets, and neuroblastoma cell lines (Komada *et al.*, 1983). At the First International Workshop on Leukocyte Differentiation Antigens in Paris, the p24 molecule was designated as a cluster of differentiation, CD9, and hereafter referred to as p24/CD9 (Bernard *et al.*, 1984).

ACTIVATION OF PLATELETS BY ANTI-CD9 MONOCLONAL ANTIBODIES

All anti-CD9 mAbs described cause platelet activation and subsequent platelet aggregation (Boucheix *et al.*, 1987; Jennings *et al.*, 1990). Anti-CD9 mAb-induced platelet aggregation was first reported by Boucheix *et al.* (1983) who produced a set of monoclonal antibodies against a cALL cell line. The ALB6 mAb showed a pattern of reactivity against various tissues that was similar to the BA-2 mAb (Boucheix *et al.*, 1985) and immunoprecipitated a 24kDa protein from platelets (Boucheix *et al.*, 1983). To analyze the possible role of p24/CD9 within the platelet membrane, the effects of ALB6 on platelet function were examined (Boucheix *et al.*, 1983). ALB6 induced platelet aggregation of normal platelets and platelets from patients with Bernard Soulier syndrome, which lack a functional glycoprotein Ib-IX (GP Ib-IX) on their platelet membrane surface (Boucheix *et al.*, 1983). The addition of ALB6 mAb to stirred platelet suspensions of both platelet-rich plasma (PRP) as well as washed platelets without the addition of fibrinogen resulted in an equivalent aggregation response. ALB6 mAb induced platelet aggregation required intact bivalent IgG antibody as ALB6 Fab fragments could not induce aggregation (Boucheix *et al.*, 1983). Also, ALB6 Fab fragments inhibited platelet aggregation induced by ADP, collagen, thrombin, intact ALB6 IgG, and A23187 ionophore (Boucheix *et al.*, 1983). ALB6 induced platelet aggregation was inhibited by chlorpromazine, diltiazem, verapamil, EDTA (ethylene diamine tetraacetic acid), and prostaglandin E_1 (PGE_1), suggesting that ALB6 induced platelet aggregation could involve an interaction with a putative calcium channel or a membrane protein involved in calcium (Ca^{2+}) flux (Boucheix *et al.*, 1983).

Studies by Enouf *et al.* (1985) also suggested an involvement of p24/CD9 in the control of calcium fluxes across lipid bilayers. They reported an increased uptake of calcium into isolated membrane vesicles containing p24/CD9 in the presence of increasing concentrations of ALB6. Azzarone *et al.* (1985) reported that ALB6 addition could restore the fibrin clot retraction efficiency of post-confluent human embryonic fibroblasts. Moreover, intact ALB6 IgG overcame the inhibitory effect of the calcium-channel blocking drugs, diltiazem and verapamil, on fibroblast-mediated retraction of a fibrin clot (Azzarone *et al.*, 1985). These results implicated p24/CD9 in the regulation of calcium-dependent interactions of adhesive proteins with their receptors (Azzarone *et al.*, 1985).

Rendu *et al.* (1987) demonstrated that ALB6 induced platelet aggregation was similar to that of thrombin, inducing a transient decrease in polyphosphoinositides, a synthesis of phosphotidic acid, and phosphorylation of platelet proteins. Jennings *et al.* (1990) first characterized the mechanism of anti-CD9 platelet activation by using anti-CD9 mAb, mAb 7. Platelet aggregation induced by mAb 7 was primarily mediated by the activation of phospholipase C through a guanine nucleotide-binding protein (G protein) and was accompanied by the formation of the second messengers, diacylglycerol (DAG) and inositol phosphate(s), calcium mobilization and protein phosphorylation (Jennings *et al.*, 1989; Jennings *et al.*, 1990; Carroll *et al.*, 1990). Kroll *et al.* (1992) confirmed the activation of phospholipase C by an anti-p24/CD9 antibody. MAb 7-induced platelet activation also initiated a platelet cytoskeletal reorganization similar to that observed on thrombin activation of platelets (Jennings *et al.*, 1989) and caused the incorporation of the fibrinogen receptor, GP IIb-IIIa, into cytoskeletons isolated from aggregating platelets (Jennings *et al.*, 1990). Seehafer and Shaw (1991) suggested that CD9 was involved in signal transduction processes since a CD9 monoclonal antibody immunoprecipitated proteins from platelets in the apparent molecular weight range, 25 to 26kDa, that bound ^{32}P-GTP.

Platelet activation via anti-CD9 mAbs requires both the participation of the antigen binding region and the Fc portion of the intact antibody (Boucheix *et al.*, 1983; Gorman *et al.*, 1985; Miller *et al.*, 1986). Gorman *et al.* (1985) showed that three anti-CD9 mAbs (FMC 8, FMC 48, and FMC56) as F(ab')$_2$ and Fab' fragments did not induce platelet aggregation. In contrast to ALB6 fragments which inhibited platelet aggregation (Boucheix *et al.*, 1983), these F(ab')$_2$ and Fab' fragments primed the aggregation response of platelets to threshold levels of ADP and collagen. Miller *et al.* (1986) reported that the anti-CD9 antibody AG-1 as Fab fragments did not cause platelet activation, but rather inhibited the activation induced by intact AG-1 mAb (Miller, *et al.*, 1986). Worthington *et al.* (1990) demonstrated that anti-p24/CD9 mAb induced platelet activation and subsequent platelet aggregation required binding of the mAb to both its antigen (p24/CD9) and binding of the Fc portion of the intact mAb to the platelet low affinity FcγRII receptor (CD32). These data are supported by three observations. Firstly, pre-incubation with the mAb IV-3, an anti-FcγRII receptor specific mAb (Rosenfeld *et al.*, 1985), inhibits the anti-p24/CD9 mAb induced aggregation responses (Worthington *et al.*, 1990; Jennings *et al.*, 1994). Secondly, anti-p24/CD9 Fab' or F(ab')$_2$ fragments do not cause platelet activation or aggregation (Boucheix *et al.*, 1983b; Gorman *et al.*, 1985; Miller *et al.*, 1986; Worthington *et al.*, 1990). Third, Worthington *et al.* (1990) demonstrated that the Fc portion of an intact secondary antibody restored the capacity of CD9 mAb F(ab')$_2$ fragments to induce platelet activation and subsequent platelet aggregation. In contrast, the cross-linking of anti-CD9 F(ab')$_2$ with F(ab')$_2$ fragments of a secondary antibody did not induce platelet aggregation (Worthington *et al.*, 1990). These data would suggest that the ability of anti-p24/CD9 mAbs to induce platelet activation and subsequent platelet aggregation may be related in part to the proximity of a selective epitope on the p24/CD9 antigen to the Fc receptor and/or the mobility of these specific membrane components within the lipid bilayer. Although most evidence suggests that the Fc receptor is required for platelet activation by anti-CD9 mAbs, F(ab')2 fragments of one anti-CD9 were reported to induce signaling through activation of p72^{syk}and the association of p60c-src with p72^{syk}. This signaling was not sufficient to cause a full-scale platelet activation response (Ozaki *et al.*, 1995).

In an attempt to evaluate CD9 function without the confounding effect of Fc receptor ligation, Griffith *et al.* (1991) immobilized anti-CD9 F(ab')$_2$ onto polystyrene latex beads, and showed these beads induced platelet aggregation. Platelet activation and aggregation response to latex bead-immobilized anti-CD9 F(ab')$_2$ could be competitively inhibited by soluble anti-CD9 F(ab')$_2$. An aggregation dose-response relationship was evident by titrating the number of anti-CD9 F(ab')$_2$ beads added to the platelets. The platelet aggregation response to immobilized anti-CD9 F(ab')$_2$ stimulus was not inhibited by the preincubation with the FcγRII receptor specific blocking IV-3 mAb, confirming that the immobilization of CD9 alone was able to initiate the aggregation response. Griffith *et al.* (1991) hypothesized that CD9 might be a component of a receptor system responsive to a rigid stimulus such as the extracellular matrix, since immobilization of other platelet membrane proteins did not cause the generation of a platelet activation signal. Recent studies showed that when the FcγRII contribution to platelet activation was reduced by thrombin degranulation of platelets, signaling directly attributed to CD9 ligation by mAb SYB-1 could be measured (Slupsky *et al.*, 1997).

THE INTERACTION OF CD9 WITH OTHER ADHESIVE CELL SURFACE RECEPTORS

Slupsky *et al.* (1989) reported that the anti-p24/CD9 mAbs, 50H.19 and ALB$_6$, caused a physical association between p24/CD9 and the platelet membrane GPIIb-IIIa complex. Iodinated platelets were incubated with or without mAbs before the addition of the crosslinker, DSP (dithiobis(succinimidyl propionate). In platelets exposed to activating concentrations of anti-CD9 mAbs prior to treatment with crosslinker, CD9 was found to be associated with the fibrinogen receptor, GPIIb-IIIa. In contrast, the priming of platelets with two different GPIIb-IIIa complex specific mAbs failed to induce a detectable reciprocal association of CD9 with the GPIIb-IIIa complex. These data imply that upon anti-p24/CD9 mAb binding and platelet activation, the p24/CD9 antigen becomes proximal to GPIIb-IIIa. The contribution of the Fc receptor in this process and the functional relationship between p24/CD9 and GPIIb-IIIa is still unclear. Masellis-Smith *et al.* (1990) reported the homotypic aggregation of nucleated cells induced by CD9 specific antibodies. Pre-B lymphoblastoid cell lines, NALM-6 and HOON, stimulated with anti-CD9 mAbs formed large homotypic aggregates that were stable and resistant to mechanical disruption. The anti-CD9 mAb-induced homotypic aggregation required the expenditure of metabolic energy, a physiologic temperature of 37°C, a functionally intact cytoskeleton, and an absolute requirement for Ca^{2+} ions. The antibody induced aggregation did not appear to be integrin mediated, since it was not blocked by RGD-containing peptides. The role of Fc receptor (CD32) in these responses was not investigated. Crossno *et al.* (1990) found that while intact anti-CD9 mAb 7 caused homotypic pre-B ALL NALM-1 cell aggregation, the number of cell aggregates were diminished by 50% when the cells were preincubated with the FcγRII receptor specific IV-3 mAb. Further, the observed increase in NALM-1 cell cytosolic calcium induced by mAb 7 was completely inhibited by preincubation of the NALM-1 cells with FcγRII receptor specific IV-3 mAb (Crossno *et al.*, 1990). Thus, the activation and aggregation response of pre-B ALL cells appears to parallel the platelet response to intact, bivalent CD9 specific antibodies. Letarte *et al.* (1993) reported that the level of CD9 surface expression may determine the level of β_1 integrin-regulated homotypic adhesion, and thus the adhesive phenotype of pre-B leukemia cells. The β_1 integrins expressed on pre-B leukemia cells are $\alpha_4\beta_1$ and $\alpha_5\beta_1$, with $\alpha_4\beta_1$ predominating. Letarte *et al.* (1993) demonstrated that anti-α_4 and anti-β_1 integrin mAbs induced strong homotypic aggregation which was Fc-receptor independent and that the extent of aggregation correlated with the level of CD9 expression.

In a subsequent study, Rubinstein *et al.* (1994) compared the ability of the anti-CD9 mAbs, ALB6 and SYB-1, to promote aggregation of NALM-6 and HEL (human erythroleukemia) cell lines. The anti-CD9 mAb induced aggregation of both cell lines was independent of Fc receptor (CD32) interactions and was not mediated by fibronectin or by or $\alpha_5\beta_1$. The anti-CD9 induced aggregation was hypothesized to be mediated by CD9 acting directly as a cell adhesion molecule or through its association with another cell adhesion molecule (Rubinstein *et al.*, 1994). Immunoprecipitation analysis of both the NALM-6 and HEL cell lines with anti-CD9 mAbs performed using the mild detergent CHAPS (3-[(3-Cholamidopropyl)dimethylammonio]-1-proane-s ulfonate) revealed that the anti-CD9 SYB-1 mAb co-precipitated molecules of the same molecular weight as the α_4 and β_1 integrin subunits. However, CD9 was not co-precipitated with mAbs specific for $\alpha_4\beta_1$.

Immunoprecipitation experiments using the anti-CD9 SYB-1 mAb in which cells were first solubilized in 1% CHAPS, followed by a more stringent wash to elute co-precipitated proteins, demonstrated the presence of $\alpha_4\beta_1$ and $\alpha_5\beta_1$. However, CD9 was never visualized as co-precipitating with mAbs for $\alpha_4\beta_1$, $\alpha_5\beta_1$ or β_1. Although, the failure to demonstrate a reciprocal association of CD9 in immunoprecipitations with mAbs specific for $\alpha_4\beta_1$, $\alpha_5\beta_1$ or β_1 weakens evidence for their proposed association, studies by Berditchevski *et al.* (1996) have confirmed β_1 integrin/CD9 complexes by immunoprecipitation of solubilized cells under non-stringent detergent conditions and by crosslinking analysis. Recent studies by Longhurst *et al.* (1997) demonstrated that CD9 is in complex with platelet GPIIb-IIIa in the absence of crosslinking agents or pre-bound CD9 mAb. Furthermore, these CD9-GPIIb-IIIa complexes also consist of the Integrin-Associated Protein 50 (IAP-50) and another yet unidentified surface protein of 190kDa (manuscript submitted).

In another model system, Forsyth (1991) reported that human umbilical vein endothelial cells (HUVEC) expressed large amounts of surface CD9. Incubation of HUVEC with anti-CD9 mAbs increased neutrophil adherence to the endothelial cells. This adherence was not inhibited in the presence of protein synthesis inhibitors, or by antibodies specific for GP IIb-IIIa ($\alpha_{IIb}\beta_3$) or β_2 integrin (CD18). It was postulated that the increase in neutrophil adherence was due either to an activation change within the endothelial cell, expression of a pre-formed adhesive ligand, or an activation of a constituitively expressed ligand on the endothelial cell surface rendering it functional (Forsyth, 1991).

CLONING OF CD9

In 1991, the successful cloning, sequencing and stable expression of the full length cDNA for CD9 was reported (Lanza *et al.*, 1991). A full-length 1.3kbp (kilobase pair) cDNA clone was isolated from a λ gt10 transformed fibroblast cDNA library. It contained the entire coding sequence for p24/CD9 as well as the 5′- and 3′-untranslated regions and the poly(A) tail. The 1330bp cDNA sequence had an open reading frame of 684 nucleotides that encoded a mature protein of 228 amino acids with a calculated molecular mass of

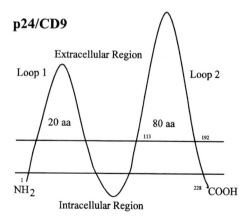

Figure 9.1 Proposed model for platelet p24/CD9 structure on the cell surface. CD9 is a single polypeptide chain of 228 amino acids with four hydrophobic sequences in the mature protein, which are believed to be transmembrane regions and two extracellular loop domains. The amino- and carboxyl- termini are postulated to be located intracellularly and the first and second extracellular loop domains are 20 and 80 amino acids in length, respectively.

25414kDa and a deduced pI of 7.2. The mature p24/CD9 protein contains 10 cysteine residues, six as juxtaposed pairs, with one potential asparagine-linked glycosylation site at residue 52 (Lanza *et al.*, 1991; Boucheix *et al.*, 1991). A hydrophilicity profile, using the method of Kyte and Doolittle (1982), showed four putative transmembrane domains (Lanza *et al.*, 1991 (Figure 9.1)). Studies of Triton X-114 solubilized platelets demonstrated that p24/CD9 partitioned in the detergent phase verifying its hydrophobic properties.

IDENTIFICATION OF CD9 AS A TETRASPANIN (TM4)

The predicted amino acid sequence of the mature p24/CD9 protein exhibits sequence homology to sixteen other cell surface proteins (Horejsi and Vlcek, 1991; Wright and Tomlinson, 1994); see Table 9.1. These homologues have a sequence identity ranging from

Table 9.1   Tetraspanin family members

CD9*	Jennings *et al.*, 1991
CD37	Classon *et al.*, 1989
CD53	Amiot, 1990
CD63/ME491*	Hotta *et al.*, 1988; Metzelaar *et al.*, 1991
CD81/TAPA-1	Oren *et al.*, 1990
CD82 (R2, C33, IA4, 4F9)	Gaugitsch *et al.*, 1991; Wright and Tomlinson,1994
CO-029	Szala *et al.*, 1990
PETA-3*	Fitter *et al.*, 1995
Sm23 and Sj23	Wright *et al.*, 1990; Davern *et al.*, 1991
A15	Emi *et al.*, 1993
TI-1	Kallin *et al.*, 1992
Uroplankins Ia and Ib	Yu *et al.*, 1994
L6	Marken *et al.*, 1992
periphren/RDS	Travis *et al.*, 1991
rom-1	Bascom *et al.*, 1992

* platelet-specific antigens.

20% to 45%, similar extracellular domain structures and dispersion within the membrane (Horejsi and Vlcek, 1991; Wright and Tomlinson, 1994). Three of the members are found in platelets: CD9, CD63/ME491, and PETA-3. CD9 and PETA-3 are expressed on the surface of resting, intact platelets whereas CD63 is located within platelet granules.

Collectively, these sixteen proteins define a new family of cell surface proteins which have been termed tetraspanins or the TM4 superfamily (Horejsi and Vlcek, 1991; Wright and Tomlinson, 1994). Characteristic features of all these homologues are a single polypeptide chain containing four putative membrane-spanning regions with both the N- and C-termini located intracellularly (Horejsi and Vlcek, 1991; Wright and Tomlinson, 1994). The highest degree of sequence identity between members of this family is observed within the putative transmembrane regions (Horejsi and Vlcek, 1991; Wright and Tomlinson, 1994). Other highly conserved features are six cysteine residues and some of the amino acids flanking them. It is not presently known how many of these or other cysteine residues are involved in stabilizing the structure of these molecules by forming cystine bridges (Horejsi and Vlcek, 1991; Wright and Tomlinson, 1994). In contrast, there are relatively large differences in the sizes of the putative major extracellular loops (Horejsi and Vlcek, 1991; Wright and Tomlinson, 1994). It is a reasonable hypothesis that the conserved transmembrane domains perform similar effector functions whereas the less conserved extracellular regions are involved in a tissue-specific function such as ligand recognition.

IDENTIFICATION OF CELL TYPES EXPRESSING CD9

Flow cytometric analyses using authentic anti-CD9 (mAb7) and Northern blot analysis under conditions of high stringency have been used as two independent methods to measure the relative expression level of CD9 in different tissue types. The highest CD9 surface expressing cells were human umbilical vein endothelium. CD9N3 CHO cells, human foreskin fibroblasts, and human metastatic melanoma cell line HS 294T were equivalent in expression. Mock-transfected CHO cells displayed no fluorescent surface labeling with the anti-CD9 antibody. The human neuroblastoma SK-N-SH, as well as the NALM-1 and HEL leukemia cell lines, demonstrated the least surface expression (Table 9.2).

Table 9.2 CD9 surface expression on cell lines.

Cell Line	Mean Fluorescent Index, MFI*
Mock CHO cells	0.00
CD9 N3 CHO cells	42.70
Fibroblasts	36.42
HUVEC	121.39
SK-N-SH	11.36
HS 294T	39.29
NALM-1	3.39
HEL	3.50

*MFI-Mean Fluorescent Index is determined by maximum binding of mAb 7 versus isotype matched control $MsIgG_{1k}$.

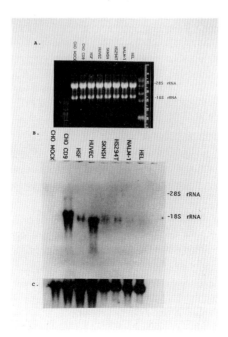

Figure 9.2   Northern blot analysis of CD9 surface expressing cell lines. Total cellular RNA was isolated from eight different cell lines, 15μg of total RNA was electrophoresed on denaturing formaldehyde agarose gels, and transferred to nitrocellulose membranes. Northern blots were hybridized with the 1.33kb cDNA for platelet p24/CD9 or the 2.0kb cDNA for chicken β-actin that was ^{32}P α dCTP labeled using a random primer DNA labeling kit. The blots were then washed at high stringency and analyzed by autoradiography. Illustrated are (A) ethidium bromide stained duplicate gel (B) autoradiogram of Northern blot hybridized with 1.3kb cDNA for platelet p24/CD9, and (C) autoradiogram of Northern blot hybridized with 2.0kb cDNA for chicken β-actin.

Northern blot analysis of reported CD9 surface expressing cell lines probed with CD9 cDNA demonstrated hybridization of a single mRNA transcript with all the reported CD9 expressing cell lines (Figure 9.2). The positive control CD9 N3 CHO cell line exhibited a single mRNA transcript with a large signal at high stringency. The CD9 N3 CHO cell mRNA CD9 transcript was slightly larger than the mRNA transcripts in the other cell lines, probably due to an incomplete processing of the extra poly-A tail on the mRNA in the CHO cell transfectants. At high stringency, the cell lines with the greatest hybridization signal were the HUVECs and human foreskin fibroblasts. Autoradiography of Northern blot hybridization with 2.0kb cDNA for chicken β-actin demonstrated equal basal RNA amounts in all eight cell lines (Figure 9.2C).

LOCALIZATION OF CD9 ON THE CELL SURFACE

Immunoelectron microscopy of spread, activated platelets probed with mAb-7-gold showed that CD9 was clustered in regions of platelet-platelet contacts (Figure 9.3), (L.K. Jennings

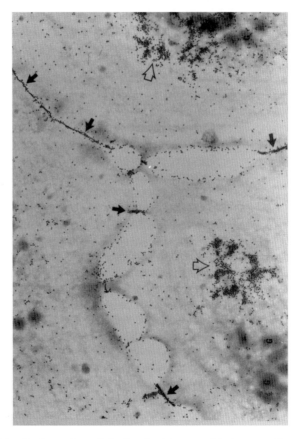

Figure 9.3   CD9 is localized at regions of platelet-platelet contact. Platelets were allowed to adhere to form-var coated, carbon stabilized grids and then probed with gold labeled mAb 7 or mouse IgG for 30 min at 37 °C. After processing, the samples were visualized by electron microscope. The gold label was primarily located at the centralized granulomere (open arrows) and at platelet margins. At points of platelet-platelet interaction, there was a dramatic concentration of mAb 7-gold (closed arrows). In some areas, the probe appears to overlie granules (G). However, several granules can be identified that are not associated with heavy label. Magnification = 33 462X.

and J. Mattson, manuscript in preparation). These studies also suggest that CD9 moves away from the periphery of the cell upon contact activation. The gold label was primarily located at the centralized granulomere (open arrows) and at platelet margins. At points of platelet-platelet interaction, there was a dramatic concentration of mAb 7-gold (closed arrows). CD9 localization at the leading edge of activated, spread platelets is compatible with the proposed role of tetraspanin proteins in cell migration. Furthermore, Cramer *et al.* (1994) showed that on resting normal platelets, CD9 was also located in the luminal face of the open canalicular system and on the α-granule limiting membrane.

EFFECTS OF CD9 EXPRESSION ON CELL PHENOTYPE

Evidence has been presented that the expression of CD9 on cells has a direct effect on the cell phenotype. In marked contrast to subconfluent, mock-transfected CHO cells which grow as loosely adherent, round or spindle shaped cells, surface expressing CD9-transfected CHO cells are more adherent at subconfluency and form colonies with polygonal boundaries. If cell cultures grow to confluency, the surface expressing CD9N3 CHO cells display a distinctive cobblestone, endothelial cell-like growth pattern at confluency (Figure 9.4a) suggesting that the expression of the CD9 protein on the CHO cell surface influences the phenotype of these clones. Mock-transfected CHO cells exhibit a swirled, fibroblast-like growth pattern similar to untransfected naive CHO cells (Figure 9.4b).

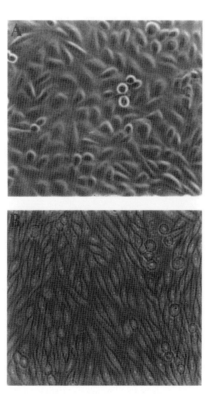

Figure 9.4   Expression of p24/CD9 in CHO cells results in a distinctive growth pattern. (A) CD9 N3 surface expressing CD9-transfected CHO cells exhibit a distinctive cobblestone, endothelial cell-like growth pattern. (B) Mock-transfected CHO cells exhibit a swirled, fibroblast-like growth pattern.

CD9-MEDIATED CELL SPREADING AND CELL MOTILITY

CHO cell secondary spreading response appears to be limited to fibronectin as compared to rat tail collagen type I, mouse collagen type IV, mouse laminin, human plasma fibrinogen, or poly-D-lysine. Substrata binding is specific and selective because both mock- and CD9-transfected CHO cells had secondary spreading solely on human plasma fibronectin (Figure 9.5). These results are not unexpected since both CHO cell lines express $\alpha_5\beta_1$, the "classic" fibronectin receptor on their cell surface. Adherent mock-transfected CHO cells exhibit a round, or elongated spindle shaped pattern, and display some degree of spreading on the fibronectin substrata (Figure 9.5a). However, the adherent CD9 N3 surface expressing CHO cells exhibit an enhanced spreading (approximately two-fold greater mean surface area, $p < 0.0001$) to the fibronectin coated wells (Figure 9.5b) similar to untransfected naive CHO

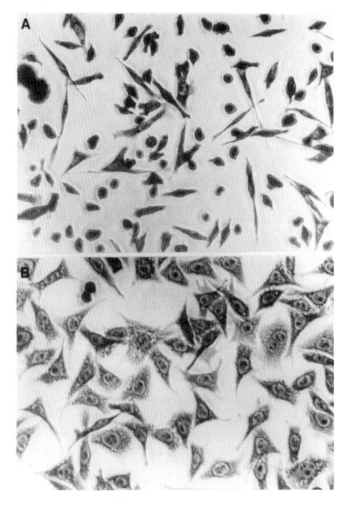

Figure 9.5   Adhesion of CHO cells to fibronectin. (A) mock-transfected CHO cells or (B) CD9 N3 surface expressing CD9-transfected CHO cells were trypsinized, washed and resuspended in adhesion buffer. Cells (1×10^6) were seeded in triplicate wells coated with 10 mg/ml fibronectin and incubated for 3 hours at 37°C. Non-adherent cells were removed and the wells washed with adhesion buffer three times. Adherent cells were then stained with modified Wright Giemsa stain and photographed under an inverted light microscope with a 40X objective. Total print magnification = 666X.

cells. These observations imply that CHO cell adherence to fibronectin is dramatically enhanced and augmented by the surface expression of CD9 on the CHO cell surface. Furthermore, an ELISA assay demonstrated that fibronectin bound to purified platelet derived CD9 or recombinant CD9, His_6-rCD9, with a K_d of 80 nm (Jennings *et al.*, unpublished data).

CD9 EXPRESSION ENHANCES HAPTOTACTIC MOTILITY OF CELLS

One of the most dramatic effects of CD9 expression and function on cells is on cell motility. Measurements of motility of mock- or CD9-transfected CHO cells migrating from the upper well of the Boyden chamber through a filter to the fibronectin-coated underside showed that CD9 expression has a direct effect on cell migration to fibronectin coated surfaces (Figure 9.6). The haptotactic motility of the CHO cells through the filter pores to the ligand coated underside is a time dependent phenomenon. There was greater than a four-fold increase in the number of CD9-transfected CHO cells adhering to the fibronectin coated filters compared to mock-transfected cells. These observations suggest the surface expression of CD9 enhanced the haptotactic motility of cells to fibronectin. Such observations would suggest that CD9 facilitates the spreading and movement of cells on fibronectin containing matrices.

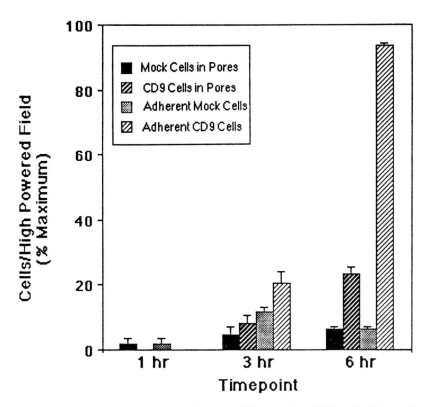

Figure 9.6 **Haptotactic motilities of mock-transfected and CD9-transfected CHO cells to fibronectin coated filters.** Using a modified Boyden chamber, the motility of mock- and CD9-transfected CHO cells migrating from the upper well of the chamber through a filter to its fibronectin-coated underside was assayed. Two types of cells were counted: filter pore cells and adherent cells to the fibronectin-coated underside were counted at 1, 3, and 6 hour time points. The results (*n* = 3) are expressed as a percent maximum of cells counted in 15 high powered fields.

CD9 AND CELL PROLIFERATION

Although not relevant to platelets, CD9 has been implicated as performing a significant role in proliferation and maturation in both normal and malignant cell populations. Previous studies have shown that CD9-transfected CHO cells proliferate 1.5–3 times faster than mock-transfected CHO cells (Jennings *et al.*, 1994). These results were confirmed by performing proliferation assays on cell cycle synchronized CHO cell lines and determining ^{3}H-thymidine incorporation. There was a two-fold increase in CD9 N3 proliferation rate versus the mock-transfected CHO cell negative control ($p < 0.0001$) (Table 9.3). Whether CD9 expression has a significant role in megakaryocyte growth and differentiation has not been determined.

Table 9.3   CD9-mediated proliferation of CHO cells

Clone	dpm ± SEM	P value
Mock CHO	31 404.50 ± 3090.31	
		p < 0.0001
CD9 N3	65 882.17 ± 4244.45	

Confluent cultures of each CHO cell line were trypsinized with 0.125% trypsin, washed with RPMI 1640 supplemented with 10% FBS by centrifugation at 800 x g, aliquoted into twelve wells of a 96 well microtiter plate at a density of 2 x 10^4/well in complete media. Cells were pulsed with 1 μCi ^{3}H-thymidine and placed in an 37 °C 5% humidified CO_2 incubator for eighteen hours. Cells were harvested onto glass fiber filters and ^{3}H-thymidine incorporation determined by counting on a Matrix 96 direct ionization β-counter (Packard Instrument Co., Meridan CT).

THE PHYSIOLOGICAL SIGNIFICANCE OF CD9 IN ADHESION PROCESSES

On the basis of these recent data, we believe that p24/CD9 functions as a cell adhesion molecule (CAM). This conclusion is derived from studies detailed above performed on Chinese Hamster Ovary (CHO) cells that have stable expression of recombinant protein on their cell surface. The major findings of these studies using the transfected CHO cell system as a model for understanding platelet CD9 are (1) surface expression of CD9 in CHO cells induces a change in the phenotypic growth pattern; (2) CD9 is localized at regions of cell to cell contact in adherent CD9 N3 surface expressing CD9-transfected CHO cells and human platelets; (3) CD9 N3 surface expressing CD9-transfected CHO cells have enhanced spreading on human plasma fibronectin; (4) surface expression of CD9 enhances the haptotactic motility of CHO cells to fibronectin; and (5) surface expression of CD9 increases the CHO cell proliferation rate.

The expression of cell adhesion molecules (CAMs) on the surface of transfected cells which do not normally express these proteins is a means to study CAM functions. This system has been used to confirm that potential CAMs mediate adhesion and promote changes in cell shape and behavior related to their function (Edelman *et al.*, 1987; Hatta *et al.*, 1988; Matsuzaki *et al.*, 1990). The surface expressing CD9-transfected CHO cell clone, CD9 N3, exhibited a distinctive cobblestone, endothelial cell-like cell growth pattern as compared to the swirled, fibroblastic growth pattern of naive untransfected and mock-transfected CHO cell lines. These results were supported by SEM of spread, surface activated platelets which showed that the highest concentration of immunogold labeled CD9 was observed at filopodial contact sites of adjacent platelets. Together, these results suggest a direct role of surface expression of p24/CD9 in platelet adhesive functions.

Cells are able to adhere to and migrate on a variety of extracellular matrix (ECM) proteins by virtue of the surface expression of several families of cell adhesion receptors (Albelda and Buck, 1990; Bevilacqua, 1993). CHO CD9-N3 cells expressing CD9 display an enhanced spreading and an increased haptotactic motility on human plasma fibronectin compared to mock-transfected CHO cells. The ability of CHO cells to adhere and migrate on a fibronectin (Fn) substrata is mediated in part by the "classic" integrin fibronectin receptor (FnR), $\alpha_5\beta_1$ (Hynes, 1992). This FnR is thought to exclusively bind Fn and mediate the adhesion and motility of cells on surfaces coated with this protein (Brown and Juliano, 1985; Pytela *et al.*, 1985; Straus *et al.*, 1989; and Bauer *et al.*, 1992). In other studies, CHO cells with very high expression of $\alpha_5\beta_1$ integrins (FnR) have a reduced motility and enhanced adhesion on fibronectin (Giancotti and Ruoslahti, 1990). Our data indicate that the directional haptotactic motility of CHO CD9-N3 cells on Fn coated filters was increased and augmented by the surface expression of CD9 on the CHO cell surface. Explanations for these observations are that CD9 may interact not only with fibronectin but also modulate or affect the functional activity of the $\alpha_5\beta_1$ integrin (FnR) or the β_1 subunit itself (Letarte *et al.*, 1993; Rubinstein *et al.*, 1994).

The increased spreading of surface expressing CD9 N3 CHO cells on fibronectin substratum as compared to the mock-transfected CHO cells could be due to a stabilization of the peripheral FnR pool to the cell surface as a result of CD9 expression. Also, CD9 could enhance the stabilization of peripheral FnR to promote the enhanced spreading of surface expressing CD9-transfected CHO cells. It remains to be determined whether CD9 may have a similar function in platelets. Evidence for the association of CD9 with integrins has been reported for $\alpha_{IIb}\beta_3$ in platelets (Slupsky *et al.*, 1989) as well as $\alpha_4\beta_1$ and $\alpha_5\beta_1$ (Masellis-Smith and Shaw, 1994; Rubinstein *et al.*, 1994). Further, co-immunoprecipitation of a homologue of CD9, CD63/ME491, a member of the tetraspanin family of cell surface proteins has been reported with $\alpha_3\beta_1$ and $\alpha_6\beta_1$ integrins (Berditchevski *et al.*, 1995). The melanoma associated antigen CD63 was shown to indirectly coprecipitate with $\alpha_3\beta_1$ and $\alpha_6\beta_1$, but not $\alpha_2\beta_1$ and $\alpha_5\beta_1$ integrins (Berditchevski *et al.*, 1995). These results collectively suggest that platelet CD9 and CD63 could be integrin-associated proteins.

The surface expression of CD9 also increased the CHO cell proliferative rate, compared to mock-transfected controls, by a factor of two (p value < 0.0001). These results confirm previous observations of an increased proliferative rate of CD9 expressing CHO cells (Jennings *et al.*, 1994), and support reports of CD9 homologues TAPA-1 (Oren *et al.*, 1990), CD37 (Ledbetter *et al.*, 1987), and CD63/ME491 (Hotta *et al.*, 1988) which have been implicated in the control of cell proliferation. The question that still remains is the mechanism by which these cell surface integral membrane proteins transmit signals intracellularly to control the regulation of gene expression and growth.

The results of these studies support the hypothesis that CD9 represents a newly discovered CAM, due to the change in the phenotypic CHO cell growth pattern secondary to CD9 surface expression and its location at regions of cell-cell contact. The identification of the ECM adhesive protein fibronectin as a ligand for CD9 reinforces the suggestion of CD9 as a new CAM. Adhesion and haptotactic motility studies demonstrate a difference between CHO cell adhesion, spreading, and motility based on the surface expression of CD9 and in the presence of active peptide regions of CD9. An increased proliferation rate in CD9 expressing CHO cells suggests that this new CAM can transmit intracellular signals to promote growth and differentiation in nucleated cells.

In summary, the question still remains as to the exact function of CD9 and other platelet TM4 proteins (PETA-3 and CD63) in platelet physiology. Our data strongly suggest that CD9 is a new adhesion molecule that mediates cell spreading and motility on the matrix fibronectin. Thus, it is possible that CD9 in platelets may have a role in the regulation of platelet adhesion and spreading directly or in combination with cell surface integrins. This hypothesis opens new and exciting avenues for the interaction of CD9 and other tetraspanins with molecules involved in vascular hemostasis.

Acknowledgements

This work was supported by the American Heart Association Grant-in-Aid and an Established Investigatorship Award, the National Heart, Lung and Blood Institute of the National Institutes of Health (HL53514) and the University of Tennessee Medical Group.

References

Albelda, S.M. and Buck, C.A. (1990). Integrins and other cell adhesion molecules. *FASEB J.*, **4**:2868–2800.

Amiot, M. (1990). Identification and analysis of cDNA clones encoding CD53: a pan-leukocyte antigen related to membrane transport proteins. *J. Immunol.*, **145**:4322–4328.

Azzarone, B., Krief, P., Soria, J. and Boucheix, C.(1985). Modulation of fibroblast-induced clot retraction by calcium channel blocking drugs and the monoclonal antibody ALB6. *J. Cell. Physiol.*, **125**:420–426.

Bascom, R.A., Manara, S., Collins, L., Molday, R.S., Kalnims, V.I. and McInnes, R.R. (1992). Cloning of the cDNA for the novel photoreceptor membrane protein (rom-1) identifies a disk rim protein protein family implicated in human retinopathies. *Neuron*, **8**:1171–84.

Bauer, J.S., Schreiner, C.L., Giancotti, F.G., Ruoslahti, E. and Juliano, R.L. 1992). Motility of fibronectin receptor-deficient cells on fibronectin and vitronectin: Collaborative interactions among integrins. *J. Cell Biol.*, **116**:477–487.

Berditchevski, F., Zutter, M.M. and hemler, M.E. (1996). Characterization of novel complexes on the cell surface between integrins and proteins with 4 transmembrane domains (TM4)proteins). *Mol. Biol. Cell.*, **7**:193–207.

Bernard, A., Boumseu, L., Dausrot, J., Milstein, C. and Schlossman, S.F. (1984). Cluster of B2 protocol. In *Leukocyte Typing* Vol I, ed. A. Bernard, pp. 61–81.

Bevilacqua, M.P. (1993). Endothelial-leukocyte adhesion molecules. *Ann. Rev. Immunol.*, **11**:767–804.

Boucheix, C., Benoit, P., Kreif, P., Billard, M., Mishal. Z., Azzarone, B. *et al.* (1987). Platelet aggregation induced by CD9 Mabs. Mechanisms and comparison with platelet aggregating properties of mabs directed against other membrane antigens. In *Leukocte Typing III*, (ed. A.J. McMichael, pp. 780–782. Oxford University Press.

Boucheix, C., Benoit, P., Frachet, P. Billard, M., Worthington, R.E., Gagnon, J. and Uzan, G. (1991). Molecular cloning of the CD9 antigen. *J. Biol. Chem.*, **266**:117–122.

Boucheix, C., Perrot, J.Y., Mirshami, M., Giannini, F., Billard, M., Bernadai, A. and Rosenfeld, C. (1985). A new set of monoclonal antibodies against acut lymphoblastic leukemia. *Leuk. Res.*, **9**:597–604.

Boucheix, C., Soria, C., Mirshahi, M., Soria, J., Perrot, J.Y., Fournier, M., Billard, M. and Rosenfeld, C. (1983). Characteristics of platelet aggregation induced by the monoclonal antibody ALB6 (acute lymphoblastic leukemia antigen p24). *FEBS Lett.*, **161**:2, 289–295.

Brown, P.J. and Juliano, R.L. (1985). Monoclonal antibodies to a cell surface glycoprotein selectively inhibit fibronectin mediated cell adhesion. *Science*, **228**:1448–1451.

Carroll, R.C., Worthington, R.E. and Boucheix, C. (1990). Stimulus-response coupling in human platelets activated by monoclonal antibodies to the CD9 antigen, a 24 kDa surface-membrane glycoprotein. *Biochem. J.*, **266**:527–535.

Classon, B.J., Williams, A.F., Willis, A.C., Seed, B. and Stamenkovic, I. (1989). The primary structure of the human leukocyte antigen CD37, a species homologue of the rat MRC OX-44 antigen. *Exp. Med.*, **172**: 1007.

Crossno, J.T., Jr., Fox, C.F., Dockter, M.E. and Jennings, L.K. (1990). The p24/CD9 antigen in pre-B cell lines and platelets: Unique properties of a new family of multiple membrane spanning proteins. *Blood*, **76** (10):1804a.

Cramer, E.M., Berger, G. and Berndt, M.C. (1994). Platelet α-granule and plasma membrane share two new components: CD9 and PECAM-1. *Blood*, **84**:1722–30.

Davern, K.M., Wright, M.D., Herrmann, U.R. and Mitchell, G.F. (1991). Further characterization of the *Schistosoma japonicum* protein Sj23, a target antigen of an immunodiagnostic monoclonal antibody. *Mol. Biochem. Parasitol.*, **48**:67–76.

Deng, C-T, Terasaki, P.I., Iwaki, Y., Hofman, F.M., Koeffler, P., Cuhan, L., Awar, N. and Billing, R. (1983). A monoclonal antibody cross-reactive with human platelets, megakaryocytes, and common acute lymphocytic leukemia cells. *Blood*, **61**:759–764.

Edelman, G.M., Murray, B.A., Mege, R.M., Cunningham, B.A. and Gallin, W.J. (1987). Cellular expression of liver and neural adhesion cell molecules after transfection with cDNAs results in cell-cell binding. *Proc. Natl. Acad. Sci.USA*, **84**:8502–8506.

Emi, N., Kitaori, K., Seto, M., Ueda, R., Saito, H. and Takahashi, T. (1993). Isolation of a novel cDNA clone showing marked similarity to ME491/CD63 superfamily. *Immunogenetics*, **37**:193–198.

Enouf, J., Bredoux, R., Boucheix, C., Mirshahi, M., Soria, C. and Levy-Toledano, S. (1985). Possible involvement of two proteins (phosphoprotein and CD9 (p24)) in regulation of platelet calcium fluxes. *FEBS Lett.*, **183**:398–402.

Fitter, S., Tetaz, T.J.,Berndt, M.C. and Ashman, L.K. (1995). Molecular cloning of cDNA encoding a novel platelet-endothelial cell tetra-span antigen, PETA-3. *Blood*, **86**:1348–1355.

Forsyth, K.D. (1991). Anti-CD9 antibodies augment neutrophil adherence to endothelium. *Immunology*, **72**:292–6.

Gaugitsch, H.W., Hofer, E., Huber, N.E., Schnabi, E. and Baumruker, T. (1991). A new superfamily of lymphoid and melanoma cell cell proteins with extensive homolgy to Schistoma nmansoni antigen Sm23. *Eur. J. Immunol.*, **21**:377–383.

Giancotti, F.G. and Ruoslahti, E. (1990). Elevated levels of the $\alpha_v\beta_1$ fibronectin receptor suppress the transformed phenotype of Chinese Hamster ovary cells. *Cell*, **60**:849– 859.

Gorman, D.J., Castaldi, P.A., Zola, H. *et al.* (1985). Preliminary functional characterization of a 24 000 dalton platelet surface protein involved in platelet activation. *Nouv. Rev. Fr. Hematol.*, **27**:255–259.

Greaves, M.F., Verbi, W., Kemshead, J. *et al.* (1980). A monoclonal antibody identifying a cell surface antigen shared by a common acute lymphoblastic leukemias and B cell antigens. *Blood*, **56**:1141–1144.

Griffith, L., Slupsky, J., Seehafer, J., Boshkov, L. and Shaw, A.R.E. (1991). Platelet activation by immobilized monoclonal antibody: Evidence for a CD9 proximal signal. *Blood*, **78**:1753–1759.

Hato, T., Ikeda, K., Yasukawa, M., Watanabe, A. and Kobayashi, Y. (1988). Exposure of platelet fibrinogen receptors by a monoclonal antibody to CD9 antigen. *Blood*, **72**:224–229.

Hatta, K., Nose, A., Nagafuchi, A. and Takeichi, M. (1988). Cloning and expression of cDNA encoding a neural Ca-dependent cell adhesion molecule: Its identity in the cadherin gene family. *J. Cell Biol.* **106**:873–881.

Hercend, T., Nadler, L.M. Pesando, J.M. *et al.* (1981). Expression of a 26 000-Dalton glycoprotein on activated human T cells. *Cell Immunol.*, **64**:192 -199.

Horejsi, V. and Vlcek, C. (1991). Novel structurally distinct family of leukocyte surface glycoproteins including CD9, CD37, CD53, CD63. *FEBS Lett.*, **288**:1–5.

Hotta, H., Ross, A.H., Huebner, K., Isobe, M., Wendeborn, S., Chao, M.V., Ricciardi, R.P., Tsujimoto, Y., Croce, C.M. and Koprowski, H. (1988). Molecular cloning and characterization of an antigen associated with early stages of melanoma tumor progression. *Cancer Res.*, **48**:2955–2962.

Hurwitz, R., Hozier, J., LeBien, T., Minowada, J., Gajl-Peczalska, K.J., Kubomishi, I. and Kersey, J. (1979). Characterization of a leukemia cell line of the pre-B phenotype. *Int. J. Cancer*, **23**:174–181.

Hynes, R.O. (1992). Integrins: versatility, modulation, and signaling in cell adhesion. *Cell*, **69**:11–25.

Jennings, L.K., Fox, C.F. and Kouns, W.C.(1989). Platelet activation mediated by the surface protein p24/CD9. *Blood*, **74** (17):1091a.

Jennings, L.K., Fox, C.F., Kouns, W.C., McKay, C.P., Ballou, L.R. and Schultz, H.E. (1990). The activation of human platelets mediated by anti-human platelet p24/CD9 monoclonal antibodies. *J. Biol. Chem.*, **265**:3815–3822.

Jennings, L.K., Crossno, J.T., Jr., Fox, C.F., White, M.M. and Green, C.A. (1994). Platelet p24/CD9, A member of the tetraspanin family of proteins. *Ann. N. Y. Acad. Sci.*, **714**:175–184.

Jennings, L.K., Wilkinson, D.A. and Crossno, J.T., Jr. (1995). CD9 is a cell adhesion molecule on human platelets. *Blood*, **86**:282a.

Jones, N.H., Borowitz, M.J. and Metzgar, R.S. (1982). Characterization and distribution of a 24,000-molecular weight antigen defined by a monoclonal antibody (DU-ALL-1) elicited to comon acute lymphoblastic leukemia (cALL) cells. *Leuk. Res.*, **6**:449–454.

Kallin, B., DeMartin, R., Etzold, T., Sorrentino, U. and Philipson, L. (1992). Cloning of a growth arrest-specific and transforming growth Factor B-regulated gene TI-1 from the epithelial cell line. *Mol. Cell. Biol.*, **11**:5338–5345.

Kersey, J.H., LeBien, T. W., Abramson, C.S. *et al.* (1981). p24: A human leukemia-associated and lympho-hemopoietic progenitor cell surface structure identified with monoclonal antibody. *J. Exp. Med.*, **153**: 726–731.

Komada, Y., Peiper, S.C., Melvin, S.L. *et al.* (1983). A monoclonal antibody (SJ-9A4) to p24 present on common ALLS, neuroblastomas and platelets-I. Characterization and development of an unique radioimmunoassay. *Leuk. Res.*, **7**:487–498.

Kroll, M.H., Mendelsohn, M.E., Miller, J.L., Ballew, K.K., Hebolica, J.K. and Schafer, A.I. (1992). Monoclonal antibody AG-1 initiates platelet activation by a pathway dependent on glycoprotein IIb-IIIa and extracellular calcium. *Biochem. Biophys. Acta.*, **1137**:248–256.

Kyte, J. and Doolittle (1982). A simple method for displaying the hydropathic character of a protein. *J. Mol. Biol.*, **157**:105–132.

Lanza, F., Wolf, D., Fox, Caro, F., Kieffer, N., Seyer, J.M., Fried, V.A., Coughlin, S.R., Phillips, D.R. and Jennings, L.K. (1991). cDNA cloning and expression of platelet p24/CD9: Evidence for a new family of multiple membrane spanning proteins. *J. Biol. Chem.*, **266**:10638–45.

Longhurst, C.M., Wilkinson, D. and Jennings, L.K. (1997). Characterization of a noncovalent association between CD9, GPIIb/IIIa ($\alpha_{IIb}\beta_3$ integrin) and integrin associated protein (IAP50) isolated from human platelets. *J. Cell Biol*, Vol. 8 Abstr 1679.

Letarte, M., Seehafer, J.G., Greaves, A., Masellis-Smith, A. and Shaw, A.R. (1993). Homotypic aggregation of pre-B leukemic cell lines by antibodies to VLA integrins correlates with their expression of CD9. *Leukemia*, **7**:93–103.

Marken, J.S., Schieven, G.L., Hellstrom, I., Hellstrom, K.E. and Aruffo, A. (1992). Cloning and expression of the tumor-associated antigen L6. *Proc. Natl. Acad. Sci., USA*, **89**:3503–3507.

Masellis-Smith, A., Jensen, G.S., Seehafer, J.G., Slupsky, J.R. and Shaw, A.R.E. (1990). Anti-CD9 monoclonal antibodies induce homotypic adhesion of pre-B cell lines by a novel mechanism. *J. Immunol.*, **144**:1607–1613.

Masellis-Smith, A. and Shaw, A.R. (1994). CD9-regulated adhesion. Anti-CD9 monoclonal antibody induce pre-B cell adhesion to bone marrow fibroblasts through de novo recognition of fibronectin. *J. Immunol.*, **152**:2768–77.

Matsuzaki, F., Mege, R.M., Jaffe, S.M., Friedlander, D.R., Gallin, W.J., Goldberg, J.I., Cunningham, B.A. and Edelman, G.M. (1990). cDNAs of cell adhesion molecules of different specificity induce changes in cell shape and border formation in cultured S180 cells. *J. Cell Biol.*, **110**:1239–1252.

Metzelaar, M.J., Wijingaard, P.L.J., Peters, L.J., Sixma, J.J., Nieuwenhuis, H.K. and Clevers, H.C. (1991). CD63 antigen. A novel lysosomal membrane glycoprotein, cloned by a screening procedure for intracellular antigens in eukaryotic cells. *J. Biol. Chem.*, **266**; 3239–3245.

Miller, J.L., Kupinski, J.M. and Hustad, K.O. (1986). Characterization of a platelet membrane protein of low molecular weight associated with platelet activation following binding by monoclonal antibody AG-1. *Blood*, **68**:743–751.

Momoi, M., Kennett, R.H. and Glick, M.C. (1980). A membrane glycoprotein from human neuroblastoma cells isolated with the use of a monoclonal antibody. *J. Biol. Chem.*, **255**:11914–11921.

Oren, R., Takahashi, C., Doss, R., Levy, R. and Levy, S. (1990). TAPA-1, the target of an anti-proliferative antibody, defines a new family of transmembrane proteins. *Mol. Cell. Biol.*, **10**:4007–4015.

Ozaki, Y., Satoh, K., Kuroda, K., Qi, R., Yatomi, Y., Yanagi, S., Sada, K., Yamamura, H., Yanabu, M., Nomura, S., *et. al.* (1995). Anti-CD9 monoclonal antibody activates p72syk in human platelets. *J. Biol. Chem.*, **270**:15119–24.

Pytela, R., Pierschbacher, M.D. and Ruoslahti, E. (1985). Identification and isolation of a 140 kd cell surface glycoprotein with properties expected of a fibronectin receptor. *Cell*, **40**:191–198.

Rendu, F., Boucheix, C., Lebreit, M., Bourdeau, N., Benoit, P., Maclouf, J., Soria C. and Levy-Toledano, S. (1987). Mechanisms of the mAb ALB6 (CD9) induced human platelet activation: Comparison with thrombin. *Biochem. Biophys. Res. Comm.*, **146**:1397–1404.

Rosenfeld, S.I., Looney, R.J., Phipps, D.C., Abraham, G.N. and Anderson, C.L. (1985). Human platelet Fc receptor for immunoglobulin G. Identification as a 40,000-molecular weight membrane protein shared by monocytes. *J. Clin. Invest.*, **76**:2317–2322.

Rubinstein E., Le Naour, F., Billard, M., Prenant, M. and Boucheix, C. (1994). CD9 antigen is an accessory subunit of the VL integrin complexes. *Eur. J. Immunol.*, **24**:3005–13.

Seehafer, J.G. and Shaw, A.R.E. (1991). Evidence that the signal-initiating membrane protein CD9 is associated with small GTP-binding proteins. *Biochem.Biophys. Res. Commun.*, **179**:401–406.

Slupsky, J. R., Seehafer, J.G., Tang, S-C, Masellis-Smith, A. and Shaw, A.R.E. (1989). Evidence that monoclonal antibodies against CD9 antigen induce specific association between CD9 and the platelet glycoprotein IIb-IIIa complex. *J. Biol. Chem.*, **264**:12289–12293.

Slupsky, J.R., Cawley, J.C., Kaplan, C. and Zuzel, M. (1997). Analysis of CD9, CD32 and p67 signalling: use of degranulated platelets indicates direct involvement of CD9 and p67 in integrin activation. *Br. J. Haematol.*, **96**:275–296.

Straus (1989). Mechanisms of fibronectin-mediated cell migration: Dependence or independence of cell migration susceptibility on RGDS-directed receptor (integrin). *Exp. Cell Res.*, **183**:129–139.

Szala, S., Kasai, Y., Steplewski, S., Rodeck, U. and Koprowski, H. (1990). Molecular cloning of cDNA for the human associated antigen CO-029 and identification of related transmembrane antigens. *Proc. Natl. Acad. Sci. U.S.A.*, **87**, 6833–6837.

Travis, G.H., Christerson, L., Danielson, P.E., Klisak, I., Sparkes, R.S., Hahn, L.B., Dryja, T.P. and Sutcuffe, J.G.. (1991). The human retinal slow (RDS) gene: Chromosome assignment and structure of the mRNA. *Genomics*, **10**:733–9.

Worthington, R.E., Carroll, R.C. and Boucheix, C. (1990). Platelet activation by CD9 monoclonal antibodies is mediated by the FcγII receptor. *Br. J. of Haematol.*, **74**:216–222.

Wright, M.D., Henkle, K.J. and Mitchell, G.F. (1990). An immunogenic Mr 23,000 integral membrane protein of Schistosoma mansoni worms that closely resembles a human tumor associated antigen. *J. Immunol.*, **144**:3195–3200.

Wright, M.D. and Tomlinson, M.G. (1994). The ins and outs of the transmembrane 4 superfamily. *Immunology Today*, **15**:430–436.

Yu, J., Lin, J.H., Wu, X.R. and Sun, T.T. (1994). Uroplakins Ia and Ib, two major differentiation products of bladder epithelium, belong to a family of four transmembrane domain (4TM) proteins. *Cell*, **125**(1):171–182.

10 Neutrophil-Platelet Interactions

Dominique Pidard*, Mustapha Si-Tahar and Michel Chignard

*Unité de Pharmacologie Cellulaire/INSERM U.485, Institut Pasteur,
25, rue du Docteur Roux, 75724 Paris Cédex 15, France*

Polymorphonuclear neutrophil leukocytes are major cellular contributors in the process of inflammation in response to infections. These cells represent one of the first lines of defence against microbial invasion. Influx of neutrophils into inflammatory sites results in phagocytosis, killing and removal of pathogenic micro-organisms, and can also participate in healing and tissue repair (Smolen and Boxer, 1995). However, uncontrolled accumulation of neutrophils, and inadequate responses of these cells can initiate acute or chronic inflammatory disorders (Smith, 1994). One of the first steps in the inflammatory process is the release of various substances from host cells as well as from the pathogens, which will then attract and activate circulating neutrophils. Such chemoattractants include N-formylated peptides derived from bacterial walls, the complement-derived C5a, membrane lipid metabolites, and the more recently described chemokines (Sandborg and Smolen, 1988; Schall and Bacon, 1994). Chemotaxis implies that neutrophils will have to cross cellular barriers such as the vascular endothelium and possibly various epithelia, and to progress through connective tissues and extracellular matrices (Nourshargh and Williams, 1995). Major metabolic events then occur to initiate phagocytosis when neutrophils contact the offending organism. During phagocytosis, different harmful substances aimed at destroying the pathogens and which are normally restricted to internal phagolysosomes, can also be exteriorized. Then, they affect the host cells and extracellular matrices, contributing to tissue damage. The releasable granule products include lysosomal hydrolases, myeloperoxidase and serine proteinases. Cytotoxic agents are also formed from the plasma membrane, *i.e.* highly reactive oxygen metabolites such as the superoxide anion ($O_2^{\cdot-}$), hydrogen peroxide (H_2O_2), and hydroxyl radical and its chlorinated derivatives (Witko-Sarsat and Descamps-Latscha, 1994). Three other types of neutrophil-derived mediators, although noncytotoxic by themselves, participate to and can amplify the inflammatory process, namely the platelet-activating factor (PAF), chemokines, and arachidonate metabolites known as eicosanoids.

Concerning the involvement of platelets in the process of inflammation, these cells can be regarded as potential protagonists as they release numerous pro-inflammatory substances. These include vasoactive amines such as serotonin and histamine; growth factors with potential chemotactic activities such as the platelet-derived growth factor (PDGF) and

---

Corresponding author: Unité de Pharmacologie Cellulaire/INSERM U.485, Institut Pasteur, 25, rue du Docteur Roux, 75724 Paris Cédex 15, France. Tel: (33)-01-45-68-86-83, Fax: (33)-01-45-68-87-03 email: dpidard@mailhost.pasteur.fr.

transforming growth factor β (TGFβ); membrane lipid derivatives such as eicosanoids and PAF, all with a large spectrum of biological activities. Likewise, platelets contain specific proteins which can act as chemokines for neutrophils, such as the platelet factor 4 (PF4) and neutrophil-activating peptide 2 (NAP-2) (Page, 1989; Mannaioni *et al.*, 1997). Moreover, platelets express membrane receptors for the Fc fragment of IgG (FcγRII) and of IgE (FcεRII) which are similar to those found on leukocyte inflammatory cells, and can be activated by IgG- or IgE-containing immune complexes (Page, 1989; Rubinstein *et al.*, 1995). Finally, platelets can interact physically with neutrophils and monocytes (Rinder *et al.*, 1991a), and can be thus potentially recruited within the vascular bed in inflamed tissues. Despite these *in vitro* findings, *in vivo* evidence remains scarce for platelets as primary cellular effectors in inflammation. Moreover, inflammation, coagulation, and thrombosis and/or haemorrhage may be such intricate events that it is difficult to decipher the proper role of platelets as compared to leukocytes, mainly neutrophils, in the overall process. Nonetheless, a number of observations indicate that platelets may be endowed with a crucial role in initiating and/or maintaining allergic inflammation, as well as some nonallergenic acute inflammation reactions (Margaretten and McKay, 1971; Kunkel, 1983; Page, 1989; Mannaioni *et al.*, 1997).

In fact, platelets are usually considered in the context of haemostasis and thrombosis. The formation of a platelet aggregate at a site of haemorrhage constitutes the primary haemostatic response. Following a rupture in the vessel wall, platelets immediately adhere and spread on the subendothelial matrix through membrane receptors belonging to the integrin or to the leucine-rich glycoprotein families (Coller, 1992; Blockmans *et al.*, 1995). Adherence activates platelets, which release substances stored in the intracellular dense granules, including adenosine diphosphate (ADP) and serotonin, produce and secrete various lipid mediators, such as thromboxane A_2 (TxA$_2$) and PAF, while the plasma membrane acquire procoagulant properties leading to the formation of α-thrombin. All these soluble substances are able to trigger activation of nearby circulating platelets, which thus undergo the same sequence of events (Blockmans *et al.*, 1995). In addition, activated platelets extrude the content of their α-granules, releasing a further pool of adhesive proteins such as fibrinogen, von Willebrand factor and thrombospondin-1 (TSP-1), which will bind on adjacent platelets to receptors such as the fibrinogen receptor, the $\alpha_{IIb}\beta_3$ integrin. This allows platelets to firmly aggregate to each other, leading to the formation of a thrombus at the site of vascular damage, which is stabilized by formation of insoluble fibrin. The whole process is normally tightly regulated, in a way to avoid platelet agonists to freely diffuse in the circulation, and to make circulating platelets refractory to activation outside the area of thrombus formation. However, if not properly controlled at sites of vascular lesions, thrombus formation, growth and embolization can lead to severe occlusive vascular diseases (Coller, 1992).

The participation of neutrophils in the thrombotic process has received more attention in recent years, even if controversies remain (Henson, 1990; Bazzoni *et al.*, 1991; Marcus and Safier, 1993; Nash, 1994). For instance, Nash (1994) pointed out that neutropenia does not seem to impair normal haemostasis. By contrast, emerging data indicate a role for neutrophils in thrombotic and vaso-occlusive diseases. Thus, neutrophils have been proposed as one cellular component involved in the initiation and propagation of deep venous thrombosis (Stewart, 1993). During extracorporeal circulation, both platelets and neutrophils can be activated and form circulating mixed aggregates, which can be responsible for the plugging of microvessels (Rinder *et al.*, 1992; Gawaz *et al.*, 1994). Finally, increasing evidence indicates that neutrophils can play an important role in the deposition and aggregation of platelets on an injured arterial wall, leading to ischemic tissue damages (Bednar *et al.*, 1985; Merhi *et al.*, 1994).

All these above considerations have allowed the emergence of the concept that neutrophils and platelets are blood cells endowed with strong interactive capacities (Henson, 1990). Although beyond the scope of this review, a third, major partner must be considered when dealing with platelet-neutrophil cooperation, *i.e.* the vascular endothelium. Activated or injured endothelial cells, or the subendothelial matrix provide the surface on which platelets and neutrophils adhere and are activated (Coller, 1992; Korthuis and Granger, 1994). Whether platelets adhere and spread first to form a pro-aggregatory area in the process of thrombus formation, or whether neutrophils adhere to the endothelium before their extravasation at an inflammatory site, each cell type has the capacity to recruit and modulate the biological functions of the other. Over the past decades, considerable investigations have been conducted in order to characterise the biochemical nature of the molecular mediators involved in platelet-neutrophil interactions, and to understand how they influence cell functions. This review will summarise our present knowledge concerning the main interaction pathways which have been explored *in vitro*, *i.e.* (i) the generation of oxygen derivatives by neutrophils; (ii) the transcellular metabolism of membrane lipid metabolites, namely eicosanoids and PAF; (iii) the release and cellular effects of neutrophil serine proteinases; (iv) the production of neutrophil chemoattractants by platelets, and (v) the involvement of adhesive molecules and adherence membrane receptors in the physical interactions between the two cell types. It should be emphasised at this point that, if numerous investigations have focused on the cross-activation of platelets and neutrophils, evidence has been provided that each cell type can also, in some circumstances, down-regulate the reactivity of the other. Examples of this latter type of interaction will be provided within each of the following sections.

GENERATION OF OXYGEN DERIVATIVES

Production of oxidants by activated neutrophils, through the NADPH oxidase pathway, is one major intracellular process providing these cells with a potent bactericidal activity (Witko-Sarsat and Descamps-Latscha, 1994). However, oxidants are also released in the extracellular environment, where they can modulate the functions of surrounding cells and participate in tissue damage. Alternatively, cells interacting with neutrophils can modulate the generation of oxidants. An example of such a cellular cross-talk is provided by platelet-neutrophil interactions.

Two decades ago, it was shown that platelet aggregation could be reduced when neutrophils were present, and that the addition of a H_2O_2 scavenger such as catalase suppressed this inhibition (Levine *et al.*, 1976). The actual activity of this compound on platelets remains far from clear, however, since some later reports showed that at low concentrations, it induces platelet activation, potentially through its transformation into hydroxyl radicals and chlorinated derivatives (Iuliano *et al.*, 1994). $O_2^{\cdot-}$ is yet another major product generated during the respiratory burst of activated neutrophils, which has been shown to initiate *per se* platelet activation and to potentiate aggregation in response to other platelet agonists (Handin *et al.*, 1977; Renesto *et al.*, 1994b). Besides, nonactivated neutrophils decrease the platelet responses to specific stimuli in a process which can be blocked by oxyhaemoglobin, suggesting the inhibitory factor as nitric oxide (NO), indeed a powerful inhibitor of platelet activation (Radomski *et al.*, 1987; Salvemini *et al.*, 1989). It thus appears that the activity of neutrophil-derived oxygen metabolites on platelets may depend on a subtle balance between the basal production of NO, which inhibits platelet activation, and the release of oxygen radicals by activated cells, which will promote platelet activation and also inactivate NO (Faint, 1992).

In fact, the reverse situation, *i.e.* modulation by platelets of the production of toxic oxygen radicals from neutrophils, has driven some more attention. Major platelet mediators in this case appear to be purine nucleotides released from dense granules. Thus, production of $O_2^{\cdot-}$ by activated human neutrophils was found to be enhanced in the presence of platelets, and this effect was related to the release of platelet ATP and ADP (Ward *et al.*, 1988). However, neutrophils express a membrane ectonucleotidase which can convert ATP and ADP into AMP and adenosine, two compounds which actually inhibit neutrophil $O_2^{\cdot-}$ generation, as they inhibit platelet activation (Ward *et al.*, 1988; Faint, 1992). Extracellular adenosine may also originate from the neutrophils themselves, through a process stimulated by contacts with nonactivated platelets (Bengtsson *et al.*, 1996). Again, a balance may thus exist between inhibitory adenosine and AMP coming from neutrophils and/or generated by the catabolism of low levels of released ADP and ATP, and activation due to the massive extrusion of the latter nucleotides by fully activated platelets. Finally, resting platelets may also limit the extracellular diffusion of neutrophil oxidants by acting as scavengers for these compounds, likely through the glutathione cycle (Faint, 1992).

Regardless of the management of toxic oxidants during *in vivo* interactions between platelets and neutrophils, these experimental observations underline the complex and ambiguous relationships which may exist between the two cell populations.

BIOSYNTHESIS OF EICOSANOIDS

Intracellular metabolites of arachidonic acid (AA), known as eicosanoids, are secretable autocoid cellular effectors with a wide range of physiological functions, and notably implicated in thrombosis and inflammation. Eicosanoids include prostaglandins, thromboxanes, leukotrienes, and lipoxins. These substances are produced by activated vascular and blood cells, upon release of membrane phospholipid-bound AA by phospholipase A_2 and processing of AA through an enzymatic cascade. The nature of the major end-product(s) depends on the particular enzyme(s) expressed by one given cell type. Thus, AA is metabolised by activated neutrophils mostly through the 5-lipoxygenase (5-LO) pathway, to produce leukotriene (LT) B_4 and 5-hydroxyeicosatetraenoic acid (5-HETE). In platelets, AA is mostly processed into TxA_2 through a cyclooxygenase pathway, and into 12-HETE through a 12-LO pathway (Maclouf, 1993; Marcus and Hajjar, 1993). LTB_4 is a very potent pro-inflammatory eicosanoid, showing a high chemotactic activity for neutrophils (Ford-Hutchinson, 1990), whereas TxA_2 is a potent amplifying stimulus for activated platelets, as well as a vasoconstrictor (FitzGerald, 1991). TxA_2 also appears to upregulate neutrophil adherence (Spagnuolo *et al.*, 1980).

The most striking situation resides in the so-called transcellular metabolism of eicosanoids between platelets and neutrophils (Maclouf, 1993; Marcus and Hajjar, 1993). Thus, in mixed activated cell suspensions, it has been shown that AA and/or its metabolites released by platelets can be taken up by neutrophils and processed into LTB_4, whose production remains lower in the absence of platelets (Maclouf *et al.*, 1982; Palmantier and Borgeat, 1991). In this situation, not only is LTB_4 produced in increased amounts, but 5-HETE and 12-HETE are formed, as well as a new metabolite, 5S,12S-di-HETE. It is remarkable that none of the two cell types alone is able to produce 5S,12S-di-HETE, and that they have to cooperate to do so (Maclouf, 1993; Marcus and Hajjar, 1993). While both the 5-HETE and 12-HETE behave as chemoattractants for neutrophils, in addition to

LTB_4, formation of 5S,12S-di-HETE, when platelets and neutrophils are both present, may limit the production of these chemotactic eicosanoids (Marcus and Hajjar, 1993).

As do neutrophils with platelet-derived AA, platelets can incorporate AA generated by neutrophils, and this has been associated with an increased production of both TxA_2 and 12-HETE (McCulloch *et al.*, 1992; Maugeri *et al.*, 1994). Other situations have been described in which platelets use neutrophil-specific eicosanoid intermediates to generate new end-products that neutrophils cannot form. Thus, it was shown that co-stimulated neutrophil-platelet suspensions can form LTC_4, a potent vaso- and broncho-constrictor which is derived from the neutrophil short-lived product LTA_4 further processed through the platelet glutathione-S-transferase pathway (Maclouf, 1993). From LTA_4, platelets can also generate lipoxin (LX) A_4, a process which depends on the 12-LO activity (Edenius *et al.*, 1994). While it has no apparent activity on platelets, LXA_4 markedly limits chemotaxis, adhesiveness and migration of neutrophils (Lefer *et al.*, 1988; Papayianni *et al.*, 1996).

Based on the experimental *in vitro* observations mentioned above, it may be a difficult task to reconcile apparently disparate data into a synthetic functional scheme concerning the transcellular metabolism of eicosanoids between platelets and neutrophils. Moreover, whether the balance will be shifted towards the net production of pro- or anti-inflammatory, and/or pro- or anti-thrombotic compounds, must depend on several factors. For example, the level of metabolic activation of the two cell types may dictate the enzymatic pathway in which each cell will utilise an exogenous eicosanoid precursor or intermediate. In addition, platelets and neutrophils are unlikely to interact in a simple two-cell type system, but rather in a multicellular process involving endothelial cells, and in a complex network of soluble cell effectors such as cytokines (Maclouf, 1993; Marcus and Hajjar, 1993).

BIOSYNTHESIS OF PLATELET-ACTIVATING FACTOR

Platelet-activating factor (PAF) is an autocoid lipid mediator produced by many cells upon their activation, which is endowed with numerous biological activities, including activation of neutrophils and platelets. As such, it is thought to be implicated in various physiological and pathophysiological conditions, and particularly in inflammatory and thrombotic states (Handley *et al.*, 1990; Korthuis and Granger, 1994). As far as platelet-neutrophil interactions are concerned, an important feature is that both cell types can produce PAF through a similar intracellular metabolic pathway. Upon cell activation, deacylation of membrane ether-linked glyceryl-phosphorylcholine by PLA_2 yields lyso-PAF, and allows the release of a free fatty acid, which is preferentially AA, another important lipid mediator (see above). Lyso-PAF is a stable but biologically inactive compound which is in turn acetylated into PAF (1-*O*-alkyl-2-acetyl-*sn*-glycero-3-phosphocholine) by an acetyltransferase (Prescott *et al.*, 1990). However, this biosynthetic route differs in potency for each cell type. Indeed, it was shown that lyso-PAF is much more quantitatively processed into PAF by neutrophil than by platelet acetyltransferase. As a consequence, platelets produce around an hundred-fold more lyso-PAF than PAF (Chignard *et al.*, 1987). However, here again, a transcellular metabolism may take place in order to increase the overall production of PAF. Thus, when autologous neutrophils and platelets are co-incubated and each cell type challenged by a specific agonist, the total production of PAF is largely above the simple addition of the amounts formed by each cell type activated separately, because platelet lyso-PAF can be taken up by neutrophils and processed through the activated acetyltransferase, resulting in an increased PAF synthesis (Coëffier *et al.*, 1990).

One major point to be considered is whether neutrophils and/or platelets can indeed release in the extracellular milieu the PAF they have synthesised intracellularly (Lorant *et al.*, 1995). From the observation mentioned above, it may seem that at least lyso-PAF can be released from platelets. However, if initial studies performed with rabbit platelets established the potential for a release of PAF, activation of human platelets with physiologic agonists resulted in a lower production of PAF compared to rabbit cells, and release of the mediator was not unequivocally established (Chignard *et al.*, 1987). In fact, recent invest-igations have shown that, from the PAF which could be released by human platelets activated with a combination of thrombin and collagen, approximately 90% appears to remain asso-ciated with the membrane of platelet-derived microparticles (Iwamoto *et al.*, 1996), which, interestingly, have been shown to bind to neutrophils and to induce their aggregation (Jy *et al.*, 1995). This is reminiscent of what has been described for the neutrophils, as upon activation, less than 10% of the synthesised PAF is released, the major part of the mediator remaining associated with the cells (Sisson *et al.*, 1987).

The next, and more important step is thus to question whether released or membrane-bound PAF can participate in a cross-activation between platelets and neutrophils. In fact, most of the *in vitro* investigations concluding to a role for PAF in such a process have been conducted with rabbit platelets. The same applies to *in vivo* or *ex vivo* animal models of acute thrombo-inflammatory reactions in which PAF was shown to be a major effector in the cooperation between platelets and neutrophils (Issekutz and Szpejda, 1986; Alloatti *et al.*, 1992). Whether PAF plays the same role in the case of human cells is questionable, since human platelets appear to be less responsive to PAF than rabbit platelets (Chignard *et al.*, 1987), and require to be first primed by another compound secretable by activated neutrophils, the serine proteinase cathepsin G, in order to fully respond to low or moderate concentrations of PAF (Renesto *et al.*, 1992). Finally, in experiments using autologous, mixed human platelet-neutrophil suspensions in which neutrophils have been activated with a specific agonist, the subsequent activation of platelets was shown to depend mostly on the release of leukocyte serine proteinases rather than PAF (see below) (Chignard *et al.*, 1986; de Gaetano *et al.*, 1990).

ROLE OF NEUTROPHIL SERINE PROTEINASES

Neutrophils contain a number of neutral serine proteinases and metalloproteinases which play important roles in the normal functions of these cells. In the quiescent cells, these proteinases are stored in intracellular granules. During the exit of activated neutrophils from the vasculature, part of these proteinases can be extruded and are found in the vicin-ity of the cells or bound to the plasma membrane. They can then participate to the migra-tion of leukocytes within tissues, because of their strong proteolytic activity on many extracellular matrix proteins. Once the neutrophil contacts a pathogen, proteinases also concur in phagocytosis and killing of the infectious agent. A number of tissue or plasma polypeptides with strong and specific antiproteinase activities normally insure that neu-trophil proteinases will only be active in the pericellular space, and will not diffuse in the tissues or in the circulation (Witko-Sarsat and Descamps-Latscha, 1994; Owen and Camp-bell, 1995).

Special attention has been focused for years on the serine proteinases stored in the azurophilic granules of neutrophils, *i.e.* neutrophil elastase (NE), cathepsin G (CG), and proteinase 3 (PR3). Indeed, these enzymes have been implicated in the pathogenesis of inflam-matory diseases, as well as in various thrombotic and/or haemorrhagic states associated with infections or malignancies (Janoff, 1985; Movat and Wasi, 1985; Eldanasouri *et al.*, 1990). This relies notably on the ability of these proteinases to markedly cleave matrix proteins and to affect coagulation and fibrinolysis (Eldanasouri *et al.*, 1990; Pintucci *et al.*, 1993; Altieri, 1995; Owen and Campbell, 1995). However, growing evidence show that these enzymes are also highly effective in modulating the activity of cell membrane receptors, mostly through proteolytic pathways. They can be thus regarded as mediators in cell-cell interactions, and this appears to be particularly true for the interactions between neutrophils and platelets.

As mentioned above, several investigations have shown that when human autologous platelets and neutrophils are mixed in a ratio similar to that found in the circulation, activation of neutrophils with a specific agonist promptly results in a potent activation of surrounding platelets, leading to formation of mixed aggregates. This process was found to depend on the release by the neutrophils of a compound which had the characteristics of a serine proteinase (Chignard *et al.*, 1986), and using a panel of specific antiproteinases and antibodies, CG was found to be responsible for platelet activation, PAF and oxidants acting as amplifying stimuli (Selak *et al.*, 1988; de Gaetano *et al.*, 1990; Ferrer-Lopez *et al.*, 1990; Evangelista *et al.*, 1991; Renesto *et al.*, 1994b). Under optimal conditions of degranulation of neutrophils, the concentration of CG present in the extracellular milieu can reach 200 to 250 nmoles/L (Evangelista *et al.*, 1991; Renesto and Chignard, 1993). However, the actual amount of CG released must be higher, since part of it can bind to platelets (see below), whereas another fraction remains associated and active at the surface of activated neutrophils (Owen *et al.*, 1995). At these concentrations, purified CG is indeed a very potent platelet agonist which reproduces the activity obtained with a stimulated neutrophil-conditioned medium (Bykowska *et al.*, 1983; Selak *et al.*, 1988; Evangelista *et al.*, 1991). Cathepsin G is similar to α-thrombin in terms of signal transduction, as being able to activate phospholipase C, PLA_2 and protein kinase C pathways and to increase the intracellular concentration of Ca^{2+} (Molino *et al.*, 1992; Selak, 1993; Si-Tahar *et al.*, 1996). Likewise, both the binding of these enzymes to the platelet surface as well as an intact enzymic activity appear mandatory for platelet activation (Selak *et al.*, 1988; Selak and Smith, 1990). The thrombin receptor belongs to the family of the seven transmembrane domain receptors coupled to G-proteins, and is a member of the subclass of the so-called proteinase-activated receptors (PAR), which require to be cleaved by their own enzyme ligand(s) in order to be activated (Coughlin, 1994). In view of the above-mentioned similarities between α-thrombin and CG as platelet agonists, it is tempting to speculate that CG also acts through a PAR. However, this specific, putative receptor should be distinct from the main thrombin receptor (PAR-T) (Selak, 1994) which is in fact down-regulated by CG (see below).

Neutrophil CG has some major effects on three platelet membrane adherence receptors implicated in cell-cell and cell-matrix interactions, and consequently on platelet activation. On the one hand, CG upregulates the activity of the platelet fibrinogen receptor, the $\alpha_{IIb}\beta_3$ integrin. Like α-thrombin, this involves two mechanisms (Shattil, 1995): (i) the shift of the plasma membrane integrin from an inactive to an active conformer during cell stimulation, a process under the control of metabolic pathways; (ii) the translocation of the α-granule-associated $\alpha_{IIb}\beta_3$ integrins to the plasma membrane during exocytosis, a process which approximately doubles the number of active receptors expressed on the cell surface (Molino *et al.*, 1993; LaRosa *et al.*, 1994b; Si-Tahar *et al.*, 1997). On the other hand, CG downregulates the biological activity of the platelet membrane glycoprotein (GP) Ib-V-IX complex. On the platelet surface, this multimeric complex operates as a receptor for von Willebrand factor (vWf), a plasma and matrix adhesive protein implicated in adherence of platelets to the vascular subendothelium, and in platelet activation and aggregation. GPIb-V-IX also contains a high affinity binding site for α-thrombin, and may be thus implicated in platelet activation by this agonist. Binding sites for vWf and for α-thrombin are located in the mid-portion of the extracellular domain within the largest subunit of the receptor, GPIbα. They are not strictly identical, but overlap within a sequence of approximately 80 amino acids which contains two intricate disulphide loops and a negatively charged carboxyterminal flanking region (Lopez, 1994). Downregulation of GPIb-V-IX by CG occurs through two different processes. Firstly, CG specifically cleaves the GPIbα subunit within the sequence carboxyterminal to the disulphide loops, thus eliminating the binding domains for α-thrombin and vWf (Molino *et al.*, 1993; Pidard *et al.*, 1994; Ward *et al.*, 1996). Secondly, CG induces the redistribution of the GPIb-V-IX complex from the external plasma membrane towards internal vacuolar structures early during platelet activation (LaRosa *et al.*, 1994a; Pidard *et al.*, 1994). Finally, through exocytosis of α-granules, CG has a strong potential to induce the expression of P-selectin on the platelet surface (LaRosa *et al.*, 1994b).

NE is the other serine proteinase stored in the azurophilic granules of neutrophils which, upon cell degranulation, can be recovered in the extracellular milieu at a concentration of ≈ 400 nmoles/L (Renesto and Chignard, 1993), whereas a fraction binds in an active form to the neutrophil membrane (Owen *et al.*, 1995). Using mixed platelet-neutrophil suspensions, it was shown that NE alone plays a little role in the neutrophil-dependent activation of platelets (Evangelista *et al.*, 1991). Indeed, incubation of isolated platelets with purified NE up to ≈ 1 μmole/L failed to trigger intracellular signalling pathways and to induce cell activation (Bykowska *et al.*, 1983; Si-Tahar *et al.*, 1997). However, several platelet membrane receptors are substrates for the enzymic activity of NE, which markedly affects their functions. Thus, exposure of isolated platelets to low concentrations of NE (<100 nmoles/L) for only a few minutes was found to markedly reduce the binding of α-thrombin to the platelet surface and the ensuing platelet exocytosis and aggregation, and to inhibit the GPIb-V-IX-dependent interaction of platelets with vWf. These effects can be ascribed to a proteolytic downregulation of the activity of the GPIb-V-IX complex. Indeed, NE specifically cleaves the GPIbα subunit slightly downstream to the cleavage site described for CG, thus eliminating both the vWf and the high-affinity α-thrombin binding sites (Brower *et al.*, 1985; Wicki and Clemetson, 1985; Pidard *et al.*, 1994). NE also appears to cleave the GPV subunit and to release much of its extracellular domain, but the potential functional significance of this proteolysis remains unknown (Wicki and Clemetson, 1985). On another hand, despite its inability to trigger an intracellular activation, NE at concentrations ≤ 200 nmoles/L was shown to rapidly induce a binding of fibrinogen to the platelet surface through a non-metabolic activation of the $\alpha_{IIb}\beta_3$ integrin (Kornecki *et al.*, 1988). In this case, activation appears to result from a very limited proteolytic event, i.e. the cleavage and release of a short glycosylated amino acid sequence located at the carboxyterminal tail of the extracellular heavy chain of the α_{IIb} subunit. This cleavage results in a conformational change of the integrin mimicking that induced by metabolic activation, leading to an enhanced affinity for fibrinogen (Si-Tahar *et al.*, 1997). A major feature is that this proteolytic activation of $\alpha_{IIb}\beta_3$ strongly potentiates the aggregation induced by a low concentration of CG (Selak, 1992; Renesto and Chignard, 1993; Si-Tahar *et al.*, 1997). Actually, this NE-dependent potentiation of platelet aggregation also applies to nonproteinase platelet agonists such as collagen and a TxA$_2$ analogue (Renesto and Chignard, 1993), and also to ADP (Pidard, unpublished observation). However, exposure of platelets to high concentrations of NE (≥ 1 μmole/L) for several minutes before addition of CG or another agonist actually results in a marked inhibition of platelet activation and aggregation (Renesto *et al.*, 1993). The mechanism of this inhibition is not yet elucidated. It may rely to the capacity of NE, under these conditions, to further proteolyse the $\alpha_{IIb}\beta_3$ integrin, with the cleavage of a large aminoterminal region of the β_3 subunit, which contains a binding site for fibrinogen (Kornecki *et al.*, 1988; Niewiarowski *et al.*, 1989; Pidard *et al.*, unpublished results). It may also depend on the capacity of NE to proteolyse platelet agonist receptors (see below).

Interestingly, the third neutrophil proteinase released from the azurophilic granules, i.e. PR3, has also been shown to not be able to induce platelet activation by itself while it potentiates the aggregation induced by CG and other platelet agonists (Renesto *et al.*, 1994a). The mechanism by which PR3 acts on platelets remains to be determined but seems to exclude a proteolytic modification of the $\alpha_{IIb}\beta_3$ integrin (Pidard *et al.*, unpublished observation).

A final mention must be made concerning the interactions of these three proteinases with membrane receptors. Recent investigations have indeed shown that both NE, CG and PR3, when present at high concentrations (≥ 400 nmoles/L), can rapidly cleave at specific and distinct sites, the first aminoterminal extracellular domain of the thrombin receptor PAR-T. These cleavages occur downstream of the thrombin cleavage site, thus preventing any subsequent activation by α-thrombin of PAR-T-bearing cells, including platelets and endothelial cells (Molino *et al.*, 1995; Renesto *et al.*, 1997).

Taken as a whole, these observations suggest that, once released from neutrophils, the serine proteinases CG, NE and PR3 have potent but ambivalent effects on surrounding platelets. Even when present at low concentrations, they are clearly endowed with most of the potency of neutrophils to activate platelets with, as consequences, an increased expression and activity of the $\alpha_{IIb}\beta_3$ integrin and the P-selectin, thus favouring platelet-to-platelet and platelet-to-neutrophil interactions (see below). On another hand, their capacity to downregulate both the GPIb-V-IX receptor and the thrombin receptor PAR-T, through cellular redistribution and/or proteolysis, should limit cell-cell or cell-matrix adherence and platelet activation. The balance between these quite opposite effects may depend on the relative accessibility and susceptibility of the platelet membrane glycoprotein substrates, as well as on the local concentrations and persistence of active enzymes. Plasma anti-proteinases specific for NE and CG, such as α_1-proteinase inhibitor and α_1-antichymo-trypsin, must be thus essential regulatory elements in these processes. However, the capacity of platelets to tightly attach to neutrophils (see below) may create a microenvironment between adjacent cell plasma membranes, in which proteinases released by or bound to neutrophils largely escape an inhibition by anti-proteinases (Evangelista *et al.*, 1991, 1993; Owen and Campbell, 1995; Owen *et al.*, 1995). Under these conditions, CG can indeed activate platelets and markers of platelet membrane proteolysis can be detected (Evangelista *et al.*, 1991; Aziz *et al.*, 1995). Activation of platelets may then reinforce the physical interactions with the neutrophils, through overexpression and/or activation of P-selectin and the $\alpha_{IIb}\beta_3$ integrin, and prolong the effects of the proteinases. There is so far no direct *in vivo* evidence for such proteinase-dependent platelet-neutrophil interactions. However, it can be speculated from the *in vitro* observations that these processes could be operative in the adult respiratory distress syndrome (ARDS), a severe lung inflammation pathology in which neutrophils and platelets both participate, and in which neutrophil chemoattractants such as C5a, interleukin-8 (IL-8) and Tumor Necrosis Factor-α (TNFα), all mediators able to strongly release neutrophil proteinases, are implicated (Demling, 1990; Chignard and Renesto, 1994). A potential role of neutrophil proteinases in *in vivo* thrombogenesis is emphasized by studies in animal models showing that a substance with antithrombotic activity, defibrotide, is a CG inhibitor (Evangelista *et al.*, 1992).

The neutrophil serine proteinases are also able to affect neutrophil functions, most likely through proteolytic modifications of membrane receptors, in an autocrine or paracrine manner, although NE and CG by themselves are not true agonists for neutrophils. Important in the context of neutrophil interactions with other cells are the observations that NE and CG have major effects on the adherence properties of neutrophils. Thus, NE can proteolyse leukosialin/sialophorin (CD43), a leukocyte membrane sialoglycoprotein providing a high negative charge to the cell membrane which appears to regulate the capacity of neutrophils to be activated and spread on various substrates (Nathan *et al.*, 1993; Remold-O'Donnell and Parent, 1995). More recently, data have been provided indicating a surprising interaction between neutrophil serine proteinases and the integrin $\alpha_M\beta_2$. This integrin is present on the surface of neutrophils, and its biological activity is upregulated during cell activation, allowing the firm attachment and spreading of neutrophils to activated endothelial cells, mostly through the binding of its endothelial ligand intercellular adhesion molecule-1 (ICAM-1) (Nourshargh and Williams, 1995). Because activated $\alpha_M\beta_2$ can bind fibrinogen (Altieri, 1995), this integrin may also be involved in the adherence of neutrophils to a (sub)endothelial fibrin(ogen) deposit, as well as to activated platelets bearing fibrin(ogen) on their surface (see below). Concerning CG, exposure of neutrophils to this proteinase was found to decrease their adherence to activated endothelial cells or to a fibrinogen substrate, whereas NE had no effect. However, the surface expression and activation of $\alpha_M\beta_2$ in CG-treated neutrophils was unchanged, and the integrin showed no obvious proteolytic modification although an intact enzymic activity of CG was a prerequisite (Renesto *et al.*, 1996). Besides, NE (and possibly PR3) has been shown to be a ligand for $\alpha_M\beta_2$ and to compete with fibrinogen for binding to the integrin in such a way that, once expressed at the surface of activated neutrophils, NE may allow cell detachment from an adhesive substrate and thus favour chemotaxis (Cai and Wright, 1996). These *in vitro* observations appear in agreement with *in vivo* studies using intravital microscopy, showing that NE may be an important factor for modulating the activity of $\alpha_M\beta_2$ on neutrophils and reducing cell adherence and extravasation (Woodman *et al.*, 1993). Altogether, these data indicate that neutrophil serine proteinases can modulate the adherence properties of their own producing cells in a very subtle manner, both through proteolytic and possibly nonproteolytic mechanisms.

ROLE OF α-GRANULE PLATELET PROTEINS IN NEUTROPHIL CHEMOTAXIS

Platelet α-granules contain an astonishing number of different proteins, among which the most important may be adhesive proteins, coagulation and fibrinolysis factors, and growth factors (Harrison and Martin-Cramer, 1993). Some of these proteins, which are actively released into the extracellular milieu upon platelet degranulation, have chemotactic and activating properties towards neutrophils.

Thus, platelet-derived growth factor (PDGF), one major platelet mitogenic protein, has been initially described as a chemotactic substance for neutrophils (Tzeng *et al.*, 1984). More recent investigations using recombinant PDGF indicate, however, that this growth factor is not active on human neutrophils (Qu *et al.*, 1995), and that PDGF may be rather a chemotactic and activating factor for other leukocytes such as monocytes and eosinophils (Mannaioni *et al.*, 1997).

More important are the platelet-derived neutrophil chemoattractants of the chemokine family. These include the platelet factor 4 (PF4) and the neutrophil-activating peptide 2 (NAP-2), which both belong to the so-called Cys-X-Cys subgroup of chemokines (Schall and Bacon, 1994), and are specific for the megakaryocyte/platelet lineage (Harrison and Martin-Cramer, 1993). PF4 has been described to be chemotactic for neutrophils and monocytes, but had to be used at rather high concentrations to exert its activity, and remained controversial as a chemokine (Bebawy *et al.*, 1986; Walz *et al.*, 1990). In fact, it appears now that PF4 has such an activity only when neutrophils are concomitantly exposed to the proinflammatory cytokine TNF-α (Peterson *et al.*, 1996). An important point is that, once released from platelets, PF4 can bind to glycosaminoglycans on the surface of activated platelets themselves, or on endothelial cells (Poncz, 1990). In so doing, it may be an important link between thrombosis and inflammation, because of its local accumulation on a vascular site on which platelets have formed a thrombus and where it may participate in the accretion and activation of neutrophils. Furthermore, PF4 has the capacity to increase the activity of NE on some macromolecular substrates (Lonky and Wohl, 1981). NAP-2 has a strong sequence homology with PF4, as well as with IL-8 (or NAP-1). In fact, NAP-2 derives from a platelet-specific protein called platelet basic protein (PBP) which, through various proteolytic cleavages, generates products such as (from the longer to the shorter) the connective tissue activating peptide III (CTAP-III), β-thromboglobulin, and various isoforms of NAP-2. Of these products, only NAP-2 has a chemotactic activity on neutrophils, as potent as that of IL-8 or C5a (Walz *et al.*, 1990). Until recently, it was believed that NAP-2 *per se* was not present in platelets nor was it generated by a platelet proteinase (Walz *et al.*, 1990). However, recent data indicate that some of the active isoforms of NAP-2 are constitutively present in and/or released by platelets (Piccardoni *et al.*, 1996). Besides, it was shown that neutrophils can generate by proteolysis NAP-2 from CTAP-III released from platelets, and the active proteinase is CG (Cohen *et al.*, 1992). NAP-2 has thus the potential for allowing a strong positive feedback process in the costimulation of platelets and neutrophils (Chignard and Renesto, 1994). However, the process could be controlled by the capacity of NE to proteolyse the chemokine precursors in a way which impairs further processing into active NAP-2 (Cohen *et al.*, 1992).

INVOLVEMENT OF ADHERENCE RECEPTORS AND ADHESIVE MOLECULES

The importance of physical interactions between platelets and neutrophils for their metabolic cooperation has been already stressed in the preceding sections. Specific cellular adherence through membrane receptors expressed on the two cell types not only allows this cooperation, but it may generate intracellular signallings by itself, and is the basis for the formation of circulatory mixed aggregates and for the accumulation of both cells at sites of vascular injury.

Formation of mixed platelet-neutrophil conjugates or aggregates have been observed *in vivo* upon activation of platelets both in experimental animal models and in clinical settings in humans (Rinder *et al.*, 1992; Lehr *et al.*, 1994). *In vitro*, using anticoagulated whole blood or isolated cell suspensions, attachment of inactivated platelets to inactivated neutrophils in suspension appears to be low, although not insignificant. By contrast, the activation of platelets by specific agonists or under high shear forces leads to an increased conjugation of platelets to neutrophils and formation of mixed clumps (Evangelista *et al.*, 1991, 1993, 1996; Rinder *et al.*, 1991b). In all these circumstances, it was established that expression of P-selectin on the surface of activated platelets is a major determinant in the initial cell attachment (Larsen *et al.*, 1989). P-selectin (CD62-P) is a transmembrane adherence receptor expressed by platelets and endothelial cells. In resting platelets, it is restricted to the membrane of internal α-granules, whereas in activated platelets, it is translocated to the plasma membrane following exocytosis of α-granules (McEver, 1991). Like other selectins, P-selectin contains an extracellular, aminoterminal carbohydrate recognition domain, which specifically binds sialylated and fucosylated oligosaccharides. The major P-selectin counter-receptor on neutrophils appears to be a mucin-like protein called P-selectin glycoprotein ligand-1 (PSGL-1), but L-selectin, a selectin specifically expressed by leukocytes, may also serve as a ligand for P-selectin (Sako *et al.*, 1993; McEver *et al.*, 1995). Since both PSGL-1 and L-selectin are constitutively expressed on neutrophils, P-selectin-mediated attachment of activated platelets does not require prior activation of leukocytes. However, this attachment appears to be a signalling event which activates the neutrophils, thus reinforcing cell interactions (Nagata *et al.*, 1993; Cooper *et al.*, 1994; Evangelista *et al.*, 1996). As mentioned above, physical interactions between the cooperating cells may also strongly favour the transcellular metabolism (Maclouf, 1993). Indeed, it was shown that an antibody directed against P-selectin, which blocked the formation of mixed neutrophil-platelet aggregates, could reduce the synthesis of TxA_2 and LTC_4 by platelets using AA and LTA_4 released by activated neutrophils (Maugeri *et al.*, 1994). Activated platelets tethered to neutrophils through P-selectin may also stimulate the leucocytes through high surface concentrations of neutrophil agonists such as PAF and various chemokines (Lorant *et al.*, 1995).

The activated $\alpha_M\beta_2$ integrin on neutrophils stimulated by bound platelets is essential for firm attachment between the two cell populations, and appears to be the receptor involved in the attachment of nonactivated platelets to activated neutrophils (Evangelista *et al.*, 1996). In mixed cell suspensions, the ligand for $\alpha_M\beta_2$ on platelets remains uncertain. A good candidate could be fibrinogen, since this adhesive protein is expressed on the surface of activated platelets bound to the $\alpha_{IIb}\beta_3$ integrin (Coller, 1992) and is a ligand for $\alpha_M\beta_2$ (Altieri, 1995). However, data reported by various laboratories show discrepancies: in some studies, the use of antibodies against $\alpha_{IIb}\beta_3$, or of peptides blocking the binding of fibrinogen to $\alpha_{IIb}\beta_3$ or to $\alpha_M\beta_2$, had no effect on platelet-neutrophil adherence (Rinder *et al.*, 1991b; Evangelista *et al.*, 1996), whereas other investigations have concluded that fibrinogen bound to $\alpha_{IIb}\beta_3$ on activated platelets may be an essential component in their interactions with neutrophils, and could be involved in platelet-dependent neutrophil activation (Spangenberg *et al.*, 1995; Ruf and Patscheke, 1995). In the latter, based on the use of a panel of antibodies and inhibitory fibrinogen peptides, the neutrophil receptors proposed to be involved in fibrinogen recognition excluded $\alpha_M\beta_2$ and were the integrin $\alpha_X\beta_2$ and/or the complex made of the Leukocyte-Response Integrin (LRI), an integrin of the β_3 subfamily, and the Integrin-Associated Protein (IAP, or CD47) (Gresham *et al.*, 1992; Spangenberg *et al.*, 1995; Ruf and Patscheke, 1995). CD47 is a transmembrane signalling molecule associated with β_3 integrins and regulating the integrin functions, and as such implicated in neutrophil migration through the endothelium (Lindberg *et al.*, 1993; Cooper *et al.*, 1995). CD47 is also endowed with adherence receptor activity as a binding site for TSP-1 (Gao *et al.*, 1996). In fact, TSP-1 might be another adhesive molecule involved in platelet-neutrophil interactions, since both cell types contain TSP-1 in intracellular granules, release it upon activation, and bind it on their membrane through several potential receptors, including CD47. TSP-1 is thus essential for platelet aggregation, and participates in neutrophil activation and motility (Frazier, 1991; Suchard *et al.*, 1991). It appears however that anti-TSP-1 antibodies have no effect on platelet-neutrophil interactions in cell suspensions (Rinder *et al.*, 1991b; Evangelista *et al.*, 1993).

Surprisingly, formation of platelet-neutrophil clumps in mixed cell suspensions was found to be transient and disaggregation occurs within minutes. Because neutrophils activated for minutes before exposure to platelets are poorly adherent, rapid modifications at the neutrophil membrane level can be considered to be the cause (Evangelista *et al.*,

1996). Such modifications may include shedding of L-selectin, redistribution of PSGL-1 (McEver *et al.*, 1995), as well as deactivation of the $\alpha_M\beta_2$ integrin (Evangelista *et al.*, 1996; Sheikh and Nash, 1996). However, it can be speculated that proteolysis by secreted NE of the fibrinogen binding site located at the aminoterminus of the β_3 integrin subunit (Kornecki *et al.*, 1988), and/or the proteolysis of fibrinogen by NE and CG (Plow, 1982; Rabhi-Sabile *et al.*, 1996) may also play a role in loosening platelet-neutrophil interactions.

The molecular basis for the interaction between platelets and neutrophils at the vascular wall shares similarities with that proposed for the interaction of circulating neutrophils with the activated endothelium at sites of inflammation (Nourshargh and Williams, 1995). Thus, *in vitro* as well as *in vivo* studies have shown that circulating neutrophils can roll and arrest on a monolayer of adherent, spread platelets, and can be actively incorporated into a platelet thrombus. This phenomenon initially depends on the ligation of P-selectin expressed on activated platelets by constitutive counter-receptor(s) on neutrophils, including L-selectin (Palabrica *et al.*, 1992; Buttrum *et al.*, 1993). Arrest and spreading of neutrophils on adherent platelets depends on cell activation, which can result from contact with platelet-derived chemoattractants, and appears to be mainly mediated through the $\alpha_M\beta_2$ integrin (Sheikh and Nash, 1996; Weber and Springer, 1997). Surprisingly, not only neutrophils can roll and spread on a platelet monolayer, but they can migrate through it in response to a chemotactic signal, again *via* the $\alpha_M\beta_2$ integrin (Diacovo *et al.*, 1996). One major ligand for the $\alpha_M\beta_2$ integrin in these models appears to be the fibrin(ogen) expressed on the adherent platelets or deposited on the vascular subendothelium (Kuijper *et al.*, 1997; Weber and Springer, 1997).

CONCLUSION

The statement made by Henson (1990) that "the fledgling nature of the field [interactions between neutrophils and platelets] is again emphasised by the different conflicting results, as is the emphasis on what could happen rather than on what does" may appear to be still valid, when one considers the experimental data accumulated over the last decade. The proposal made by Bazzoni and colleagues (1991) could be regarded, however, as an appropriate basis to try to understand how these interactions can have a physiological or pathophysiological relevance. Thus, circulating platelets and neutrophils, either resting or in a low state of activation, may physically interact in a continuous although transient manner, and repress their activation. This could occur for instance through the capacity of platelets to limit oxidants production, or to process some leukotrienes into products with no or inhibitory activity on neutrophils. Conversely, neutrophils may limit platelet activation through the expression of a membrane ectonucleotidase metabolising ADP, basal production of NO, or low level of membrane expression of serine proteinases able to down-regulate a platelet receptor such as GPIb-V-IX. In fact, even when one cell type is stimulated to participate in its proper physiological function, the other cell population may still have a down-regulating activity. This is very simply but clearly illustrated by the lower capacity of platelets to aggregate in response to specific agonists when autologous neutrophils are present (Zatta *et al.*, 1990). Active incorporation of neutrophils into a thrombus, through adherence to activated platelets, and their potential subsequent activation, may not necessarily facilitate thrombus growth, but rather control it, in view of the potent proteolytic activity of CG and NE on platelet adhesive proteins such as fibrinogen and TSP-1. Nonetheless, if the two cell populations are strongly activated and can accumulate locally, they tend to firmly adhere to each other through engagement of selectin and integrin receptors. In so doing, they create a pericellular environment highly enriched in cell agonists with autocrine and/or paracrine activities, including oxidants, lipid metabolites, adenine nucleotides, chemokines and proteinases, which can be protected from or can overwhelm their potential antagonists. Alternatively, the deposit of platelets or of neutrophils at primary sites of thrombosis or inflammation can initiate vascular and tissue damages creating the conditions for the recruitment of the other cell type. In all circumstances, a major element allowing platelet-neutrophil interactions to degenerate into a thrombo-inflammatory reaction leading to tissue destruction may well be the endothelium, whose anti-thrombotic, anti-inflammatory, and vasomotor properties can be profoundly affected by adherent platelets, neutrophils, and their products (Siminiak *et al.*, 1995).

Future investigations should consider two major issues. One is to establish animal models of human pathologies involving platelets and neutrophils (such as ARDS and more generally multiple organ failure syndrome, or organ ischaemia/reperfusion injury) in species in which cell interactions are mediated through molecules active *in vitro* in human cell systems. Then, the relative role of each class of effectors (*e.g.* adherence receptors, chemokines, eicosanoids, proteinases) in the platelet-neutrophil cooperation and in the thrombo-inflammatory process will have to be evaluated by using specific drugs interfering with these molecules, in *in vivo* situations corresponding to the real level of complexity which is to be expected in processes of multiple cell interactions.

Acknowledgements

Financial supports to the authors are provided by the Institut Pasteur de Paris, the Institut National de la Santé et de la Recherche Médicale (INSERM), the Centre National de la Recherche Scientifique (CNRS), the Ministère de la Recherche et de la Technologie (MRT), and the Association pour la Recherche sur le Cancer (ARC), France.

References

Alloatti, G., Montrucchio, G., Emanuelli, G. and Camussi, G. (1992). Platelet-activating factor (PAF) induces platelet/neutrophil cooperation during myocardial reperfusion. *Journal of Molecular and Cellular Cardiology*, **24**:163–171.

Altieri, D.C. (1995). Inflammatory cell participation in coagulation. *Seminars in Cell Biology*, **6**:269–274.

Aziz, K.A., Cawley, J.C., Kamiguti, A.S. and Zuzel, M. (1995). Degradation of platelet glycoprotein Ib by elastase released from primed neutrophils. *British Journal of Haematology*, **91**:46–54.

Bazzoni, G., Dejana, E. and Del Maschio, A. (1991). Platelet-neutrophil interactions. Possible relevance in the pathogenesis of thrombosis and inflammation. *Haematologica*, **76**:491–499.

Bednar, M., Smith, B., Pinto, A. and Mullane, K.M. (1985). Neutrophil depletion supress ^{111}In-labeled platelet accumulation in infarcted myocardium. *Journal of Cardiovascular Pharmacology*, **7**:906–912.

Bengtsson, T., Zalavary, S., Stendahl, O. and Grenegard, M. (1996). Release of oxygen metabolites from chemoattractant-stimulated neutrophils is inhibited by resting platelets: role of extracellular adenosine and actin polymerisation. *Blood*, **87**:4411–4423.

Bebawy, S.T., Gorka, J., Hyers, T.M. and Webster, R.O. (1986). *In vitro* effects of platelet factor 4 on normal human neutrophil functions. *Journal of Leukocyte Biology*, **39**:423–434.

Blockmans, D., Deckmyn, H. and Vermylen, J. (1995). Platelet activation. *Blood Reviews*, **9**:145–156.

Brower, M.S., Levin, R.I. and Garry, K. (1985). Human neutrophil elatase modulates platelet function by limited proteolysis of membrane glycoproteins. *Journal of Clinical Investigation*, **75**:657–666.

Buttrum, S.M., Hatton, R. and Nash, G.B. (1993). Selectin-mediated rolling of neutrophils on immobilized platelets. *Blood*, **82**:1165–1174.

Bykowska, K., Kaczanowska, J., Karpowicz, M., Stachurska, J. and Kopec, M. (1983). Effect of neutral proteases from blood leukocytes on human platelets. *Thrombosis and Haemostasis*, **50**:768–772.

Cai, T.-Q. and Wright, S.D. (1996). Human leukocyte elastase is an endogenous ligand for the integrin CR3 (CD11b/CD18, Mac-1, $\alpha_M\beta_2$) and modulates polymorphonuclear leukocyte adhesion. *Journal of Experimental Medicine*, **184**:1213–1223.

Chignard, M., Lalau-Keraly, C., Nunez, D., Coëffier, E. and Benveniste, J. (1987). PAF-acether and platelets. In *Platelets in Biology and Pathology III*, edited by D.E MacIntyre and J.L. Gordon, pp. 289–315. Amsterdam: Elsevier.

Chignard, M. and Renesto, P. (1994). Proteinases and cytokines in neutrophil and platelet interactions *in vitro*. Possible relevance to the adult respiratory distress syndrome. *Annals of the New York Academy of Sciences*, **725**:309–322.

Chignard, M., Selak, M.A. and Smith, J.B. (1986). Direct evidence for the existence of a neutrophil-derived platelet activator (neutrophilin). *Proceedings of the National Academy of Sciences of the United States of America*, **83**:8609–8613.

Coëffier, E., Delautier, D., Le Couedic, J.P., Chignard, M., Denizot, Y. and Benveniste, J. (1990). Cooperation between platelets and neutrophils for paf-acether (platelet-activating factor) formation. *Journal of Leukocyte Biology*, **47**:234–243.

Cohen, A.B., Stevens, M.D., Miller, E.J., Atkinson, M.A. and Mullenbach, G. (1992). Generation of the neutrophil activating peptide-2 by cathepsin G and cathepsin G-treated human platelets. *American Journal of Physiology*, **263**:L249–L256.

Coller, B.S. (1992). Platelets in cardiovascular thrombosis and thrombolysis. In *The Heart and Cardiovascular System,* Second Edition, edited by H.A. Fozzard *et al.*, pp. 219–273. New York: Raven Press, Ltd.

Cooper, D., Butcher, C.M., Berndt, M.C. and Vadas, M.A. (1994). P-Selectin interacts with a β_2-integrin to enhance phagocytosis. *Journal of Immunology*:**153**:3199–3209.

Cooper, D., Lindberg, F.P., Gamble, J.R., Brown, E.J. and Vadas, M.A. (1995). Transendothelial migration of neutrophils involves integrin-associated protein (CD47). *Proceedings of the National Academy of Sciences of the United States of America*, **92**:3978–3982.

Coughlin, S.R. (1994). Protease-activated receptors start a family. *Proceedings of the National Academy of Sciences of the United States of America*, **91**:9200–9202.

de Gaetano, G., Evangelista, V., Rajtar, G., Del Maschio, A. and Cerletti, C. (1990). Activated polymorphonuclear leukocytes stimulate platelet function. *Thrombosis Research*, Suppl. XI. 25–32.

Demling, R.H. (1990). Current concepts on the adult respiratory distress syndrome. *Circulatory Shock*, **30**:297–309.

Diacovo,T.G., Roth, S.J., Buccola, J.M., Bainton, D.F. and Springer, T.A. (1996). Neutrophil rolling, arrest, and transmigration across activated, surface-adherent platelets via sequential action of P-selectin and the β_2-integrin CD11b/CD18. *Blood*, **88**:146–157.

Edenius, C., Tornhamre, S. and Lindgren, J.A. (1994). Stimulation of lipoxin synthesis from leukotriene A4 by endogenously formed 12-hydroperoxyeicosatetraenoic acid in activated platelets. *Biochimica et Biophysica Acta*, **1210**:361–367.

Eldanasouri, N., Seitz, R., Wolf, M., Egbring, R., Shams Elden, A. and Havemann, K. (1990). Cytochemichal determination of intracellular polymorphonuclear leukocyte elastase content in patients with severe bacterial infection and septicaemia correlated with coagulation parameters. *Annals of Clinical Biochemistry*, **27**:575–580.

Evangelista, V., Manarini, S., Rotondo, S., Martelli, N., Polischuk, R., McGregor, J.L. *et al.* (1996). Platelet/polymorphonuclear leukocyte interaction in dynamic conditions: evidence of adhesion cascade and cross talk between P-selectin and the β_2 integrin CD11b/CD18. *Blood*, **88**:4183–4194.

Evangelista, V., Piccardoni, P., de Gaetano, G. and Cerletti, C. (1992). Defibrotide inhibits platelet activation by cathepsin G released from stimulated polymorphonuclear leucocytes. *Thrombosis and Haemostasis*, **67**:660–664.

Evangelista, V., Piccardoni, P., White, J.G., de Gaetano, G. and Cerletti, C. (1993). Cathepsin G-dependent platelet stimulation by activated polymorphonuclear leucocytes and its inhibition by antiproteinases: role of P-selectin-mediated cell-cell adhesion. *Blood*, **81**:2947–2957.

Evangelista, V., Rajtar, G., de Gaetano, G., White, J.G. and Cerletti, C. (1991). Platelet activation by fMLP-stimulated polymorphonuclear leukocytes: the activity of cathepsin G is not prevented by antiproteinases. *Blood*, **77**:2379–2388.

Faint, R.W (1992). Platelet-neutrophil interactions: their significance. *Blood Reviews*, **6**:83–91.

Ferrer-Lopez, P., Renesto, P., Schattner, M., Bassot, S., Laurent, P. and Chignard, M. (1990). Activation of human platelets by C5a-stimulated neutrophils: a role for cathepsin G. *American Journal of Physiology*, **258**:C1100–C1107.

FitzGerald, G.A. (1991). Mechanisms of platelet activation: thromboxane A_2 as an amplifying signal for other agonists. *American Journal of Cardiology*, **68**:11B–15B.

Ford-Hutchinson, A.W. (1990). LTB$_4$ in inflammation. *Critical Review of Immunology*, **10**:1–12.

Frazier, W.A. (1991). Thrombospondins. *Current Opinion in Cell Biology*, **3**:792–799.

Gao, A.-G., Lindberg, F.P., Finn, M.A., Blystone, S.D., Brown, E.J. and Frazier, W.A. (1996). Integrin-associated protein is a receptor for the C-terminal domain of thrombospondin. *Journal of Biological Chemistry*, **271**:21–24.

Gawaz, M.P., Mujais, S.K., Schmidt, B. and Gurland, H.J. (1994). Platelet-leukocyte aggregation during hemodialysis. *Kidney International*, **46**:489–495.

Gresham, H.D., Adams, S.P. and Brown, E.J. (1992). Ligand binding specificity of the leukocyte response integrin expressed by human neutrophils. *Journal of Biological Chemistry*, **267**:13895–13902.

Handin, R.I., Karabin, R. and Boxer, G.J. (1977). Enhancement of platelet function by superoxide anion. *Journal of Clinical Investigation*, **59**:959–963.

Handley, D.A., Saunders, R.N., Houlihan, W.J. and Tomesch, J.C. (1990). *Platelet-Activating Factor in Endotoxin and Immune Diseases*. New York: Marcel Dekker, Inc.

Harrison, P. and Martin-Cramer, E. (1993). Platelet α-granules. *Platelets*, **7**:52–62.

Henson, P.M. (1990). Interactions between neutrophils and platelets. *Laboratory Investigation*, **62**:391–393.

Issekutz, A.C. and Szpejda, M. (1986). Evidence that platelet activating factor may mediate some acute inflammatory responses. Studies with the platelet-activating factor antagonists, CV 3988. *Laboratory Investigation*, **54**:275–281.

Iuliano, L., Pedersen, J.Z., Praticò, D., Rotilio, G. and Violi, F. (1994). Role of hydroxyl radicals in the activation of human platelets. *European Journal of Biochemistry*, **221**:695–704.

Iwamoto, S., Kawasaki, T., Kambayashi, J., Ariyoshi, H. and Monden, M. (1996). Platelet microparticles: a carrier of platelet-activating factor? *Biochemical and Biophysical Research Communications*, **218**:940–944.

Janoff, A. (1985). Elastase and emphysema. Current assessment of the protease-antiprotease hypothesis. *American Review of Respiratory Diseases*, **132**:417–433.

Jy, W., Mao, W.-W., Horstman, L.L., Tao, J. and Ahn, Y.S. (1995). Platelet microparticles bind, activate and aggregate neutrophils *in vitro*. *Blood Cells, Molecules, and Diseases*, **21**:217–231.

Kornecki, E., Yigal, H.E., Egbring, R., Gramse, M., Seitz, R., Eckardt, A. *et al.* (1988). Granulocyte-platelet interactions and platelet fibrinogen receptor exposure. *American Journal of Physiology*, **255**:H6561–H6568.

Korthuis, R.J. and Granger, D.N. (1994). Pathogenesis of ischemia/reperfusion: role of neutrophil-endothelial cell adhesion. In *The Handbook of Immunopharmacology. Adhesion Molecules*, edited by C.G. Wegner, pp. 163–190. London: Academic Press Ltd.

Kuijper, P.H.M., Gallardo-Torres, H.I., Lammers, J.W.J, Sixma, J.J., Koenderman, L. and Zwaginga, J.J. (1997). Platelet and fibrin deposition at the damaged vessel wall. Cooperative substrates for neutrophil adhesion under flow conditions. *Blood*, **89**:166–175.

Kunkel, S.L. (1983). Generalized Schwartzman reaction: an enigmatic model for therapeutic agents? *Laboratory Investigation*, **6**:653–655.

LaRosa, C.A., Rohrer, M.J., Benoit, S.E., Barnard, M.R. and Michelson, A.D. (1994a). Neutrophil cathepsin G modulates the platelet surface expression of the glycoprotein (GP) Ib-IX complex by proteolysis of the von Willebrand factor binding site on GPIbα and by cytoskeletal-mediated redistribution of the remainder of the complex. *Blood*, **84**:158–168.

LaRosa, C.A., Rohrer, M.J., Benoit, S.E., Rodino, L.J., Barnard, M.R. and Michelson, A.D. (1994b). Human neutrophil cathepsin G is a potent platelet activator. *Journal of Vascular Surgery*, **19**:306–319.

Larsen, E., Celi, A., Gilbert, G.E., Furie, B.C., Erban, J.K., Bonfanti, R. *et al.* (1989). PADGEM protein: a receptor that mediates the interaction of activated platelets with neutrophils and monocytes. *Cell*, **59**:305–312.

Lefer, A.M., Stahl, G.L., Lefer, D.J., Brezinski, M.E., Nicolaou, K.C., Veale, C.A. *et al.* (1988). Lipoxin A_4 and B_4: comparison of eicosanoids having bronchoconstrictor and vasodilator actions but lacking platelet aggregatory activity. *Proceedings of the National Academy of Sciences of the United States of America*, **85**:8340–8344.

Lehr, H.-A., Olofsson, A.M., Carew, T.E., Vajkoczy, P., Von Andrian U.H., Hübner, C. *et al.* (1994). P-selectin mediates the interaction of circulating leukocytes with platelets and microvascular endothelium in response to oxidized lipoprotein *in vivo*. *Laboratory Investigation*, **71**:380–386.

Levine, P.H., Weinger, R.S., Simon, J., Scoon, K.L. and Krimsky, N.I. (1976). Leukocyte-platelet interaction. Release of hydrogen peroxide by granulocytes as a modulator of platelet reaction. *Journal of Clinical Investigation*, **57**:955–963.

Lindberg, F.P., Gresham, H.D., Schwartz, E. and Brown, E.J. (1993). Molecular cloning of integrin-associated protein: an immunoglobulin family member with multiple membrane-spanning domains implicated in $\alpha_v\beta_3$-dependent ligand binding. *Journal of Cell Biology*, **123**:485–496.

Lonky, S.A. and Wohl, H. (1981). Stimulation of human leukocyte elastase by platelet factor 4. Physiologic, morphologic and biochemical effects on hamster lungs *in vitro*. *Journal of Clinical Investigation*, **67**:817–826.

Lopez, J.A. (1994). The platelet glycoprotein Ib-IX complex. *Blood Coagulation and Fibrinolysis*, **5**:97–119.

Lorant, D.E., Zimmerman, G.A., McIntyre, T.M. and Prescott, S.M. (1995). Platelet-activating factor mediates procoagulant activity on the surface of endothelial cells by promoting leukocyte adhesion. *Seminars in Cell Biology*, **6**:295–303.

Maclouf, J. (1993). Transcellular biosynthesis of arachidonic acid metabolites: from *in vitro* investigations to *in vivo* reality. *Baillière's Clinical Haematology*, **6**:593–608.

Maclouf, J., Fruteau de Laclos, B. and Borgeat, P. (1982). Stimulation of leukotriene biosynthesis in human blood leukocytes by platelet-derived 12-hydroperoxy-eicosatetraenoic acid. *Proceedings of the National Academy of Sciences of the United States of America*, **79**:6042–6046.

Mannaioni, P.F., Di Bello, M.G. and Masini, E. (1997). Platelets and inflammation: role of platelet-derived growth factor, adhesion molecules and histamine. *Inflammation Research*, **46**:4–18.

Marcus, A.J. and Hajjar, D.P. (1993). Vascular transcellular signaling. *Journal of Lipid Research*, **34**:2017–2030.

Marcus, A.J. and Safier, L.B. (1993). Thromboregulation: multicellular modulation of platelet reactivity in hemostasis and thrombosis. *FASEB Journal*, **7**:516–522.

Margaretten, W. and McKay, D.G. (1971). The requirement for platelets in the active Arthus reaction. *American Journal of Pathology*, **64**:257–270.

Maugeri, N., Evangelista, V., Celardo, A., Dell'Elba, G., Martelli, N., Piccardoni *et al*. (1994). Polymorpho-nuclear leukocyte-platelet interaction: role of P-selectin in thromboxane B$_2$ and leukotriene C$_4$ cooperative synthesis. *Thrombosis and Haemostasis*, **72**:450–456.

McCulloch, R.K., Croft, K.D. and Vandongen, R. (1992). Enhancement of platelet 12-HETE production in the presence of polymorphonuclear leukocytes during calcium ionophore stimulation. *Biochimica et Biophysica Acta*, **1133**:142–146.

McEver, R.P. (1991). Selectins: novel receptors that mediate leukocyte adhesion during inflammation. *Thrombosis and Haemostasis*, **65**:223–228.

McEver, R.P., Moore, K.L. and Cummings, R.D. (1995). Leukocyte trafficking mediated by selectin-carbohydrate interactions. *Journal of Biological Chemistry*, **270**:11025–11028.

Merhi, Y., Lacoste, L.L. and Lam, J.Y.T. (1994). Neutrophil implications in platelet deposition and vasoconstric-tion after deep arterial injury by angioplasty in pigs. *Circulation*, **90**:997–1002.

Molino, M., Blanchard, N., Belmonte, E., Tarver, A.P., Abrams, C., Hoxie, J.A. *et al*. (1995). Proteolysis of human platelet and endothelial cell thrombin receptor by neutrophil-derived cathepsin G. *Journal of Biological Chemistry*, **270**:11168–11175.

Molino, M., Di Lallo, M., de Gaetano, G. and Cerletti, C. (1992). Intracellular Ca^{2+} rise in human platelets induced by polymorphonuclear leucocyte-derived cathepsin G. *Biochemical Journal*, **288**:741–745.

Molino, M., Di Lallo, M., Martelli, N., de Gaetano, G. and Cerletti, C. (1993). Effects of leukocyte-derived cathepsin G on platelet membrane glycoprotein Ib-IX and IIb-IIIa complexes: a comparison with thrombin. *Blood*, **82**:2442–2451.

Movat, H.Z. and Wasi, S. (1985). Severe microvascular injury induced by lysosomal releasates of human poly-morphonuclear leukocytes. Increase in vasopermeability, hemorrhage, and microthrombosis due to degradation of subendothelial and perivascular matrices. *American Journal of Pathology*, **121**:404–417.

Nagata, K., Tsuji, T., Todoroki, N., Katagiri, Y., Tanoue, K., Yamazaki, H. *et al*. (1993). Activated platelets induce superoxide anion release by monocytes and neutrophils through P-selectin (CD62). *Journal of Immunology*, **151**:3267–3273.

Nash, G.B. (1994). Adhesion between neutrophils and platelets: a modulator of thrombotic and inflammatory events? *Thrombosis Research*, **74** (Suppl.1):S3–S11.

Nathan, C., Xie, Q.-W., Halbwachs-Mecarelli, L. and Jin, W.W. (1993). Albumin inhibits neutrophil spreading and hydrogen peroxide release by blocking the shedding of CD43 (sialophorin, leukosialin). *Journal of Cell Biology*, **122**:243–256.

Niewiarowski, S., Norton, K.J., Eckardt, A., Lukasiewicz, H., Holt, J.C. and Kornecki, E. (1989). Structural and functional characterization of major platelet membrane components derived by limited proteolysis of glyco-protein IIIa. *Biochimica et Biophysica Acta*, **983**:91–99.

Nourshargh, S. and Williams, T.J. (1995). Molecular and cellular interactions mediating granulocyte accumula-tion *in vivo*. *Seminars in Cell Biology*, **6**:317–326.

Owen, C.A. and Campbell, E.J. (1995). Neutrophil proteinases and matrix degradation. The cell biology of peri-cellular proteolysis. *Seminars in Cell Biology*, **6**:367–376.

Owen, C.A., Campbell, M.A., Sannes, P.L., Boukedes, S.S. and Campbell, E.J. (1995). Cell surface-bound elastase and cathepsin G on human neutrophils: a novel, non-oxidative mechanism by which neutrophils focus and preserve catalytic activity of serine proteinases. *Journal of Cell Biology*, **131**:775–789.

Page, C.P. (1989). Platelets as inflammatory cells. *Immunopharmacology*, **17**:51–59.

Palabrica, T., Furie, B.C., Aronowitz, M., Benjamin, C., Hsu, Y.-M., Sajer, S.A. *et al*. (1992). Leukocyte accumu-lation which promotes fibrin deposition is mediated *in vivo* by P-selectin (CD62) on adherent platelets. *Nature*, **359**:848–851.

Palmantier, R. and Borgeat, P. (1991). Thrombin-activated platelets promote leukotriene B$_4$ synthesis in polymor-phonuclear leukocytes stimulated by physiological agonists. *British Journal of Pharmacology*, **103**:1909–1916.

Papayianni, A., Serhan, C.N. and Brady, H.R. (1996). Lipoxin A$_4$ and B$_4$ inhibit leukotriene-stimulated interac-tions of human neutrophils and endothelial cells. *Journal of Immunology*, **156**:2264–2272.

Peterson, F., Ludwig, A., Flad, H.D. and Brandt, E. (1996). TNF-α renders human neutrophils responsive to plate-let factor 4. Comparison of PF-4 and IL-8 reveals different activity profiles of the two chemokines. *Journal of Immunology*, **156**:1954–1962.

Piccardoni, P., Evangelista, V., Piccoli, A., de Gaetano, G., Walz, A. and Cerletti, C. (1996). Thrombin-activated human platelets release two NAP-2 variants that stimulate polymorphonuclear leukocytes. *Thrombosis and Haemostasis*, **76**:780–785.

Pidard, D., Renesto, P., Berndt, M.C., Rabhi, S., Clemetson, K.J. and Chignard, M. (1994). Neutrophil proteinase cathepsin G is proteolytically active on the human platelet glycoprotein Ib-IX receptor: characterization of the cleavage sites within the glycoprotein Ibα subunit. *Biochemical Journal*, **303**:489–498.

Pintucci, G., Iacoviello, L., Castelli, M.P., Amore, C., Evangelista, V., Cerletti, C. *et al*. (1993). Cathepsin G-induced release of PAI-1 in the culture medium of endothelial cells: a new thrombogenic role for polymorphonuclear leukocytes? *Journal of Laboratory and Clinical Medicine*, **122**:69–79.

Plow, E.F. (1982). Leukocyte elastase release during blood coagulation. A potential mechanism for activation of the alternative fibrinolytic pathway. *Journal of Clinical Investigation*, **69**:564–572.

Poncz, M. (1990). Molecular biology of the megakaryocyte-specific gene platelet factor 4. In *Molecular and Cellular Biology of Cytokines*, edited by J.J. Oppenheim, M.C. Powanda, M.J. Kluger, and C.A. Dinarello, pp. 65–76. New York: Wiley-Liss, Inc.

Prescott, S.M., Zimmerman, G.A. and McIntyre, T.M. (1990). Platelet-activating factor. *Journal of Biological Chemistry*, **265**:17381–17384.

Qu, J., Condliffe, A.M., Lawson, M., Plevin, R.J., Riemersma, R.A., Barclay, G.R. *et al.* (1995). Lack of effect of recombinant platelet-derived growth factor on human neutrophil function. *Journal of Immunology*, **154**:4133–4141.

Rabhi-Sabile, S., Pidard, D., Lawler, J., Renesto, P., Chignard, M. and Legrand, C. (1996). Proteolysis of thrombospondin during cathepsin G-induced platelet aggregation: functional role of the 165-kDa carboxy-terminal fragment. *FEBS Letters*, **386**:82–86.

Radomski, M.W., Palmer, R.M.J. and Moncada, S. (1987). Comparative pharmacology of endothelium-derived relaxing factor, nitric oxide and prostacyclin in platelets. *British Journal of Pharmacology*, **92**:181–187.

Remold-O'Donnell, E. and Parent, D. (1995). Specific sensitivity of CD43 to neutrophil elastase. *Blood*, **86**:2395–402.

Renesto, P., Balloy, V. and Chignard, M. (1993). Inhibition by human leukocyte elastase of neutrophil-mediated platelet activation. *European Journal of Pharmacology*, **248**:151–155.

Renesto, P. and Chignard, M. (1993). Enhancement of cathepsin G-induced platelet activation by leukocyte elastase: consequence for the neutrophil-mediated platelet activation. *Blood*, **28**:139–144.

Renesto, P., Halbwachs-Mecarelli, L., Bessou, G., Balloy, V. and Chignard, M. (1996). Inhibition of neutrophil-endothelial cell adhesion by a neutrophil product, cathepsin G. *Journal of Leukocyte Biology*, **59**:855–863.

Renesto, P., Halbwachs-Mecarelli, L., Nusbaum, P., Lesavre, P. and Chignard, M. (1994a). Proteinase 3. A neutrophil proteinase with activity on platelets. *Journal of Immunology*, **152**:4612–4617.

Renesto, P., Kadiri, C. and Chignard, M. (1992). Combined activation of platelets by cathepsin G and platelet activating factor, two neutrophil-derived agonists. *British Journal of Haematology*, **80**:205–213.

Renesto, P., Si-Tahar, M. and Chignard, M. (1994b). Modulation by superoxide anions of neutrophil-mediated platelet activation. *Biochemical Pharmacology*, **47**:1401–1404.

Renesto, P., Si-Tahar, M., Moniatte, M., Balloy, V., Van Dorsselaer, A., Pidard, D. *et al.* (1997). Specific inhibition of thrombin-induced cell activation by the neutrophil proteinases elastase, cathepsin G and proteinase 3. Evidence for distinct cleavage sites within the aminoterminal domain of the thrombin receptor. *Blood*, **89**:1944–1953.

Rinder, H.M., Bonan, J.L., Rinder, C.S., Ault, K.A. and Smith, B.R. (1991a). Dynamics of leukocyte-platelet adhesion in whole blood. *Blood*, **78**:1730–1737.

Rinder, H.M., Bonan, J.L., Rinder, C.S., Ault, K.A. and Smith, B.R. (1991b). Activated and unactivated platelet adhesion to monocytes and neutrophils. *Blood*, **78**:1760–1769.

Rinder, C.S., Bonan, J.L., Rinder, H.M., Mathew, J., Hines, R. and Smith, B.R. (1992). Cardiopulmonary bypass induces leukocyte-platelet adhesion. *Blood*, **79**:1201–1205.

Rubinstein, E., Boucheix, C., Worthington, R.E. and Carroll, R.C. (1995). Anti-platelet antibody interactions with Fcγ receptor. *Seminars in Thrombosis and Hemostasis*, **21**:10–22.

Ruf, A. and Patscheke, H. (1995). Platelet-induced neutrophil activation: platelet-expressed fibrinogen induces the oxidative burst in neutrophils by an interaction with CD11c/CD18. *British Journal of Haematology*, **90**:791–796.

Sako, D., Chang, X.-J., Barone, K.M., Vachino, G., White, H.M., Shaw, G. *et al.* (1993). Expression cloning of a functional glycoprotein ligand for P-selectin. *Cell*, **75**:1179–1186.

Salvemini, D., De Nucci, G., Gryglewski, R.J. and Vane, J.R. (1989). Human neutrophils and mononuclear cells inhibit platelet aggregation by releasing a nitric oxide-like factor. *Proceedings of the National Academy of Sciences of the United States of America*, **86**:6328–6332.

Sandborg, R.R. and Smolen, J.E. (1988). Biology of disease. Early biochemical events in leukocyte activation. *Laboratory Investigation*, **59**:300–320.

Schall, T.J. and Bacon, K.B. (1994). Chemokines, leukocyte trafficking, and inflammation. *Current Opinion in Immunology*, **6**:865–873.

Selak, M.A. (1992). Neutrophil elastase potentiates cathepsin G-induced platelet activation. *Thrombosis and Haemostasis*, **68**:570–576.

Selak, M.A. (1993). Cathepsin G activates platelets in the presence of plasma and stimulates phosphatidic acid formation and lysosomal enzyme release. *Platelets*, **4**:85–89.

Selak, M.A. (1994). Cathepsin G and thrombin: evidence for two different platelet receptors. *Biochemical Journal*, **297**:269–275.

Selak, M.A., Chignard, M. and Smith, J.B. (1988). Cathepsin G is a strong platelet agonist released by neutrophils. *Biochemical Journal*, **251**:293–299.

Selak, M.A. and Smith, J.B. (1990). Cathepsin G binding to platelets. Evidence for a specific receptor. *Biochemical Journal*, **266**:55–62.

Shattil, S.J. (1995). Function and regulation of the β_3 integrins in hemostasis and vascular biology. *Thrombosis and Haemostasis*, **74**:149–155.

Sheikh, S. and Nash, G.B. (1996). Continuous activation and deactivation of integrin CD11b/CD18 during *de novo* expression enables rolling neutrophils to immobilize on platelets. *Blood*, **87**:5040–5050.

Siminiak, T., Flores, N.A. and Sheridan, D.J. (1995). Neutrophil interactions with endothelium and platelets: possible role in the development of cardiovascular injury. *European Heart Journal*, **16**:160–170.

Sisson, J.H., Prescott, S.M., McIntyre, T.M. and Zimmerman, G.A. (1987). Production of platelet-activating factor by stimulated human polymorphonuclear leucocytes. Correlation of synthesis with release, functional events, and leukotriene B_4 metabolism. *Journal of Immunology*, **138**:3918–3926.

Si-Tahar, M., Pidard, D., Balloy, V., Moniatte, M., Van Dorsselaer, A. and Chignard, M. (1997). Human neutrophil elastase proteolytically activates the platelet $\alpha_{IIb}\beta_3$ integrin through cleavage of the carboxyterminus of the α_{IIb} subunit heavy chain. Involvement in the potentiation of platelet aggregation. *Journal of Biological Chemistry*, **272**:11636–11647.

Si-Tahar, M., Renesto, P., Falet, H., Rendu, F. and Chignard, M. (1996). The phospholipase C/protein kinase C pathway is involved in cathepsin G-induced human platelet activation: comparison with thrombin. *Biochemical Journal*, **313**:401–408.

Smith, J.A. (1994). Neutrophils, host defense, and inflammation: a double-edged sword. *Journal of Leukocyte Biology*, **56**:672–686.

Smolen, J.E. and Boxer, L.A. (1995). Functions of neutrophils. In *Hematology,* Fifth Edition, edited by E. Beutler *et al.*, pp. 779–798. New York: McGraw Hill, Inc.

Spagnuolo, P.J., Ellner, J.J., Hassid, A. and Dunn, M.J. (1980). Thromboxane A_2 mediates augmented polymorphonuclear leukocyte adhesiveness. *Journal of Clinical Investigation*, **66**:406–414.

Spangenberg, P., Redlich, H., Haferkorn, R., Götzrath, M., Lösche, W., Kehrel, B., *et al.* (1995). Adhesion of platelets to polymorphonuclear leukocyte (PMNL): PMNL-counterreceptors of platelet-bound fibrinogen and activation of platelets or PMNL during adhesion. *Thrombosis and Haemostasis*, **73**:1080.

Stewart, G.J. (1993). Neutrophils and deep venous thrombosis. *Haemostasis*, **23**(suppl.1):127–140.

Suchard, S.J., Burton, M.J., Dixit, V.M. and Boxer, L.A. (1991). Human neutrophil adherence to thrombospondin occurs through a CD11/CD18-independent mechanism. *Journal of Immunology*, **146**:3945–3952.

Tzeng, D.Y., Deuel, T.F., Huang, J.S., Senior, R.M., Boxer, L.A. and Baehner, R.L. (1984). Platelet-derived growth factor promotes polymorphonuclear leukocyte activation. *Blood*, **64**:1123–1128.

Walz, A., Dewald, B. and Baggiolini, M. (1990). Formation and biological activity of NAP-2, a neutrophil-activating peptide derived from platelet α-granule precursors. In *Molecular and Cellular Biology of Cytokines*, edited by J.J. Oppenheim, M.C. Powanda, M.J. Kluger, and C.A. Dinarello, pp. 363–368. New York: Wiley-Liss, Inc.

Ward, C.M., Andrews, R.K., Smith, A.I. and Berndt, M.C. (1996). Mocarhagin, a novel cobra venom metalloproteinase, cleaves the platelet von Willebrand factor receptor glycoprotein Ibα. Identification of the sulfated tyrosine/anionic sequence Tyr-276-Glu-282 of glycoprotein Ibα as a binding site for von Willebrand factor and α-thrombin. *Biochemistry*, **35**:4929–4938.

Ward, P.A., Cunningham, T.W., McCulloch, K.K., Phan, S.H., Powell, J. and Johnson, K.J. (1988). Platelet enhancement of O_2- responses in stimulated human neutrophils. Idenfication of platelet factor as adenine nucleotide. *Laboratory Investigation*, **58**:37–47.

Weber, C. and Springer, T.A. (1997). Neutrophil accumulation on activated, surface-adherent platelets in flow is mediated by interaction of Mac-1 with fibrinogen bound to $\alpha_{IIb}\beta_3$ and stimulated by platelet-activating factor. *Journal of Clinical Investigation*, **100**:2085–2093.

Wicki, A.N. and Clemetson, K.J. (1985). Structure and function of platelet membrane glycoproteins Ib and V. Effects of leukocyte elastase and other proteases on platelets response to von Willebrand factor and thrombin. *European Journal of Biochemistry*, **153**:1–11.

Witko-Sarsat, V. and Descamps-Latscha, B. (1994). Neutrophil-derived oxidants and proteinases as immunomodulatory mediators in inflammation. *Mediators of Inflammation*, **3**:257–273.

Woodman, R.C., Reinhardt, P.H., Kanwar, S., Johnston, F.L. and Kubes, P. (1993). Effects of human neutrophil elastase (HNE) on neutrophil function *in vitro* and in inflamed microvessels. *Blood*, **82**:2188–2195.

Zatta, A., Prosdocimi, M., Bertelé, V., Bazzoni, G. and Del Maschio, A. (1990). Inhibition of platelet function by polymorphonuclear leukocytes. *Journal of Laboratory and Clinical Medicine*, **116**:651–660.

11 P-selectin

Rodger P. McEver*

W.K. Warren Medical Research Institute and Department of Medicine and Biochemistry and Molecular Biology, University of Oklahoma Health Sciences Center, and Cardiovascular Biology Research Program, Oklahoma Medical Research Foundation, Oklahoma City, OK 73104, USA

INTRODUCTION

P-, E-, and L-selectin are structurally related membrane glycoproteins that initiate the attachment of flowing leukocytes to platelets, endothelial cells, or other leukocytes at sites of inflammation or tissue injury. This chapter focuses on P-selectin, the only selectin expressed on platelets. I will review aspects of its structure, the regulation of its expression, the mechanisms by which it mediates multicellular interactions under shear stress, and its role in physiological and pathological forms of inflammation and hemostasis.

SUBCELLULAR AND TISSUE DISTRIBUTION OF P-SELECTIN

P-selectin (CD62P) was discovered following the development of monoclonal antibodies (mAbs) that react with activated platelets but not with unactivated platelets (McEver and Martin, 1984; Hsu-Lin *et al.*, 1984). The mAbs bind to a platelet membrane glycoprotein with an apparent Mr of 140000, which was originally named GMP-140 (granule membrane protein of 140 kDa) (Stenberg *et al.*, 1985) or PADGEM (platelet-activation-dependent granule-external membrane protein) (Berman *et al.*, 1986). Immunogold analysis of ultra-thin sections of human platelets indicated that P-selectin is localized to the membranes of α granules of resting cells (Stenberg *et al.*, 1985; Berman *et al.*, 1986). Within seconds after activation of platelets with thrombin or other agonists, P-selectin is redistributed to the cell surface through fusion of granule membranes with the plasma membrane. *In vitro*, P-selectin remains on the activated cell surface for at least one hour (George *et al.*, 1986). These properties have led to the widespread use of mAbs to P-selectin as probes for activated platelets. The initial radioligand-binding assays of isolated platelets (McEver and Martin, 1984; Hsu-Lin *et al.*, 1984) have been replaced by flow cytometric analyses of unfractionated platelets in anticoagulated whole blood (reviewed in

---

Corresponding author: W.K. Warren Medical Research Institute, University, of Oklahoma Health Sciences Center, 825 N.E. 13th Street, Oklahoma City, OK 73104, USA. Tel: 405-271-6480, Fax: 405-271-3137, email: rodger-mcever@ouhsc.edu.

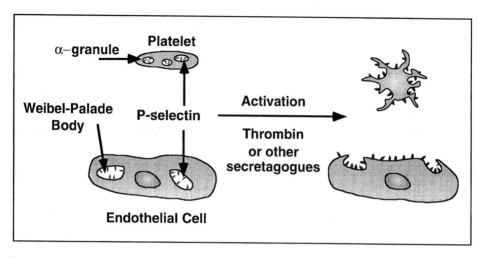

Figure 11.1 Subcellular location of P-selectin. The protein is concentrated in the membranes of secretory granules: the α granules of platelets and the Weibel-Palade bodies of endothelial cells. Upon activation of these cells with thrombin or other secretagogues, P-selectin is rapidly translocated to the cell surface as the granule membranes fuse with the plasma membrane.

Michelson, 1996). *In vivo*, anti-P-selectin mAbs may also be useful probes for radionuclide imaging of thrombi (Palabrica *et al.*, 1989; Miller *et al.*, 1991).

Immunoperoxidase analysis of normal human tissues indicated that P-selectin is present in endothelial cells as well as in platelets and their precursors, megakaryocytes. The endothelial cell protein is concentrated in postcapillary venules, the major site of leukocyte extravasation during inflammation (McEver *et al.*, 1989). In endothelial cells, P-selectin is localized to the membranes of Weibel-Palade bodies, the secretory storage granules that store large multimers of von Willebrand factor (McEver *et al.*, 1989; Bonfanti *et al.*, 1989). Within minutes after activation of these cells with thrombin, histamine, or complement components, P-selectin is translocated to the cell surface (McEver *et al.*, 1989; Hattori *et al.*, 1989a; Hattori *et al.*, 1989b). Unlike expression on activated platelets, P-selectin expression on activated endothelial cells is transient, peaking within 3–10 minutes and then declining to basal levels within 30 minutes as a result of endocytosis (Hattori *et al.*, 1989a).

Thus, P-selectin is stored in secretory granules of both platelets and endothelial cells, where it is rapidly mobilized to the cell surface by thrombin and other agonists that induce degranulation (Figure 11.1).

STRUCTURE AND FUNCTION OF P-SELECTIN

P-selectin is synthesized by cultured endothelial cells and by HEL cells, a human cell line with features of megakaryocytes (McEver *et al.*, 1989; Johnston *et al.*, 1989b). Core high-mannose N-linked glycans are attached to the nascent protein chain. These glycans are converted to complex forms during passage through the Golgi complex. The mature glycoprotein contains 30% carbohydrate by weight (Johnston *et al.*, 1989b).

The cDNA-derived amino acid sequence of P-selectin predicts that it is an elongated protein composed of a tandem array of modular, cysteine-rich domains (Johnston *et al.*, 1989a). The extracellular region contains nine potential attachment sites for N-glycans, most of which are probably occupied. Following the signal peptide, there is an N-terminal

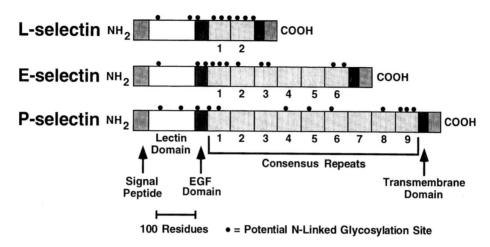

Figure 11.2 **Schematic diagram of the three human selectins.** Each selectin has an N-terminal carbohydrate-recognition domain like those in C-type lectins, followed by an EGF-like domain, a series of short consensus repeats, a transmembrane domain, and a short cytoplasmic tail. Each protein has many potential attachment sites for N-glycans, most of which are utilized.

carbohydrate-recognition domain characteristic of Ca^{2+}-dependent (C-type lectins), then an epidermal growth factor (EGF)-like motif, a series of short consensus repeats (SCRs) like those in complement-regulatory proteins, a transmembrane domain, and a short cytoplasmic tail (Figure 11.2).

The human gene for P-selectin is located on the long arm of chromosome 1 at bands q21–24 (Watson *et al.*, 1990). It spans over 50 kb and contains 17 exons, most of which encode structurally distinct domains (Johnston *et al.*, 1990). Both platelets and endothelial cells contain an alternatively spliced form of P-selectin mRNA that lacks the exon encompassing the transmembrane domain (Johnston *et al.*, 1989a; Johnston *et al.*, 1990). Chinese hamster ovary (CHO) cells transfected with a cDNA encoding this transcript secrete a soluble monomeric form of P-selectin (Ushiyama *et al.*, 1993). Furthermore, P-selectin antigen is present in human plasma (Dunlop *et al.*, 1992; Ushiyama *et al.*, 1993; Katayama *et al.*, 1993), and some of this material is clearly derived from the alternatively spliced transcript (Ishiwata *et al.*, 1994). However, the plasma levels of P-selectin are very low, raising questions as to its biological significance. Plasma P-selectin levels are modestly increased in patients with inflammatory and thrombotic disorders, raising the possibility that such measurements could be used as markers of disease activity (Katayama *et al.*, 1993; Takeda *et al.*, 1994; Ikeda *et al.*, 1994; Chong *et al.*, 1994; Kaikita *et al.*, 1995; Ikeda *et al.*, 1995; Sakamaki *et al.*, 1995).

The cDNAs encoding E-selectin (Bevilacqua *et al.*, 1989) and L-selectin (Lasky *et al.*, 1989; Siegelman *et al.*, 1989; Tedder *et al.*, 1989) were reported virtually simultaneously to the report of the cDNA encoding P-selectin (Figure 11.2). All three proteins have similar domain organizations, although the number of SCRs vary. The selectins share significant sequence similarity, particularly in the lectin and EGF domains which are over 60% identical at both the nucleotide and amino acid levels. The genes for the three selectins are clustered on chromosome 1 in both mice and humans (Watson *et al.*, 1990). They also have similar exon/intron boundaries, supporting the concept that they evolved through exon rearrangement and duplication (Collins *et al.*, 1991; Johnston *et al.*, 1990; Ord *et al.*, 1990).

The structural similarities of the selectins immediately led to studies that explored their functional relationships. L-selectin, which is expressed on most leukocytes, was originally described as a homing receptor that mediates the adhesion of lymphocytes to specialized high endothelial venules (HEV) of peripheral lymph nodes (Gallatin *et al.*, 1983). E-selectin was shown to be transiently synthesized by endothelial cells that are stimulated by tumor necrosis factor α (TNF-α), interleukin-1 (IL-1), or lipopolysaccharide (LPS), where it mediates adhesion of leukocytes (Bevilacqua *et al.*, 1987). P-selectin was shown to mediate adhesion of leukocytes to platelets and endothelial cells stimulated with thrombin, histamine, or other secretagogues (Larsen *et al.*, 1989; Hamburger and McEver, 1990; Geng *et al.*, 1990). Many studies have confirmed the importance of the selectins in mediating multicellular interactions of leukocytes in response to immunological or inflammatory challenges (McEver *et al.*, 1995; Tedder *et al.*, 1995). There is general consensus that the selectins play a pivotal proximal role in a multistep cascade of activation and adhesion events during leukocyte recruitment (Springer, 1995; Butcher and Picker, 1996). Under shear stress, the selectins promote the tethering and transient rolling adhesion of leukocytes to activated endothelial cells or platelets. The relatively slow velocities of the rolling leukocytes allows them to encounter locally generated chemoattractants and lipid autacoids that activate the leukocytes. The activated leukocytes then use integrins for shear-resistant firm adhesion and for later events such as cell spreading or emigration into the underlying tissues. This chapter will emphasize the functions of P-selectin in leukocyte recruitment, but will include some relevant comparisons to the functions of L- and E-selectin.

SELECTIN-LIGAND INTERACTIONS

The selectins mediate cell adhesion through interactions of the N-terminal C-type lectin domains with glycoconjugates on the surfaces of target cells. All three selectins bind sialylated and fucosylated glycans, of which the prototype is sialyl Lewis x (sLe^x, Neu Ac$\alpha_{2,3}$Gal$\beta_{1,4}$[Fuc$\alpha_{1,3}$]GlcNAcβ_{1-R}), a terminal component of oligosaccharides attached to many proteins and lipids on leukocytes and on some endothelial cells (Varki, 1994; McEver *et al.*, 1995). Importantly, P- and L-selectin, but not E-selectin, also bind sulfated structures such heparin and sulfatides (Varki, 1994; McEver *et al.*, 1995). A central question has been whether the selectins bind preferentially to specific glyconjugates on cell surfaces, and if so, whether such binding is required for optimal cell adhesion.

P-selectin binds to a limited number of relatively high affinity sites on human leukocytes (Moore *et al.*, 1991; Skinner *et al.*, 1991). Treatment of the leukocytes with sialidases eliminates the binding sites, indicating that they must be sialylated to be recognized by P-selectin (Moore *et al.*, 1991). Treatment with proteases also eliminates binding, indicating that the binding sites are on glycoprotein(s) rather than on glycolipids (Moore *et al.*, 1991). Leukocyte adhesion to immobilized P-selectin is eliminated by treatment of the cells with sialidase, or by preincubation with fluid-phase sLe^x or antibodies to sLe^x (Polley *et al.*, 1991). This result is consistent with the notion that the high affinity binding sites for P-selectin contain sLe^x. However, P-selectin binds with much lower affinity to CHO cells transfected with an α1,3/1,4 fucosyltransferase that express large quantities of sLe^x, demonstrating that this tetrasaccharide is not sufficient to confer high affinity binding (Zhou *et al.*, 1991). Heparin prevents binding of P-selectin to leukocytes, providing an early clue that sulfation might be an additional feature of the high affinity binding sites for P-selectin (Skinner *et al.*, 1991).

As assessed by blotting assays or by affinity chromatography, P-selectin binds to only a single glycoprotein in lysates of human myeloid cells (Moore *et al.*, 1992). This protein, now termed P-selectin glycoprotein ligand-1 (PSGL-1), is a disulfide-bonded homodimer

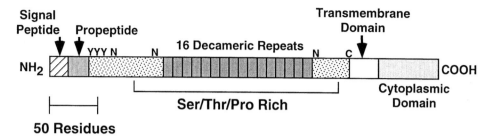

Figure 11.3 Schematic diagram of human PSGL-1. "N" indicates one of the three potential sites for addition of N-glycans. "Y" indicates one of the three clustered tyrosines near the N-terminus of the mature protein that are within a consensus sequence for tyrosine sulfation. "C" indicates the single cysteine in the extracellular domain that is the putative site for disulfide linkage of the two identical subunits of PSGL-1.

with two 120-kDa subunits. PSGL-1 has at most two or three N-glycans, but has many clustered, sialylated O-glycans that render the protein susceptible to cleavage with O-sialoglycoprotein endopeptidase (OSGE) (Moore *et al.*, 1992; Norgard *et al.*, 1993). Treatment of intact leukocytes with OSGE eliminates the high affinity binding sites for P-selectin, suggesting that they correspond to PSGL-1 (Norgard *et al.*, 1993; Ushiyama *et al.*, 1993). PSGL-1 contains sLe^x, but this constitutes only a small fraction of the total sLe^x on the surface of human neutrophils (Norgard *et al.*, 1993). Sialidase treatment of PSGL-1 eliminates binding to P-selectin, whereas enzymatic removal of the N-glycans does not; this suggests that sialylated O-glycans are required for binding (Moore *et al.*, 1992). A cDNA encoding PSGL-1 was isolated from a human promyelocytic HL-60 cell library by expression cloning in COS cells transfected with an 1,3/1,4 fucosyltransferase, establishing that PSGL-1 must also be fucosylated to bind P-selectin (Sako *et al.*, 1993).

The cDNA-derived amino acid sequence indicates that PSGL-1 is a type 1 membrane protein (Figure 11.3). It has an N-terminal signal peptide, followed by a putative propeptide. The extracellular domain of the mature protein extends from residues 42 to 309, and is followed by a 25-residue transmembrane domain and a 69-residue cytoplasmic tail. The extracellular domain has the characteristic features of a mucin; it is rich in serines, threonines, and prolines, and it includes 15 decameric consensus repeats. There are only three potential sites for attachment of N-glycans, consistent with the relatively small quantities of N-glycans observed on native PSGL-1. There is a single extracellular cysteine near the junction with the transmembrane domain; this is the putative site for disulfide linkage between two identical subunits of PSGL-1. Near the N-terminus are three clustered tyrosines at residues 46, 48, and 51 within an acidic consensus motif for tyrosine sulfation. The coding sequence of PSGL-1 is contained within a single exon in the human gene (Veldman *et al.*, 1995). The gene from human leukocytes encodes 16 decameric repeats, whereas the gene from HL-60 cells, from which the cDNA for PSGL-1 was originally isolated, contains only 15 repeats (Veldman *et al.*, 1995; Moore *et al.*, 1995). A cDNA for murine PSGL-1 predicts a protein of similar size to the human molecule (Yang *et al.*, 1996). The murine protein has a signal peptide, a propeptide, and a single cysteine near the transmembrane domain. The sequences of the transmembrane domain and cytoplasmic domain are very similar to those of human PSGL-1, suggesting conserved functions. Interestingly, the murine extracellular domain has only five decameric repeats and has little sequence similarity to the human protein other than having a high percentage of serines, threonines, and prolines. However, it does have two tyrosines within a consensus sequence for tyrosine sulfation near the N-terminus.

Studies with glycosidases indicate that PSGL-1 from human neutrophils has O-glycans with sialylated and fucosylated polylactosamine (Moore *et al.*, 1994). Extension with

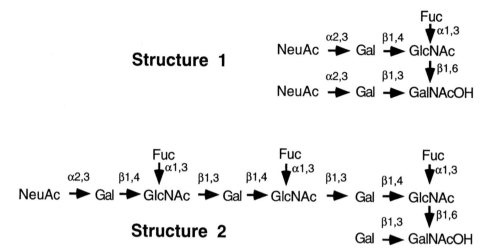

Figure 11.4 Structures of the fucosylated O-glycans of PSGL-1. Both structures have a core-2 backbone created by addition of an N-acetylglucosamine $\beta_{1,6}$-linked to N-acetylgalactosamine. Structure 1 is much less common than structure 2. Both fucosylated glycans constitute only 14% of the total O-glycans of PSGL-1 from human HL-60 cells. One or both of these structures comprises part of the recognition site(s) on PSGL-1 for selectins.

polylactosamine is thought to occur only on the $\beta_{1,6}$ branch of core-2 O-glycans (Maemura and Fukuda, 1992). CHO cells synthesize only simple core-1 O-glycans because they do not express the core 2 $\beta_{1,6}$ N-acetylglucosaminyltransferase (C2GnT) that is required to create the core-2 structure (Sasaki *et al.*, 1987; Maemura and Fukuda, 1992). In order to bind P-selectin, PSGL-1 expressed on transfected CHO cells must be co-expressed with both an $\alpha_{1,3}$ fucosyltransferase and C2GnT; thus, PSGL-1 requires sialylated and fucosylated, core-2 O-glycans to interact with P-selectin (Li *et al.*, 1996b). The structures of the O-glycans on native PSGL-1 from human HL-60 cells have been determined (Wilkins *et al.*, 1996). Remarkably, only 14% of the O-glycans are fucosylated, which occur as two distinct core-2 species (Figure 11.4). The more common of the two has a novel structure with a sialylated, trifucosylated polylactosamine extension on the $\beta_{1,6}$ branch. The O-glycans of CD43, a sialomucin of similar size expressed on HL-60 cells, have virtually no fucosylated structures and almost no polylactosamine (Maemura and Fukuda, 1992; Wilkins *et al.*, 1996). Both PSGL-1 and CD43 have many short core-2 O-glycans, indicating that formation of the core-2 branch is not sufficient for polylactosamine extension or fucosylation. These data suggest that some fucosylated O-glycans are constructed only at specific sites on specific proteins. The paucity of fucosylated O-glycans on PSGL-1 indicates that the protein is not merely a "scaffold" for attachment of many clustered sLe^x-bearing O-glycans that increase the avidity of binding to P-selectin.

   L-selectin binds preferentially to several sialomucins on endothelial cells, including GlyCAM-1, CD34, MAdCAM-1, and a 200-kDa molecule (Baumhueter *et al.*, 1993; Berg *et al.*, 1993; Lasky *et al.*, 1992; Hemmerich *et al.*, 1994). These proteins must be modified with O-glycans that are sulfated as well as sialylated and fucosylated to be recognized by L-selectin (Imai *et al.*, 1991; Imai *et al.*, 1993; Hemmerich *et al.*, 1994). Like the L-selectin glycoprotein ligands, PSGL-1 is sulfated (Wilkins *et al.*, 1995; Pouyani and Seed, 1995; Sako *et al.*, 1995). However, the O-glycans of PSGL-1 are not sulfated (Wilkins *et al.*, 1996); instead, the sulfate is present exclusively as tyrosine sulfate (Wilkins *et al.*, 1995). Sulfation occurs on one or more of the three clustered tyrosines at residues 46, 48, and 51 of PSGL-1 (Pouyani and Seed, 1995; Sako *et al.*, 1995; Li *et al.*, 1996b). Enzymatic removal of sulfate (Wilkins *et al.*, 1995), blockade of sulfate synthesis (Pouyani and Seed, 1995; Sako *et al.*, 1995), proteolytic removal of an N-terminal fragment containing the three clustered tyrosines (De Luca *et al.*, 1995), or replacement of the tyrosines with phenylalanines (Pouyani and Seed, 1995; Sako *et al.*, 1995; Li *et al.*, 1996b) abolishes the ability of PSGL-1 to interact with P-selectin. PL1, a mAb to an epitope spanning residues 49–62 of PSGL-1 (Li *et al.*, 1996a), blocks binding of purified PSGL-1 to P-selectin (Moore *et al.*, 1995). An Ig chimera containing residues 42–60 of PSGL-1, when co-expressed in COS cells with an $\alpha_{1,3}$ fucosyltransferase, also binds P-selectin, although perhaps not as well as the full-length molecule (Sako *et al.*, 1995). Collectively, these data suggest that P-selectin binds preferentially to an N-terminal, composite recognition site on PSGL-1 that includes at least one tyrosine sulfate and at least one sialylated and fucosylated, core-2 O-glycan. The relative orientations of the tyrosine sulfate(s) and O-glycan(s) required for optimal binding remain to be determined. Thr-44 and Thr-57 are the potential attachment sites for O-glycans that are closest to the clustered tyrosines. Alanine substitutions of these residues diminish binding of truncated or chimeric forms of recombinant PSGL-1 to P-selectin (Pouyani and Seed, 1995; Sako *et al.*, 1995). However, it has not been established that O-glycans at either position are required for full-length PSGL-1 to bind P-selectin.

The anti-PSGL-1 mAb PL1 completely blocks binding of fluid-phase P-selectin to leukocytes, and prevents adhesion of leukocytes to P-selectin under both static and shear conditions (Moore *et al.*, 1995). Therefore, the N-terminal region of PSGL-1 accounts for all the high affinity binding sites for P-selectin on leukocytes, and is required for adhesion of leukocytes to P-selectin under a variety of conditions. It should be noted that PSGL-1 also binds to L- and E-selectin. Binding of PSGL-1 to L-selectin is blocked by mAb PL1 (Guyer *et al.*, 1996; Walcheck *et al.*, 1996; Tu *et al.*, 1996; Spertini *et al.*, 1996), by proteolytic removal of the N-terminal clustered tyrosines (Spertini *et al.*, 1996), or by preventing sulfate synthesis (Spertini *et al.*, 1996). These results suggest that L- and P-selectin bind to a similar N-terminal region of PSGL-1 that includes both tyrosine sulfate and O-glycan(s). In contrast, binding of PSGL-1 to E-selectin is not inhibited by these maneuvers. This suggests that E-selectin binds to one or more additional sites on PSGL-1 that do not require sulfation.

PSGL-1 is expressed on the surface of virtually all leukocytes, although it is expressed at lower levels on B cells (Moore *et al.*, 1995; Laszik *et al.*, 1996). P-selectin binds to PSGL-1 on all myeloid cells (Moore *et al.*, 1995). However, it binds to PSGL-1 on only a subset of T cells (Moore *et al.*, 1995; Vachino *et al.*, 1995); most of these have the memory phenotype (Moore and Thompson, 1992) and they may be predominantly γ/δ cells (Diacovo *et al.*, 1996c). This suggests that differentiating T cells alter the post-translational modifications of PSGL-1 in order to confer P-selectin recognition. PSGL-1 is expressed on circulating dendritic cells, on tissue monocyte-derived dendritic cells, and on dendritic cells in lymphoid organs; the function of the protein in these cells has not been established (Laszik *et al.*, 1996). P-selectin binds to hematopoietic stem cells in the bone marrow, probably through PSGL-1 (Zannettino *et al.*, 1995). This interaction could potentially regulate hematopoetic cell differentiation. However, P-selectin-deficient mice have no obvious defects in hematopoiesis, although the kinetics of neutrophil clearance may be altered (Mayadas *et al.*, 1993; Johnson *et al.*, 1995).

P-selectin binds to CD24, a small, O-glycosylated, glycosylphosphatidylinositol-linked protein on leukocytes (Aigner *et al.*, 1995). It also binds to a 160-kDa glycoprotein on leukocytes that appears to have N-glycans but not O-glycans (Lenter *et al.*, 1994). The physiologic significance of these interactions is unknown, as neither CD24 nor the 160-kDa protein contributes measurably to the total number of high affinity binding sites for P-selectin on leukocytes or to adhesion of leukocytes to P-selectin.

Like all C-type lectins, the selectins bind in a Ca^{2+}-dependent manner to their carbo-hydrate ligands (McEver *et al.*, 1995). Ca^{2+} binds directly to P-selectin, altering the con-formation of the lectin domain (Geng *et al.*, 1991). Molecular modeling suggests that the structures of the lectin and EGF domains of P-selectin (Erbe *et al.*, 1993; Hollenbaugh *et al.*, 1993) are similar to those of E-selectin, which have been determined by X-ray crys-tallography (Graves *et al.*, 1994). The structural data, in conjunction with early results gen-erated from site-directed mutagenesis of P- or E-selectin (Graves *et al.*, 1994; Erbe *et al.*, 1993; Hollenbaugh *et al.*, 1993), suggest that sLe^x-like glycans bind to a small region that overlaps a single Ca^{2+} coordination site on the opposite face from where the EGF domain is attached. One mutagenesis study of P-selectin suggests that the binding site for sulfatides overlaps the binding site for sLe^x-like glycans (Bajorath *et al.*, 1994). However, more recent mutagenesis studies cast doubts on the assumptions as to the specific residues required for P-selectin to bind to either sLe^x-like glycans or to sulfated structures such as sulfatides (Revelle *et al.*, 1996b; Revelle *et al.*, 1996a). The crystal structure of E-selectin suggests very little contact between the lectin and EGF domains (Graves *et al.*, 1994). However, chimeric selectins in which EGF domains and/or SCRs have been exchanged exhibit altered ligand specificity in some studies (Kansas *et al.*, 1994; Tu *et al.*, 1996). Even substitutions of individual residues in the EGF domain can alter ligand specificity (Revelle *et al.*, 1996a). Collectively, these data suggest that residues in both lectin and EGF domains contribute to ligand specificity. However, only co-crystallization of a selectin with a biolo-gically relevant ligand is likely to resolve the true nature of the molecular contacts.

LEUKOCYTE INTERACTIONS WITH P-SELECTIN UNDER HYDRODYNAMIC FLOW

Many *in vitro* and *in vivo* studies have documented that leukocytes use selectins to attach to and roll on the vessel wall under the shear forces characteristic of postcapillary venules, where most leukocyte recruitment takes place (reviewed in McEver *et al.*, 1995; Ley and Tedder, 1995). Neutrophils were originally shown to tether to and roll on purified P-selectin incorporated into planar membranes in a parallel-plate flow chamber (Lawrence and Springer, 1991). Subsequent studies have shown that leukocytes roll on P-selectin expres-sed on activated human endothelial cells (Jones *et al.*, 1994) and on monolayers of activated platelets (Buttrum *et al.*, 1993; Yeo *et al.*, 1994; Diacovo *et al.*, 1996b). Leukocytes also roll on P-selectin immobilized on plastic or expressed on transfected CHO cells (Moore *et al.*, 1995). Selectin-ligand interactions must form rapidly to facilitate tethering, and then dissociate rapidly to facilitate rolling. Furthermore, selectin-ligand bonds must have signi-ficant mechanical strength; that is, shear forces must not significantly accelerate the rate of dissociation. To continue rolling, new selectin-ligand bonds must form at the leading edge of the cell to replace those bonds that dissociate at the trailing edge. The rolling velocity of the cell reflects the balance between the rates of association and dissociation of these bonds. Studies of the transient tethering of neutrophils on very low densities of P-selectin under flow support the concept that P-selectin-ligand bonds have fast association and dis-sociation rates with high tensile strength (Alon *et al.*, 1995a). These studies assume that the cellular kinetic measurements reflect the intrinsic rates of association and dissociation of selectin-ligand bonds. However, this assumption must be confirmed by measuring the kin-etics of binding of purified selectins to their ligands.

Studies with blocking antibodies or other inhibitors suggest that PSGL-1 on all classes of leukocytes must interact with P-selectin to mediate adhesion under either static or shear conditions (Moore *et al.*, 1995; Weyrich *et al.*, 1995b; Weyrich *et al.*, 1996; Alon *et al.*, 1994; Diacovo *et al.*, 1996c). This requirement probably reflects, in part, the superior binding of PSGL-1 to P-selectin as measured by biochemical assays (see above). The orientations of both PSGL-1 and P-selectin on the cell surface may also enhance their interactions under hydrodynamic flow. P-selectin is a highly extended molecule, which allows its N-terminal lectin domain to extend approximately 40 nm above the cell surface (Ushiyama *et al.*, 1993). When expressed on CHO cells, shortened P-selectin constructs with less than six SCRs mediate adhesion of neutrophils normally under static conditions, but are much less effective in supporting attachment and rolling of neutrophils under shear forces (Patel *et al.*, 1995). Those cells that do attach roll at significantly higher velocities and are much more susceptible to detachment by increasing shear forces. The shorter constructs function better, however, when expressed in glycosylation-defective Lec8 CHO cells, which have less glycocalyx. These data suggest that the lectin domain of P-selectin must extend a sufficient distance above the cell membrane to facilitate rapid contact with PSGL-1 on flowing leukocytes, and to minimize repulsion between the apposing glycocalyces. Like most sialomucins, PSGL-1 is also a highly extended molecule, which projects its N-terminal P-selectin-binding domain approximately 50 nm above the cell surface (Li *et al.*, 1996a). Thus, the binding domains of both P-selectin and PSGL-1 extend significantly higher above the cell surface than most other glycoproteins and glycolipids. Furthermore, PSGL-1 is concentrated on the tips of microvilli (Moore *et al.*, 1995), a location that enhances the ability of adhesion molecules on leukocytes to mediate attachment under flow conditions (Von Andrian *et al.*, 1995). Taken together, these data suggest that the orientations of PSGL-1 and P-selectin are ideally suited to optimize their opportunities for interactions under shear stresses (Figure 11.5). It should be noted that activated leukocytes do not

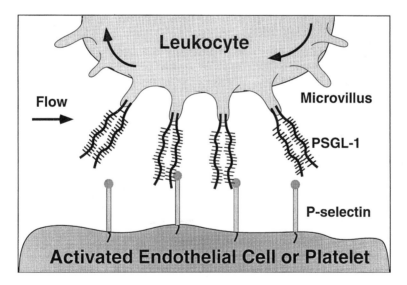

**Figure 11.5 Interactions of PSGL-1 on flowing leukocytes with P-selectin on activated platelets or endo-
thelial cells.** Both glycoproteins are highly extended, which projects their N-terminal binding sites 40–50 nm
above the plasma membrane. This distance is higher than that of most other cell-surface glycoproteins and glycolipids that constitute the glycocalyx. PSGL-1 is also concentrated on the tips of microvilli, the regions on leukocytes that first contact the vessel wall. These physical factors enhance the opportunities for rapid encounters of
PSGL-1 with P-selectin under hydrodynamic flow. The drawing is designed to emphasize the long lengths of PS-
GL-1 and P-selectin. The lengths of the microvilli are actually much longer than those of the molecules themselves. Platelets are much smaller than either leukocytes or endothelial cells.

adhere efficiently to P-selectin, even under static conditions (Lawrence and Springer, 1991; Lorant *et al.*, 1995; Doré *et al.*, 1996; Rinder *et al.*, 1994). This lessened adhesion is associated with a cytoskeletally-mediated redistribution of PSGL-1 to the uropods of the activated cells (Lorant *et al.*, 1995; Doré *et al.*, 1996). This redistribution may provide a method to transfer adhesive control from selectins to integrins to allow migration of leukocytes through endothelial cells into inflammatory tissues.

Leukocytes also attach to and roll on P-selectin expressed by activated platelets or endothelial cells *in vivo*. Mice rendered genetically deficient in P-selectin develop normally and have no obvious abnormalities in the absence of infectious challenge (Mayadas *et al.*, 1993). However, leukocytes roll poorly on postcapillary venules of exteriorized tissues under conditions where P-selectin is normally expressed (Mayadas *et al.*, 1993; Ley *et al.*, 1995). At later time periods, leukocytes resume rolling to some extent, apparently because of the induction of L-selectin ligands on the venular endothelium (Ley *et al.*, 1995). Studies with blocking mAbs to P-selectin also demonstrate that leukocytes roll on P-selectin expressed on postcapillary venules, usually within the first hour after tissue trauma or after superfusion of secretagogues such as histamine (Doré *et al.*, 1993; Kubes and Kanwar, 1994; Asako *et al.*, 1994; Nolte *et al.*, 1994; Ley, 1994). The anti-PSGL-1 mAb PL1 blocks rolling of human neutrophils infused into rat venules expressing P-selectin, suggesting that PSGL-1 must interact with P-selectin for leukocytes to roll on P-selectin *in vivo* as well as *in vitro* (Norman *et al.*, 1995). In a primate model of thrombus formation, leukocytes accumulate on activated adherent platelets. The accumulation of leukocytes is associated with increased generation of fibrin, both of which are blocked by anti-P-selectin antibodies (Palabrica *et al.*, 1992). Using wild-type and P-selectin-deficient mice, it has been observed that unstimulated platelets use an unidentified ligand to roll on P-selectin expressed on activated endothelial cells (Frenette *et al.*, 1995). In another mouse model, circulating activated platelets use P-selectin to attach to high endothelial venules of lymph nodes, probably by interacting with ligands that L-selectin also recognizes. Furthermore, the activated platelets bind to circulating lymphocytes, indirectly recruiting the lymphocytes to the vessel surface (Diacovo *et al.*, 1996a). Thus, P-selectin mediates multicellular interactions of leukocytes because of its expression on both activated platelets and endothelial cells, and because of its ability to bind to ligands on leukocytes and, in some cases, on endothelial cells or even platelets (Figure 11.6).

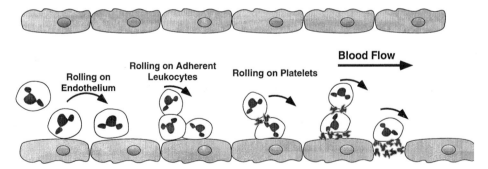

Figure 11.6   Multicellular interactions mediated by PSGL-1 and P-selectin under shear stress. PSGL-1 on leukocytes mediates tethering and rolling of leukocytes on P-selectin expressed on activated endothelial cells during inflammation and on adherent activated platelets during hemorrhage. Activated platelets use P-selectin to bridge adjacent leukocytes, indirectly recruiting them to the cell surface. Activated platelets use P-selectin to roll on L-selectin ligands on the endothelial cell, and nonactivated platelets use an unidentified ligand to roll on P-selectin expressed on activated endothelial cells. Under permissive shear stresses, PSGL-1 binds to L-selectin, initiating leukocyte-leukocyte contacts that amplify leukocyte recruitment.

SIGNALING OF LEUKOCYTES ATTACHED TO P-SELECTIN

In the multistep model of leukocyte recruitment, leukocytes rolling on the endothelium through selectins encounter locally presented signaling molecules that activate the leukocytes; the activated leukocytes then develop firm, shear-resistant adhesion through integrins. Specific combinations of adhesion and signaling molecules favor the recruitment of particular classes of leukocytes (Springer, 1995; Butcher and Picker, 1996; Zimmerman *et al.*, 1992). A prototype of such combinations is the transient co-expression of P-selectin and platelet-activating factor (PAF) on the surface of endothelial cells stimulated with thrombin or histamine (Lorant *et al.*, 1991). P-selectin tethers neutrophils to the endothelial cell surface, allowing PAF to activate the neutrophil. The activated neutrophil then adheres more firmly through interactions of β_2 integrins with counter-receptors on the endothelial cell surface. This juxtacrine presentation of adhesion and activating molecules promotes the sequential tethering, activation, and firm adhesion of various leukocyte classes to endothelial cells that are stimulated by various mediators.

Leukocytes attached to P-selectin on activated platelets also encounter juxtacrine activators, although they are less well characterized. Neutrophils rolling on monolayers of activated platelets are frequently activated, which results in their abrupt arrest through interaction of the $\alpha_M\beta_2$ integrin on neutrophils with an unknown ligand on the platelets (Yeo *et al.*, 1994; Diacovo *et al.*, 1996b). Neutrophils adherent to activated platelets may produce superoxide (Nagata *et al.*, 1993). Antibodies to P-selectin prevent both integrin-dependent arrest and superoxide production, indicating that tethering through P-selectin is required for the subsequent activation event.

In addition to positioning the leukocyte near a juxtacrine activator such as PAF or a chemokine, P-selectin might directly signal the adherent leukocyte through PSGL-1. Some reports suggest that P-selectin may transduce a negative signal that renders neutrophils refractory to stimulation (Gamble *et al.*, 1990; Wong *et al.*, 1991). However, this proposed mechanism is not consistent with the need for efficient activation of leukocytes tethered through selectins. In contrast, other reports suggest that P-selectin is sufficient to transmit proinflammatory signals to leukocytes. Thus, P-selectin may stimulate neutrophils to bind unopsonized zymosan (Cooper *et al.*, 1994) or may cause monocytes to synthesize very small quantities of tissue factor (Celi *et al.*, 1994). Other studies indicate that P-selectin transduces a signal which cooperates with suboptimal signals transduced through conventional agonists to generate a specific leukocyte response. Binding of P-selectin to neutrophils may increase the degree of integrin activation in response to PAF (Lorant *et al.*, 1991). Monocytes mobilize the transcription factor NF-κB and synthesize the cytokines TNF-α and IL-8 when they adhere to immobilized P-selectin and PAF, but not to a surface displaying either P-selectin or PAF alone (Weyrich *et al.*, 1995b). Monocytes secrete a different profile of cytokines when they adhere to activated platelets. This reponse is reproduced by incubating monocytes with P-selectin and RANTES, a chemokine secreted by activated platelets, but is not reproduced with either P-selectin or RANTES alone (Weyrich *et al.*, 1996). In this study, neither activated platelets nor purified P-selectin plus RANTES caused monocytes to synthesize tissue factor (Weyrich *et al.*, 1996). Taken together, these studies suggest that P-selectin transmits signals through PSGL-1 that are usually insufficient to activate leukocytes, as measured by changes in integrin function, production of superoxides, synthesis of tissue factor, or secretion of cytokines. However, signals transmitted through PSGL-1 may cooperate with suboptimal signals generated through conventional activators to produce such responses. The nature of the response may depend on the density of P-selectin on the activated platelet or endothelial cell, the duration of adhesion, and the specific co-activating chemokine or lipid autacoid presented on the cell surface.

PHYSIOLOGICAL EXPRESSION OF P-SELECTIN

As discussed earlier, P-selectin is stored in the membranes of secretory storage granules: the α granules of platelets and the Weibel-Palade bodies of endothelial cells. After passing through the Golgi complex, newly synthesized P-selectin is probably sorted in the trans-Golgi network for delivery to secretory granules (Disdier *et al.*, 1992). Thrombin, histamine, or other secretagogues rapidly mobilize P-selectin from these storage compartments to the cell surface. P-selectin remains on the surface of activated platelets for prolonged periods (George *et al.*, 1986), which may explain its ability to form stable platelet-leukocyte conjugates (Rinder *et al.*, 1991). In contrast, P-selectin is rapidly internalized from the surface of activated endothelial cells (Hattori *et al.*, 1989a). Once in endosomes, P-selectin may recycle to the cell surface, as do many internalized membrane glycoproteins such as the low density lipoprotein receptor. It may be transported from endosomes to the trans-Golgi network to be resorted into newly forming Weibel-Palade bodies (Subramaniam *et al.*, 1993). However, the major destination for endosomal P-selectin is the lysosome, where it is rapidly degraded; thus, P-selectin has a short half-life if it is not sorted into the regulated secretory pathway (Green *et al.*, 1994).

The sequence of the 35-residue cytoplasmic domain of P-selectin is highly conserved in different species, suggesting that it has important functions (Setiadi *et al.*, 1995). Indeed, the cytoplasmic domain contains signals that mediate sorting of P-selectin into secretory granules (Disdier *et al.*, 1992), endocytosis into clathrin-coated pits (Setiadi *et al.*, 1995), rapid delivery from endosomes to lysosomes (Green *et al.*, 1994), and perhaps movement from endosomes to synaptic-like vesicles (Norcott *et al.*, 1996). Mutational analysis indicates that many residues in the cytoplasmic domain contribute to the endocytosis of P-selectin (Setiadi *et al.*, 1995). This suggests that at least some sorting signals depend on the entire conformation of the cytoplasmic domain rather than on a short linear stretch of amino acids. The cytoplasmic domain is acylated and phosphorylated, which might affect its conformation (Fujimoto *et al.*, 1993; Fujimoto and McEver, 1993; Crovello *et al.*, 1993; Crovello *et al.*, 1995). The signals in the cytoplasmic tail presumably interact with sorting molecules in the cytoplasm. Although still uncharacterized, these molecules may be related to adaptins or other protein components of clathrin- or coatomer-coated vesicles, which mediate most vesicular transport events (Rothman and Wieland, 1996).

The steady-state distribution of P-selectin in megakaryocytes/platelets and endothelial cells reflects the balance between its rate of synthesis and its efficiency of sorting into various subcellular compartments. At normal levels of synthesis, a significant percentage of P-selectin molecules probably move directly from the trans-Golgi network into secretory granules. Those molecules that are not immediately sorted travel instead to the plasma membrane. However, in endothelial cells and probably in megakaryocytes (but not platelets), P-selectin is quickly cleared from the cell surface by rapid endocytosis and efficient delivery from endosomes to lysosomes. Thus, the basal levels of cell-surface P-selectin are low. Secretagogues induce rapid redistribution of a larger cohort of P-selectin molecules from secretory granules to the plasma membrane. This pool of P-selectin is sufficient to mediate leukocyte adhesion. In endothelial cells, however, the adhesion is transient because P-selectin is again endocytosed and degraded. This rapid but transient expression of P-selectin appears to account for the importance of the protein during the earliest phases of neutrophil recruitment in murine models of acute inflammation (Mayadas *et al.*, 1993; Johnson *et al.*, 1995; Bullard *et al.*, 1995). On the other hand, increased synthesis may saturate the sorting pathway into secretory granules, diverting more P-selectin to the plasma membrane (Hahne *et al.*, 1993; Yao *et al.*, 1996). This results in constitutive cell-surface expression of P-selectin despite its continued endocytosis and degradation, and may explain the importance of P-selectin in mononuclear cell recruitment in some models of chronic inflammation (Subramaniam *et al.*, 1995; Tipping *et al.*, 1996). Subsequent challenge of the "primed" endothelium with thrombin, histamine, or other secretagogues mobilizes even more P-selectin to the cell surface (Hahne *et al.*, 1993; Yao *et al.*, 1996).

In cultured rodent and bovine endothelial cells, TNF-α, IL-1β, or LPS transiently increase steady-state levels of P-selectin mRNA as well as E-selectin mRNA (Weller *et al.*, 1992). *In vivo*, administration of these mediators to mice or rats also increases P-selectin mRNA and protein (Sanders *et al.*, 1992; Weller *et al.*, 1992; Auchampach *et al.*, 1994; Gotsch *et al.*, 1994; Mayadas *et al.*, 1993). However, TNF-α, IL-1β, or LPS do not increase transcription of P-selectin mRNA in cultured human endothelial cells (Burns *et al.*, 1995; Yao *et al.*, 1996). Furthermore, injection of LPS into the dermis of non-human primates does not increase expression of P-selectin in postcapillary venules (Silber *et al.*, 1994). These data suggest that there may be important differences in the mechanisms for inducible transcription of P-selectin between primates and other species. Unlike TNF-α, IL-1β, or LPS, the cytokines IL-4 or oncostatin M markedly increase transcription of P-selectin in cultured human endothelial cells (Yao *et al.*, 1996). P-selectin mRNA rises only after 7 hours, reflecting a requirement for new protein synthesis. This suggests that the cytokines first activate one or more genes whose products then activate the P-selectin gene. Accumulation of mRNA reaches a peak by 10–24 hours and then persists for at least 72 hours. The elaboration of IL-4 or oncostatin M may contribute to the sustained apical expression of P-selectin observed on human venular endothelium in chronic or allergic forms of inflammation (Grober *et al.*, 1993; Symon *et al.*, 1994; Johnson-Tidey *et al.*, 1994). IL-3 may also increase P-selectin mRNA in human endothelial cells (Khew-Goodall *et al.*, 1996), although its effects are less consistent and less pronounced than those of IL-4 or oncostatin M (Yao *et al.*, 1996).

The mechanisms that regulate transcription of the P-selectin gene require further study. The 5′ flanking region of the human gene has been sequenced (Pan and McEver, 1993). There is no canonical TATA box, and transcription is initiated at multiple sites. The first 249 base pairs upstream of the translational initiation site are sufficient to confer constitutive expression of a reporter gene in transfected endothelial cells. A GATA element is required for optimal constitutive expression. There is also a novel κB element that binds homodimers of p50 or p52, but not heterodimers containing p65 (Pan and McEver, 1995). Co-transfection experiments indicate that interactions of Bcl-3 and p52 homodimers with the κB element augment expression of a reporter gene in endothelial cells. In contrast, co-expression of p50 homodimers with Bcl-3 represses expression, although the repression is prevented by co-expression of Bcl-3. These data suggest that differential interactions of Bcl-3 with p50 or p52 homodimers help regulate the expression of P-selectin in humans. The pathways by which IL-4 or oncostatin M affect transcription of P-selectin are unknown. It is also not understood why TNF-α, IL-1β, and LPS increase expression of P-selectin in the endothelium of mice but not of humans.

The physiological significance of platelet-leukocyte interactions mediated by P-selectin is not fully appreciated. As mentioned earlier, activated platelets may deliver leukocytes to the vessel wall by simultaneously binding to L-selectin ligands on endothelial cells and to PSGL-1 on circulating leukocytes (Diacovo *et al.*, 1996a). Leukocytes may also carry platelets into sites of inflammation (Issekutz *et al.*, 1983), a mechanism for delivery of platelet factor 4, platelet-derived growth factor, RANTES, and other platelet-derived chemokines. Close contact between platelets and leukocytes may facilitate transcellular metabolism of leukotrienes and lipoxins (Murphy *et al.*, 1991; Serhan and Sheppard, 1990; Marcus, 1990). Some of these products induce translocation of P-selectin to the surface of endothelial cells (Datta *et al.*, 1995). Adhesion of activated platelets to leukocytes enhances secretion of cytokines, tissue factor, or other mediators by the leukocytes (Weyrich *et al.*, 1995b; Weyrich *et al.*, 1996; Celi *et al.*, 1994; Nagata *et al.*, 1993). Such interactions may amplify both inflammation and hemostasis, which are frequently linked responses.

PATHOLOGICAL EXPRESSION OF P-SELECTIN

Excessive accumulation of leukocytes is responsible for a variety of inflammatory disorders such as ischemia-reperfusion injury, Gram-negative shock, and rheumatoid arthritis (Albelda *et al.*, 1994; Sharar *et al.*, 1995). Tissue injury results from release of oxygen-derived radicals, proteases, and other mediators, to which the endothelium is particularly vulnerable (Ward and Varani, 1990). Dysregulated expression of P-selectin has been implicated in several forms of leukocyte-mediated tissue injury.

Activated complement and oxygen radicals, which are frequently present during the early stages of sepsis or ischemia-reperfusion syndromes, mobilize P-selectin to the surface of endothelial cells *in vitro* (Hattori *et al.*, 1989b; Patel *et al.*, 1991; Foreman *et al.*, 1994; Vischer *et al.*, 1995). Oxygen radicals prolong the expression of P-selectin on the cell surface, perhaps by inhibiting endocytosis (Patel *et al.*, 1991). Endothelial dysfunction decreases formation of nitric oxide, a oxygen-radical scavenger that may normally dampen the expression of P-selectin (Murohara *et al.*, 1995). Hypoxia *per se* also translocates P-selectin to the surface of endothelial cells (Rainger *et al.*, 1995; Pinsky *et al.*, 1996). Consistent with these observations, ischemia-reperfusion induces expression of P-selectin on endothelial cells *in vivo* (Winn *et al.*, 1993; Weyrich *et al.*, 1995a; Okada *et al.*, 1994). Furthermore, mAbs and other P-selectin inhibitors significantly reduce both neutrophil accumulation and tissue injury in many ischemia-reperfusion models (Weyrich *et al.*, 1993; Winn *et al.*, 1993; Davenpeck *et al.*, 1994; Mulligan *et al.*, 1993; Gauthier *et al.*, 1994; Chen *et al.*, 1994; Winn *et al.*, 1994; Lee *et al.*, 1995; Lefer *et al.*, 1996). Antibodies to P-selectin decrease neutrophil accumulation in the tissues of rats injected with LPS (Coughlan *et al.*, 1994) and in some models of acute lung injury in rats (Mulligan *et al.*, 1992).

Oxidized low density lipoprotein activates both platelets and endothelial cells, promoting P-selectin-dependent platelet-leukocyte aggregates and leukocyte adhesion to the arterial endothelium *in vitro* and *in vivo* (Lehr *et al.*, 1994b). Cigarette smoke, which causes release of oxygen radicals, produces similar effects *in vivo* (Lehr *et al.*, 1994a). Some viral infections prolong the expression of P-selectin on the surface of cultured endothelial cells (Etingin *et al.*, 1992). These insults may allow monocytes and other leukocytes to emigrate beneath the endothelium during the early stages of atherosclerosis. In the later stages of atherosclerosis, P-selectin is observed on the apical surface of the endothelium (Johnson-Tidey *et al.*, 1994), perhaps because of the local synthesis of IL-4 or oncostatin M by subendothelial macrophages or T cells (Yao *et al.*, 1996). The endothelially expressed P-selectin may promote recruitment of additional monocytes, particularly in areas of arterial bifurcation where shear stresses are lower. Rupture of advanced atherosclerotic plaques promotes platelet aggregation and thrombin formation. Leukocytes accumulating on adherent platelets may express tissue factor, further augmenting thrombin and fibrin generation.

Consistent with this notion, mAbs to P-selectin accelerate pharmacological thrombolysis in a primate model of arterial thrombosis (Toombs *et al.*, 1995).

These data suggest that inhibitors of P-selectin function or expression might be effective therapeutics in some inflammatory or thrombotic disorders. A potential risk of such agents is interference with the physiological recruitment of leukocytes required to combat infections. However, antibodies to P-selectin do not significantly increase infections in some experimental models, suggesting that other adhesion molecules may suffice for this purpose (Sharar *et al.*, 1993). Furthermore, infections are relatively uncommon in patients with leukocyte adhesion deficiency 2, a congenital disorder of fucose metabolism that causes defective synthesis of all selectin ligands (Etzioni *et al.*, 1992). The limited number of infections may result from the ability of α_4 integrins to tether flowing mononuclear leukocytes and eosinophils to the endothelium (Alon *et al.*, 1995b; Berlin *et al.*, 1995).

Metastasis of some cancer cells may occur through usurpation of adhesion molecules normally used for leukocyte or platelet adhesion. Platelets promote metastasis of tumor cells in some experimental models, perhaps by favoring the formation of platelet-tumor cell aggregates. Some tumors use integrins for these adhesive interactions (Boukerche *et al.*, 1989). However, P-selectin mediates adhesion of activated platelets to neuroblastoma and small cell lung cancer cells *in vitro* (Stone and Wagner, 1993). The P-selectin ligand(s) on these and other tumor cells appear to be sialomucins, although they are likely to be distinct from PSGL-1 (Crottet *et al.*, 1996). Tumor cells might attach to P-selectin on both activated endothelial cells and platelets, forming multicellular aggregates on the surface of the microvasculature. These tumor cells would be well positioned to emigrate into the extravascular tissues.

References

Aigner, S., Ruppert, M., Hubbe, M., Sammar, M., Sthoeger, Z., Butcher, E.C. *et al.* (1995). Heat stable antigen (mouse CD24) supports myeloid cell binding to endothelial and platelet P-selectin. *Int. Immunol.*, **7**:1557–1565.

Albelda, S.M., Smith, C.W. and Ward, P.A. (1994). Adhesion molecules and inflammatory injury. *FASEB J.*, **8**:504–512.

Alon, R., Rossiter, H., Wang, X., Springer, T.A. and Kupper, T.S. (1994). Distinct cell surface ligands mediate T lymphocyte attachment and rolling on P- and E-selectin under physiological flow. *J. Cell Biol.*, **127**:1485–1495.

Alon, R., Hammer, D.A. and Springer, T.A. (1995a). Lifetime of the P-selectin: carbohydrate bond and its response to tensile force in hydrodynamic flow. *Nature*, **374**:539–542.

Alon, R., Kassner, P.D., Carr, M.W., Finger, E.B., Hemler, M.E. and Springer, T.A. (1995b). The integrin VLA-4 supports tethering and rolling in flow on VCAM-1. *J. Cell Biol.*, **128**:1243–1253.

Asako, H., Kurose, I., Wolf, R., DeFrees, S., Zheng, Z.-L., Phillips, M.L. *et al.* (1994). Role of H1 receptors and P-selectin in histamine-induced leukocyte rolling and adhesion in postcapillary venules. *J. Clin. Invest.*, **93**:1508–1515.

Auchampach, J.A., Oliver, M.G., Anderson, D.C. and Manning, A.M. (1994). Cloning, sequence comparison and *in vivo* expression of the gene encoding rat P-selectin. *Gene*, **145**:251–255.

Bajorath, J., Hollenbaugh, D., King, G., Harte, W., Jr., Eustice, D.C., Darveau, R.P. *et al.* (1994). CD62/P-selectin binding sites for myeloid cells and sulfatides are overlapping. *Biochemistry*, **33**:1332–1339.

Baumhueter, S., Singer, M.S., Henzel, W., Hemmerich, S., Renz, M., Rosen, S.D. *et al.* (1993). Binding of L-selectin to the vascular sialomucin CD34. *Science*, **262**:436–438.

Berg, E.L., McEvoy, L.M., Berlin, C., Bargatze, R.F. and Butcher, E.C. (1993). L-selectin-mediated lymphocyte rolling on MAdCAM-1. *Nature*, **366**:695–698.

Berlin, C., Bargatze, R.F., Campbell, J.J., Von Andrian, U.H., Szabo, M.C., Hasslen, S.R. *et al.* (1995). α_4 integrins mediate lymphocyte attachment and rolling under physiologic flow. *Cell*, **80**:413–422.

Berman, C.L., Yeo, E.L., Wencel-Drake, J.D., Furie, B.C., Ginsberg, M.H. and Furie, B. (1986). A platelet α granule membrane protein that is associated with the plasma membrane after activation. *J. Clin. Invest.*, **78**:130–137.

Bevilacqua, M.P., Pober, J.S., Mendrick, D.L., Cotran, R.S. and Gimbrone, M.A., Jr. (1987). Identification of an inducible endothelial-leukocyte adhesion molecule. *Proc. Natl. Acad. Sci., USA*, **84**:9238–9242.

Bevilacqua, M.P., Stengelin, S., Gimbrone, M.A., Jr. and Seed, B. (1989). Endothelial leukocyte adhesion molecule 1: an inducible receptor for neutrophils related to complement regulatory proteins and lectins. *Science*, **243**:1160–1165.

Bonfanti, R., Furie, B.C., Furie, B. and Wagner, D.D. (1989). PADGEM (GMP 140) is a component of Weibel-Palade bodies of human endothelial cells. *Blood*, **73**:1109–1112.

Boukerche, H., Berthier-Vergnes, O., Tabone, E., Dore, J.-F., Leung, L.L.K. and McGregor, J.L. (1989). Platelet-melanoma cell interaction is mediated by the glycoprotein IIb-IIIa complex. *Blood*, **74**:658–663.

Bullard, D.C., Qin, L., Lorenzo, I., Quinlin, W.M., Doyle, N.A., Bosse, R. *et al.* (1995). P-selectin/ICAM-1 double mutant mice: Acute emigration of neutrophils into the peritoneum is completely absent but is normal into pulmonary alveoli. *J. Clin. Invest.*, **95**: 782–1788.

Burns, S.A., DeGuzman, B.J., Newburger, J.W., Mayer, J.E., Jr., Neufeld, E.J. and Briscoe, D.M. (1995). P-selectin expression in myocardium of children undergoing cardiopulmonary bypass. *J. Thorac. Cardiovasc. Surg.*, **110**:924–933.

Butcher, E.C. and Picker, L.J. (1996). Lymphocyte homing and homeostasis. *Science*, **272**:60–66.

Buttrum, S.M., Hatton, R. and Nash, G.B. (1993). Selectin-mediated rolling of neutrophils on immobilized platelets. *Blood*, **82**:1165–1174.

Celi, A., Pellegrini, G., Lorenzet, R., De Blasi, A., Ready, N., Furie, B.C. *et al.* (1994). P-selectin induces the expression of tissue factor on monocytes. *Proc. Natl. Acad. Sci., USA*, **91**:8767–8771.

Chen, L.Y., Nichols, W.W., Hendricks, J.B., Yang, B.C. and Mehta, J.L. (1994). Monoclonal antibody to P-selectin (PB1.3) protects against myocardial reperfusion injury in the dog. *Cardiovasc. Res.*, **28**:1414–1422.

Chong, B.H., Murray, B., Berndt, M.C., Dunlop, L.C., Brighton, T. and Chesterman, C.N. (1994). Plasma P-selectin is increased in thrombotic consumptive platelet disorders. *Blood*, **83**:1535–1541.

Collins, T., Williams, A., Johnston, G.I., Kim, J., Eddy, R., Shows, T. *et al.* (1991). Structure and chromosomal location of the gene for endothelial-leukocyte adhesion molecule 1. *J. Biol. Chem.*, **266**:2466–2473.

Cooper, D., Butcher, C.M., Berndt, M.C. and Vadas, M.A. (1994). P-selectin interacts with a β_2-integrin to enhance phagocytosis. *J. Immunol.*, **153**:3199–3209.

Coughlan, A.F., Hau, H., Dunlop, L.C., Berndt, M.C. and Hancock, W.W. (1994). P-selectin and platelet-activating factor mediate initial endotoxin-induced neutropenia. *J. Exp. Med.*, **179**:329–334.

Crottet, P., Kim, Y.J. and Varki, A. (1996). Subsets of sialylated, sulfated mucins of diverse origins are recognized by L-selectin. Lack of evidence for unique oligosaccharide sequences mediating binding. *Glycobiology*, **6**:191–208.

Crovello, C.S., Furie, B.C. and Furie, B. (1993). Rapid phosphorylation and selective dephosphorylation of P-selectin accompanies platelet activation. *J. Biol. Chem.*, **268**:14590–14593.

Crovello, C.S., Furie, B.C. and Furie, B. (1995). Histidine phosphorylation of P-selectin upon stimulation of human platelets: A novel pathway for activation-dependent signal transduction. *Cell*, **82**:279–286.

Datta, Y.H., Romano, M., Jacobson, B.C., Golan, D.E., Serhan, C.N. and Ewenstein, B.M. (1995). Peptido-leukotrienes are potent agonists of von Willebrand factor secretion and P-selectin surface expression in human umbilical vein endothelial cells. *Circulation*, **92**:3304–3311.

Davenpeck, K.L., Gauthier, T.W., Albertine, K.H. and Lefer, A.M. (1994). Role of P-selectin in microvascular leukocyte-endothelial interaction in splanchnic ischemia-reperfusion. *Am. J. Physiol. Heart Circ. Physiol.*, **267**:H622–H630.

De Luca, M., Dunlop, L.C., Andrews, R.K., Flannery, J.V., Jr., Ettling, R., Cumming, D.A. *et al.* (1995). A novel cobra venom metalloproteinase, mocarhagin, cleaves a 10-amino acid peptide from the mature N terminus of P-selectin glycoprotein ligand receptor, PSGL-1, and abolishes P-selectin binding. *J. Biol. Chem.*, **270**:26734–26737.

Diacovo, T.G., Puri, K.D., Warnock, R.A., Springer, T.A. and Von Andrian, U.H. (1996a). Platelet-mediated lymphocyte delivery to high endothelial venules. *Science*, **273**:252–255.

Diacovo, T.G., Roth, S.J., Buccola, J.M., Bainton, D.F. and Springer, T.A. (1996b). Neutrophil rolling, arrest, and transmigration across activated, surface-adherent platelets via sequential action of P-selectin and the β_2-integrin CD11b/CD18. *Blood*, **88**:146–157.

Diacovo, T.G., Roth, S.J., Morita, C.T., Rosat, J.P., Brenner, M.B. and Springer, T.A. (1996c). Interactions of human α/β and γ/δ T lymphocyte subsets in shear flow with E-selectin and P-selectin. *J. Exp. Med.*, **183**:1193–1203.

Disdier, M., Morrissey, J.H., Fugate, R.D., Bainton, D.F. and McEver, R.P. (1992). Cytoplasmic domain of P-selectin (CD62) contains the signal for sorting into the regulated secretory pathway. *Mol. Biol. Cell*, **3**:309–321.

Doré, M., Korthuis, R.J., Granger, D.N., Entman, M.L. and Smith, C.W. (1993). P-selectin mediates spontaneous leukocyte rolling *in vivo*. *Blood*, **82**:1308–1316.

Doré, M., Burns, A.R., Hughes, B.J., Entman, M.L. and Smith, C.W. (1996). Chemoattractant-induced changes in surface expression and redistribution of a functional ligand for P-selectin on neutrophils. *Blood*, **87**:2029–2037.

Dunlop, L.C., Skinner, M.P., Bendall, L.J., Favaloro, E.J., Castaldi, P.A., Gorman, J.J. *et al.* (1992). Characterization of GMP-140 (P-selectin) as a circulating plasma protein. *J. Exp. Med.*, **175**:1147–1150.

Erbe, D.V., Watson, S.W., Presta, L.G., Wolitzky, B.A., Foxall, C., Brandley, B.K. *et al.* (1993). P- and E-selectin use common sites for carbohydrate ligand recognition and cell adhesion. *J. Cell Biol.*, **120**:1227–1235.

Etingin, O.R., Silverstein, R.L. and Hajjar, D.P. (1992). Identification of a monocyte receptor on herpesvirus-infected endothelial cells. *Proc. Natl. Acad. Sci., USA*, **88**:7200–7203.

Etzioni, A., Frydman, M., Pollack, S., Avidor, I., Phillips, M.L., Paulson, J.C. *et al.* (1992). Brief report: recurrent severe infections caused by a novel leukocyte adhesion deficiency. *N. Engl. J. Med.*, **327**:1789–1792.

Foreman, K.E., Vaporciyan, A.A., Bonish, B.K., Jones, M.L., Johnson, K.J., Glovsky, M.M. *et al.* (1994). C5a-induced expression of P-selectin in endothelial cells. *J. Clin. Invest.*, **94**:1147–1155.

Frenette, P.S., Johnson, R.C., Hynes, R.O. and Wagner, D.D. (1995). Platelets roll on stimulated endothelium *in vivo*: An interaction mediated by endothelial P-selectin. *Proc. Natl. Acad. Sci., USA*, **92**:7450–7454.

Fujimoto, T. and McEver, R.P. (1993). The cytoplasmic domain of P-selectin is phosphorylated on serine and threonine residues. *Blood*, **82**:1758–1766.

Fujimoto, T., Stroud, E., Whatley, R.E., Prescott, S.M., Muszbek, L., Laposata, M. *et al.* (1993). P-selectin is acylated with palmitic acid and stearic acid at cysteine 766 through a thioester linkage. *J. Biol. Chem.*, **268**:11394–11400.

Gallatin, W.M., Weissman, I.L. and Butcher, E.C. (1983). A cell surface molecule involved in organ-specific homing of lymphocytes. *Nature*, **304**:30–34.

Gamble, J.R., Skinner, M.P., Berndt, M.C. and Vadas, M.A. (1990). Prevention of activated neutrophil adhesion to endothelium by soluble adhesion protein GMP140. *Science*, **249**:414–417.

Gauthier, T.W., Davenpeck, K.L. and Lefer, A.M. (1994). Nitric oxide attenuates leukocyte-endothelial interaction via P-selectin in splanchnic ischemia-reperfusion. *Am. J. Physiol. Gastrointest. Liver Physiol.*, **267**:G562–G568.

Geng, J.-G., Bevilacqua, M.P., Moore, K.L., McIntyre, T.M., Prescott, S.M., Kim, J.M. *et al.* (1990). Rapid neutrophil adhesion to activated endothelium mediated by GMP-140. *Nature*, **343**:757–760.

Geng, J.-G., Moore, K.L., Johnson, A.E. and McEver, R.P. (1991). Neutrophil recognition requires a Ca^{2+}-induced conformational change in the lectin domain of GMP-140. *J. Biol. Chem.*, **266**:22313–22318.

George, J.N., Pickett, E.B., Saucerman, S., McEver, R.P., Kunicki, T.J., Kieffer, N. *et al.* (1986). Platelet surface glycoproteins. Studies on resting and activated platelets and platelet membrane microparticles in normal subjects, and observations in patients during adult respiratory distress syndrome and cardiac surgery. *J. Clin. Invest.*, **78**:340–348.

Gotsch, U., Jager, U., Dominis, M. and Vestweber, D. (1994). Expression of P-selectin on endothelial cells is upregulated by LPS and TNF-α *in vivo*. *Cell Adhes. Commun.*, **2**:7–14.

Graves, B.J., Crowther, R.L., Chandran, C., Rumberger, J.M., Li, S., Huang, K.-S. *et al.* (1994). Insight into E-selectin/ligand interaction from the crystal structure and mutagenesis of the lec/EGF domains. *Nature*, **367**:532–538.

Green, S.A., Setiadi, H., McEver, R.P. and Kelly, R.B. (1994). The cytoplasmic domain of P-selectin contains a sorting determinant that mediates rapid degradation in lysosomes. *J. Cell Biol.*, **124**:435–448.

Grober, J.S., Bowen, B.L., Ebling, H., Athey, B., Thompson, C.B., Fox, D.A. *et al.* (1993). Monocyte-endothelial adhesion in chronic rheumatoid arthritis: *in situ* detection of selectin and integrin-dependent interactions. *J. Clin. Invest.*, **91**:2609–2619.

Guyer, D.A., Moore, K.L., Lynam, E., Schammel, C.M.G., Rogelj, S., McEver, R.P. *et al.* (1996). P-selectin glycoprotein ligand-1 (PSGL-1) is a ligand for L-selectin in neutrophil aggregation. *Blood*, **88**:2415–2421.

Hahne, M., Jäger, U., Isenmann, S., Hallmann, R. and Vestweber, D. (1993). Five tumor necrosis factor-inducible cell adhesion mechanisms on the surface of mouse endothelioma cells mediate the binding of leukocytes. *J. Cell Biol.*, **121**:655–664.

Hamburger, S.A. and McEver, R.P. (1990). GMP-140 mediates adhesion of stimulated platelets to neutrophils. *Blood*, **75**:550–554.

Hattori, R., Hamilton, K.K., Fugate, R.D., McEver, R.P. and Sims, P.J. (1989a). Stimulated secretion of endothelial von Willebrand factor is accompanied by rapid redistribution to the cell surface of the intracellular granule membrane protein GMP-140. *J. Biol. Chem.*, **264**:7768–7771.

Hattori, R., Hamilton, K.K., McEver, R.P. and Sims, P.J. (1989b). Complement proteins C5b-9 induce secretion of high molecular weight multimers of endothelial von Willebrand factor and translocation of granule membrane protein GMP-140 to the cell surface. *J. Biol. Chem.*, **264**:9053–9060.

Hemmerich, S., Butcher, E.C. and Rosen, S.D. (1994). Sulfation-dependent recognition of high endothelial venules (HEV)-ligands by L-selectin and MECA 79, an adhesion-blocking monoclonal antibody. *J. Exp. Med.*, **180**:2219–2226.

Hollenbaugh, D., Bajorath, J., Stenkamp, R. and Aruffo, A. (1993). Interaction of P-selectin (CD62) and its cellular ligand: Analysis of critical residues. *Biochemistry*, **32**:2960–2966.

Hsu-Lin, S-C., Berman, C.L., Furie, B.C., August, D. and Furie, B. (1984). A platelet membrane protein expressed during platelet activation and secretion. Studies using a monoclonal antibody specific for thrombin-activated platelets. *J. Biol. Chem.*, **259**:9121–9126.

Ikeda, H., Nakayama, H., Oda, T., Kuwano, K., Muraishi, A., Sugi, K. *et al.* (1994). Soluble form of P-selectin in patients with acute myocardial infarction. *Coronary Artery Dis.*, **5**:515–518.

Ikeda, H., Takajo, Y., Ichiki, K., Ueno, T., Maki, S., Noda, T. *et al.* (1995). Increased soluble form of P-selectin in patients with unstable angina. *Circulation*, **92**:1693–1696.

Imai, Y., Singer, M.S., Fennie, C., Lasky, L.A. and Rosen, S.D. (1991). Identification of a carbohydrate-based endothelial ligand for a lymphocyte homing receptor. *J. Cell Biol.*, **113**:1213–1222.

Imai, Y., Lasky, L.A. and Rosen, S.D. (1993). Sulphation requirement for GlyCAM-1, an endothelial ligand for L-selectin. *Nature*, **361**:555–557.

Ishiwata, N., Takio, K., Katayama, M., Watanabe, K., Titani, K., Ikeda, Y. *et al.* (1994). Alternatively spliced isoform of P-selectin is present *in vivo* as a soluble molecule. *J. Biol. Chem.*, **269**:23708–23715.

Israels, S.J., Gerrard, J.M., Jacques, Y.V., McNicol, A., Cham, B., Nishibori, M. *et al.* (1992). Platelet dense granule membranes contain both granulophysin and P-selectin (GMP-140). *Blood*, **80**:143–152.

Issekutz, A.C., Ripley, M. and Jackson, J.R. (1983). Role of neutrophils in the deposition of platelets during acute inflammation. *Lab. Invest.*, **49**:716–724.

Johnson, R.C., Mayadas, T.N., Frenette, P.S., Mebius, R.E., Subramaniam, M., Lacasce, A. *et al.* (1995). Blood cell dynamics in P-selectin-deficient mice. *Blood*, **86**:1106–1114.

Johnson-Tidey, R.R., McGregor, J.L., Taylor, P.R. and Poston, R.N. (1994). Increase in the adhesion molecule P-selectin in endothelium overlying atherosclerotic plaques. Coexpression with intercellular adhesion molecule-1. *Am. J. Pathol.*, **144**:952–961.

Johnston, G.I., Cook, R.G. and McEver, R.P. (1989a). Cloning of GMP-140, a granule membrane protein of platelets and endothelium: sequence similarity to proteins involved in cell adhesion and inflammation. *Cell*, **56**:1033–1044.

Johnston, G.I., Kurosky, A. and McEver, R.P. (1989b). Structural and biosynthetic studies of the granule membrane protein, GMP-140, from human platelets and endothelial cells. *J. Biol. Chem.*, **264**:1816–1823.

Johnston, G.I., Bliss, G.A., Newman, P.J. and McEver, R.P. (1990). Structure of the human gene encoding granule membrane protein-140, a member of the selectin family of adhesion receptors for leukocytes. *J. Biol. Chem.*, **265**:21381–21385.

Jones, D.A., Abbassi, O., McIntire, L.V., McEver, R.P. and Smith, C.W. (1994). P-selectin mediates neutrophil rolling on histamine-stimulated endothelial cells. *Biophys. J.*, **65**:1560–1569.

Kaikita, K., Ogawa, H., Yasue, H., Sakamoto, T., Suefuji, H., Sumida, H. *et al.* (1995). Soluble P-selectin is released into the coronary circulation after coronary spasm. *Circulation*, **92**:1726–1730.

Kansas, G.S., Saunders, K.B., Ley, K., Zakrzewicz, A., Gibson, R.M., Furie, B.C. *et al.* (1994). A role for the epidermal growth factor-like domain of P-selectin in ligand recognition and cell adhesion. *J. Cell Biol.*, **124**:609–618.

Katayama, M., Handa, M., Araki, Y., Ambo, H., Kawai, Y., Watanabe, K. *et al.* (1993). Soluble P-selectin is present in normal circulation and its plasma level is elevated in patients with thrombotic thrombocytopenic purpura and haemolytic uraemic syndrome. *Br. J. Haematol.*, **84**:702–710.

Khew-Goodall, Y., Butcher, C.M., Litwin, M.S., Newlands, S., Korpelainen, E.I., Noack, L.M. *et al.* (1996). Chronic expression of P-selectin on endothelial cells stimulated by the T-cell cytokine, interleukin-3. *Blood*, **87**:1432–1438.

Kubes, P. and Kanwar, S. (1994). Histamine induces leukocyte rolling in post-capillary venules: A P-selectin-mediated event. *J. Immunol.*, **152**:3570–3577.

Larsen, E., Celi, A., Gilbert, G.E., Furie, B.C., Erban, J.K., Bonfanti, R. *et al.* (1989). PADGEM protein: A receptor that mediates the interaction of activated platelets with neutrophils and monocytes. *Cell* **59**:305–312.

Lasky, L.A., Singer, M.S., Yednock, T.A., Dowbenko, D., Fennie, C., Rodriguez, H. *et al.* (1989). Cloning of a lymphocyte homing receptor reveals a lectin domain. *Cell*, **56**:1045–1055.

Lasky, L.A., Singer, M.S., Dowbenko, D., Imai, Y., Henzel, W.J., Grimley, C. *et al.* (1992). An endothelial ligand for L-selectin is a novel mucin-like molecule. *Cell*, **69**:927–938.

Laszik, Z., Jansen, P.J., Cummings, R.D., Tedder, T.F., McEver, R.P. and Moore, K.L. (1996). P-selectin glycoprotein ligand-1 is broadly expressed in cells of myeloid, lymphoid, and dendritic lineage and in some non-hematopoietic cells. *Blood*, **88**:3010–3021.

Lawrence, M.B. and Springer, T.A. (1991). Leukocytes roll on a selectin at physiologic flow rates: Distinction from and prerequisite for adhesion through integrins. *Cell*, **65**:859–873.

Lee, W.P., Gribling, P., De Guzman, L., Ehsani, N. and Watson, S.R. (1995). A P-selectin-immunoglobulin G chimera is protective in a rabbit ear model of ischemia-reperfusion. *Surgery*, **117**:458–465.

Lefer, D.J., Flynn, D.M. and Buda, A.J. (1996). Effects of a monoclonal antibody directed against P-selectin after myocardial ischemia and reperfusion. *Am. J. Physiol.*, **39**:H88–H98.

Lehr, H.-A., Frei, B. and Arfors, K.-E. (1994a). Vitamin C prevents cigarette smoke-induced leukocyte aggregation and adhesion to endothelium *in vivo*. *Proc. Natl. Acad. Sci., USA*, **91**:7688–7692.

Lehr, H.-A., Olofsson, A.M., Carew, T.E., Vajkoczy, P., Von Andrian, U.H., Hübner, C. *et al.* (1994b). P-selectin mediates the interaction of circulating leukocytes with platelets and microvascular endothelium in response to oxidized lipoprotein *in vivo*. *Lab. Invest.*, **71**:380–386.

Lenter, M., Levinovitz, A., Isenmann, S. and Vestweber, D. (1994). Monospecific and common glycoprotein ligands for E- and P-selectin on myeloid cells. *J. Cell Biol.*, **125**:471–481.

Ley, K. (1994). Histamine can induce leukocyte rolling in rat mesenteric venules. *Am. J. Physiol. Heart Circ. Physiol.*, **267**:H1017–H1023.

Ley, K., Bullard, D.C., Arbonés, M.L., Bosse, R., Vestweber, D., Tedder, T.F. *et al.* (1995). Sequential contribution of L- and P-selectin to leukocyte rolling *in vivo*. *J. Exp. Med.*, **181**:669–675.

Ley, K. and Tedder, T.F. (1995). Leukocyte interactions with vascular endothelium: New insights into selectin-mediated attachment and rolling. *J. Immunol.*, **155**:525–528.

Li, F., Erickson, H.P., James, J.A., Moore, K.L., Cummings, R.D. and McEver, R.P. (1996a). Visualization of P-selectin glycoprotein ligand-1 as a highly extended molecule and mapping of protein epitopes for monoclonal antibodies. *J. Biol. Chem.*, **271**:6342–6348.

Li, F., Wilkins, P.P., Crawley, S., Weinstein, J., Cummings, R.D. and McEver, R.P. (1996b). Post-translational modifications of recombinant P-selectin glycoprotein ligand-1 required for binding to P- and E-selectin. *J. Biol. Chem.*, **271**:3255–3264.

Lorant, D.E., Patel, K.D., McIntyre, T.M., McEver, R.P., Prescott, S.M. and Zimmerman, G.A. (1991). Coexpression of GMP-140 and PAF by endothelium stimulated by histamine or thrombin: A juxtacrine system for adhesion and activation of neutrophils. *J. Cell Biol.*, **115**:223–234.

Lorant, D.E., McEver, R.P., McIntyre, T.M., Moore, K.L., Prescott, S.M. and Zimmerman, G.A. (1995). Activation of polymorphonuclear leukocytes reduces their adhesion to P-selectin and causes redistribution of ligands for P-selectin on their surfaces. *J. Clin. Invest.*, **96**:171–182.

Maemura, K. and Fukuda, M. (1992). Poly-*N*-acetyllactosaminyl *O*-glycans attached to leukosialin: The presence of sialyl Le^x structures in *O*-glycans. *J. Biol. Chem.*, **267**:24379–24386.

Marcus, A.J. (1990). Eicosanoid interactions between platelets, endothelial cells, and neutrophils. *Methods Enzymol.*, **187**:585–598.

Mayadas, T.N., Johnson, R.C., Rayburn, H., Hynes, R.O., and Wagner, D.D. (1993). Leukocyte rolling and extravasation are severely compromised in P selectin-deficient mice. *Cell*, **74**:541–554.

McEver, R.P., Beckstead, J.H., Moore, K.L., Marshall-Carlson, L. and Bainton, D.F. (1989). GMP-140, a platelet α-granule membrane protein, is also synthesized by vascular endothelial cells and is localized in Weibel-Palade bodies. *J. Clin. Invest.*, **84**:92–99.

McEver, R.P., Moore, K.L. and Cummings, R.D. (1995). Leukocyte trafficking mediated by selectin-carbohydrate interactions. *J. Biol. Chem.*, **270**:11025–11028.

McEver, R.P. and Martin, M.N. (1984). A monoclonal antibody to a membrane glycoprotein binds only to activated platelets. *J. Biol. Chem.*, **259**:9799–9804.

Michelson, A.D. (1996). Flow cytometry: a clinical test of platelet function. *Blood*, **87**:4925–4936.

Miller, D.D., Boulet, A.J., Tio, F.O., Garcia, O.J., McEver, R.P., Palmaz, J.C. *et al.* (1991). *In vivo* ^{99m}Technetium S12 antibody imaging of platelet α-granules in rabbit endothelial neointimal proliferation after angioplasty. *Circulation*, **83**:224–236.

Moore, K.L., Varki, A. and McEver, R.P. (1991). GMP-140 binds to a glycoprotein receptor on human neutrophils: Evidence for a lectin-like interaction. *J. Cell Biol.*, **112**:491–499.

Moore, K.L., Stults, N.L., Diaz, S., Smith, D.L., Cummings, R.D., Varki, A. *et al.* (1992). Identification of a specific glycoprotein ligand for P-selectin (CD62) on myeloid cells. *J. Cell. Biol.*, **118**:445–456.

Moore, K.L. and Thompson, L.F. (1992). P-selectin (CD62) binds to subpopulations of human memory T lymphocytes and natural killer cells. *Biochem. Biophys. Res. Commun.*, **186**:173–181.

Moore, K.L., Eaton, S.F., Lyons, D.E., Lichenstein, H.S., Cummings, R.D. and McEver, R.P. (1994). The P-selectin glycoprotein ligand from human neutrophils displays sialylated, fucosylated, *O*-linked poly-*N*-acetyllactosamine. *J. Biol. Chem.*, **269**:23318–23327.

Moore, K.L., Patel, K.D., Bruehl, R.E., Fugang, L., Johnson, D.A., Lichenstein, H.S. *et al.* (1995). P-selectin glycoprotein ligand-1 mediates rolling of human neutrophils on P-selectin. *J. Cell Biol.*, **128**:661–671.

Mulligan, M.S., Polley, M.J., Bayer, R.J., Nunn, M.F., Paulson, J.C. and Ward, P.A. (1992). Neutrophil-dependent acute lung injury. Requirement for P-selectin (GMP-140). *J. Clin. Invest.*, **90**:1600–1607.

Mulligan, M.S., Paulson, J.C., De Frees, S., Zheng, Z.-L., Lowe, J.B. and Ward, P.A. (1993). Protective effects of oligosaccharides in P-selectin-dependent lung injury. *Nature*, **364**:149–151.

Murohara, T., Parkinson, S.J., Waldman, S.A. and Lefer, A.M. (1995). Inhibition of nitric oxide biosynthesis promotes P-selectin expression in platelets — Role of protein kinase C. *Arterioscler. Thromb. Vasc. Biol.*, **15**:2068–2075.

Murphy, R.C., Maclouf, J. and Henson, P.M. (1991). Interaction of platelets and neutrophils in the generation of sulfidopeptide leukotrienes. *Adv. Exp. Med. Biol.*, **314**:91–101.

Nagata, K., Tsuji, T., Todoroki, N., Katagiri, Y., Tanoue, K., Yamazaki, H. *et al.* (1993). Activated platelets induce superoxide anion release by monocytes and neutrophils through P-selectin (CD62). *J. Immunol.*, **151**:3267–3273.

Nolte, D., Schmid, P., Jäger, U., Botzlar, A., Roesken, F., Hecht, R. *et al.* (1994). Leukocyte rolling in venules of striated muscle and skin is mediated by P-selectin, not by L-selectin. *Am. J. Physiol. Heart Circ. Physiol.*, **267**:H1637–H1642.

Norcott, J.P., Solari, R. and Cutler, D.F. (1996). Targeting of P-selectin to two regulated secretory organelles in PC12 cells. *J. Cell Biol.*, **134**:1229–1240.

Norgard, K.E., Moore, K.L., Diaz, S., Stults, N.L., Ushiyama, S., McEver, R.P. *et al.* (1993). Characterization of a specific ligand for P-selectin on myeloid cells. A minor glycoprotein with sialylated *O*-linked oligosaccharides. *J. Biol. Chem.*, **268**:12764–12774.

Norman, K.E., Moore, K.L., McEver, R.P. and Ley, K. (1995). Leukocyte rolling *in vivo* is mediated by P-selectin glycoprotein ligand-1. *Blood*, **86**:4417–4421.

Okada, Y., Copeland, B.R., Mori, E., Tung, M.-M., Thomas, W.S. and Del Zoppo, G.J. (1994). P-selectin and intercellular adhesion molecule-1 expression after focal brain ischemia and reperfusion. *Stroke*, **25**:202–210.

Ord, D.C., Ernst, T.J., Zhou, L.-J., Rambaldi, A., Spertini, O., Griffin, J. *et al.* (1990). Structure of the gene encoding the human leukocyte adhesion molecule-1 (TQ1, Leu-8) of lymphocytes and neutrophils. *J. Biol. Chem.*, **265**:7760–7767.

Palabrica, T., Lobb, R., Furie, B.C., Aronovitz, M., Benjamin, C., Hsu, Y.-M. *et al.* (1992). Leukocyte accumulation promoting fibrin deposition is mediated *in vivo* by P-selectin on adherent platelets. *Nature*, **359**:848–851.

Palabrica, T.M., Furie, B.C., Konstam, M.A., Aronovitz, M.J., Connolly, R., Brockway, B.A. *et al.* (1989). Thrombus imaging in a primate model with antibodies specific for an external membrane protein of activated platelets. *Proc. Natl. Acad. Sci., USA,* **86**:1036–1040.

Pan, J. and McEver, R.P. (1993). Characterization of the promoter for the human P-selectin gene. *J. Biol. Chem.,* **268**:22600–22608.

Pan, J. and McEver, R.P. (1995). Regulation of the human P-selectin promoter by Bcl-3 and specific homodimeric members of the NF-κB/Rel family. *J. Biol. Chem.,* **270**:23077–23083.

Patel, K.D., Zimmerman, G.A., Prescott, S.M., McEver, R.P. and McIntyre, T.M. (1991). Oxygen radicals induce human endothelial cells to express GMP-140 and bind neutrophils. *J. Cell Biol.,* **112**:749–759.

Patel, K.D., Nollert, M.U. and McEver, R.P. (1995). P-selectin must extend a sufficient length from the plasma membrane to mediate rolling of neutrophils. *J. Cell Biol.,* **131**:1893–1902.

Pinsky, D.J., Naka, Y., Liao, H., Oz, M.C., Wagner, D.D., Mayadas, T.N. *et al.* (1996). Hypoxia-induced exocytosis of endothelial cell Weibel-Palade bodies. A mechanism for rapid neutrophil recruitment after cardiac preservation. *J. Clin. Invest.,* **97**:493–500.

Polley, M.J., Phillips, M.L., Wayner, E., Nudelman, E., Singhal, A.K., Hakomori, S. *et al.* (1991). CD62 and endothelial cell-leukocyte adhesion molecule 1 (ELAM-1) recognize the same carbohydrate ligand, sialyl-Lewis x. *Proc. Natl. Acad. Sci., USA,* **88**:6224–6228.

Pouyani, T. and Seed, B. (1995). PSGL-1 recognition of P-selectin is controlled by a tyrosine sulfation consensus at the PSGL-1 amino terminus. *Cell,* **83**:333–343.

Rainger, G.E., Fisher, A., Shearman, C. and Nash, G.B. (1995). Adhesion of flowing neutrophils to cultured endothelial cells after hypoxia and reoxygenation *in vitro. Am. J. Physiol. Heart Circ. Physiol.,* **269**:H1398–H1406.

Revelle, B.M., Scott, D. and Beck, P.J. (1996a). Single amino acid residues in the E- and P-selectin epidermal growth factor domains can determine carbohydrate binding specificity. *J. Biol. Chem.,* **271**:16160–16170.

Revelle, B.M., Scott, D., Kogan, T.P., Zheng, J.H. and Beck, P.J. (1996b). Structure-function analysis of P-selectin-sialyl Lewis^x binding interactions — Mutagenic alteration of ligand binding specificity. *J. Biol. Chem.,* **271**:4289–4297.

Rinder, H.M., Bonan, J.L., Rinder, C.S., Ault, K.A. and Smith, B.R. (1991). Dynamics of leukocyte-platelet adhesion in whole blood. *Blood,* **78**:1730–1737.

Rinder, H.M., Tracey, J.L., Rinder, C.S., Leitenberg, D. and Smith, B.R. (1994). Neutrophil but not monocyte activation inhibits P-selectin-mediated platelet adhesion. *Thromb. Haemost.,* **72**:750–756.

Rothman, J.E. and Wieland, F.T. (1996). Protein sorting by transport vesicles. *Science,* **272**:227–234.

Sakamaki, F., Ishizaka, A., Handa, M., Fujishima, S., Urano, T., Sayama, K. *et al.* (1995). Soluble form of P-selectin in plasma is elevated in acute lung injury. *Am. J. Respir. Crit. Care Med.,* **151**:1821–1826.

Sako, D., Chang, X.-J., Barone, K.M., Vachino, G., White, H.M., Shaw, G. *et al.* (1993). Expression cloning of a functional glycoprotein ligand for P-selectin. *Cell,* **75**:1179–1186.

Sako, D., Comess, K.M., Barone, K.M., Camphausen, R.T., Cumming, D.A. and Shaw, G.D. (1995). A sulfated peptide segment at the amino terminus of PSGL-1 is critical for P-selectin binding. *Cell,* **83**:323–331.

Sanders, W.E., Wilson, R.W., Ballantyne, C.M. and Beaudet, A.L. (1992). Molecular cloning and analysis of *in vivo* expression of murine P-selectin. *Blood,* **80**:795–800.

Sasaki, H., Bothner, B., Dell, A. and Fukuda, M. (1987). Carbohydrate structure of erythropoetin expressed in Chinese hamster ovary cells by a human erythropoetin cDNA. *J. Biol. Chem.,* **262**:12059–12076.

Serhan, C.N. and Sheppard, K.-A. (1990). Lipoxin formation during human neutrophil-platelet interactions. Evidence for the transformation of leukotriene α_4 by platelet 12-lipoxygenase *in vitro. J. Clin. Invest.,* **85**:772–780.

Setiadi, H., Disdier, M., Green, S.A., Canfield, W.M. and McEver, R.P. (1995). Residues throughout the cytoplasmic domain affect the internalization efficiency of P-selectin. *J. Biol. Chem.,* **270**:26818–26826.

Sharar, S.R., Sasaki, S.S., Flaherty, L.C., Paulson, J.C., Harlan, J.M. and Winn, R.K. (1993). P-selectin blockade does not impair leukocyte host defense against bacterial peritonitis and soft tissue infection in rabbits. *J. Immunol.,* **151**:4982–4988.

Sharar, S.R., Winn, R.K. and Harlan, J.M. (1995). The adhesion cascade and anti-adhesion therapy: An overview. *Springer Semin. Immunopathol.,* **16**:359–378.

Siegelman, M.H., van de Rijn, M. and Weissman, I.L. (1989). Mouse lymph node homing receptor cDNA clone encodes a glycoprotein revealing tandem interaction domains. *Science,* **243**:1165–1172.

Silber, A., Newman, W., Reimann, K.A., Hendricks, E., Walsh, D. and Ringler, D.J. (1994). Kinetic expression of endothelial adhesion molecules and relationship to leukocyte recruitment in two cutaneous models of inflammation. *Lab. Invest.,* **70**:163–175.

Skinner, M.P., Lucas, C.M., Burns, G.F., Chesterman, C.N. and Berndt, M.C. (1991). GMP-140 binding to neutrophils is inhibited by sulfated glycans. *J. Biol. Chem.,* **266**:5371–5374.

Spertini, O., Cordey, A.-S., Monai, N., Giuffre, L. and Schapira, M. (1996). P-selectin glycoprotein ligand-1 (PSGL-1) is a ligand for L-selectin on neutrophils, monocytes and CD34$^+$ hematopoietic progenitor cells. *J. Cell Biol.,* **135**:523–531.

Springer, T.A. (1995). Traffic signals on endothelium for lymphocyte recirculation and leukocyte emigration. *Annu. Rev. Physiol.,* **57**:827–872.

Stenberg, P.E., McEver, R.P., Shuman, M.A., Jacques, Y.V. and Bainton, D.F. (1985). A platelet α-granule membrane protein (GMP-140) is expressed on the plasma membrane after activation. *J. Cell Biol.,* **101**:880–886.

Stone, J.P. and Wagner, D.D. (1993). P-selectin mediates adhesion of platelets to neuroblastoma and small cell lung cancer. *J. Clin. Invest.*, **92**:804–813.

Subramaniam, M., Koedam, J.A. and Wagner, D.D. (1993). Divergent fates of P- and E-selectins after their expression on the plasma membrane. *Mol. Biol. Cell*, **4**:791–801.

Subramaniam, M., Saffaripour, S., Watson, S.R., Mayadas, T.N., Hynes, R.O. and Wagner, D.D. (1995). Reduced recruitment of inflammatory cells in a contact hypersensitivity response in P-selectin-deficient mice. *J. Exp. Med.*, **181**:2277–2282.

Symon, F.A., Walsh, G.M., Watson, S.R. and Wardlaw, A.J. (1994). Eosinophil adhesion to nasal polyp endothelium is P-selectin-dependent. *J. Exp. Med.*, **180**:371–376.

Takeda, I., Kaise, S., Nishimaki, T. and Kasukawa, R. (1994). Soluble P-selectin in the plasma of patients with connective tissue diseases. *Int. Arch. Allergy Immunol.*, **105**:128–134.

Tedder, T.F., Isaacs, C.M., Ernst, T.J., Demetri, G.D., Adler, D.A. and Disteche, C.M. (1989). Isolation and chromosomal localization of cDNAs encoding a novel human lymphocyte cell surface molecule, LAM-1. Homology with the mouse lymphocyte homing receptor and other human adhesion proteins. *J. Exp. Med.*, **170**:123–133.

Tedder, T.F., Steeber, D.A., Chen, A. and Engel, P. (1995). The selectins: Vascular adhesion molecules. *FASEB J.*, **9**:866–873.

Tipping, P.G., Huang, X.R., Berndt, M.C. and Holdsworth, S.R. (1996). P-selectin directs T lymphocyte-mediated injury in delayed-type hypersensitivity responses: Studies in glomerulonephritis and cutaneous delayed-type hypersensitivity. *Eur. J. Immunol.*, **26**:454–460.

Toombs, C.F., DeGraaf, C.L., Martin, J.P., Geng, J.G., Anderson, D.C. and Shebuski, R.J. (1995). Pretreatment with a blocking monoclonal antibody to P-selectin accelerates pharmacological thrombolysis in a primate model of arterial thrombosis. *J. Pharmacol. Exp. Ther.*, **275**:941–949.

Tu, L., Chen, A., Delahunty, M.D., Moore, K.L., Watson, S.R., McEver, R.P. *et al.* (1996). L-selectin binds to PSGL-1 on leukocytes: Interactions between the lectin, EGF and consensus repeat domains of the selectins determine ligand binding specificity. *J. Immunol.*, **157**:3995–4004.

Ushiyama, S., Laue, T.M., Moore, K.L., Erickson, H.P. and McEver, R.P. (1993). Structural and functional characterization of monomeric soluble P-selectin and comparison with membrane P-selectin. *J. Biol. Chem.*, **268**:15229–15237.

Vachino, G., Chang, X-J., Veldman, G.M., Kumar, R., Sako, D., Fouser, L.A. *et al.* (1995). P-selectin glycoprotein ligand-1 is the major counter-receptor for P-selectin on stimulated T cells and is widely distributed in non-functional form on many lymphocytic cells. *J. Biol. Chem.*, **270**:21966–21974.

Varki, A. (1994). Selectin ligands. *Proc. Natl. Acad. Sci., USA*, **91**:7390–7397.

Veldman, G.M., Bean, K.M., Cumming, D.A., Eddy, R.L., Sait, S.N.J. and Shows, T.B. (1995). Genomic organization and chromosomal localization of the gene encoding human P-selectin glycoprotein ligand. *J. Biol. Chem.*, **270**:16470–16475.

Vischer, U.M., Jornot, L., Wollheim, C.B. and Theler, J.-M. (1995). Reactive oxygen intermediates induce regulated secretion of von Willebrand factor from cultured human vascular endothelial cells. *Blood*, **85**:3164–3172.

Von Andrian, U.H., Hasslen, S.R., Nelson, R.D., Erlandsen, S.L. and Butcher, E.C. (1995). A central role for microvillous receptor presentation in leukocyte adhesion under flow. *Cell*, **82**:989–999.

Walcheck, B., Moore, K.L., McEver, R.P. and Kishimoto, T.K. (1996). Neutrophil-neutrophil interactions under hydrodynamic shear stress involve L-selectin and PSGL-1: a mechanism that amplifies initial leukocyte accumulation on P-selectin *in vitro*. *J. Clin. Invest.*, **98**:1081–1087.

Ward, P.A. and Varani, J. (1990). Mechanisms of neutrophil-mediated killing of endothelial cells. *J. Leukocyte Biol.*, **48**:97–102.

Watson, M.L., Kingsmore, S.F., Johnston, G.I., Siegelman, M.H., Le Beau, M.M., Lemons, R.S. *et al.* (1990). Genomic organization of the selectin family of leukocyte adhesion molecules on human and mouse chromosome 1. *J. Exp. Med.*, **172**:263–272.

Weller, A., Isenmann, S. and Vestweber, D. (1992). Cloning of the mouse endothelial selectins. Expression of both E- and P-selectin is inducible by tumor necrosis factor. *J. Biol. Chem.*, **267**:15176–15183.

Weyrich, A.S., Ma, X., Lefer, D.J., Albertine, K.H. and Lefer, A.M. (1993). *In vivo* neutralization of P-selectin protects feline heart and endothelium in myocardial ischemia and reperfusion injury. *J. Clin. Invest.*, **91**:2620–2629.

Weyrich, A.S., Buerke, M., Albertine, K.H. and Lefer, A.M. (1995a). Time course of coronary vascular endothelial adhesion molecule expression during reperfusion of the ischemic feline myocardium. *J. Leukocyte Biol.*, **57**:45–55.

Weyrich, A.S., McIntyre, T.M., McEver, R.P., Prescott, S.M. and Zimmerman, G.A. (1995b). Monocyte tethering by P-selectin regulates monocyte chemotactic protein-1 and tumor necrosis factor-α secretion. *J. Clin. Invest.*, **95**:2297–2303.

Weyrich, A.S., Elstad, M.R., McEver, R.P., McIntyre, T.M., Moore, K.L., Morrissey, J.H. *et al.* (1996). Activated platelets signal chemokine synthesis by human monocytes. *J. Clin. Invest.*, **97**:1525–1534.

Wilkins, P.P., Moore, K.L., McEver, R.P. and Cummings, R.D. (1995). Tyrosine sulfation of P-selectin glycoprotein ligand-1 is required for high affinity binding to P-selectin. *J. Biol. Chem.*, **270**:22677–22680.

Wilkins, P.P., McEver, R.P. and Cummings, R.D. (1996). Structures of the O-glycans on P-selectin glycoprotein ligand-1 from HL-60 cells. *J. Biol. Chem.*, **271**:18732–18742.

Winn, R.K., Liggitt, D., Vedder, N.B., Paulson, J.C. and Harlan, J.M. (1993). Anti-P-selectin monoclonal anti-body attenuates reperfusion injury to the rabbit ear. *J. Clin. Invest.*, **92**:2042–2047.

Winn, R.K., Paulson, J.C. and Harlan, J.M. (1994). A monoclonal antibody to P-selectin ameliorates injury associated with hemorrhagic shock in rabbits. *Am. J. Physiol. Heart Circ. Physiol.*, **267**:H2391–H2397.

Wong, C.S., Gamble, J.R., Skinner, M.P., Lucas, C.M., Berndt, M.C. and Vadas, M.A. (1991). Adhesion protein GMP140 inhibits superoxide anion release by human neutrophils. *Proc. Natl. Acad. Sci., USA*, **88**:2397–2401.

Yang, J., Galipeau, J., Kozak, C.A., Furie, B.C. and Furie, B. (1996). Mouse P-selectin glycoprotein ligand-1: Molecular cloning, chromosomal localization, and expression of a functional P-selectin receptor. *Blood*, **87**:4176–4186.

Yao, L., Pan, J., Setiadi, H., Patel, K.D. and McEver, R.P. (1996). Interleukin 4 or oncostatin M induces a prolonged increase in P-selectin mRNA and protein in human endothelial cells. *J. Exp. Med.*. **184**:81–92.

Yeo, E.L., Sheppard, J.-A.I. and Feuerstein, I.A. (1994). Role of P-selectin and leukocyte activation in polymorphonuclear cell adhesion to surface adherent activated platelets under physiologic shear conditions (an injury vessel wall model). *Blood*, **83**:2498–2507.

Zannettino, A.C.W., Berndt, M.C., Butcher, C., Butcher, E.C., Vadas, M.A. and Simmons, P.J. (1995). Primitive human hematopoietic progenitors adhere to P-selectin (CD62P). *Blood*, **85**:3466–3477.

Zhou, Q., Moore, K.L., Smith, D.F., Varki, A., McEver, R.P. and Cummings, R.D. (1991). The selectin GMP-140 binds to sialylated, fucosylated lactosaminoglycans on both myeloid and nonmyeloid cells. *J. Cell Biol.*, **115**:557–564.

Zimmerman, G.A., Prescott, S.M. and McIntyre, T.M. (1992). Endothelial cell interactions with granulocytes: tethering and signaling molecules. *Immunol. Today*, **13**:93–100.

12 Role of PECAM-1 in Vascular Biology

Robert K. Andrews and Michael C. Berndt

*Hazel and Pip Appel Vascular Biology Laboratory, Baker Medical Research Institute,
Commercial Road, Prahran, VIC 3181, Australia, Tel: (61)-3-9522-4333,
Fax: (61)-3-9521-1362*

One of the most interesting questions in vascular biology is how cells circulating under high shear flow in the bloodstream can selectively target a particular site on the blood vessel wall and fulfil their specialized functions. In the inflammatory response, neutrophils, monocytes or lymphocytes roll on activated endothelial cells before sticking, and then transmigrating through the endothelium to reach the site of damaged tissue. In haemostasis, platelets adhere to the subendothelial matrix of a damaged vessel wall, spread over the surface and recruit more platelets to form a platelet aggregate or thrombus. Cell adhesion is also critical in the pathological processes of atherogenesis and thrombosis. Each stage of vascular cell adhesion is regulated by specific receptors on the cell surface that control cell-cell and cell-matrix interactions. Many of these adhesive receptors not only regulate contact adhesion, but also initiate signals that activate the cell and regulate post-adhesion cellular events. These include the cytoskeletal rearrangements involved in cell shape change and motility, secretion from storage organelles, and expression of additional activation-dependent receptors. PECAM-1 (platelet-endothelial cell adhesion molecule-1; CD31) is an adhesive receptor found on endothelial cells, platelets and leukocytes that plays a key role in regulating leukocyte migration through endothelial cell junctions. Recent evidence also suggests that PECAM-1 regulates the adhesive function of integrin receptors on leukocytes, and that cytoplasmic phosphorylation of PECAM-1 in platelets is associated with activation-dependent rearrangement of cytoskeletal actin filaments. This review will examine the structure of PECAM-1 and its role in vascular biology. Interestingly, many characteristics of PECAM-1 structure, function and regulation are shared by other adhesive receptors found on vascular cells.

STRUCTURE OF PECAM-1

The cloning of PECAM-1 from human endothelial cell or HL60 cell libraries in 1990 (Newman *et al.*, 1990; Simmons *et al.*, 1990; Stockinger *et al.*, 1990) sparked renewed interest in the biological function of this protein, previously known as the hec7 or CD31 myeloid cell differentiation antigen. The 130-kiloDalton glycoprotein was homologous to members of the immunoglobulin (Ig) superfamily (Table 12.1). The 711-amino acid mature protein consists of six tandem repeats of the Ig domain in the N-terminal extracellular region, a transmembrane domain and an ~ 118-residue cytoplasmic domain. There are nine consensus Asn-linked glycosylation sites in the extracellular region, and a consensus

Table 12.1 PECAM-1 and some other members of the immunoglobulin superfamily found on vascular cells

Protein Structure[a]	Protein	Distribution[b]	Ligand(s)
	PECAM-1[c]	P, E, M, L	PECAM-1, GAGs, $\alpha_v\beta_3$
	sPECAM-1[d]	Plasma	
	ICAM-1	E	LFA-1 $(\alpha_L\beta_2)$ Mac-1 $(\alpha_M\beta_2)$
	cICAM-1[d]	Plasma	
	ICAM-2	E, L, P	LFA-1 $(\alpha_L\beta_2)$
	ICAM-3	N, M, L	LFA-1 $(\alpha_L\beta_2)$
	ALCAM	M, L	CD6
	VCAM-1	E	VLA-4 $(\alpha_4\beta_1)$ LPAM-1 $(\alpha_4\beta_7)$
	VCAM-1[d]	E	
	MadCAM-1[e]	E	L-selectin LPAM-1 $(\alpha_4\beta_7)$
	KDR (Flk-1)[f] FLT-1[f]	E E, M	VEGF VEGF, PlGF
	sFLT-1	Plasma	
	PDGFR (A and B)[f]	E	PDGF
	FGIR[f]	E	FGF-1, -2, -4
	CD22	L	Sialylated ligand
	CD22[d]	L	Sialylated ligand

[a] The stippled bar represents the plasma membrane, the extracellular side is to the right; ▨, transmembrance domain; ⌒ immunoglobulin domain.

[b] P, platelets; E, endothelial cells; M, monocytes; N, neutrophils; L, lymphocytes (subset).

[c] Consensus, ℗ phosphorylation ● and , N-linked glycosylation site(s).

[d] Alternatively-spliced form.

[e] Vertical lines represent glycosylation.

[f] Shaded ovals, tyrosine kinase-like domains.

tyrosine phosphorylation site within the cytoplasmic domain at Tyr-686 (Newman *et al.*, 1990). The Ig domain is a conserved sequence of ~ 80–100 amino acids, commonly containing a single disulfide bridge. There are three major subgroups of the Ig superfamily, C1, C2 and V, and secondary structure variability between Ig domains of different groups.

Five of the six PECAM-1 Ig domains are of the C2 type, with seven anti-parallel β strands folded into a two-layer β-sheet sandwich stabilized by the disulfide bond; the fifth domain is also of the C2 type, but lacks one of the predicted β sheets (Newman *et al.*, 1990). There may be from ~ 15% up to >45% amino acid identity between Ig domains of the same or different proteins. The human PECAM-1 gene, localized to chromosome 17, consists of a total of ~ 65 kilobases (Kirschbaum *et al.*, 1994). A pseudogene with 76% identity to the PECAM-1 gene was identified on chromosome 3. The mature protein is encoded by 16 exons, with each of the six Ig domains encoded by an individual exon. The cytoplasmic domain of ~ 118 amino acid residues is encoded by eight different exons, providing ample opportunity for alternative-splicing variations within this domain. Minor differences in the predicted amino acid sequence of the mature protein reported from different laboratories may reflect some polymorphism in the PECAM-1 gene.

The Ig superfamily of proteins, already considerable in number ten years ago (Hunkapiller and Hood, 1986), has continued to expand. As well as the heavy and light chains of immunoglobulins, Ig domains are found in T- and B-cell receptors and accessory proteins, such as those involved in MHC recognition (CD4 and CD8), in the neural cell adhesion molecules, N-CAM-1, neural-glial (Ng)-CAM, L1, fascilin II, amalgam and contactin, and in carcinoembryonic antigen, CEA (Hunkapiller and Hood, 1986; Newman *et al.*, 1990). In the vasculature, Ig domain-containing proteins include the adhesive proteins PECAM-1, intercellular adhesion molecule-1 (ICAM-1), ICAM-2 and ICAM-3, vascular cell adhesion molecule-1 (VCAM-1), activated leukocyte cell adhesion molecule (ALCAM), and Mad-CAM-1 (Table 12.1). Endothelial cells also express the signalling receptor proteins KDR and FLT-1, that bind vascular endothelial growth factor (VEGF) or placental growth factor (PlGF); the platelet-derived growth factor (PDGF) A and B receptors; and the fibroblast growth factor (FGF) receptors (Table 12.1). These latter proteins involved primarily in angiogenesis have Ig domains more closely related to the V subclass, and contain a tyrosine kinase-like domain within the cytoplasmic tail (Mustonen and Alitalo, 1995; Thomas, 1996). FLT-1 is also upregulated on activated monocytes where it is involved in VEGF-induced chemotaxis (Barleon *et al.*, 1996). Interestingly, although PECAM-1 does not contain a tyrosine kinase-like domain, there is some amino acid sequence homology within the cytoplasmic domains of PECAM-1 and the signalling PDGF A and B receptors (Newman *et al.*, 1990) hinting at a role for PECAM-1 in cell signalling (discussed below). Another recently described member of the Ig superfamily in the vasculature is the B lymphocyte cell adhesion molecule, CD22 (Table 12.1). Found on subsets of B lymphocytes, CD22 mediates interactions between B cells through an interaction with sialic acid-containing structures (Wilson *et al.*, 1991; van der Merwe *et al.*, 1996). CD22 contains seven extracellular Ig domains, including an N-terminal V-type domain followed by six C2-type domains. An alternatively-spliced form of CD22 lacks the fourth and fifth domains.

DISTRIBUTION OF PECAM-1 ON VASCULAR CELLS

PECAM-1 is found on a number of different vascular cell types including endothelial cells at all locations throughout the vasculature, platelets, monocytes, granulocytes, subpopulations of lymphocytes and populations of primitive haematopoietic progenitor cells (Stockinger *et al.*, 1990; Ashman *et al.*, 1991; Ruco *et al.*, 1992; Watt *et al.*, 1993; Leavesley *et al.*, 1994). PECAM-1 is also found on solid tumour cell lines, albeit with a different pattern of glycosylation compared with the platelet protein (Tang *et al.*, 1993). Transcription and surface expression of PECAM-1 on different cell types may be modulated by inflammatory stimuli. For example, as a consequence of cell activation by TNFα, IFNγ or phytohemagglutinin, synthesis of PECAM-1 is downregulated in lymphocytes and endothelial cells associated with decreased levels of PECAM-1 mRNA transcripts (Zehnder *et al.*, 1992; Stewart *et al.*, 1996). In contrast, stimulation of the monocytic cell line, U937, with TGFβ results in increased PECAM-1 mRNA levels and surface expression (Lastres *et al.*, 1994). Distribution on such a wide range of vascular cells makes PECAM-1 fairly unusual among vascular cell adhesion receptors, which are generally restricted to fewer myeloid cell lineages. Further, this distribution pattern makes PECAM-1 a candidate for mediating a myriad of interactions between different vascular cells in the (patho)physiological processes of inflammation and thrombosis, some of which will be discussed in following sections.

CELL INTERACTIONS MEDIATED BY PECAM-1

An initial clue to one of the physiological functions of PECAM-1 came from immunohistochemical studies localizing PECAM-1 to endothelial cell-cell junctions (Newman *et al.*, 1990; Stockinger *et al.*, 1990; Metzelaar *et al.*, 1991; Muller *et al.*, 1993; Ayalon *et al.*, 1994). This suggested that PECAM-1 may regulate formation and stabilization of the endothelial cell monolayer. In cultured endothelial cells, immunohistochemical localization showed that PECAM-1 was concentrated at cell borders, at distinct sites from receptors of the cadherin family (Ayalon *et al.*, 1994). The cadherins form Ca^{2+}-dependent homotypic interactions that regulate specific cell-cell adhesion during development and in mature tissue (Jaffe *et al.*, 1990). However, unlike the cadherin receptors, PECAM-1-mediated cell-cell interactions were Ca^{2+}-independent (Ayalon *et al.*, 1994). In transfected COS-7 or 3T3 cell lines, PECAM-1 was concentrated at cell-cell junctions formed between fluorescently-labeled cells transfected with full length PECAM-1 cDNA, but these junctions did not occur between transfected and mock transfected cells (Albelda *et al.*, 1991). In the latter case, the expressed PECAM-1 remained diffusely distributed over the cell surface. Recent studies confirmed homotypic interactions with purified PECAM-1 in phospholipid vesicles (Sun *et al.*, 1996b). PECAM-1 expressed on placental microvessels or cultured endothelial cells following stimulation by histamine or inflammatory cytokines, respectively, shifted from the characteristic localization at cell-cell borders and redistributed over the cell surface (Leach *et al.*, 1995; Romer *et al.*, 1995). This redistribution corresponded to a widening of distance between adjacent endothelial membranes at tight junctions (from 4.1 nm to 6.1 nm), and increased microvessel permeability (Leach *et al.*, 1995). In a similar vein, mild peroxide treatment of cultured endothelial cells, a situation mimicking oxidative conditions of some inflammatory diseases, resulted in redistribution of PECAM-1 from cell borders to the basal surface (Bradley *et al.*, 1995). There were related topographical changes with respect to other adhesive receptors, including ICAM-1 and ICAM-2. A recent study using recombinant PECAM-1 suggested that specific residues within the first Ig domain (Asp-11, Asp-33, Lys-89, Lys-50 and Asp-51) may be involved in homotypic PECAM-1 interaction (Newton *et al.*, 1997). Interestingly, Fab fragments of a monoclonal antibody against Cys-496-Gly-501 (CAVNEG) within the sixth Ig domain augmented homophilic interactions between PECAM-1-containing proteoliposomes and PECAM-1-expressing cells (Sun *et al.*, 1996b).

In addition to homotypic interactions involving PECAM-1, there is evidence for heterotypic interactions between PECAM-1 and glycosaminoglycan structures (GAGs) regulating cell-cell binding. L cells transfected with PECAM-1 cDNA adhered to other PECAM-1 transfected cells or mock transfected cells to a similar extent (Muller *et al.*, 1992). Interactions of PECAM-1 transfected L cells, unlike homotypic interactions, were Ca^{2+}-dependent, and were inhibited by soluble sulfated glycosaminoglycans such as heparin and chondroitin sulfate, and by enzymatic removal of cell surface GAGs (DeLisser *et al.*, 1993). This indicated that separate binding sites on PECAM-1 were involved in homotypic *versus* heterotypic binding. Similarly, neural cell adhesion receptors of the Ig superfamily, N-CAM-1 and Ng-CAM, also engage in heterotypic interactions with GAG-like ligands as well as homotypic interactions (Rao *et al.*, 1992). Mapping of anti-PECAM-1 monoclonal antibodies showed that antibodies recognizing epitopes within the second or sixth Ig domains specifically blocked heterotypic adhesion of PECAM-1-transfected L cells (Yan *et al.*, 1995b). The second Ig domain of PECAM-1 contains an amino acid sequence, Leu-Lys-Arg-Glu-Lys-Asn (LKREKN), corresponding to a sequence in N-CAM-1 that binds GAGs, and representing a GAG-binding consensus recognition sequence. A synthetic peptide based on LKREKN inhibited PECAM-1-mediated aggregation of transfected L cells (DeLisser *et al.*, 1993). In this same study, cells transfected with a mutant form of PECAM-1 lacking the second Ig domain failed to aggregate. Another sequence at the C-terminal end of the sixth Ig domain (Asn-552-Lys-574) mediates lymphocyte adhesion and is proximal to the membrane-spanning domain (Zehnder *et al.*, 1995). Both a monoclonal antibody that maps to this site and a synthetic peptide based on the sequence inhibit B cell/T cell adhesion involving heterotypic PECAM-1 interactions. Further, a monoclonal antibody that maps to Cys-496-Gly-501 at the N-terminal end of the sixth Ig domain, also inhibits heterotypic PECAM-1-dependent adhesion (Yan *et al.*, 1995b). This antibody augmented homotypic binding (above). Two likely divalent cation binding sites have been identified, corresponding to clusters of negatively-charged residues within Asp-443-Glu-446 and Glu-487-Glu-542 of the fifth and sixth Ig domains, respectively (Jackson *et al.*, 1997a).

Deleting the entire cytoplasmic domain of recombinant PECAM-1 in L cells abolished heterotypic, Ca^{2+}-dependent aggregation involving GAGs, and, in contrast to wild-type PECAM-1-transfected cells, there was no localization of the mutant PECAM-1 to cell-cell junctions (DeLisser *et al.*, 1994). However, L cells expressing PECAM-1 containing either one-third or two thirds of the cytoplasmic tail demonstrated homotypic aggregation in related experiments without the requirement for Ca^{2+}. These results suggested that PECAM-1 on one cell can bind either to itself or to GAGs expressed on another cell, and that cells have the potential to alter the adhesive phenotype of PECAM-1 through its cytoplasmic domain. These studies further show that the cytoplasmic domain of PECAM-1 not only regulates the ligand binding properties of the protein, but at the same time alters the divalent cation dependence of PECAM-1-mediated cell adhesion (DeLisser *et al.*, 1994; Sun *et al.*, 1996a). Additional studies identified a region of the cytoplasmic domain of murine PECAM-1 that directs ligand-binding specificity (Yan *et al.*, 1995a). It was observed in the developing mouse embryo that from the earliest manifestation of the circulatory system, cells expressed multiple isoforms of PECAM-1 as a result of alternative splicing of exons encoding the cytoplasmic domain (Baldwin *et al.*, 1994). This implied that PECAM-1 may play an important role in angiogenesis, and that the cytoplasmic region of the receptor was regulating PECAM-1 function. Recombinant murine PECAM-1 cDNA containing various combinations of the cytoplasmic domain exons was therefore expressed in L cells to determine the effect of cytoplasmic sequences on ligand-binding function (Yan *et al.*, 1995a). In this expression system, the presence of the amino acid sequence corresponding to exon 14 was associated with Ca^{2+}-dependent heterotypic interactions, whereas mutant forms of PECAM-1 lacking exon 14 sequence mediated Ca^{2+}-independent homotypic binding. The mechanism for this profound effect of the exon 14-related sequence, LGTRATETVYSEIRKVDP (Figure 12.1),

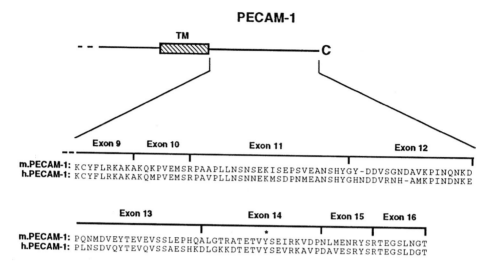

Figure 12.1   Amino acid sequence of the cytoplasmic domain of human PECAM-1 (Newman *et al.*, 1990) compared with the murine sequence (Xie and Muller, 1993). The hatched box represents the transmembrane (TM) domain. The amino acid sequences corresponding to exons 9 to 16 comprising the cytoplasmic domain of murine PECAM-1 are also indicated. The asterisk marks the consensus phosphorylation site at Tyr-686 in human PECAM-1.

is not clear, but may involve differences in phosphorylation at Tyr-686, alterations in cytoskeletal association and/or conformational changes of the receptor (Yan *et al.*, 1995a).

A further heterotypic ligand for PECAM-1 is the vitronectin receptor of the integrin superfamily, $\alpha_v\beta_3$ (Piali *et al.*, 1995). $\alpha_v\beta_3$ is found on several vascular cell types, including endothelial cells, T lymphocytes and platelets. It binds adhesive proteins of the plasma or extracellular matrix such as vitronectin, fibronectin, laminin, fibrinogen and von Willebrand Factor *via* an Arg-Gly-Asp (RGD) amino acid sequence in the ligand. PECAM-1 does not contain an RGD sequence. Binding to $\alpha_v\beta_3$ involves the second Ig domain, as determined by functional analysis of recombinant deletion mutants of PECAM-1 (Piali *et al.*, 1995). Other members of the Ig superfamily on vascular cells, ICAM-1, ICAM-2, ICAM-3, VCAM-1 and MadCAM-1 also bind to integrin receptors (Table 12.2).

THE ROLE OF PECAM-1 IN LEUKOCYTE TRAFFICKING

A major function for PECAM-1, ICAM-1 and VCAM-1 of the Ig superfamily, in concert with adhesive receptors from the selectin and integrin families, is to deliver circulating leukocytes from the bloodstream through the vessel wall to sites of infection. The sequence of events involved in neutrophil migration through the endothelium is outlined in Figure 12.2A. Some of the receptors involved in different stages of leukocyte adhesion and migration are shown in Table 12.2. In the case of neutrophils and monocytes, this process is initiated by endothelial cell activation by cytokines or other inflammatory stimuli released from infected tissue. The transcriptional regulation of many of the endothelial adhesion receptors expressed in response to cytokine stimulation is linked, and involves the nuclear factor-κB (NF-κB) system (Collins *et al.*, 1995; Ledebur and Parks, 1995). Similar stages of transmigration occur in normal lymphocyte circulation, from blood vessels to extravascular fluid and the lymphatic system (Springer, 1995).

Table 12.2 Comparison of PECAM-1 with other vascular cell adhesive receptors

Physiological process	Stage	Cell[a]	Receptor	Family	Ligand[b]	Expression[c]	Soluble form[d]	Cytoskeletal association/signalling
Leukocyte/endothelial cell interactions	1. Initial rolling	N	PSGL-1	Sialomucin	P-, E-selectin*	C		Cytoplasmic tail mediates adhesive function; linked to p56^{lck}
		N, M, L	L-selectin	Selectin	MadCAM-1*[e]	C	Prot.	Cytoplasmic tail phosphorylated in activated platelets
	2. Tight Adhesion	E	P-selectin	Selectin	PSGL-1*	A	Alt. S.	
		E	E-selectin	Selectin	PSGL-1/ESL-1*	I		
		E	MadCAM-1[e]	Ig Superfamily	L-selecin*, LPAM-1	C		
		N, M, L	LFA-1	$\alpha_L\beta_2$ Integrin	ICAM-1/ICAM-2	C		Induces p130^{cas} phosphorylation; β_2 phosphorylated
		N, M	Mac-1	$\alpha_M\beta_2$ Integrin	ICAM-1	C		
		N, M	p150/95	$\alpha_X\beta_2$ Integrin	?	C		Stimulates CD19-associated tyrosine kinase pathway
		M, L	VLA-4	$\alpha_4\beta_1$ Integrin	VCAM-1	C		
		L	LPAM-1	$\alpha_4\beta_7$ Integrin	VCAM-1, MadCAM-1	C		
		E, L	ICAM-1	Ig Superfamily	LFA-1/Mac-1	C/I	Alt. S.	Cytoplasmic tail associated with cytoskeletal α-actinin
		E, L, P	ICAM-2	Ig Superfamily	LFA-1	C		Activates p59^{fyn} and p56^{lyk} tyrosine kinases
		N, M, L	ICAM-3[f]	Ig Superfamily	LFA-1	C		
		E	VCAM-1	Ig Superfamily	VLA-4, LPAM-1	I		
	3. Transmigration	E, M, L, P	PECAM-1	Ig Superfamily	PECAM-?, $\alpha_v\beta_3$, GAGs*	C	Alt. S.	Cytoplasmic tail phosphorylated; associated with shp-2 and src
		E, L, P	$\alpha_v\beta_3$	Integrin	Collagen	C		
Platelet thrombus formation	1. Adhesion to matrix[g]	P	GP Ib-IX-V	Leu-rich/Sialomucin	vWF	C	Prot.	GP Ibα binds 14.3.3 (zeta) and cytoskeletal ABP
		P	GP Ia-IIa	$\alpha_2\beta_1$ Integrin	Collagen	C		Activates focal adhesion kinase, pp125^{fak}
		P	GP IV	–	Collagen/TSP	C		Activates pp60^{fyn}, pp62^{yes} and pp54/58^{lyn}
		P	GP VI	–	Collagen	C		Activates pp60^{src} and p72^{syk}
	2. Platelet aggregation	P	GP IIb-IIa	$\alpha_{IIb}\beta_3$ Integrin	Fibrinogen*	A		Activates pp125^{fak}
	3. Platelet/leukocyte interactions	P, M, L, E	PECAM-1?	Ig Superfamily	PECAM-1, $\alpha_v\beta_3$, GAGs*	C	Alt. S.	Cytoplasmic tail phosphorylated; associated with shp-2 and src
		P	P-selectin	Selectin	PSGL-1*	A	Alt. S.	Cytoplasmic tail phosphorylated on activated platelets
		N	PSGL-1	Sialomucin	P-selectin*	C		
		P, L, E	ICAM-2	Ig Superfamily	LFA-1	C		

[a] N, neutrophils; M, monocytes; L, lymphocytes (subset); E, endothelial cells; P, platelets. [b](*) denotes Ca^{2+}-dependent ligand binding. [c] C, constitutive surface expression; A, activation-dependent expression (rapid); I, induced by cytokines (several hours). [d] Prot., proteolytic soluble fragment; Alt. S., alternatively-spliced soluble form. [e] L-selectin also binds related mucin-like proteins CD34 and GlyCAM-1. [f] Leukocyte receptor involved in immune cell interactions. [g] ABP, actin-binding protein; TSP, thrombospondin; vWF, von Willebrand Factor.

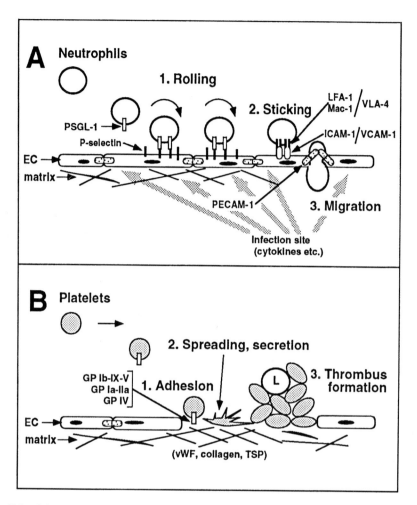

Figure 12.2   Schematic representation of the stages involved in (A) neutrophil trafficking and (B) platelet adhesion and thrombus formation (L, leukocyte).

The first stage of leukocyte trafficking is the rolling of circulating cells along the endothelial cell monolayer lining the vessel wall. This process is primarily mediated by selectins and their carbohydrate/sialomucin counter-receptors (Varki, 1994; McEver *et al.*, 1995; Springer, 1995). Proteins of the selectin family contain an N-terminal Ca^{2+}-dependent (C-type) lectin domain, an adjacent epidermal growth factor (EGF)-like domain, a variable number of complement regulatory protein-like domains, a transmembrane domain and a cytoplasmic tail. P-selectin is stored on intracellular granule membranes and rapidly expressed on the surface of activated endothelial cells, while E-selectin expression is induced in response to inflammatory stimuli. The lectin domain and elements of the adjacent EGF domain are both involved in Ca^{2+}-dependent binding of P-selectin to P-selectin glycoprotein ligand-1 (PSGL-1) on neutrophils (Kansas *et al.*, 1994). PSGL-1 is a membrane-spanning disulfide-linked homodimer consisting of a cytoplasmic domain, a transmembrane domain, an extracellular sialomucin domain and an N-terminal anionic sequence containing 1–3 sulfated tyrosine residues (Sako *et al.*, 1993; 1995; Pouyani and Seed, 1995). E-selectin also binds to PSGL-1 in a Ca^{2+}-dependent manner, as well as to E-selectin ligand-1 (ESL-1) and other glycoproteins. The third member of the selectin family, L-selectin, is constitutively expressed on leukocytes and binds to carbohydrate structures on CD34, MadCAM-1 or GlyCAM-1 expressed on endothelium at peripheral lymph nodes, or transiently expressed on endothelial cells at inflammatory sites (Varki *et al.*, 1994; McEver *et al.*, 1995; Springer, 1995). Mice deficient in selectin expression show markedly diminished leukocyte accumulation in response to inflammatory stimuli, demonstrating the importance of the rolling stage in initiating subsequent stages of the transmigration pathway (Mayadas *et al.*, 1993; Kunkel *et al.*, 1996). After the initial rolling stage, selectin receptors may be down-regulated by shedding of an extracellular proteolytic fragment or altered cytoskeletal association (Kansas *et al.*, 1993; Kahn *et al.*, 1994; Preece *et al.*, 1996).

In the second stage of trafficking, rolling cells are more tightly bound to the endothelium by receptors from the integrin family on leukocytes binding to counter-receptors of the Ig superfamily, ICAM-1, ICAM-2, or VCAM-1 expressed on endothelial cells (Table 12.2). ICAM-1 is synthesized and expressed on the surface of stimulated endothelial cells over several hours, and binds to the leukocyte integrins, LFA-1 ($\alpha_L\beta_2$) and Mac-1 ($\alpha_M\beta_2$). VCAM-1 is similarly up-regulated on endothelial cells and binds to the leukocyte integrin, VLA-4 ($\alpha_4\beta_1$) or the related integrin LPAM-1 ($\alpha_4\beta_7$). Endothelial ICAM-1 and VCAM-1 also support leukocyte rolling to some extent, but not to the same degree as the selectins (Alon *et al.*, 1995; Kunkel *et al.*, 1996). The related Ig superfamily receptor, ICAM-3, is expressed on leukocytes, where it also binds LFA-1 and is involved in regulation of the immune response (Holness *et al.*, 1995). The importance of LFA-1 and Mac-1 in leukocyte function is evident in the congenital disease, leukocyte adhesion deficiency (LAD), where defective β_2 expression is associated with ineffectual neutrophil extravasation (Springer, 1995). The β_2 integrins are generally constitutively expressed, however ligand binding may be increased following cell activation. It appears as if all of these integrins transmit intracellular signals when engaged by ligand, thus regulating subsequent cellular events (Table 12.2) (Juliano and Haskill, 1993; Clark and Brugge, 1995; Petruzzelli *et al.*, 1996; Xiao *et al.*, 1996).

The third stage of leukocyte trafficking is the migration of adhered cells through the endothelial cell monolayer. This partly involves breaking and forming of integrin-dependent interactions on the advancing cell. F(ab′)$_2$ fragments of a monoclonal antibody against integrin-associated protein (CD47) on neutrophils and endothelial cells also blocks activated neutrophil migration, suggesting CD47 plays a part in this process (Cooper *et al.*, 1995). PECAM-1 also has a recently defined role in leukocyte migration. A number of studies have shown that anti-PECAM-1 antibodies or recombinant soluble forms of PECAM-1 inhibit leukocyte migration. Muller and coworkers (1993) used an *in vitro* model of leukocyte migration featuring cultured endothelial cell monolayers grown on hydrated collagen gels. An anti-PECAM-1 monoclonal antibody and a recombinant soluble form of PECAM-1 lacking the transmembrane domain both blocked monocyte migration through the endothelium by as much as 70–90%. Pretreatment of either monocytes or endothelial cells with the antibody was similarly effective, consistent with a homotypic mechanism of PECAM-1-mediated adhesion in this system. Migration of neutrophils through cytokine-activated endothelium was inhibited to the same extent by the anti-PECAM-1 antibody or fluid-phase PECAM-1 (Muller *et al.*, 1993). Furthermore, where migration was blocked, scanning electron microscopy of the endothelium showed that the cells remained tightly bound to the apical surface of the endothelial cell junctions (Muller *et al.*, 1993). Anti-PECAM-1 antibodies also block neutrophil accumulation at inflammatory sites in a variety of animal models *in vivo* (Vaporciyan *et al.*, 1993; Bogen *et al.*, 1994). One model was the accumulation of neutrophils in the rat peritoneal cavity induced by glycogen. Intravenous injection of polyclonal anti-PECAM-1 Fab fragments blocked neutrophil migration to the peritoneum by ~ 75% (Vaporciyan *et al.*, 1993). There was similar inhibition of leukocyte accumulation in the lung induced by IgG immune complexes when rats were infused with anti-PECAM-1 F(ab′)$_2$ fragment. In a third model involving transplanted normal human skin onto immuno-deficient mice as inducer of an inflammatory response, anti-PECAM-1 Fab fragments also reduced the number of leukocytes reaching the dermal interstitium from the microvascula-ture (Vaporciyan *et al.*, 1993). In another murine peritonitis model, intravenous injection of an anti-PECAM-1 monoclonal antibody markedly inhibited neutrophil accumulation in response to thioglycollate stimulation for up to 48 hours (Bogen *et al.*, 1994). In this case, neutrophils were observed to be adherent to the luminal surface of mesenteric venules, but there was limited transmigration. PECAM-1 has also been associated with lymphocyte transmigration following antigenic stimulation (Bogen *et al.*, 1992, 1994), while homing of transplanted hematopoietic stem cells across vascular endothelium was at least partially inhibited by blockade of lymphocyte integrins as well as PECAM-1 (Yong *et al.*, 1998). All of these studies are consistent with a role for PECAM-1 in the migration phase of leukocyte trafficking, however there is also a pool of PECAM-1 expressed on the apical surface of endothelial cells adjacent to the cell-cell border (Muller *et al.*, 1993), and there may not be a clear distinction between the function of PECAM-1 in tight adhesion and migration. In addition, PECAM-1 not only functions in migration of monocytes across endothelium, but also mediates passage of cells through the subendothelial matrix (Liao *et al.*, 1995). Mono-clonal antibodies mapping into the first and second Ig domains of PECAM-1 inhibited dia-pedesis by ~ 70%, whereas a monoclonal antibody against the sixth Ig domain blocked only "interstitial" migration through the extracellular matrix. In the latter case, monocytes accumulated at the sublaminal surface of the cultured endothelial cells. This study sug-gested that migration through endothelium and matrix involves separate binding domains on PECAM-1, and provided evidence for heterotypic interactions involving PECAM-1 and a noncellular matrix-associated ligand (Liao *et al.*, 1995).

SOLUBLE FORMS OF PECAM-1 AND OTHER ADHESIVE RECEPTORS

The inflammatory response may be regulated to some extent *in vivo* by soluble forms of receptors present in the circulation. This may be especially relevant in minimizing cell adhesion and activation at locations downstream of inflammatory sites. Selectins circulate in the plasma as a result of either proteolytic cleavage of L-selectin at a site near the mem-brane, or by secretion of an alternately-spliced soluble form of P-selectin containing cyto-plasmic sequence but lacking the transmembrane domain (Dunlop *et al.*, 1992; Kahn *et al.*, 1994; Ishiwata *et al.*, 1994; Preece *et al.*, 1996). Soluble P-selectin has been shown to pre-vent activated neutrophil adhesion to activated endothelial cells (Gamble *et al.*, 1990). Human plasma contains ~ 0.175 μg/mL (male) or ~ 0.251 μg/mL (female) of soluble P-selectin. A circulating form of ICAM-1 (cICAM-1) is present in normal human serum at similar levels, 0.1–0.2 μg/mL, and becomes elevated up to 0.7 μg/mL in patients with congenital defi-ciency of ICAM-1-binding β_2 integrins (Rothlein *et al.*, 1991). An alternatively-spliced form of PECAM-1 lacking the transmembrane domain, but containing the cytoplasmic domain has been identified in the media of cultured cells and in plasma at 0.01–0.025 μg/mL (Gold-berger *et al.*, 1994). This soluble form of PECAM-1 is functional since it blocks PECAM-1-dependent cell adhesion. Plasma also contains another soluble form of PECAM-1 lacking both the cytoplasmic and transmembrane domains, that may arise from proteolysis of cell surface or alternatively-spliced PECAM-1 (Goldberger *et al.*, 1994). Like selectins, soluble forms of PECAM-1 potentially regulate leukocyte migration, either by competing for adhesive sites with membrane-bound PECAM-1 or by modulating the adhesive activity of β_2 integrins or other receptors (discussed below). While precise physiological roles for these functional, soluble extracellular forms of adhesive receptors has yet to be established, it is interesting to note that a functional proteolytic fragment of the platelet adhesion receptor, the glycoprotein (GP) Ib-IX-V complex, also circulates in the plasma, as does an alternatively-spliced extracellular fragment of the non-adhesive, signalling receptor of the Ig superfamily, FLT-1 (Table 12.1).

A POTENTIAL ROLE FOR PECAM-1 IN THROMBUS FORMATION

To prevent the loss of blood from a damaged blood vessel, platelets in flowing blood first adhere to the vessel wall matrix, and then form a platelet aggregate or thrombus. Haemostasis is a coordinated sequence of events involving specific cell-surface adhesive receptors (Figure 12.2B), not unlike the process of leukocyte trafficking (Figure 12.2A). Not surprisingly, some of the receptors involved in thrombosis are structurally and functionally related to the leukocyte/endothelial cell receptors (Table 12.2). For example, both neutrophil PSGL-1 and the ligand-binding subunit (GP Ibα) of the platelet GP Ib-IX-V complex that mediate initial cell contact with the vessel wall are sialomucins, with an elongated highly-glycosylated region elevating the ligand binding domain from the cell surface; an analogous anionic/sulfated tyrosine sequence is involved in ligand recognition of both receptors (De Luca *et al.*, 1995; Andrews *et al.*, 1997). Different integrin receptors — $\alpha_L\beta_2$, $\alpha_M\beta_2$ and $\alpha_4\beta_1$ on neutrophils and $\alpha_{IIb}\beta_3$ on platelets — regulate secondary stages of cell-cell interactions after initial contact with the vessel wall. Like leukocytes, platelets also roll along activated endothelium, an interaction mediated by P-selectin expressed on endothelial cells (Frenette *et al.*, 1995). A further connection between the haemostatic and inflammatory responses of vascular cells is the intriguing recent finding that neutrophils adhere more efficiently to adhered activated platelets under flow conditions than to activated endothelial cells (Kuijper *et al.*, 1996). Activated platelets may therefore accentuate leukocyte accumulation at sites of injury. In these circumstances, neutrophil adhesion is dependent upon selectins and β_2 integrins. As discussed for leukocyte trafficking above, PECAM-1 may play a comparable role in later stages of thrombus formation. Alternatively, PECAM-1 may be one of the receptors involved in interactions of platelets with activated endothelial cells, a process critical for the initial stages of atherogenesis (Ross, 1995; Springer, 1995).

In the first stage of haemostasis at high physiological shear, the platelet membrane GP Ib-IX-V complex binds to von Willebrand Factor (vWF), an adhesive glycoprotein in the subendothelial matrix (Weiss, 1995; Andrews *et al.*, 1997) (see Chapter 3). At low shear, other platelet receptors such as the collagen receptors, GP IV, GP VI and the $\alpha_2\beta_1$ integrin (GP Ia-IIa), may mediate platelet contact with the vessel wall (Staatz *et al.*, 1989; Tandon *et al.*, 1989; Moroi *et al.*, 1996). GP Ib-IX-V, GP Ia-IIa, GP IV and GP VI are all capable of activating platelets when engaged by ligand via an association of their cytoplasmic domains with specific signalling molecules (Table 12.2) (Huang *et al.*, 1991; Shattil and Brugge, 1991; Bull *et al.*, 1994; Du *et al.*, 1994; Clark and Brugge, 1995; Ichinohe *et al.*, 1995; Asazuma *et al.*, 1996). Platelet activation triggered by adhesion receptors, or by agonists such as thrombin or ADP, leads to Ca^{2+} mobilization, Ca^{2+}-dependent activation of the platelet membrane fibrinogen receptor, the $\alpha_{IIb}\beta_3$ integrin GP IIb-IIIa, and cytoskeletal rearrangements involved in shape change, spreading and secretion (Lipfert *et al.*, 1992; Fox, 1993; Clemetson, 1995; Andrews *et al.*, 1997).

The physiological importance of leukocytes and erythrocytes in the rate and extent of thrombus formation is poorly understood, however, there is some evidence that metabolic cross-talk between leukocytes and platelets may be a significant factor in thrombosis (Marcus, 1994; Furie and Furie, 1995) (see Chapter 10). It is not clear which adhesive receptors are directly involved in mediating platelet-leukocyte interactions. Some receptors expressed on platelets, P-selectin, ICAM-2 and PECAM-1, are known to mediate the interaction of endothelial cells with leukocytes. P-selectin, stored within platelet α-granules and expressed on the surface of activated platelets, could conceivably mediate binding of platelets to PSGL-1 on neutrophils. In support of this, an anti-P-selectin antibody reportedly blocks the interaction of human platelets and neutrophils *in vitro* (Valles *et al.*, 1993), however studies with platelets from P-selectin null mice suggest the platelet receptor may have a more limited role (Frenette *et al.*, 1998). There are ~ 3000 copies of ICAM-2 on a resting platelet, and the copy number is unchanged upon activation (Diacovo *et al.*, 1994). Platelet ICAM-2 mediates binding of platelets to either purified LFA-1 or stimulated T lymphocytes *in vitro*, interactions inhibitable by anti-ICAM-2 or anti-LFA-1 monoclonal antibodies (Diacovo *et al.*, 1994). PECAM-1 has no defined role in platelet aggregation, however, it could play a physiological role in thrombosis, by binding PECAM-1, GAGs or integrins on leukocytes. Like P-selectin, PECAM-1 is found on platelet α-granule membranes as determined by immunohistochemical studies (Cramer *et al.*, 1994), that would enable increased surface expression of PECAM-1 following platelet activation. There is reportedly an increase in surface expression of PECAM-1 from ~ 7760 copies on a resting platelet to ~ 14500 copies on a thrombin-activated platelet, as assessed by monoclonal antibody binding (Metzelaar *et al.*, 1991). Engagement of PECAM-1 by anti-PECAM-1 monoclonal antibodies was recently shown to augment both platelet activation induced by other agonists, and adhesion and aggregation under low shear flow (Varon *et al.*, 1998).

OTHER POSSIBLE ROLES FOR PECAM-1

Recent studies have provided at least two other potential roles for PECAM-1 in the vasculature. Firstly, blocking PECAM-1 function with specific antibodies inhibited rodent endothelial tube formation in *in vitro* models of angiogenesis (DeLisser *et al.*, 1997). Furthermore, murine vasculogenesis was shown to be accompanied by Tyr-686 phosphorylation of the PECAM-1 cytoplasmic tail (Pinter *et al.*, 1997). Secondly, PECAM-1 has recently been implicated as one of the receptors involved in adhesion of *Plasmodium falciparum* infected erythrocytes to endothelium (Treutiger *et al.*, 1997).

INTERACTION OF PECAM-1 WITH THE CYTOSKELETON

The contractile actin filaments making up the cytoskeleton play an essential part in cell signalling, secretion and shape changes involved in cell motility. Many adhesion receptors on vascular cells (Table 12.2) and cell adhesion molecules such as N-CAM-1 and the cadherins on other cell types (Nagafuchi and Takeichi, 1988; Jaffe *et al.*, 1990), interact with the cytoskeleton. These interactions often control cytoskeletal changes and regulate the adhesive activity of the receptor (Jaffe *et al.*, 1990; Fox, 1993; Kansas *et al.*, 1993). Recent evidence suggests that PECAM-1 also interacts with the cytoskeleton in platelets and other cells. In platelets, PECAM-1 is differentially associated with the cytoskeleton before and after platelet activation, coinciding with protein kinase C-dependent phosphorylation of serine residues within its cytoplasmic domain (Newman *et al.*, 1992; Zehnder *et al.*, 1992; Hoyt and Lerea, 1995). The relative amount of PECAM-1 that co-isolated with cytoplasmic actin filaments increased from ~ 1% in the unactivated platelet to >60% after activation (Newman *et al.*, 1992). Phosphorylation of the cytoplasmic tail of PECAM-1 on a time scale corresponding to major cytoskeletal reorganization in activated platelets would be consistent with a role for PECAM-1 in regulating at least some of these changes. For instance, cAMP-dependent phosphorylation at Ser-166 within the GP Ib β-chain of the GP Ib-IX-V complex is known to regulate cytoskeletal rearrangement by inhibiting actin polymerization (Andrews *et al.*, 1997). PECAM-1 is also phosphorylated in endothelial cells and a lymphocytic cell line following activation (Zehnder *et al.*, 1992), and in TGFβ-stimulated monocytic U937 cells (Lastres *et al.*, 1994). In the monocytic cell line, PECAM-1 phosphorylation also results in increased association with the cytoskeleton (Lastres *et al.*, 1994).

A positively-charged stretch of amino acids, Arg-Gln-Arg-Lys-Ile-Lys-Lys-Tyr-Arg (RQRKIKKYR), within the cytoplasmic domain of ICAM-1 at a site close to the membrane reportedly mediates binding of ICAM-1 to the cytoskeletal protein, α-actinin (Carpén *et al.*, 1992). α-Actinin forms a homodimer that cross-links actin filaments. It also binds to the cytoplasmic domain of some integrins and to PI-3-kinase, suggesting it may play both a structural and signalling role (Clark and Brugge, 1995). PECAM-1 contains an analogous positively-charged sequence Arg-Lys-Ala-Lys-Ala-Lys-Gln-Lys (RKAKAKQK) proximal to the cytoplasmic face of the membrane (Newman *et al.*, 1990), implying that PECAM-1 might also bind α-actinin and play a part in cytoskeletal attachment. This sequence is highly conserved in the murine PECAM-1 sequence (RKAKAKQM) (Xie and Muller, 1993). In the integrin β_1 subunit, residues within the sequence Arg-Arg-Glu-Phe-Ala-Lys-Phe-Glu-Lys-Glu-Lys (RREFAKFEKEK) adjacent to the cytoplasmic face of the membrane participate in binding α-actinin (Otey *et al.*, 1993).

PECAM-1-INDUCED CELL SIGNALLING

As discussed above, many of the adhesion receptors on vascular cells are now known to induce signal transduction when engaged by ligand (Table 12.2). Signalling by adhesion molecules regulates the transition of cells from one adhesive phenotype to another, allowing progression through different stages of transmigration or aggregation, by modulating receptor expression/affinity and re-organizing the cytoskeleton. The following evidence would suggest that PECAM-1 is itself able to induce cell signalling: (1) Binding of certain anti-PECAM-1 monoclonal antibodies to PECAM-1 on leukocytes, endothelial cells and platelets has been shown to upregulate the adhesive function of other receptors. On distinct subsets of T cells, binding of anti-PECAM-1 antibodies results in an amplification of $\alpha_4\beta_1$ integrin (VLA-4) mediated adhesion (Tanaka *et al.*, 1992). There was a smaller but related effect on $\alpha_L\beta_2$ integrin (LFA-1) dependent adhesion. Other anti-PECAM-1 antibodies were without effect on VLA-4 or LFA-1 dependent adhesion. For some of the activating antibodies, the degree of amplification was increased by using polyvalent antibody complexes, whereas Fab fragments had only minimal effect, consistent with cross-linking of two or more receptors as a mechanism for PECAM-1 dependent cell signalling. Upregulation of VLA-4 dependent adhesion to VCAM-1 by anti-PECAM-1 antibodies was confirmed in human hematopoietic progenitor cells (Leavesley *et al.*, 1994). Monovalent Fab

fragments of another anti-PECAM-1 monoclonal antibody increased LFA-1-dependent
adhesion of monocytes to endothelial cells, and IgG or polyvalent cross-linked Fab fragments of the same antibody showed even greater binding (Berman and Muller, 1995). A
monoclonal antibody against murine PECAM-1 also specifically increased LFA-1-
dependent adhesion of lymphokine-activated killer cells to endothelium (Piali *et al.*, 1993;
Berman *et al.*, 1996). These killer cells migrate through the endothelial layer and accumulate at sites of tumour lesions. The adhesive function of β_2 integrins on monocytes and neutrophils is known to be upregulated by surface receptors other than PECAM-1, including
FcγRIII and the urokinase receptor, CD87 (Zhou and Brown, 1994; Sitrir *et al.*, 1996).
Finally, in platelets, bivalent but not monovalent anti-PECAM-1 monoclonal antibodies
that mapped into the sixth Ig domain enhanced activation-dependent platelet adhesion and
aggregation at low shear flow (Varon *et al.*, 1998). (2) Engagement of PECAM-1 on monocytes by an anti-PECAM-1 monoclonal antibody induced secretion of the cytokines TNFα,
IL-1 and IL-8 (Chen *et al.*, 1994). This apparent PECAM-1-dependent signal transduction
required co-ligation of PECAM-1 and the FcγRII receptor. In an earlier study, an anti-
PECAM-1 monoclonal antibody was also reported to stimulate a respiratory burst in monocytes *via* an Fcγ receptor-dependent mechanism (Stockinger *et al.*, 1990). In endothelial
cells, certain anti-PECAM-1 antibodies stimulated elevation of intracellular Ca^{2+} and prostacyclin release (Gurubhagavatula *et al.*, 1997). (3) The cytoplasmic domain of PECAM-1
contains a consensus tyrosine phosphorylation site at Tyr-686, located within the exon 14-
related cytoplasmic sequence of PECAM-1 (Figure 12.1) implicated in regulating heterotypic ligand binding (Yan *et al.*, 1995a; Famiglietti *et al.*, 1997). Tyrosine phosphorylation
of PECAM-1 has been demonstrated in endothelial cells by phosphoamino acid analysis
and reactivity towards an anti-phosphotyrosine antibody (Lu *et al.*, 1996). In this study, the
level of PECAM-1 tyrosine phosphorylation was decreased following ligand binding to β_1
integrin receptors during cell migration, concomitant with increased phosphorylation of
pp125^{fak}. Mutation of Tyr-686 to Phe at least partially reversed PECAM-1-mediated
inhibition of endothelial cell migration (Lu *et al.*, 1996). The B lymphocyte receptor CD22,
another member of the Ig superfamily, is also known to be phosphorylated on a cytoplasmic tyrosine residue, and subsequently binds and activates shp, a protein tyrosine phosphatase that attenuates cell signalling through membrane immunoglobulin receptors on
lymphocytes (Doody *et al.*, 1995). Although there is no demonstrable tyrosine phosphorylation in thrombin-activated platelets (Newman *et al.*, 1992), tyrosine phosphorylation
was reported to occur in aggregated platelets (Jackson *et al.*, 1997b; Varon *et al.*, 1998).
Phosphorylation of specific tyrosine residues within the PECAM-1 cytoplasmic domain in
platelets and endothelial cells has recently been linked to enhanced association with the
src-homology 2 (SH2) domains of signalling molecules such as shp-2 and src (Jackson
et al., 1997b; Lu *et al.*, 1997; Masuda *et al.*, 1997). (4) Although PECAM-1 does not
contain a tyrosine kinase-like domain, its cytoplasmic tail shows some homology to the
C-terminal cytoplasmic region of the PDGF A and B tyrosine-kinase receptors of the Ig
superfamily (Newman *et al.*, 1990). (5) ICAM-3, another receptor from the Ig superfamily
on leukocytes, induces signalling in a T cell line when engaged by ligand (Juan *et al.*,
1994). This signalling requires receptor cross-linking and is associated with Ca^{2+} mobilization, activation of tyrosine kinases p59^{fyn} and p56^{lck}, and increased β_1 and β_2 integrin-
dependent cell adhesion. Ig superfamily receptors on neural cells, such as N-CAM-1,
induce signal transduction by a pathway involving activation of the src-related tyrosine
kinase, p59^{fyn} (Beggs *et al.*, 1994). The known signalling function of ICAM-3, CD22,
N-CAM-1 and the PDGF receptors demonstrates the capacity of adhesive receptors of the
Ig superfamily to transmit intracellular signals in response to ligand binding.

SPECIFICITY OF INTERACTIONS INVOLVING IG DOMAINS

The structural differences between ICAM-1, VCAM-1 and PECAM-1 that determine their ligand specificity remains an interesting and largely unanswered question. However, recent structure-function analysis of a number of Ig superfamily receptors provides insight into the molecular determinants of ligand specificity. The major form of VCAM-1 on endothelial cells contains two prospective binding sites for VLA-4, within the conserved first and fourth Ig domains (Osborn *et al.*, 1992). A less prevalent alternatively-spliced form of VCAM-1 lacking the fourth Ig domain (Table 12.1) still binds VLA-4, but with markedly lower affinity. This splice variant shows different affinity and divalent cation dependency towards the integrin heterodimers, $\alpha_4\beta_1$ and $\alpha_4\beta_7$, compared with the full length receptor (Kilger *et al.*, 1995). Studies with recombinant VCAM-1 have identified the human sequence encompassing Arg-Thr-Gln-Ile-Asp-Ser-Pro-Leu (^{36}RTQIDSPL^{43}), conserved within the first and fourth Ig domains, as a motif that mediates, at least in part, binding of VCAM-1 to VLA-4 (Osborn *et al.*, 1994; Renz *et al.*, 1994; Vonderheide *et al.*, 1994). In addition, adjacent amino acid residues may also regulate VLA-4 binding (Newham *et al.*, 1997). The X-ray crystal structure of a VLA-4-binding fragment of VCAM-1 (comprising the first and second Ig domains) resolved to 1.8 Å showed that the RTQIDSPL sequence is within the loop structure linking the C and D β strands of the first Ig domain, and is part of a binding site accessible on the exposed surface of the Ig domain β sheet (Jones *et al.*, 1995). On the lymphocyte receptor, CD22, a sialic acid-binding region has been localized to a cluster of amino acid residues within the first Ig domain by site-directed mutagenesis and by homology to other sialoadhesins of the Ig superfamily (van der Merwe *et al.*, 1996). The position of the binding site on the tertiary structure of the domain is largely superimposable on the intergrin-binding site on the first Ig domain of VCAM-1. Further, a sequence within the third Ig domain of chicken N-CAM-1, KYSFNYDGSE, has been implicated in homotypic N-CAM-1-dependent adhesion (Rao *et al.*, 1992), and is located at the predicted C′ β strand of the Ig domain. Likewise, the LKREKN sequence of PECAM-1 that represents a GAG-binding recognition sequence (DeLisser *et al.*, 1993) is located in the vicinity of the D β strand of the second Ig domain (Newman *et al.*, 1990).

Other structural regions of the Ig domain have also been implicated as ligand-binding sites. On ICAM-3, amino acid residues affecting LFA-1 binding are within the C, C′, F and G β strands that together make up a face of the Ig domain β sheet (Holness *et al.*, 1996). On ALCAM, targeted mutagenesis suggests that the ligand (CD6) binding site is comprised of residues clustered on the A′GFCC′C″ β-sheet face of its N-terminal Ig domain (Skonier *et al.*, 1996). A sequence towards the N-terminal end of the sixth Ig domain of PECAM-1, Cys-Ala-Val-Asn-Glu-Gly (CAVNEG), incorporating the epitope for a monoclonal antibody that inhibits heterotypic PECAM-1-dependent adhesion but augments homotypic adhesion (Yan *et al.*, 1995b; Sun *et al.*, 1996b), is situated at the B β strand (Newman *et al.*, 1990). Another sequence at the C-terminal end of the sixth Ig domain of PECAM-1 (NHASSVPRSKILTVRVILAPWKK) that mediates heterotypic adhesion (Zehnder *et al.*, 1995) forms part of the predicted G β strand. Structure-function analysis of the first Ig domain which is necessary for homotypic PECAM-1 interaction has identified five functional residues located on both faces of the Ig fold that were potentially involved in homotypic adhesion (Asp-11 and Asp-33 on the predicted A and B strands, Lys-89 within the F-G loop, and Lys-50 and Asp-51 on the C-D loop) (Newton *et al.*, 1997).

Like modular domains on other proteins, Ig domains of the Ig superfamily receptors appear to have evolved to utilize a conserved structural motif with variable regions of amino acid sequence modified to interact with different ligands. In this respect, it is interesting to examine some of the ligands for the Ig domain-containing proteins. The α_L and α_M subunits of LFA-1 and Mac-1, respectively, both contain a conserved 200-amino acid insert domain (I or A domain) in the extracellular region, and this domain is directly involved in binding of ICAM-1 (Randi and Hogg, 1994; Huang and Springer, 1995; Kamata *et al.*, 1995). Studies mapping epitopes of inhibitory monoclonal antibodies against the α_L I domain provided evidence that distinct sequences within the domain were involved in binding ICAM-3 as opposed to ICAM-1 (Binnerts *et al.*, 1996). The α_4 subunit of VLA-4, on the other hand, does not contain an I domain, so its Ig domain-containing ligand, VCAM-1, must bind to other structures on the $\alpha_4\beta_1$ integrin. Similarly, there is no I domain in the α_v subunit of $\alpha_v\beta_3$ that functions as a PECAM-1 ligand (Piali *et al.*, 1995). The I domain structure is widely distributed among other vascular proteins, and constitutes a ligand-binding site in both the α_2 subunit of the $\alpha_2\beta_1$ integrin and vWF. The I domain in $\alpha_2\beta_1$ binds collagen type I and the A1 domain of vWF binds the platelet membrane receptor, GP Ib-IX-V (Table 12.2) (Kamata *et al.*, 1994; Andrews *et al.*, 1997). One of the regions within the vWF A1 domain identified as binding to GP Ib-IX-V (Asp-514-Glu-542) is an α-helical-loop structure that corresponds to an overlapping region of the LFA-1 α_L subunit that mediates binding to Ig superfamily ligands (Champe *et al.*, 1995; Andrews *et al.*, 1997). This suggests there is a common ligand-recognition region within the I domain structure. Interactions between I domains and Ig domains may be relatively selective, or be promiscuous in that receptors cross-react with more than one counter-receptor. For example, ICAM-1 binds both of the β_2 integrins, LFA-1 and Mac-1, on leukocytes, while LFA-1 interacts with either ICAM-1, ICAM-2 or ICAM-3. Finally, other factors such as glycosylation, phosphorylation or interactions of the cytoplasmic tail of the Ig superfamily receptors, in addition to the primary sequence of Ig domains themselves, clearly play a part in determining ligand-binding specificity (Newman *et al.*, 1992; Zehnder *et al.*, 1992; Yan *et al.*, 1995a).

CONCLUSIONS AND FUTURE DIRECTIONS

PECAM-1 is an adhesive receptor of the Ig superfamily that maintains the integrity of the vascular endothelial cell monolayer at cell-cell junctions, and plays a crucial role in leukocyte migration from the circulation through the endothelium. Recent studies show that PECAM-1 can function as a Ca^{2+}-independent homotypic receptor or a Ca^{2+}-dependent heterotypic receptor for glycosaminoglycans; the nature of PECAM-1 adhesion can apparently be regulated by its cytoplasmic domain. In platelets, phosphorylation of the cytoplasmic domain increases the cytoskeletal association of PECAM-1, suggesting a role for this protein in regulating the cytoskeletal rearrangements involved in cell shape change and motility. Some interesting questions on the structure-function relationships of PECAM-1 remain to be answered. What are the differences between PECAM-1 and the other Ig superfamily adhesive receptors, ICAM-1 and VCAM-1, that confer specificity towards different ligands? How is the ligand-binding phenotype of PECAM-1 regulated in different vascular cells? What is the function of platelet PECAM-1? Defining the precise physiological roles for PECAM-1 on platelets, endothelial cells and leukocytes has the potential for selective therapeutic intervention of inflammatory, and perhaps thrombotic disease.

References

Albelda, S.M., Muller, W.A., Buck, C.A. and Newman, P.J. (1991). Molecular and cellular properties of PECAM-1 (endoCAM/CD31): a novel vascular cell-cell adhesion molecule. *The Journal of Cell Biology*, **114**:1059–1068.

Alon, R., Kassner, P.D., Woldemar Carr, M., Finger, E.B., Hemler, M.E. and Springer, T.A. (1995). The integrin VLA-4 supports tethering and rolling in flow on VCAM-1. *The Journal of Cell Biology*, **128**:1243–1253.

Andrews, R.K., López, J.A. and Berndt, M.C. (1997). Molecular mechanisms of platelet adhesion and activation. *International Journal of Biochemistry and Cell Biology*, **29**:91–105.

Asazuma, N., Yatomi, Y., Ozaki, Y., Qi, R., Kuroda, K., Satoh, K. and Kume, S. (1996). Protein-tyrosine phosphorylation and p72^{syk} activation in human platelets stimulated with collagen is dependent upon glycoprotein Ia/IIa and actin polymerization. *Thrombosis and Haemostasis*, **75**:648–654.

Ashman, L.K., Aylett, G.W., Cambareri, A.C. and Cole, S.R. (1991). Different eitopes of the CD31 antigen identified by monoclonal antibodies: cell type-specific patterns of expression. *Tissue Antigens*, **38**:199–207.

Ayalon, O., Sabanai, H., Lampugnani, M.-G., Dejana, E. and Geiger, B. (1994). Spatial and temporal relationships between cadherins and PECAM-1 in cell-cell junctions of human endothelial cells. *The Journal of Cell Biology*, **126**:247–258.

Baldwin, H.S., Shen, H.M., Yan, H.-C., DeLisser, H.M., Chung, A., Mickanin, C., Trask, T., Kirschbaum, N.E., Newman, P.J., Albelda, S.M. and Buck, C.A. (1994). Platelet endothelial cell adhesion molecule-1 (PECAM-1/CD31): alternatively spliced, functionally distinct isoforms expressed during mammalian cardiovascular development. *Development*, **120**:2539–2553.

Barleon, B., Sozzani, S., Zhou, D., Weich, H.A., Mantovani, A. and Marmé, D. (1996). Migration of human monocytes in response to vascular endothelial growth factor (VEGF) is mediated via the VEGF receptor flt-1. *Blood*, **87**:3336–3343.

Beggs, H.E., Soriano, P. and Maness, P.F. (1994). NCAM-dependent neurite outgrowth is inhibited in neurons from *Fyn*-minus mice. *The Journal of Cell Biology*, **127**:825–833.

Berman, M.E. and Muller, W.A. (1995). Ligation of platelet/endothelial cell adhesion molecule-1 (PECAM-1/CD31) on monocytes and neutrophils increases binding capacity of leukocyte CR3 (CD11b/CD18). *Journal of Immunology*, **154**:299–307.

Berman, M.E., Xie, Y. and Muller, W.A. (1996). Roles of platelet/endothelial cell adhesion molecule-1 (PECAM-1, CD31) in natural killer cell transendothelial migration and β_2 integrin activation. *Journal of Immunology*, **156**:1515–1524.

Binnerts, M.E., van Kooyk, Y., Edwards, C.P., Champe, M., Presta, L., Bodary, S.C., Figdor, C.G. and Berman, P.W. (1996). Antibodies that selectively inhibit leukocyte function-associated antigen 1 binding intercellular adhesion molecule-3 recognize a unique epitope within the CD11a I domain. *The Journal of Biological Chemistry*, **271**:9962–9968.

Bogen, S.A., Baldwin, H.S., Watkins, S.C., Albelda, S.M. and Abbas, A.K. (1992). Association of murine CD31 with transmigrating lymphocytes following antigenic stimulation. *American Journal of Pathology*, **141**:843–854.

Bogen, S.A., Pak, J., Garifallou, M., Deng, X. and Muller, W.A. (1994). Monoclonal antibody to murine PECAM-1 (CD31) blocks acute inflammation *in vivo*. *The Journal of Experimental Medicine*, **179**:1059–1064.

Bradley, J.R., Thiru, S. and Pober, J.S. (1995). Hydrogen peroxide-induced endothelial retraction is accompanied by a loss of the normal spatial organization of endothelial cell adhesion molecules. *American Journal of Pathology*, **147**:627–641.

Bull, H.A., Brickell, P.M. and Dowd, P.M. (1994). *src*-related protein tyrosine kinases are physically associated with the surface antigen CD36 in human dermal microvascular endothelial cells. *FEBS Letters*, **351**:41–44.

Carpén, O., Pallai, P., Staunton, D.E. and Springer, T.A. (1992). Association of intercellular adhesion molecule-1 (ICAM-1) with actin-containing cytoskeleton and α-actinin. *The Journal of Cell Biology*, **118**:1223–1234.

Champe, M., McIntyre, B.W. and Berman, P.W. (1995). Monoclonal antibodies that block the inserted domain of CD11a. *The Journal of Biological Chemistry*, **270**:1388–1394.

Chen, W., Knapp, W., Majdic, C., Stockinger, H., Bohmig, G.A. and Zlabinger, G.J. (1994). Co-ligation of CD31 and FcγRII induces cytokine production in human monocytes. *Journal of Immunology*, **152**:3991–3997.

Clark, E.A. and Brugge, J.S. (1995). Integrin and signal transduction pathways: the road taken. *Science*, **268**:233–239.

Clemetson, K.J. (1995). Platelet activation: signal transduction via membrane receptors. *Thrombosis and Haemostasis*, **74**:111–116.

Collins, T., Read, M.A., Neish, A.S., Whitley, M.Z., Thanos, D. and Maniatis, T. (1995). Transcriptional regulation of endothelial cell adhesion molecules: NF-κB and cytokine-inducible enhancers. *FASEB Journal*, **9**:899–909.

Cooper, D., Lindberg, F.P., Gamble, J.R., Brown, E.J. and Vadas, M.A. (1995). Transendothial migration of neutrophils involves integrin-associated protein (CD47). *Proceedings of the National Academy of Sciences, USA*, **92**:3978–3982.

Cramer, E.M., Berger, G. and Berndt, M.C. (1994). Platelet α-granule membrane shares two new components: CD9 and PECAM-1. *Blood*, **84**:1722–1730.

DeLisser, H.M., Yan, H.C., Newman, P.J., Muller, W.A., Buck, C.A. and Albelda, S.M. (1993). Platelet/endothelial cell adhesion molecule-1 (CD31)-mediated cellular aggregation involves cell surface glycosaminoglycans. *The Journal of Biological Chemistry*, **268**:16037–16046.

DeLisser, H.M., Chilkotowsky, J., Yan, H.-C., Daise, M.L., Buck, C.A. and Albelda, S.M. (1994). Deletions in the cytoplasmic domain of platelet-endothelial cell adhesion molecule-1 (PECAM-1, CD31) result in changes in ligand binding properties. *The Journal of Cell Biology*, **124**:195–203.

DeLisser, H.M., Christofidou-Solomidou, M., Strieter, R.M., Burdick, M.D., Robinson, C.S., Wexler, R.S., Kerr, J.S., Garlanda, C., Merwin, J.R., Madri, J.A. and Albelda, S.M. (1997). Involvement of endothelial PECAM-1/CD31 in angiogenesis. *American Journal of Pathology*, **151**:671–677.

De Luca, M., Dunlop, L.C., Andrews, R.K., Flannery J.V., Jr, Ettling, R., Cumming, D.A., Veldman, G.M., and Berndt, M.C. (1995). A novel cobra venom metalloproteinase, mocarhagin, cleaves a 10-amino acid peptide from the mature N terminus of P-selectin glycoprotein ligand receptor, PSGL-1, and abolishes P-selectin binding. *The Journal of Biological Chemistry*, **270**:26734–26737.

Diacovo, T.G., deFougerolles, A.R., Bainton, D.F. and Springer, T.A. (1994). A functional integrin ligand on the surface of platelets: intercellular adhesion molecule-2. *Journal of Clinical Investigation*, **94**:1243–1251.

Doody, G.M., Justement, L.B., Delibrias, C.C., Matthews, R.J., Lin, J., Thomas, M.L. and Fearon, D.T. (1995). A role in B cell activation for CD22 and the protein tyrosine phosphatase SHP. *Science*, **269**:242–244.

Du, X., Harris, S.J., Tetaz, T.J., Ginsberg, M.H., and Berndt, M.C. (1995). Association of a phospholipase A_2 (14.3.3 protein) with the platelet glycoprotein Ib-IX complex. *The Journal of Biological Chemistry*, **269**: 18287–18290.

Dunlop, L.C., Skinner, M.P., Bendall, L.J., Favoloro, E.J., Castaldi, P.A., Gorman, J.J., Gamble, J.R., Vadas, M.A. and Berndt, M.C. (1992). Characterization of GMP-140 (P-selectin) as a circulating plasma protein. *The Journal of Experimental Medicine*, **175**:1147–1150.

Famiglietti, J., Sun, J., DeLisser, H.M. and Albelda, S.M. (1997). Tyrosine residue in exon 14 of the cytoplasmic domain of platelet endothelial cell adhesion molecule-1 (PECAM-1/CD31) regulates ligand binding specificity. *The Journal of Cell Biology*, **138**:1425–1435.

Fox, J.E.B. (1993). The platelet cytoskeleton. *Thrombosis and Haemostasis*, **70**:884–893.

Frenette, P.S., Johnson, R.C., Hynes, R.O. and Wagner, D.D. (1995). Platelets roll on stimulated endothelium *in vivo*: an interaction mediated by endothelial P-selectin. *Proceedings of the National Academy of Science, USA*, **92**:7450–7454.

Frenette, P.S., Moyna, C., Hartwell, D.W., Lowe, J.B., Hynes, R.O. and Wagner, D.D. (1998). Platelet-endothelial interactions in inflamed mesenteric venules. *Blood*, **91**:1318–1324.

Furie, B. and Furie, B.C. (1995). The molecular basis of platelet and endothelial cell interaction with neutrophils and monocytes: role of P-selectin and the P-selectin ligand, PSGL-1. *Thrombosis and Haemostasis*, **74**:224–227.

Gamble, J.R., Skinner, M.P., Berndt, M.C. and Vadas, M.A. (1990). Prevention of activated neutrophil adhesion to activated endothelium mediated by GMP-140. *Science*, **249**:414–417.

Goldberger, A., Middleton, K.A., Oliver, J.A., Paddock, C., Yan, H.-C., DeLisser, H.M., Albelda, S.M. and Newman, P.J. (1994). Biosynthesis and processing of the cell adhesion molecule PECAM-1 includes production of a soluble form. *The Journal of Biological Chemistry*, **269**:17183–17191.

Gurubhagavatula, I., Amrani, Y., Pratico, D., Ruberg, F.L., Albelda, S.M. and Panettieri R.A., Jr (1997). Engagement of human PECAM-1 (CD31) on human endothelial cells increases intracellular calcium ion concentration and stimulates prostacyclin release. *Journal of Clinical Investigation*, **101**:212–222.

Holness, C.L., Bates, P.A., Littler, A.J., Buckley, C.D., McDowall, A., Bossy, D., Hogg, N. and Simmons, D.L. (1995). Analysis of the binding site on intercellular adhesion molecule 3 for the leukocyte integrin lymphocyte function-associated antigen 1. *The Journal of Biological Chemistry*, **270**:877–884.

Hoyt, C.H. and Lerea, K.M. (1995). Aggregation-dependent signalling in human platelets is sensitive to protein serine/threonine phosphatase inhibitors. *Biochemistry*, **34**:9565–9570.

Huang C. and Springer T.A. (1995). A binding interface on the I domain of lymphocyte function-associated antigen-1 (LFA-1) is required for specific interaction with intercellular adhesion molecule 1 (ICAM-1). *The Journal of Biological Chemistry*, **270**:19008–19016.

Huang, M.-M., Bolen, J.B., Barnwell, J.W., Shattil, S.S. and Brugge, J.S. (1991). Membrane glycoprotein IV (CD36) is physically associated with the Fyn, Lyn,and Yes protein-tyrosine kinases in human platelets. *Proceedings of the National Academy of Sciences, USA*, **88**:7844–7848.

Hunkapiller, T. and Hood, L. (1986). The growing immunoglobulin gene superfamily. *Nature*, **323**:15–16.

Ichinohe, T., Takayama, H., Ezumi, Y., Yanagi, S., Yamamura, H. and Okuma, M. (1995). Cyclic AMP-insensitive activation of csrc and syk protein-tyrosine kinases through platelet membrane glycoprotein VI. *The Journal of Biological Chemistry*, **270**:28129–28036.

Ishiwata, N., Takio, K., Katayama, M., Watanabe, K., Titani, K., Ikeda, Y. and Handa, M. (1994). Alternatively spliced isoform of P-selectin is present *in vivo* as a soluble molecule. *The Journal of Biological Chemistry*, **269**:23708–23715.

Jackson, D.E., Loo, R.O., Holyst, M.T. and Newman, P.J. (1997a). Identification and characterization of functional cation coordination sites in platelet endothelial cell adhesion molecule-1. *Biochemistry*, **36**:9395–9404.

Jackson, D.E., Ward, C.M., Wang, R. and Newman, P.J. (1997b). The protein-tyrosine phosphatase shp-2 binds platelet/endothelial cell adhesion molecule-1 (PECAM-1) and forms a distinct signaling complex during platelet aggregation. Evidence for a mechanistic link between PECAM-1 and integrin-mediated cellular signaling. *The Journal of Biological Chemistry*, **272**:6986–6993.

Jaffe, S.H., Friedlander, D.R., Matsuzaki, F., Crossin, K.L., Cunningham, B.A. and Edelman, G.M. (1990). Differential effects of the cytoplasmic domains of cell adhesion molecules on cell aggregation and sorting-out. *Proceedings of the National Academy of Sciences, USA*, **87**:3589–3593.

Jones, E.Y., Harlos, K., Bottomley, M.J., Robinson, R.C., Driscoll, P.C., Edwards, R.M., Clements, J.M., Dudgeon, T.J. and Stuart, D.I. (1995). Crystal structure of an integrin-binding fragment of vascular cell adhesion molecule-1 at 1.8 Å resolution. *Nature*, **373**:539–544.

Juan, M., Viñas, O., Pino-Otín, M.R., Places, L., Martínez-Cáceres, E., Barceló, J.J., Miralles, A., Vilella, R., de la Fuente, M.A., Vives, J., Yargüe, A. and Gayà, A. (1994). CD50 (intercellular adhesion molecule 3) stimulation induces calcium mobilization and tyrosine phosphorylation through p59^{fyn} and p56^{lck} in Jurkat T cell line. *The Journal of Experimental Medicine*, **179**:1747–1756.

Juliano, R.L. and Haskill, S. (1993). Signal transduction from the extracellular matrix. *The Journal of Cell Biology*, **120**:577–585.

Kahn, J., Ingraham, R.H., Shirley, F., Migaki, G.I. and Kishimoto, T.K. (1994). Membrane proximal cleavage of L-selectin: identification of the cleavage site and a 6-kD transmembrane peptide fragment of L-selectin. *The Journal of Cell Biology*, **125**:461–470.

Kamata, T., Puzon, W. and Takada, Y. (1994). Identification of putative ligand binding sites within I domain of integrin $\alpha_2\beta_1$ (VLA-2, CD49b/CD29). *The Journal of Biological Chemistry*, **269**:9659–9663.

Kamata, T., Wright, R. and Takada, Y. (1995). Critical threonine and aspartic acid residues within the I domains of β_2 integrins for interactions with intercellular adhesion molecule 1 (ICAM-1) and C3bi. *The Journal of Biological Chemistry*, **270**:12531–12535.

Kansas, G.S., Ley, K., Munro, J.M. and Tedder, T.F. (1993). Regulation of leukocyte rolling and adhesion to high endothelial venules through the cytoplasmic domain of L-selectin. *The Journal of Experimental Medicine*, **177**:833–888.

Kansas, G.S., Saunders, K.B., Ley, K., Zakrzewicz, A., Gibson, R.M., Furie, B.C., Furie, B. and Tedder, T.F. (1994). A role for the epidermal growth factor-like domain of P-selectin in ligand recognition and cell adhesion. *The Journal of Cell Biology*, **124**:609–618.

Kilger, G., Needham, L.A., Nielsen, P.J., Clements, J., Vestweber, D. and Holzmann, B. (1995). Differential regulation of α_4 integrin-dependent binding to domains 1 and 4 of vascular cell adhesion molecule-1. *The Journal of Biological Chemistry*, **270**:5979–5984.

Kirschbaum, N.E., Gumina, R.J. and Newman, P.J. (1994). Organization of the gene for human platelet/endothelial cell adhesion molecule-1 shows alternatively spliced isoforms and a functionally complex cytoplasmic domain. *Blood*, **84**:4028–4037.

Kuijper, P.H.M., Torres, G., van der Linden, J.A.M., Lammers, J.-W.J., Sixma, J.J., Koenderman, L. and Zwaginga, J.J. (1996). Platelet-dependent primary hemostasis promotes selectin- and integrin-mediated neutrophil adhesion to damaged endothelium under flow conditions. *Blood*, **87**:3271–3281.

Kunkel, E.J., Jung, U., Bullard, D.C., Norman, K.E., Wolitzky, B.A., Vestweber, D., Beaudet, A.L. and Ley, K. (1996). Absence of trauma-induced leukocyte rolling in mice deficient in both P-selectin and intercellular adhesion molecule 1. *The Journal of Experimental Medicine*, **183**:57–65.

Lastres, P., Almendro, N., Bellon, T., Lopez-Guerrero, J.A., Eritja, R. and Bernabeu, C. (1994). Functional regulation of platelet/endothelial cell adhesion molecule-1 by TGFβ_1 in promonocytic U-937 cells. *Journal of Immunology*, **153**:4206–4218.

Leach, L., Eaton, B.M., Westcott, E.D.A. and Firth, J.A. (1995). Effect of histamine on endothelial permeability and structure and adhesion molecules of the paracellular junctions of perfused human placental microvessels. *Microvascular Research*, **50**:323–337.

Leavesley, D.I., Oliver, J.M., Swart, B.W., Berndt, M.C., Haylock, D.N. and Simmons, P.J. (1994). Signals from platelet/endothelial cell adhesion molecule enhance the adhesive activity of the very late antigen-4 integrin of human CD34$^+$ hemopoietic progenitor cells. *Journal of Immunology*, **153**:4673–4683.

Ledebur, H.C., Parks, T.P. (1995). Transcriptional regulation of the intercellular adhesion molecule-1 gene by inflammatory cytokines in human endothelial cells. *The Journal of Biological Chemistry*, **270**:933–943.

Liao, F., Huynh, H.K., Eiroa, A., Greene, T., Polizzi, E. and Muller, W.A. (1995). Migration of monocytes across endothelium and passage through extracellular matrix involve separate molecular domains of PECAM-1. *The Journal of Experimental Medicine*, **182**:1337–1342.

Lipfert, L., Haimovich, B., Schaller, M.D., Cobb, B.S., Parsons, J.T. and Brugge, J.S. (1992). Integrin-dependent phosphorylation and activation of the protein kinase pp125^{fak} in platelets. *The Journal of Cell Biology*, **119**:905–912.

Lu, T.T., Yan, L.G. and Madri, J.A. (1996). Integrin engagement mediates tyrosine dephosphorylation on platelet-endothelial cell adhesion molecule 1. *Proceedings of the National Academy of Science, USA*, **93**:11808–11813.

Lu, T.T., Barreuther, M., Davis, S. and Madri, J.A. (1997). Platelet endothelial cell adhesion molecule-1 is phosphorylatable by c-src, binds src-src homology 2 domain, and exhibits immunoreceptor tryosine-based activation motif-like properties. *The Journal of Biological Chemistry*, **272**:14442–14446.

McEver, R.P., Moore, K.L. and Cummings, R.D. (1995). Leukocyte trafficking mediated by selectin-carbohydrate interactions. *The Journal of Biological Chemistry*, **270**:11025–11028.

Marcus, A.J. (1994). Cellular interactions of platelets in thrombosis. In *Thrombosis and Hemorrhage*, edited by J. Loscalzo and A.I. Schafer, pp. 279–289. Boston: Blackwell Scientific.

Masuda, M., Osawa, M., Shigematsu, H., Harada, N. and Fujiwara, K. (1997). Platelet endothelial cell adhesion molecule-1 is a major SH-PTP2 binding protein in vascular endothelial cells. *FEBS Letters*, **408**:331–336.

Mayadas, T.N., Johnson, R.C., Rayburn, H., Hynes, R.O. and Wagner, D.D. (1993). Leukocyte rolling and extravasation are severely compromised in P-selectin-deficient mice. *Cell*, **74**:541–554.

Metzelaar, M.J., Korteweg, J., Sixma, J.J. and Nieuwenhuis, H.K. (1991). Biochemical characterization of PECAM-1 (CD31 antigen) on human platelets. *Thrombosis and Haemostasis*, **66**:700–707.

Moroi, M., Jung, S.M., Shinmyozu, K., Tomiyama, Y., Ordinas, A. and Diaz-Ricart, M. Analysis of platelet adhesion to a collagen-coated surface under flow conditions: the involvement of glycoprotein VI in the platelet adhesion. *Blood*, **88**:2081–2092.

Muller, W.A., Berman, M.E., Newman, P.J., DeLisser, H.M. and Albelda, S.M. (1992). A heterophilic adhesion mechanism for platelet/endothelial cell adhesion molecule 1 (CD31). *The Journal of Experimental Medicine*, **175**:1401–1404.

Muller, W.A., Weigl, S.A., Deng, X. and Phillips, D.M. (1993). PECAM-1 is required for transendothelial migration of leukocytes. *The Journal of Experimental Medicine*, **178**:449–460.

Mustonen, T. and Alitalo, K. (1995). Endothelial receptor tyrosine kinases involved in angiogenesis. *The Journal of Cell Biology*, **129**:895–898.

Nagafuchi, A. and Takeichi, M. (1988). Cell binding function of E-cadherin is regulated by the cytoplasmic domain. *The EMBO Journal*, **7**:3679–3684.

Newham, P., Craig, S.E., Seddon, G.N., Schofield, N.R., Rees, A., Edwards, R.M., Jones, E.Y. and Humphries, M.J. (1997). α_4 integrin binding interfaces on VCAM-1 and MadCAM-1. Integrin binding footprints identify accessory binding sites that play a role in integrin specificity. *The Journal of Biological Chemistry*, **272**:19429–19440.

Newman, P.J., Berndt, M.C., Gorski, J., White II, G.C., Lyman, S., Paddock, C. and Muller, W.A. (1990). PECAM-1 (CD31) cloning and relation to adhesion molecules of the immunoglobulin gene superfamily. *Science*, **247**:1219–1222.

Newman, P.J., Hillery, C.A., Albrecht, R., Parise, L.V., Berndt, M.C., Mazurov, A.V., Dunlop, L.C., Zhang, J. and Rittenhouse, S.E. (1992). Activation-dependent changes in human platelet PECAM-1: phosphorylation, cytoskeletal association, and surface membrane redistribution. *The Journal of Cell Biology*, **119**:239–246.

Newton, J.P., Buckley, C.D., Jones, E.Y. and Simmons, D.L. (1997). Residues on both faces of the first immunoglobulin fold contribute to homophilic binding sites of PECAM-1/CD31. *The Journal of Biological Chemistry*, **272**:20555–20563.

Osborn, L., Vassallo, C. and Benjamin, C.D. (1992). Activated endothelium binds lymphocytes through a novel binding site in the alternatively spliced domain of vascular cell adhesion molecule-1. *The Journal of Experimental Medicine*, **176**:99–107.

Osborn, L., Vassallo, C., Griffiths Browning, B., Tizard, R., Haskard, D.O., Benjamin, C.D., Dougas, I. and Kirchhausen, T. (1994). Arrangement of domains, and amino acid residues required for binding of vascular cell adhesion molecule-1 to its counter-receptor VLA-4 ($\alpha_4\beta_1$). *The Journal of Cell Biology*, **124**:601–608.

Otey, C.A., Vasquez, G.B., Burridge, K., Erickson, B.W. (1993). Mapping of the α-actinin binding site within the β_1 integrin cytoplasmic face. *The Journal of Biological Chemistry*, **268**:21193–21197.

Petruzzelli, L., Takami, M. and Herrera, R. (1996). Adhesion through the interaction of lymphocyte function-associated antigen-1 with intracellular adhesion molecule-1 induces tyrosine phosphorylation of p130cas and its association with c-CrkII. *The Journal of Biological Chemistry*, **271**:7796–7801.

Piali, L., Albelda, S.M., Baldwin, H.S., Hammel, P., Gisler, R.H., Imhof, B.A. (1993). Murine platelet endothelial cell adhesion molecule (PECAM-1/CD31) modulates β_2 integrins on lymphokine-activated killer cells. *European Journal of Immunology*, **23**:2464–2471.

Piali, L., Hammel, P., Uherek, C., Bachmann, F., Gisler, R.H., Dunon, D. and Imhof, B.A. (1995). CD31/PECAM-1 is a ligand for $\alpha_v\beta_3$ integrin involved in adhesion of leukocytes to endothelium. *The Journal of Cell Biology*, **130**:451–460.

Pinter, E., Barreuther, M., Lu, T., Imhof, B.A. and Madri, J.A. (1997). Platelet-endothelial cell adhesion molecule-1 (PECAM-1/CD31) tyrosine phosphorylation state changes during vasculogenesis in the murine conceptus. *American Journal of Pathology*, **150**:1523–1530.

Pouyani, T. and Seed, B. (1995). PSGL-1 recognition of P-selectin is controlled by a tyrosine sulfation consensus at the PSGL-1 amino terminus. *Cell*, **83**:333–343.

Preece, G., Murphy, G. and Ager, A. (1996). Metalloproteinase-mediated regulation of L-selectin levels on leukocytes. *The Journal of Biological Chemistry*, **271**:11634–11640.

Randi, A.M., Hogg, N. (1994). I domain of β_2 integrin lymphocyte function-associated antigen-1 contains a binding site for ligand intercellular adhesion molecule-1. *The Journal of Biological Chemistry*, **269**:12395–12398.

Rao, Y., Wu, X.-F., Gariepy, J., Rutishauser, U. and Siu, C.-H. (1992). Identification of a peptide sequence involved in homophilic binding in the neural cell adhesion molecule NCAM. *The Journal of Cell Biology*, **118**:937–949.

Renz, M.E., Chiu, H.H., Jones, S., Fox, J., Kim, K.J., Presta, L.G. and Fong, S. (1994). Structural requirements for adhesion of soluble recombinant murine vascular cell adhesion molecule-1 to $\alpha_4\beta_1$. *The Journal of Cell Biology*, **125**:1395–1406.

Romer, L.H., McLean, N.V., Yan, H.C., Daise, M., Sun, J. and DeLisser, H.M. (1995). IFN-γ and TNF-α induce redistribution of PECAM-1 (CD31) on human endothelial cells. *Journal of Immunology*, **154**:6582–6592.

Ross, R. (1995). Cell biology of atherosclerosis. *Annual Review of Physiology*, **57**:791–804.

Rothlein, R., Mainolfi, E.A., Czajkowski, M. and Marlin, S.D. (1991). A form of circulating ICAM-1 in human serum. *The Journal of Immunology*, **147**:3788–3793.

Ruco, L.P., Pomponi, D., Pigott, R., Gearing, A.J.H., Baiocchini, A. and Baroni, C.D. (1992). Expression and cell distribution of the intercellular adhesion molecule, vascular cell adhesion molecule, endothelial leukocyte adhesion molecule, and endothelial cell adhesion molecule (CD31) in reactive human lymph nodes and in Hodgkin's disease. *American Journal of Pathology*, **140**:1337–1344.

Sako, D., Chang, X.J., Barone, K.M., Vachino, G., White, H.M., Shaw, G.D., Veldman, G.M., Bean, K.M., Ahern, T.J., Furie, B., Cumming, D.A. and Larsen, G.R. (1993). Expression cloning of a functional glyco-protein ligand for P-selectin. *Cell*, **75**:1179–1186.

Sako, D., Comess, K.M., Barone, K.M., Camphausen, R.T., Cumming, D.A. and Shaw, G.D. (1995). A sulfated peptide segment at the amino terminus of PSGL-1 is critical for P-selectin binding. *Cell*, **83**:323–331.

Shattil, S.J. and Brugge, J.S. (1991). Protein tyrosine phosphorylation and the adhesive functions of platelets. *Current Opinion in Cell Biology*, **3**:869–879.

Simmons, D.L., Walker, C., Power, C. and Pigott, R. (1990). Molecular cloning of CD31, a putative intercellular adhesion molecule closely related to carcinoembryonic antigen. *The Journal of Experimental Medicine*, **171**:2147–2152.

Sitrir, R.G., Todd III, R.F., Petty, H.R., Brock, T.G., Shollenberger, S.B., Albrecht, E. and Gyetko, M.R. (1996). The urokinase receptor (CD87) facilitates CD11b/CD18-mediated adhesion of human monocytes. *Journal of Clinical Investigation*, **97**:1942–1951.

Skonier, J.E., Bowen, M.A., Emswiler, J., Aruffo, A. and Bajorath, J. (1996). Recognition of diverse proteins by members of the immunoglobulin superfamily: delineation of the receptor binding site in the human CD6 lig-and ALCAM. *Biochemistry*, **35**:12287–12291.

Springer, T.A. (1995). Traffic signals on endothelium for lymphocyte recirculation and leukocyte emigration. *Annual Review of Physiology*, **57**:827–872.

Staatz, W.D., Rajpara, S.M., Wayner, E.A., Carter, W.G. and Santoro, S.A. (1989). The membrane glycoprotein Ia-IIa (VLA-2) complex mediates the Mg^{++}-dependent adhesion of platelets to collagen. *The Journal of Cell Biology*, **108**:1917–1924.

Stewart, R.J., Kashxur, T.S. and Marsden, P.A. (1996). Vascular endothelial platelet endothelial adhesion molecule-1 (PECAM-1) expression is decreased by $TNF\alpha$ and $IFN\gamma$. Evidence for cytokine-induced desta-bilization of messenger ribonucleic acid transcripts in bovine endothelial cells. *Journal of Immunology*, **156**:1221–1228.

Stockinger, H., Gadd, S.J., Eher, R., Majdic, O., Schreiber, W., Kasinrerk, W., Strass, B., Schnabl, E. and Knapp, W. (1990). Molecular characterization and functional analysis of the leukocyte surface protein CD31. *The Journal of Immunology*, **145**:3889–3897.

Sun, J., Williams, J., Yan, H.-C., Amin, K.M., Albelda, S.M. and DeLisser, H.M. (1996a). Platelet endothelial cell adhesion molecule-1 (PECAM-1) homophilic adhesion is mediated by immunoglobulin-like domains 1 and 2 and depends on the cytoplasmic domain and the level of surface expression. *The Journal of Biological Chemistry*, **271**:18561–18570.

Sun, Q.-H., DeLisser, H.M., Zukowski, M.M., Paddock, C., Albelda, S.M. and Newman, P.J. (1996b). Individu-ally distinct Ig domains in PECAM-1 regulate homophilic binding and modulate receptor affinity. *The Journal of Biological Chemistry*, **271**:11090–11098.

Tanaka, Y., Albelda, S.M., Horgan, K.T., van Seventer, Y., Shimizu, Y., Newman, W., Hallam, J., Newman, P.J., Buck, C.A. and Shaw, S. (1992). CD31 expressed on distinctive T cell subsets is a preferential amplifier of β_1 integrin-mediated adhesion. *The Journal of Experimental Medicine*, **176**:245–253.

Tandon, N.N., Kralisz, U. and Jamieson, G.A. (1989). Identification of glycoprotein IV (CD36) as a primary receptor for platelet collagen adhesion. *The Journal of Biological Chemistry*, **264**:7576–7583.

Tang, D.G., Chen, Y.Q., Newman, P.J., Shi, L., Gao, X., Diglio, C.A. and Honn, K.V. (1993). Identification of PECAM-1 in solid tumor cells and its potential involvement in tumor cell adhesion to endothelium. *The Journal of Biological Chemistry*, **268**:22883–22894.

Thomas, K.A. (1996). Vascular endothelial growth factor, a potent and selective angiogenic agent. *The Journal of Biological Chemistry*, **271**:603–606.

Treutiger, C.J., Heddini, A., Fernandez, V., Muller, W.A. and Wahlgren, M. (1997). PECAM-1/CD31, an endothelial receptor for binding *Plasmodium falciparum*-infected erythrocytes. *Nature Medicine*, **3**:1405–1408.

Valles, J., Santos, M.T., Marcus, A.J., Safier, L.B., Broekman, M.J., Islam, N., Ullman, H.L. and Aznar, J. (1993). Downregulation of human platelet reactivity by neutrophils. Participation of lipoxygenase derivatives and adhesive proteins. *Journal of Clinical Investigation*, **92**:1357–1365.

van der Merwe, P.A., Crocker, P.R., Vinson, M., Barclay, A.N., Schauer, R. and Kelm, S. (1996). Localization of the putative sialic acid-binding site on the immunoglobulin superfamily cell-surface molecule CD22. *The Journal of Biological Chemistry*, **271**:9273–9280.

Vaporciyan, A.A., DeLisser, H.M., Yan, H.-C., Mendiguren, I.I., Thom, S.R., Jones, M.L., Ward, P.A. and Albelda, S.M. (1994). Involvement of platelet-endothelial cell adhesion molecule-1 in neutrophil recruit-ment *in vivo*. *Science*, **262**:1580–1582.

Varki, A. (1994). Selectin Ligands. *Proceedings of the National Academy of Science, USA*, **91**:7390–7394.

Varon, D., Jackson, D.E., Shenkman, B., Dardik, R., Tamarin, I., Savion, N. and Newman, P.J. (1998). Platelet/endothelial cell adhesion molecule-1 serves as a costimulatory agonist receptor that modulates integrin-dependent adhesion and aggregation of human platelets. *Blood*, **91**:500–507.

Vonderheide, R.H., Tedder, T.F., Springer, T.A. and Staunton, D.E. (1994). Residues within a conserved amino acid motif of domains 1 and 4 of VCAM-1 are required for binding to VLA-4. *The Journal of Cell Biology*, **125**:215–222.

Watt, S.M., Williamson, J., Genevier, H., Fawcett, J., Simmons, D.L., Hatzfeld, A., Nesbitt, S.A. and Coombe, D.R. (1993). The heparin binding PECAM-1 adhesion molecule is expressed by CD34⁺ hematopoietic precursor cells with early myeloid and B-lymphoid cell phenotypes. *Blood*, **82**:2649–2663.

Weiss, H.J. (1995). Flow-related platelet deposition on subendothelium. *Thrombosis and Haemostasis*, **74**:117–122.

Wilson, G.L., Fox, C.H., Fauci, A.S. and Kehrl, J.H. (1991). cDNA cloning of the B cell membrane protein CD22: a mediator of B-B cell interactions. *The Journal of Experimental Medicine*, **173**:137–146.

Xiao, J., Messinger, Y., Jin, J., Myers, D.E., Bolen, J.B. and Uckun, F.M. (1996). Signal transduction through the β_1 integrin family surface adhesion molecules VLA-4 and VLA-5 of human B-cell precursors activates CD19 receptor-associated protein-tyrosine kinases. *The Journal of Biological Chemistry*, **271**:7659–7664.

Xie, Y. and Muller, W.A. (1993). Molecular cloning and adhesive properties of murine platelet/endothelial cell adhesion molecule 1. *Proceedings of the National Academy of Science, USA*, **90**:5569–5573.

Yan, H.-C., Baldwin, H.S., Sun, J., Buck, C.A., Albelda, S.M. and DeLisser, H.M. (1995a). Alternative splicing of a specific cytoplasmic exon alters the binding characteristics of murine platelet/endothelial cell adhesion molecule-1 (PECAM-1). *The Journal of Biological Chemistry*, **270**:23672–23680.

Yan, H.-C., Pilewski, J.M., Zhang, Q., DeLisser, H.M., Romer, L. and Albelda, S.M. (1995b). Localization of multiple functional domains on human PECAM-1 (CD31) by monoclonal antibody mapping. *Cell Adhesion Communications*, **3**:45–66.

Yong, K.L., Watts, M., Shaun Thomas, N., Sullivan, A., Ings, S. and Linch, D.C. (1998). Transmigration of CD34+ cells across specialized and nonspecialized endothelium requires prior activation by growth factors and is mediated by PECAM-1 (CD31). *Blood*, **91**:1196–1205.

Zehnder, J.L., Hirai, K., Shatsky, M., McGregor, J.L., Levitt, L.J. and Leung, L.L.K. (1992). The cell adhesion molecule CD31 is phosphorylated after cell activation. Down-regulation of CD31 in activated T lymphocytes. *The Journal of Biological Chemistry*, **267**:5243–5249.

Zehnder, J.L., Shatsky, M., Leung, L.L.K., Butcher, E.C., McGregor, J.L. and Levitt, L.J.(1995). Involvement of CD31 in lymphocyte-mediated immune responses: importance of the membrane-proximal immunoglobulin domain and identification of an inhibiting CD31 peptide. *Blood*, **85**:551–574.

Zhou, M.-J. and Brown, E.J. (1994). CR3 (Mac-1, $\alpha_M\beta_2$, CD11b/CD18) and FcγRIII cooperate in generation of a neutrophil respiratory burst: requirement for FcγRII and tyrosine phosphorylation. *The Journal of Cell Biology*, **125**:1407–1416.

13 Role of Thrombospondins in Vascular Biology

Philip J. Hogg* and Kylie A. Hotchkiss

The Centre for Thrombosis and Vascular Biology, School of Pathology, University of New South Wales, Sydney NSW 2052, Australia

Thrombospondins are large glycoproteins secreted by most cells and activated platelets which function in cell-cell and cell-matrix interactions. This chapter focuses on the role of thrombospondins in vascular development, vascular remodelling, and vascular disease. In particular, the function of thrombospondins in angiogenesis, atherogenesis and thrombosis will be discussed. It is beyond the scope of this chapter to discuss in any detail the gene structure and the role of thrombospondins in inflammation and wound repair, and in diseases such as cancer and malaria. These subjects have been comprehensively covered in other recent reviews (Lahav, 1993a,b; Bornstein and Sage, 1994; Hogg, 1994; Bornstein, 1995; Adams *et al.*, 1995).

THE THROMBOSPONDIN FAMILY

The thrombospondins (TSP) are a family of multidomain glycoproteins. TSP-1 was first recognized more than 20 years ago as a protein released from platelet α-granules in response to activation by thrombin (Baenziger *et al.*, 1971; Baenziger *et al.*, 1972). TSP-1 is a calcium-sensitive, disulfide-bonded homo-trimer with a subunit molecular mass of 150 kDa (Lawler *et al.*, 1978; Margossian *et al.*, 1981; Lawler and Simons, 1983). The TSP-1 subunit contains a unique heparin-binding domain at the amino-terminus, followed by a connecting region that contains the cysteines that participate in interchain disulfide linkages, a procollagen-like module, three properdin-like or type 1 modules, three epidermal growth factor-like or type 2 modules, 12 unique calcium-binding loops (or seven type 3 repeats), and a unique carboxy-terminal globular domain (Lawler and Hynes, 1986) (Figure 13.1).

TSP-1 is the best studied of the TSPs and has been shown to play an important role in cell-cell and cell-matrix interactions (reviewed by Lahav, 1993; Bornstein and Sage, 1994; Bornstein, 1995; Adams *et al.*, 1995). TSP-1 is synthesized and secreted by a vast array of normal and transformed cells in culture, including fibroblasts (Jaffe *et al.*, 1983), endothelial cells (Mosher *et al.*, 1982), and vascular smooth muscle cells (VSMC) (Majack *et al.*, 1985). The TSP-1 gene is a member of the "immediate early" cellular response genes (Framson and Bornstein, 1993), and has been implicated in the control of cell cycle

---

*Corresponding author: Centre for Thrombosis and Vascular Research, School of Pathology, University of New South Wales, Sydney NSW 2052, Australia, Tel: +61-2-9385-1004, Fax: +61-2-9385-1389, e-mail: P.Hogg@unsw.EDU.AU.

Class 1

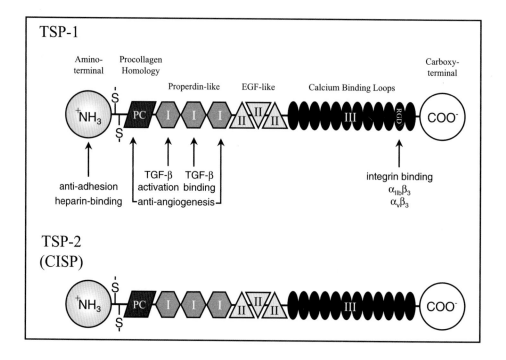

Class 2

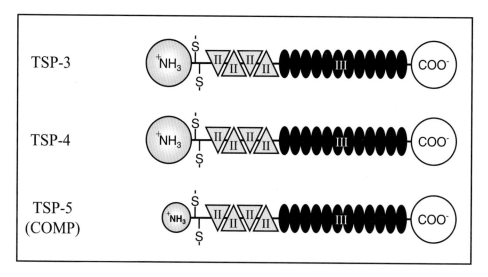

Figure 13.1 Domain structure of the thrombospondins. TSP-1 and TSP-2 belong to class 1 and are homo-trimers, while TSP-3, TSP-4 and TSP-5 belong to class 2 and are homo-pentamers.

progression (Majack *et al.*, 1988). TSP-1 produced by cells binds to their surface and influences their morphology, proliferation and migration. Different cell binding motifs in TSP-1 interact with different cell surface receptors. The Arg-Gly-Asp sequence in the eleventh Ca^{2+}-binding loop binds the integrins $\alpha_{IIb}\beta_3$ and $\alpha_v\beta_3$ (Lawler *et al.*, 1988; Lawler and Hynes, 1989; Sun *et al.*, 1992), while the Val-Thr-Cys-Gly sequence in the type 1 or properdin-like domains binds CD36 (Li *et al.*, 1993). TSP-1 also interacts with heparin sulphate proteoglycans on cell surfaces through the amino-terminal heparin-binding domain (McKeown-Longo *et al.*, 1984), and with the integrin-associated protein (CD47) via the carboxy-terminal domain (Frazier *et al.*, 1999).

The TSP family now numbers five members which are encoded by homologous but separate genes in human and mouse and fall into two distinct classes (for review see Bornstein and Sage, 1994; Adams *et al.*, 1995) (Figure 13.1). TSP-1 and TSP-2/CISP (the bovine form of TSP-2 was originally isolated as Corticotropin-Induced Secreted Protein (Pellerin *et al.*, 1993)) have the same domain structure and can be expressed as both homo- and hetero-trimers (O'Rouke *et al.*, 1992). TSP-3, TSP-4 and TSP-5/COMP (also called Cartilage Oligomeric Matrix Protein (Oldberg *et al.*, 1992; DeCesare *et al.*, 1994)) are lacking the connecting region, the procollagen homology, and the properdin-like modules of TSP-1 and TSP-2, but contain an extra epidermal growth factor-like module. In addition, the amino-terminal domain of TSP-5 is foreshortened. TSP-3, TSP-4 and TSP-5 are pentameric.

ROLE IN VASCULAR DEVELOPMENT

TSP-1 and TSP-2 are differentially expressed during development of the murine vasculature (Iruela-Arispe *et al.*, 1993; Kyriakides *et al.*, 1998a, b). Transcripts for TSP-2 were detected as early as day 11 in the dorsal aorta, and persisted in the vasculature through day 18. TSP-2 mRNA was found in large and small vessel endothelium, and was often associated with marked angiogenic activity, for example in the meninges and choroid plexus. Endothelial cells were also a major source of TSP-2 in the chick embryo (Tucker, 1993). In contrast, TSP-1 mRNA was not seen in blood vessels in the early mouse embryo, but was associated with capillaries later in development. Neither TSP-3 (Iruela-Arispe *et al.*, 1993) or TSP-4 (Tucker *et al.*, 1995) transcripts were detected near blood vessels. TSP-5 transcript has not been investigated.

These descriptive observations suggest that TSP-1 and TSP-2 are playing different roles in vascular development. They indicate that TSP-2 may be a factor that promotes or controls the development and growth of blood vessels. The finding that TSP-1 is present only when the vasculature is essentially established, is consistent with the presence of TSP-1 in the matrix surrounding blood vessels in adult human kidney and skin (Wight *et al.*, 1985). These results are in accordance with the many *in vitro* and *in vivo* observations that TSP-1 is an inhibitor of angiogenesis (see below).

ROLE IN VASCULAR REMODELING

Blood vessels develop by two processes, vasculogenesis and angiogenesis. Vasculogenesis only occurs during embryogenesis, and is the process whereby endothelial cells are born from progenitor cell types. The sprouting of new capillaries from existing vessels is termed angiogenesis, and occurs during embryogenesis and in the adult. The vasculature is quiescent in the normal adult mammal, except in the female reproductive cycles of ovulation, menstruation, implantation, and pregnancy. Outside of these cycles, angiogenesis in the adult is largely confined to pathological situations, such as wound healing and tumor growth.

The process of angiogenesis is complex (reviewed by Cockerill *et al.*, 1995; Folkman, 1995). In response to an appropriate stimulus, the basement membrane surrounding an endothelial cell tube is locally degraded, which triggers the endothelial cells underlying this disrupted matrix to change shape and invade the surrounding stroma. The invading endothelial cells proliferate and develop into a migrating column. A gradation of differentiation trails behind the proliferating cells at the leading edge of the column. The cells of the column wall stop proliferating, change shape, and tightly adhere to each other to form the lumen of the new capillary. Ultimately, the new capillaries fuse and form into loops, resulting in a circulatory system that facilitates exchange of nutrients and waste products in the region.

TSP-1 has been shown to modulate the behavior of both endothelial and non-endothelial cells involved in the angiogenic process. A number of *in vitro* observations support an important role for TSP-1 in endothelial cell biology. TSP-1 inhibits proliferation, motility and sprouting of cultured endothelial cells (Rastinejad *et al.*, 1989; Good *et al.*, 1990; Taraboletti *et al.*, 1990; Bagavandoss and Wilks, 1990; Vogel *et al.*, 1993; DiPietro *et al.*, 1994; Canfield and Schor, 1995; Tolsma *et al.*, 1997). A clue to the mechanism of this property of TSP-1 was provided by Murphy-Ullrich and Höök (1989), who showed that endothelial cells attached to TSP-1 develop fewer focal adhesion contacts than when attached to fibronectin. This focal adhesion disassembly is mediated by the amino-terminal heparin binding domain of TSP-1 (Murphy-Ullrich *et al.*, 1993) and requires cGMP-dependent protein kinase activity (Murphy-Ullrich *et al.*, 1996) and activation of phosphoinositide 3-kinase (Greenwood *et al.*, 1998) (Table 1). These cell culture effects appear to be manifested as anti-angiogenic *in vivo*.

Table 13.1   TSP-1 sequences that function in vascular biology

Domain	Sequence	Function	Reference
Amino-terminal	ELTGAARKGSGRRLVKGPD	Anti-adhesion	Murphy-Ullrich *et al.*, 1993
Procollagen homology	NGVQYRN	Anti-angiogenesis	Tolsma *et al.*, 1993
Type 1 repeats	GGWSHW	TGF-β binding	Schultz-Cherry *et al.*, 1995
	RFK	TGF-β activation	Schultz-Cherry *et al.*, 1995
	SPWDICSVTCGGGVQKRSR	Anti-angiogenesis	Tolsma *et al.*, 1993
Type 3 repeats	DXDXDGXXDXXDX	Calcium binding	Sun *et al.*, 1992
	RGD	Integrin binding	Lawler *et al.*, 1988; Sun *et al.*, 1992

The anti-angiogenic properties of TSP-1 were first recognized by Bouck and colleagues (Rastinejad *et al.*, 1989; Good *et al.*, 1990). These investigators identified a nontumorigenic hamster cell line that became tumorigenic concomitant with a mutation that inactivated a tumor suppressor gene. Comparison of the cell lines revealed that the nontumorigenic cell was releasing high levels of a truncated form of TSP-1, which was found to inhibit endothelial cell chemotaxis *in vitro* and fibroblast growth factor (FGF) stimulated corneal neovascularization *in vivo*. Intact platelet (Good *et al.*, 1990) and recombinant (Sheibani and Frazier, 1995) TSP-1 were subsequently shown to also have anti-angiogenic properties. Using proteolytic fragments and synthetic peptides, the anti-angiogenic activity of TSP-1 was mapped to the procollagen homology region (Tolsma *et al.*, 1993) and to the type 1 repeats (Tolsma *et al.*, 1993; Vogel *et al.*, 1993) (Table 13.1). Two separate anti-angiogenic activities of non-overlapping peptides derived from the type 1 repeats were identified. The activity of one of the peptides was dependent on its ability to activate transforming growth factor-β (TGF-β) (Vogel *et al.*, 1993; Schultz-Cherry *et al.*, 1995), an autocrine inhibitor of endothelial cell proliferation (see below). The activity of the other peptide was independent of TGF-β activation (Tolsma *et al.*, 1993). These observations are supported by the findings of RayChaudhury *et al.* (1994), who showed that conversion of endothelial cells to an angiogenic phenotype by viral transformation was associated with loss of TSP-1 expression. Also, overexpression of TSP-1 in stable transfectants derived from this angiogenic cell line restored the normal phenotype (Sheibani and Frazier, 1995), and down-regulation of endogenous TSP-1 expression in these stable transfectants using antisense technology increased chemotactic responses to FGF-2 and capillary morphogenesis *in vitro* (DiPietro *et al.*, 1994). In addition, transfection of TSP-1 into a human breast carcinoma cell line reduced primary growth and metastasis, which correlated with impaired neovascularization of the tumors (Weinstat-Saslow *et al.*, 1994). Similarly, transfection of a 449-amino acid N-terminal domain of TSP-1 into v-src NIH3T3 cells caused a dose-dependent suppression of tumour growth and neovascularization in nude mice (Castle *et al.*, 1997).

TSP-1 expression is regulated by the wild-type p53 tumor suppressor protein in fibroblasts (Dameron *et al.*, 1994) and epithelial cells (Volpert *et al.*, 1995). Loss of p53 function upon transformation of these cells resulted in a precipitous drop in the level of expression of TSP-1, which correlated with tumorigenic ability *in vivo*. Restoration of p53 function upregulated TSP-1 and impaired the angiogenic capability of the tumors.

Pro-angiogenic activity of TSP-1 has also been reported (BenEzra *et al.*, 1993; reviewed in Tuszynski and Nicosia, 1996). BenEzra *et al.* (1993) found that TSP-1 potentiated the angiogenic effect of FGF-2 in the rabbit cornea model, in contrast to the findings of Bouck and colleagues (Good *et al.*, 1990; Tolsma *et al.*, 1993). BenZra *et al.* (1993) used much higher concentrations of TSP-1 and FGF-2 in their assays. It was noted that the presence of TSP-1 promoted influx of leukocytes into the cornea, which may have produced sufficient angiogenic factors to account for the positive response. Nicosia and Tuszynski showed that TSP-1 promoted angiogenesis from an aortic ring culture imbedded in either collagen or fibrin (Nicosia and Tuszynski, 1994). These investigators proposed that TSP-1 acts primarily in this system by stimulating myofibroblasts, which in turn produce angiogenic factors. Therefore, it is apparent that TSP-1 can be a negative or positive modulator of angiogenesis depending on the specifics of the angiogenesis assay employed. Although, the weight of evidence is in favor of TSP-1 being a net angiogenesis inhibitor.

TSP-2 is also an angiogenesis inhibitor. Volpert *et al.* (1995) tested the ability of recombinant murine and bovine TSP-2, and bovine TSP-5, to block the migration of capillary endothelial cells towards a variety of inducers and to inhibit FGF-induced neovascularization in the rat cornea. Both TSP-2 preparations were inhibitors *in vitro* and *in vivo*, whereas TSP-5 was inactive. These results imply that the procollagen homology and the properdin-like type 1 modules of TSP-2, that it shares with TSP-1 and are missing in TSP-5, are important for the anti-angiogenic activity. This is in accordance with studies of the anti-angiogenic properties of TSP-1 fragments and peptides that have identified the procollagen domain and type 1 repeats as important for this function of TSP-1 (see above). TSP-2 expression has also been correlated with inhibition of angiogenesis and metastasis of colon cancer (Tokunaga *et al.*, 1999).

The recent availability of TSP-1 null mice have provided another useful tool for examining the role of TSP-1 in vascular remodelling. The first report of this nature comes from Iruela-Arispe and colleagues (Iruela-Arispe *et al.*, 1996), who compared the cycle of growth and involution of the vasculature during mammary gland development in TSP-1 null versus wild-type controls. TSP-1 protein was found in association with the basement membrane of blood vessels and glands during all phases of mammary development in normal mice, and was increased at stages of involution by 7–12 fold over levels detected in glands from lactating or virgin mice. There was a marked increase in the vascular density of involuting glands in TSP-1 null mice compared to control animals, which was associated with impaired endothelial apoptosis in the TSP-1 null animals. Therefore, in this system TSP-1 may regulate angiogenesis by controlling the rate of endothelial cell death.

The mechanism or mechanisms by which TSP-1 influences endothelial cell phenotype are largely unknown. A provocative hypothesis can be formulated based on recent work from Cheresh and co-workers on the requirement for appropriate ligation of the endothelial cell $\alpha_v\beta_3$ integrin for endothelial cell survival. Cheresh and co-workers have shown that antagonists of α_v integrins can block tumor induced angiogenesis by promoting unscheduled apoptosis of angiogenic endothelial cells, while leaving preexisting vascular cells intact (Brooks *et al.*, 1994). The apoptosis was associated with the ability of α_v integrin antagonists to promote p53 activity in proliferative endothelial cells (Stromblad *et al.*, 1996). The observations that endothelial cells can interact with TSP-1 in an RGD-dependent manner through the $\alpha_v\beta_3$ integrin receptor (Lawler *et al.*, 1988; Sun *et al.*, 1992) (Table 13.1), and that upregulation of p53 correlates with enhanced TSP-1 expression (Dameron *et al.*, 1994; Volpert *et al.*, 1995), suggests that TSP-1 may be a natural counterpart of the synthetic α_v integrin antagonists. That is, endothelial cell-derived TSP-1 may compete with $\alpha_v\beta_3$ ligands such as vitronectin for binding to $\alpha_v\beta_3$ integrin, which could destabilize the endothelial cell-matrix interaction resulting in cell death. The finding that TSP-1 decreases focal adhesion contacts in endothelial cells (Murphy-Ullrich and Höök, 1989) supports this hypothesis. Importantly, interaction of TSP-1 with endothelial cell $\alpha_v\beta_3$ integrin is dependent on the disulfide-bond structure of TSP-1 (Sun *et al.*, 1992; Hotchkiss *et al.*, 1998), which might provide for an intriguing control of the anti-angiogenic properties of this protein.

Each subunit of TSP-1 contains a free thiol that can reside on any one of 12 different cysteines in the carboxy-terminal Ca^{2+}-binding repeats and globular domain of TSP-1 (Speziale and Detwiler, 1990) (Table 13.1). This finding implies that there is a complex intramolecular disulfide interchange process operating in TSP-1. TSP-1 can be isolated in two different disulfide-bonded conformational states by varying the Ca^{2+} ion concentration used in the purification buffers. Sun *et al.* (1992) have shown that TSP-1 purified in buffers containing 0.1 mM Ca^{2+} (TSP-1$^{0.1}$) is a homogeneous TSP-1 population in that the free thiol is at Cys974, while the work of Speziale and Detwiler (1990) suggests that TSP-1 purified in buffers containing 2 mM Ca^{2+} (TSP-1$^{2.0}$) is probably a heterogeneous population of molecules in which the free thiol resides on one of any twelve Cys. These different isomeric forms of TSP-1 not only have different RGD-dependent cell adhesive activity (Sun *et al.*, 1992), but also inhibit neutrophil enzymes with markedly differing potencies (Hogg *et al.*, 1993a, b; Hogg, 1994), bind Ca^{2+} with different stoichiometries (Misenheimer and Mosher, 1995), and bind platelet-derived growth factor (PDGF) with very different affinities (Hogg *et al.*, 1997) (see below).

TSP-1 functions in angiogenesis bound to cell-surfaces or extracellular matrix (ECM), and there is evidence that the conformation of TSP-1 bound to cells or matrix is different from that of TSP-1 in solution. We have shown that the anti-TSP-1 monoclonal antibodies, D4.6 and A65M, interact with TSP-1$^{0.1}$ bound to plastic or ECM, but not with TSP-1$^{2.0}$ bound to these surfaces (Matthias *et al.*, 1996). Antibody D4.6 immunostained TSP-1 in the ECM and on the surface of cultured fibroblasts and VSMC (Dixit *et al.*, 1986; personal observations), and antibodies D4.6 and A65M immunostained blood vessel walls and occasional matrix cells in human rheumatoid synovial tissue (Matthias *et al.*, 1996). Moreover, antibody A65M was developed against TSP-1 in endothelial cell ECM (Underwood *et al.*, 1990), implying that TSP-1$^{0.1}$ or equivalent isoform is produced by endothelial cells.

Importantly, Majack *et al.* (1988) showed that antibodies specific for Ca^{2+}-depleted TSP-1 or TSP-$1^{0.1}$, D4.6 and A6.1, are effective in inhibiting PDGF-mediated VSMC proliferation (see below). These results indicate that different disulfide-bonded forms of TSP-1 are produced by cultured cells and in human tissue.

Disulfide interchange in a protein is usually very specific and quite slow, unless catalyzed. We have shown that the enzyme, protein disulfide isomerase (PDI), catalyzes disulfide interchange in TSP-1 and changes its enzyme inhibitory properties (Hotchkiss *et al.*, 1996), its interaction with anti-TSP-1 monoclonal antibodies (Hotchkiss *et al.*, 1996), its binding to PDGF$_{BB}$ (Hogg *et al.*, 1997), and its interaction with $\alpha_v\beta_3$ integrin on endothelial cells (Hotchkiss *et al.*, 1998). Both platelet TSP-1 and TSP-1 bound in the ECM of cultured fibroblasts were substrates for PDI (Hotchkiss *et al.*, 1996). PDI functions as a disulfide isomerase by virtue of two very reactive dithiols/disulfides in the common sequence Trp-Cys-Gly-His-Cys-Lys (for review see Freedman *et al.*, 1994). These dithiols/disulfides undergo a cycle of oxidation/reduction reactions with thiol(s) or disulfide(s) in the protein substrate resulting in formation, reduction, or isomerization of disulfide bond(s) in the substrate. Importantly, it was demonstrated that the active site disulfides of PDI must be reduced to catalyze disulfide interchange in TSP-1 (Hotchkiss *et al.*, 1996). PDI was recently identified on the platelet surface (Chen *et al.*, 1995; Essex *et al.*, 1995), and we have demonstrated PDI on the surface of endothelial, fibroblast, and VSMC in an active reduced conformation (Hotchkiss *et al.*, 1998). In addition, we have found that platelet activation triggers reduction of platelet surface PDI (unpublished observations). Therefore, PDI is present on the surface of TSP-1-secreting cells in an active reduced conformation, which makes it well positioned to regulate the structure/function of this protein. Should cell surface PDI regulate adhesion of TSP-1 to endothelial cell $\alpha_v\beta_3$ integrin, this enzyme may function to modulate the anti-angiogenic properties of TSP-1.

It is interesting to note that secretion of urokinase plasminogen activator (uPA) was decreased, and plasminogen activator inhibitor-1 increased, in cells transfected with TSP-1 (Sheibani and Frazier, 1995). Also, TSP-1 has been shown to be an inhibitor of plasmin activity (Hogg *et al.*, 1992) (see below). Because the plasminogen/plasmin system is thought to play an important role in mediating focal degradation of basement membrane matrix in angiogenesis (for review see Mignatti and Rifkin, 1993), the anti-angiogenic properties of TSP-1 may relate in part to control of this proteolytic system.

ROLE IN VASCULAR DISEASE

TSP-1 plays an important role in the vascular disease, atherosclerosis. The atherosclerotic lesion initially results from various forms of injury to the endothelium and smooth muscle of the artery wall (reviewed in Ross, 1993). These events trigger uncontrolled migration and proliferation of VSMC, which results in thickening of the vascular wall. The ultimate consequence of the disease is rupture of the atherosclerotic plaque, which promotes platelet adhesion and aggregation, culminating in thrombus formation and partial or complete occlusion of the artery. TSP-1 is thought to function in all these stages of atherogenesis.

TSP-1 is found in large amounts in atherosclerotic lesions (Wight *et al.*, 1985; Riessen *et al.*, 1998) and derives mostly from activated VSMC (Raugi *et al.*, 1990). TSP-1 is expressed as an immediate-early response to PDGF stimulation of cultured VSMC (Framson and Bornstein, 1993). PDGF is made by most cells in atherosclerotic lesions, and is generally considered to be an important factor responsible for the excessive migration and proliferation of VSMC (Ross, 1993; Khachigian *et al.*, this volume). These findings imply that TSP-1 and

PDGF co-localize extracellularly in atherosclerotic lesions. Studies using anti-TSP-1 mono-clonal antibodies to neutralize endogenously produced TSP-1 have shown that TSP-1 is a requirement for PDGF-mediated proliferation (Majack *et al.*, 1988) and migration (Yab-kowitz *et al.*, 1993) of VSMC. Majack *et al.* (1988) reported that five monoclonal antibod-ies against TSP-1 specifically inhibited VSMC growth in serum, and presented evidence that cell surface associated TSP-1 is functionally essential for the proliferation of VSMC. Yabkowitz *et al.* (1993) showed that a anti-TSP-1 monoclonal antibody negated PDGF-mediated migration of VSMC, and proposed that TSP-1 functions as an autocrine motility factor to modulate VSMC migration. Therefore, PDGF and TSP-1 activities are closely linked in the pathogenesis of atherosclerosis.

The mechanism by which TSP-1 promotes PDGF activity is not understood but it is apparent that TSP-1 must interact with the VSMC cell surface to be active. We recently reported that $PDGF_{BB}$ binds tightly and specifically to TSP-1 and that this interaction is markedly dependent on the disulfide-bonded conformation of TSP-1 (Hogg *et al.*, 1997). Also, binding of $PDGF_{BB}$ to TSP-1 did not preclude $PDGF_{BB}$ from binding to its receptor on rat aortic VSMC. It was suggested that interaction of TSP-1 with PDGF may play a role in the targeting of PDGF to the PDGF-receptor on VSMC (Hogg *et al.*, 1997). TSP-1 pro-duced by VSMC binds to the VSMC surface and ECM (Majack *et al.*, 1985, 1988). It is plausible that PDGF-TSP-1 interactions will result in high local concentrations of PDGF in the vicinity of or at the cell surface which will enhance the interaction of PDGF with its sig-naling receptors. This is analogous to the FGF/glycosaminoglycan (GAG)/FGF receptor interactions (Schlessinger *et al.*, 1995). Cell-surface GAGs have been shown to enhance the binding or targeting of FGF to its signaling receptor, probably by reducing the dimen-sionality of FGF diffusion from three dimensions in the extracellular space to two dimen-sions at the cell-surface (Schlessinger *et al.*, 1995). The net effect is an increase in the local concentration of FGF at the cell-surface which greatly enhances chances of interaction with high affinity FGF receptors. Another similarity between the FGF-GAG and PDGF-TSP-1 interactions is their ionic nature (Hogg *et al.*, 1997), which usually implies rapid dis-sociation rates. In contrast, high affinity signaling receptors such as the FGF and PDGF receptors are often characterized by rapid on rates and slow off rates (Schlessinger *et al.*, 1995; Bowen-Pope and Ross, 1995). It is possible, therefore, that rapid dissociation of PDGF from abundant TSP-1 molecules on the cell-surface will lead to formation of a more stable complex of PDGF with the less abundant high affinity PDGF receptors.

TGF-β is an autocrine inhibitor of VSMC proliferation, and therefore may play an important role in atherogenesis (for review see Rifkin *et al.*, 1993; Ross, 1993). Binding of TSP-1 to latent TGF-β activates the growth factor. This property of TSP-1 has been local-ized to the Arg-Phe-Lys peptide sequence in the type 1 repeats (Schultz-Cherry *et al.*, 1995) (Table 13.1). An adjacent sequence, Trp-Ser-His-Trp, may function in the binding of TGF-β to TSP-1. The Arg-Phe-Lys sequence is unique to TSP-1. However, the Trp-Ser-Xxx-Trp sequence is present in the type 1 repeats of TSP-2, and peptides containing this sequence or recombinant TSP-2 are competitive inhibitors of activation of latent TGF-β by purified TSP-1 (Schultz-Cherry *et al.*, 1995). Therefore, TSP-1 may be an activator of lat-ent TGF-β, while TSP-2 may regulate this activity of TSP-1. Importantly, active TFG-β also binds tightly to TSP-1 in a manner which does not perturb its biological activity (Murphy-Ullrich *et al.*, 1992). This interaction may improve the half-life of TGF-β *in vivo*, or focus the activities of this growth factor in an analogous fashion as that proposed for PDGF (see above). However, the *in vivo* relevance of activation of TGF-β by TSP-1 is con-troversial. TSP-1 overexpression in endothelial cells does not increase activation of endo-genous latent TGF-β (Sheibani and Frazier, 1995), and the majority of latent TGF-β

released from activated platelets is not activated despite being exposed to high concentrations of TSP-1. On the other hand, systemic administration of the TGF-β_1 activating peptide derived from TSP-1, KRFK (see Table 13.1), to TSP-1 null mice caused *in vivo* activation of TGF-β_1 and the lung and pancreatic abnormalities in the null mice reverted toward wild type (Crawford *et al.*, 1998). Moreover, systemic administration of a peptide that blocks activation of TGF-β_1 by TSP-1, LSKL, to wild type mice caused lung and pancreatic abnormalities similar to those observed in TGF-β_1 null mice (Crawford *et al.*, 1998). These disparate findings may relate to the disulfide-dependent structural changes in TSP-1 described above. It may be that only a small fraction of TSP-1 molecules in a given population are in the appropriate configuration to activate TGF-β. Further work is required to test this hypothesis.

The interaction of TSP-1 with platelets has been thoroughly studied. Thrombin triggers release of TSP-1 from platelet α-granules. The secreted TSP-1 binds to the platelet surface with a distribution similar to that of fibrinogen and $\alpha_{IIb}\beta_3$ (Asch *et al.*, 1985). TSP-1 promotes adhesion and aggregation of platelets, and also clot formation and clot retraction (for example see Tuszynski and Kowalski, 1991; Bacon-Baguley *et al.*, 1990). TSP-1 interacts directly with the $\alpha_{IIb}\beta_3$ (Karczewski *et al.*, 1989) and $\alpha_v\beta_3$ (Lawler and Hynes, 1989) integrin receptors on the platelet surface in an RGD and divalent cation dependent manner (Table 13.1). It has been proposed that binding of TSP-1 to fibrinogen bound to $\alpha_{IIb}\beta_3$ is responsible for promoting platelet aggregation (for review see Silverstein *et al.*, 1986), however the details of these interactions are controversial. Both the carboxy and amino-terminal domains of TSP-1 have been implicated in the facilitation of platelet aggregation by TSP-1 (Legrand *et al.*, 1994; Rabhisabile *et al.*, 1996). CD36 has also been proposed as a platelet TSP-1 receptor (see G. Burns, this volume).

In addition to interacting with the platelet surface, TSP-1 binds to other components of the nascent and mature clot (Murphy-Ullrich and Mosher, 1985; Watkins *et al.*, 1990). Bale *et al.* (1985) demonstrated that TSP-1 specifically co-polymerizes with fibrin during blood coagulation. Examination of the effects of TSP-1 on fibrin polymerization and structure revealed that TSP-1 interacts with fibrin intermediates to accelerate fiber growth (Bale and Mosher, 1986). It was proposed that TSP-1 may serve as a trifunctional branching unit during network formation. The properties of fibrin around aggregating platelets, therefore, may be influenced by secreted TSP-1.

Fibrin clot dissolution is under control of the plasminogen/plasmin enzyme system. Most components of the fibrinolytic system have been shown to interact with TSP-1 (for review see Hogg, 1994). TSP-1 forms binary and ternary complexes with plasminogen and tissue plasminogen activator, binary complexes with uPA, and binary and ternary complexes with plasminogen and histidine-rich glycoprotein (reviewed in Silverstein and Nachman, 1987). These complex interactions have been proposed to regulate plasminogen activation on the surface of fibrin, ECM or cells. In addition to regulating plasmin formation, TSP-1 also regulates plasmin activity. TSP-1 was shown to be a slow tight-binding inhibitor of plasmin and uPA (Hogg *et al.*, 1992; Stephens *et al.*, 1992), while Anonick *et al.* (1993) reported that TSP-1 is a rapid hyperbolic mixed-type plasmin inhibitor that slows the inactivation of plasmin by α_2-antiplasmin and α_2-macroglobulin. In contrast, others have reported that TSP-1 is not an inhibitor of plasmin or uPA (Silverstein *et al.*, 1986; Hosokawa *et al.*, 1993). We proposed that the reason for these conflicting reports may relate to the structural lability of TSP-1 (Hogg, 1994; see above). TSP-1 has also been shown to inhibit uPA-mediated fibrinolysis (Mosher *et al.*, 1992; Hosokawa *et al.*, 1993), but the mechanism of this action of TSP-1 is not known. These findings indicate that TSP-1 can function in clot formation and dissolution, although whether TSP-1 is an essential requirement for regulated thrombosis and hemostasis is still open to debate.

In summary, TSP-1 functions in proliferation and migration of VSMC in the diseased vascular wall, and the thrombotic events which are triggered by atherosclerotic plaque rupture. The advent of TSP-1 and TSP-2 knockout mice will be invaluable in assessing the importance of these TSPs in mouse models of atherogenesis.

CONCLUSIONS AND FUTURE DIRECTIONS

It is clear that TSP-1 and probably TSP-2 function in vascular hemostasis. TSP-2 appears to be involved in vascular development, while TSP-1 is expressed only when the vasculature is essentially established or being remodeled. TSP-3, TSP-4 and TSP-5 do not appear to play a role in vascular biology. TSP-1 and TSP-2 have been strongly implicated as inhibitors of angiogenesis, and TSP-1 is probably important for VSMC proliferation and migration in the atherosclerotic vessel wall. The requirement or otherwise for TSP-1 in thrombosis is less well defined.

Further understanding of the protein chemistry and structure of TSP-1 and TSP-2 is required before we can examine the details of their cell biological effects. There is increasing evidence that TSP-1 bound at the cell surface or in ECM has a different conformation to TSP-1 purified from activated platelet releasate, which is the source of TSP-1 used in most studies. The conformational flexibility of TSP-1, and perhaps TSP-2, is poorly understood and there are many indicators that change in TSP structure regulates TSP function. Characterization of factors which catalyze changes in TSP structure, such as PDI, will also help in elucidating how TSP function is controlled.

References

Adams, J.C., Tucker, R.P. and Lawler, J. (1995). The thrombospondin gene family, New York: Springer-Verlag.

Anonick, P.K., Yoo, J.K., Webb, D.J. and Gonias, S.L. (1993). Characterization of the antiplasmin activity of human thrombospondin-1 in solution. *Biochemical Journal*, **289**:903–909.

Asch, A.S., Leung, L.L.K. and Polley, M.J. (1985). Platelet membrane topography: colocalization of thrombospondin and fibrinogen with the glycoprotein IIb/IIIa complex. *Blood*, **66**:926–934.

Bacon-Baguley, T., Ogilvie, M.L. and Gartner, T.K. (1990). Thrombospondin binding to specific sequences within the Aα and Bβ chains of fibrinogen. *Journal of Biological Chemistry*, **265**:2317–2323.

Baenziger, N.L., Brodie, G.N. and Majerus, P.W. (1971). A thrombin-sensitive protein of human platelet membranes. *Proceedings of the National Academy of Science, USA*, **68**:240–243.

Baenziger, N.L., Brodie, G.N. and Majerus, P.W. (1972). Isolation and properties of a thrombin-sensitive protein of human platelets. *Journal of Biological Chemistry*, **247**:2723–2731.

Bagavandoss, P. and Wilks, J.W. (1990). Specific inhibition of endothelial cell proliferation by thrombospondin. *Biochemical and Biophysical Research Communications*, **170**:867–872.

Bale, M.D., Westrick, L.G. and Mosher, D.F. (1985). Incorporation of thrombospondin into fibrin clots. *Journal of Biological Chemistry*, **260**:7502–7508.

Bale, M.D. and Mosher, D.F. (1986). Effects of thrombospondin on fibrin polymerization and structure. *Journal of Biological Chemistry*, **261**:862–68.

BenEzra, D., Griffin, B.W., Maftzir, G. and Aharonov, O. (1993). Thrombospondin and *in vivo* angiogenesis induced by basic fibroblast growth factor or lipopolysaccharide. *Investigations in Opthamology and Visual Science*, **34**:3601–3608.

Bornstein, P. and Sage, E.H. (1994). Thrombospondins. *Methods in Enzymology*, **245**:62–85.

Bornstein, P. (1995). Diversity of function is inherent in matricellular proteins: an appraisal of thrombospondin 1. *Journal of Cell Biology*, **130**:503–506.

Bowen-Pope, D.F. and Ross, R. (1995). Methods for studying the platelet-derived growth factor receptor. *Methods in Enzymology*, **109**:69–100.

Brooks, P.C., Montgomery, A.M., Rosenfeld, M., Reisfeld, R.A., Hu, T., Klier, G., Cheresh, D.A. (1994). Integrin $\alpha_4\beta_3$ antagonists promote tumor regression by inducing apoptosis of angiogenic blood vessels. *Cell,* **79**:1157–1164.

Canfield, A.E. and Schor, A.M. (1995). Evidence that tenascin and thrombospondin 1 modulate sprouting of endothelial cells. *Journal of Cell Science,* **108**:797–809.

Castle V.P., Dixit, V.M. and Polverini, P.J. (1997). Thrombospondin-1 suppresses tumorigenesis and angiogenesis in serum- and anchorage-independent NIH 3T3 cells. *Laboratory Investigation,* **77**:51–61.

Chen, K., Detwiler, T.C. and Essex, D.W. (1995). Characterization of protein disulfide isomerase released from activated platelets. *British Journal of Haematology,* **90**:425–431.

Cockerill, G.W., Gamble, J.R. and Vadas, M.A. (1995). Angiogenesis: models and modulators. *International Reviews of Cytology,* **1159**:113–160.

Crawford, S.E., Stellmach, V., Murphy-Ullrich, J.E., Ribeiro, S.M.F., Lawler, J., Hynes, R.O., Biovin, G.P. and Buock, N. (1998). Thrombospondin-1 is a major activator of TGF-β_1 *in vivo. Cell,* **93**:1159–1170.

Dameron, K.M., Volpert, O.V., Tainsky, M.A. and Bouck, N. (1994). Control of angiogenesis in fibroblasts by p53 regulation of thrombospondin-1. *Science,* **265**:1582–1584.

DeCesare, P.E., Mörgelin, M., Mann, K. and Paulsson, M. (1994). Cartilage oligomeric matrix protein and thrombospondin: purification from articular cartilage, electron microscopic structure, and chondrocyte binding. *British Journal of Biochemistry,* **223**:927–937.

Dixit, V.M., Galvin, N.J., O'Rourke, K.M. and Frazier, W.A. (1986). Monoclonal antibodies that recognize calcium-dependent structures of human thrombospondin. *Journal of Biological Chemistry,* **261**, 1962–1968.

DiPietro, L.A., Nebgen, D.R. and Polverini, P.J. (1994). Downregulation of endothelial cell thrombospondin 1 enhances *in vitro* angiogenesis. *Journal of Vascular Research,* **31**:178–185.

Essex, D.W., Chen, K. and Swiatkowska, M. (1995). Localization of protein disulfide isomerase to the external surface of the platelet plasma membrane. *Blood,* **86**:2163–2173.

Framson, P. and Bornstein, P. (1993). A serum response element and a binding site for NF-Y mediate the serum response of the human thrombospondin 1 gene. *Journal of Biological Chemistry,* **268**:4989–4996.

Frazier, W.A., Gao, A.G., Dimitry, J., Chung, J., Brown, E.J., Lindberg, F.P. and Linder, M.E. (1999). The thrombospondin receptor integrin-associated protein (CD47) functionally couples to heterotrimeric Gi. *Journal of Biological Chemistry,* **274**:8554–8560.

Freedman, R.B., Hirst, T.R. and Tuite, M.F. (1994). Protein disulfide isomerase: building bridges in protein folding. *Trends in Biochemical Sciences,* **19**:331–336.

Folkman, J. (1995). Tumor angiogenesis. In *The Molecular Basis of Cancer,* ed. J. Mendelsohn, P.M. Howley, M.A. Israel and L.A. Liotta, pp. 206–232. Philadelphia: W.B. Saunders Co.

Good, D.J., Polverini, P.J., Rastinejad, F., Le Beau, M.M., Lemons, R.S., Frazier, W.A. and Bouck, N.P. (1990). A tumor suppressor-dependent inhibitor of angiogenesis is immunologically and functionally indistinguishable from a fragment of thrombospondin. *Proceedings of the National Academy of Sciences, USA,* **87**:6624–6628.

Greenwood J.A., Pallero, M.A., Theibert, A.B. and Murphy-Ullrich, J.E. (1998). Thrombospondin signaling of focal adhesion disassembly requires activation of phosphoinositide 3-kinase. *Journal of Biological Chemistry,* **273**:1755–1763.

Hogg, P.J., Stenflo, J. and Mosher, D.F. (1992). Thrombospondin is a slow tight-binding inhibitor of plasmin. *Biochemistry,* **31**:265–269.

Hogg, P.J., Owensby, D.A., Mosher, D.F., Misenheimer, T.M. and Chesterman, C.N. (1993a). Thrombospondin is a tight-binding competitive inhibitor of neutrophil elastase. *Journal of Biological Chemistry,* **268**:7139–7146.

Hogg, P.J., Owensby, D.A. and Chesterman, C.N. (1993b). Thrombospondin 1 is a tight-binding competitive inhibitor of cathepsin G: Determination of the kinetic mechanism of inhibition and localization of cathepsin G binding to the thrombospondin 1 type 3 repeats. *Journal of Biological Chemistry,* **268**:21811–21818.

Hogg, P.J. (1994). Thrombospondin 1 as an enzyme inhibitor. *Thrombosis and Haemostasis,* **72**:787–792.

Hogg, P.J., Hotchkiss, K.A., Jiménez, B.M., Stathakis, P. and Chesterman, C.N. (1997). Interaction of platelet-derived growth factor with thrombospondin-1. *Biochemical Journal,* **326**:709–716.

Hosokawa, T., Muraishi, A., Rothman, V.L., Papale, M. and Tuszynski, G.P. (1993). The effect of thrombospondin on invasion of fibrin gels by human A549 lung carcinoma. *Oncology Research,* **5**:183–189.

Hotchkiss, K.A., Chesterman, C.N. and Hogg, P.J. (1996). Catalysis of disulfide isomerization in thrombospondin 1 by protein disulfide isomerase. *Biochemistry,* **35**:9761–9767.

Hotchkiss, K.A., Matthias, L.J. and Hogg, P.J. (1998). Exposure of the cryptic Arg-Gly-Asp sequence in thrombospondin-1 by protein disulfide isomerase. *Biochimica et Biophysica Acta,* **1388**:478–488.

Iruela-Arispe, M.L., Liska, D.J., Sage, E.H. and Bornstein, P. (1993). Differential expression of thrombospondin 1, 2, and 3 during murine development. *Developmental Dynamics,* **197**:40–56.

Iruela-Arispe, M.L., Patil, S., Oikemus, S., Hynes, R.O. and Lawler, J. (1996). Mice that lack TSP-1 have a defect in vascular regression. *Journal of Vascular Research,* **33**:Supplement **1**: 40.

Jaffe, E.A., Ruggiero, J.T., Leung, L.L., Doyle, M.J., McKeown-Longo, P.J. and Mosher, D.F. (1983). Cultured fibroblasts synthesize and secrete thrombospondin and incorporate it into extracellular matrix. *Proceedings of the National Academy of Science, USA,* **80**:998–1002.

Karczewski, J., Knudsen, K.A., Smith, L., Murphy, A., Rothman, V.L. and Tuszynski, G.P. (1989). The interaction of thrombospondin with GPIIb-IIIa. *Journal of Biological Chemistry,* **264**:21322–21326.

Kyriakides, T.R., Zhu, Y.H., Yang, Z. and Bornstein, P. (1998a). The distribution of the matricellular protein thrombospondin-2 in tissues of embryonic and adult mice. *Journal of Histochemistry and Cytochemistry,* **46**:1007–1015.

Kyriakides, T.R., Zhu, Y.H., Smith L.T., Bain, S.D., Yang, Z., Lin, M.T., Danielson, K.G., Iozzo, R.V., LaMarca, M., McKinney, C.E., Ginns, E.I. and Bornstein, P. (1998b). Mice that lack thrombospondin-2 display connective tissue abnormalities that are associated with disordered collagen fibrillogenesis, an increased vascular density, and a bleeding diathesis. *Journal of Cell Biology,* **140**:419–430.

Lahav, J. (1993a). The functions of thrombospondin and its involvement in physiology and pathophysiology. *Biochimica et Biophysica Acta,* **1182**:1–14.

Lahav, J. (1993b). Thrombospondin, London: CRC Press.

Lawler, J.W., Slayter, H.S. and Coligan, J.E. (1978). Isolation and characterization of a high molecular weight glycoprotein from human blood platelets. *Journal of Biological Chemistry,* **253**:8609–8616.

Lawler, J. and Simons, E.R. (1983). Cooperative binding of calcium to thrombospondin. The effect of calcium on the circular dichroism and limited tryptic digestion of thrombospondin. *Journal of Biological Chemistry,* **258**:12098–12101.

Lawler, J. and Hynes, R.O. (1986). The structure of human thrombospondin, an adhesive glycoprotein with multiple calcium-binding sites and homologies with several different proteins. *Journal of Cell Biology,* **103**:1635–1648.

Lawler, J., Weinstein, R. and Hynes, R.O. (1988). Cell attachment to thrombospondin: the role of Arg-Gly-Asp, calcium and integrin receptors. *Journal of Cell Biology,* **107**:2351–2361.

Lawler, J. and Hynes, R.O. (1989). An integrin receptor on normal and thrombasthenic platelets that binds thrombospondin. *Blood,* **74**:2022–2027.

Legrand, C., Morandi, V., Mendelovitz, S., Shaked, H.., Hartman, J.R. and Panet, A. (1994). Selective inhibition of platelet macroaggregate formation by a recombinant heparin-binding domain of human thrombospondin. *Arteriosclerosis and Thrombosis,* **14**:1784–1791.

Li, W.-X., Howard, R.J. and Leung, L.L.K. (1993). Identification of SVTCG in thrombospondin as the conformation-dependent, high affinity binding site for its receptor, CD36. *Journal of Biological Chemistry,* **268**:16179–16184.

Majack, R.A., Cook, S.C. and Bornstein, P. (1985). Platelet-derived growth factor and heparin-like glycosaminoglycans regulate thrombospondin synthesis and deposition in the matrix by smooth muscle cells. *Journal of Cell Biology,* **101**:1059–1071.

Majack, R.A., Goodman, L.V. and Dixit, V.M. (1988). Cell surface thrombospondin is functionally essential for vascular smooth muscle cell proliferation. *Journal of Cell Biology,* **106**:415–422.

Margossian, S.S., Lawler, J.W. and Slayter, H.S. (1981). Physical characterization of platelet thrombospondin. *Journal of Biological Chemistry,* **256**:7495–7500.

Matthias, L.J., Gotis-Graham, I., Underwood, P.A., McNeil, H.P. and Hogg, P.J. (1996). Identification of monoclonal antibodies that recognize different disulfide bonded forms of thrombospondin 1. *Biochimica et Biophysica Acta,* **1296**:138–144.

McKeown-Longo, P.J., Hanning, R. and Mosher, D.F. (1984). Binding and degradation of platelet thrombospondin by cultured fibroblasts. *Journal of Cell Biology,* **98**:22–28.

Mignatti, P. and Rifkin, D.B. (1993). Biology and biochemistry of proteinases in tumor invasion. *Physiology Reviews,* **73**:161–195.

Misenheimer, T.M. and Mosher, D.F. (1995). Calcium ion binding to thrombospondin 1. *Journal of Biological Chemistry,* **270**:1729–1733.

Mosher, D.F., Doyle, M.J. and Jaffe, E.A. (1982). Synthesis and secretion of thrombospondin by cultured human endothelial cells. *Journal of Cell Biology,* **93**:343–348.

Mosher, D.F., Misenheimer, T.M., Stenflo, J. and Hogg, P.J. (1992). Modulation of fibrinolysis by thrombospondin. *Annals of the New York Academy of Science,* **667**:64–70.

Murphy-Ullrich, J.E. and Mosher, D.F. (1985). Localization of thrombospondin in clots formed *in situ. Blood,* **66**:1098–1104.

Murphy-Ullrich, J.E. and Höök, M. (1989). Thrombospondin modulates focal adhesions in endothelial cells. *Journal of Cell Biology,* **109**:1309–1319.

Murphy-Ullrich, J.E., Schultz Cherry, S. and Höök, M. (1992). Transforming growth factor-β complexes with thrombospondin. *Molecular Biology of the Cell,* **3**:181–188.

Murphy-Ullrich, J.E., Gurusiddappa, S., Frazier, W.A. and Höök, M. (1993). Heparin-binding peptides from thrombospondins 1 and 2 contain focal adhesion-labilizing activity. *Journal of Biological Chemistry,* **268**:26784–26789.

Murphy-Ullrich, J.E., Pallero, M.A., Boerth, N., Greenwood, J.A., Lincoln, T.M., and Cornwell, T.L. (1996). Cyclic GMP-dependent protein kinase is required for thrombospondin and tenascin mediated focal adhesion disassembly. *Journal of Cell Science,* **109**:2499–2508.

Nicosia, R.F. and Tuszynski, G.P. (1994). Matrix-bound thrombospondin promotes angiogenesis *in vitro. Journal of Cell Biology,* **124**:183–193.

Oldberg, A., Antonsson, P., Lindblom, K. and Heinegard, D. (1992). COMP (cartilage oligomeric matrix protein) is structurally related to the thrombospondins. *Journal of Biological Chemistry,* **267**:6132–6136.

O'Rourke, K.M., Laherty, C.D. and Dixit, V.M. (1992). Thrombospondin 1 and thrombospondin 2 are expressed as both homo- and heterotrimers. *Journal of Biological Chemistry*, **267**:24921–24924.

Pellerin, S., Lafeuillade, B., Scherrer, N., Gagnon, J., Shi, D. L., Chanbaz, E.M. and Feige, J.-J. (1993). Corticotropin-induced secreted protein, an ACTH-induced protein secreted by adrenocortical cells, is structurally related to the thrombospondins. *Journal of Biological Chemistry*, **268**:4304–4310.

Rabhisabile, S., Thibert, V. and Legrand, C. (1996). Thrombospondin peptides inhibit the secretion-dependent phase of platelet aggregation. *Blood Coagulation and Fibrinolysis*, **7**:237–240.

Rastinejad, F., Polverini, P.J. and Bouck, N.P. (1989). Regulation of the activity of a new inhibitor of angiogenesis by a cancer suppressor gene. *Cell*, **56**:345–355.

Raugi, G.J., Mullen, J.S., Bark, D.H., Okada, T. and Mayberg, M.R. (1990). Thrombospondin deposition in rat carotid artery. *American Journal of Pathology*, **137**:179–185.

RayChaudhury, A., Frazier, W.A. and D'Amore, P.A. (1994). Comparison of normal and tumorigenic endothelial cells: differences in thrombospondin production and responses to transforming growth factor-β. *Journal of Cell Science*, **107**:39–46.

Riessen, R., Kearney, M., Lawler, J. and Isner, J.M. (1998). Immunolocalization of thrombospondin-1 in human atherosclerotic and restenotic arteries. *American Heart Journal*, **135**:357–364.

Rifkin, D.B., Kojima, S., Abe, M. and Harpel, J.G. (1993). TGF-β: structure, function, and formation. *Thrombosis and Haemostasis*, **70**:177–179.

Ross, R. (1993). The pathogenesis of atherosclerosis. *Nature*, **362**, 801–809.

Schlessinger, J., Lax, I. and Lemmon, M. (1995). Regulation of growth factor activation by proteolysis: what is the role of the low affinity receptors? *Cell*, **83**:357–360.

Schultz-Cherry, S., Chen, H., Mosher, D.F., Misenheimer, T.M., Krutzsch, H.C., Roberts, D.D. and Murphy-Ullrich, J.E. (1995). Regulation of transforming growth factor-β activation by discrete sequences of thrombospondin 1. *Journal of Biological Chemistry*, **270**:7304–7310.

Sheibani, N. and Frazier, W.A. (1995). Thrombospondin 1 expression in transformed endothelial cells restores a normal phenotype and suppresses their tumorigenesis. *Proceedings of the National Academy of Sciences, USA*, **92**:6788–6792.

Silverstein, R.L., Leung, L.L.K. and Nachman, R.L. (1986). Thrombospondin: a versatile multifunctional glycoprotein. *Arteriosclerosis*, **6**:245–253.

Silverstein, R.L., Harpel, P.C. and Nachman, R.L. (1986). Tissue plasminogen activator and urokinase enhance the binding of plasminogen to thrombospondin. *Journal of Biological Chemistry*, **261**:9959–9965.

Silverstein, R.L. and Nachman, R.L. (1987). Thrombospondin-plasminogen interactions: modulation of plasmin generation. *Seminars in Thrombosis and Haemostasis*, **13**:335–342.

Speziale, M.V. and Detwiler, T.C. (1990). Free thiols of platelet thrombospondin. *Journal of Biological Chemistry*, **265**:17859–17867.

Stephens, R.W., Aumailley, M., Timpl, R., Reisberg, T., Tapiovaara, H., Myöhänen, H., Murphy-Ullrich, J. and Vaheri, A. (1992). Urokinase binding to laminin-nidogen. *European Journal of Biochemistry*, **207**:937–942.

Stromblad, S., Becker, J.C., Yebra, M., Brooks, P.C. and Cheresh, D.A. (1996). Suppression of p53 activity and p21WAF1/CIP1 expression by vascular cell integrin $\alpha_v\beta_3$ during angiogenesis. *Journal of Clinical Investigation*, **98**:426–433.

Sun, X., Skorstengaard, K. and Mosher, D.F. (1992). Disulfides modulate RGD-inhibitable cell adhesive activity of thrombospondin. *Journal of Cell Biology*, **118**:693–701.

Taraboletti, G., Roberts, D., Liotta, L.A. and Giavazzi, R. (1990). Platelet thrombospondin modulates endothelial cell adhesion, motility, and growth: a potential angiogenesis regulatory factor. *Journal of Cell Biology*, **111**:765–772.

Tokunaga, T., Nakamura, M., Oshika, Y., Abe, Y., Ozeki, Y., Fukushima, Y., Hatanaka, H., Sadahiro, S., Kijima, H., Tsuchida, T., Yamazaki, H., Tamaoki, N., Ueyama, S. (1999). Thrombospondin-2 expression is correlated with inhibition of angiogenesis and metastasis of colon cancer. *British Journal of Cancer*, **79**:354–359.

Tolsma, S.S., Volpert, O.V., Good, D.J., Frazier, W.A., Polverini, P.J. and Bouck, N. (1993). Peptides derived from two separate domains of the matrix protein thrombospondin 1 have anti-angiogenic activity. *Journal of Cell Biology*, **122**:497–511.

Tolsma, S.S., Stack, M.S. and Bouck, N. (1997). Lumen formation and other angiogenic activities of cultured capillary endothelial cells are inhibited by thrombospondin-1. *Microvascular Research*, **54**:13–26.

Tucker, R.P. (1993). The *in situ* localization of tenascin splice variants and thrombospondin 2 mRNA in the avian embryo. *Development*, **117**:347–358.

Tucker, R.P., Adams, J.C. and Lawler, J. (1995). Thrombospondin-4 is expressed by early osteogenic tissues in the chick embryo. *Developmental Dynamics*, **203**:477–490.

Tuszynski, G.P. and Kowalska, M.A. (1991). Thrombospondin-induced adhesion of human platelets. *Journal of Clinical Investigation*, **87**:1387–1394.

Tuszynski, G.P. and Nicosia, R.F. (1996). The role of thrombospondin-1 in tumor progression and angiogenesis. *BioEssays*, **18**:71–76.

Underwood, P.A., Steele, J.G., Dalton, B.A. and Bennett, F.A. (1990). Solid-phase monoclonal antibodies: a novel method of directing the function of biologically active molecules by presenting a specific orientation. *Journal of Immunological Methods*, **127**:91–101.

Vogel, T., Guo, N.H., Krutzsch, H.C., Blake, D.A., Hartman, J., Mendelovitz, S., Panet, A. and Roberts, D.D. (1993). Modulation of endothelial cell proliferation, adhesion, and motility by recombinant heparin-binding domain and synthetic peptides from the type 1 repeats of thrombospondin. *Journal of Cellular Biochemistry*, **53**:74–84.

Volpert, O.V., Stellmach, V. and Bouck, N. (1995). The modulation of thrombospondin and other naturally occurring inhibitors of angiogenesis during tumor progression. *Breast Cancer Research and Treatment*, **36**:119–126.

Volpert, O.V., Tolsma, S.S., Pellerin, S., Feige, J..-J., Chen, H., Mosher, D.F. and Bouck, N. (1995). Inhibition of angiogenesis by thrombospondin-2. *Biochemical and Biophysical Research Communications*, **217**:326–332.

Watkins, S.C., Raso, V. and Slayter, H.S. (1990). Immunoelectron-microscopic studies of human platelet thrombospondin, Von Willebrand factor, and fibrinogen redistribution during clot formation. *Histochemical Journal*, **22**:507–518.

Weinstat-Saslow, D.L., Zabrenetzky, V.S., VanHoutte, K., Frazier, W.A., Roberts, D.D. and Steeg, P.S. (1994). Transfection of thrombospondin 1 complementary DNA into a human breast carcinoma cell line reduces primary tumor growth, metastatic potential, and angiogenesis. *Cancer Research*, **54**:6504–6511.

Wight, T.N., Raugii, G.J., Mumby, S.M. and Bornstein, P. (1985). Light microscopic immunolocation of thrombospondin in human tissues. *Journal of Histochemistry and Cytochemistry*, **33**:295–302.

Yabkowitz, R., Mansfield, P.J., Ryan, U. and Suchard, S.J. (1993). Thrombospondin mediates migration and potentiates platelet-derived growth factor-dependent migration of calf pulmonary artery smooth muscle cells. *Journal of Cellular Physiology*, **157**:24–32.

14 Platelet-Derived Growth Factor: Regulation of Gene Expression and Roles in Vascular Pathobiology

Levon M. Khachigian,[1,*] Eric S. Silverman,[2] Volkhard Lindner,[3] Amy J. Williams,[2] Colin N. Chesterman,[1] and Tucker Collins[2]

[1]Centre for Thrombosis and Vascular Research, School of Pathology,
The University of New South Wales, and Department of Haematology,
Prince of Wales Hospital, Sydney, Australia
[2]Vascular Research Division, Brigham and Women's Hospital & Harvard Medical School,
Boston MA 02115, USA
[3]Department of Surgery, Maine Medical Research Institute, South Portland, ME 04106, USA

The vascular system is lined by a continuous monolayer of endothelial cells. Located between the circulating blood elements and surrounding tissues, these cells provide a non-adherent and non-thrombogenic surface, and help maintain blood flow and vascular reactivity. These cells also have the capacity to regulate cell migration and proliferation. Multiplicity of endothelial function is mediated by their ability to elaborate regulatory molecules such as anticoagulants, prothrombotic mediators, vasodilators, vasoconstrictors, growth factors, cytokines, adhesion molecules and extracellular matrix components. The integrity of the endothelium is therefore, a fundamental requirement for the maintenance of normal structure and function of the vessel wall.

In the "response to injury" hypothesis of atherogenesis (Ross and Glomset, 1973; Ross, 1986; Ross, 1993), endothelial dysfunction triggers a cascade of molecular and cellular events that lead eventually to the formation of atherosclerotic lesions. A hallmark in the development of these lesions is the "activation" of otherwise quiescent endothelium and adhesion and transmigration of monocytes and lymphocytes through interendothelial junctions. Monocytes develop into lipophilic macrophages and, together with lymphocytes, form the so-called fatty streak or Type III lesion (Stary et al., 1995), the first recognizable lesion of atherosclerosis (Ross et al., 1990). Atheroma, or Type IV lesions (Stary et al., 1995) consist of a distinct lipid core overwhich macrophages, foam cells, lymphocytes and smooth muscle cells (SMCs) reside in a proteoglycan-rich milieu beneath the endothelium. The core of the plaque contains cellular debris, cholesterol crystals, abundant lipid, foamy macrophages and calcium. These lesions may progress to the Type V fibroatheromatous lesion, in which prominent new connective tissue has formed (Stary et al., 1995). It is during this phase that SMCs migrate and proliferate in the neointima, elaborating large amounts of extracellular matrix. In human atheroma obtained after sudden cardiac death, much of the lesion

---

*Corresponding author: Centre for Thrombosis and Vascular Research, School of Pathology, The University of New South Wales, Sydney NSW 2052. Tel: +61-2-9385-2537, Fax: +61-2-9385-1389, e-mail: L. Khachigian@ unsw. EDU.AU.

mass is due to SMCs and matrix (Cliff *et al.*, 1988; Roberts, 1989). Type VI, or "complicated" lesions (Stary *et al.*, 1995) are characterized by thrombotic deposits, hematomas, hemorrhages, or disruptions of the lesion surface. Thrombotic deposits may be responsible for a reduction in lumen diameter and plaque rupture could lead to myocardial infarction or stroke.

Chemotactic and mitogenic responsiveness in the developing lesion may be a consequence of secretory products generated, initially, by activated endothelium. A growth-regulatory molecule implicated in the development of the atherosclerotic lesion is platelet-derived growth factor (PDGF). This chapter will review its regulation and involvement in this and other vascular pathologic settings.

THE PDGF LIGAND

In 1974, it was reported that the principal source of mitogenic activity in whole blood serum derived from platelets, an activity later termed PDGF (Kohler and Lipton, 1974; Ross *et al.*, 1974). PDGF is a mitogen and chemoattractant for cells of mesenchymal origin, such as fibroblasts and SMCs. It was first purified from outdated platelets in two laboratories using conventional chromatographic techniques (Antoniades *et al.*, 1979; Heldin *et al.*, 1979). PDGF is a glycoprotein with a pI of 10.2 (Deuel *et al.*, 1981). It has a relative molecular mass of approximately 26–30kDa and consists of two polypeptide chains, designated A and/or B, held together by disulfide bonds (Deuel *et al.*, 1981; Andersson *et al.*, 1992). The A-chain is approximately 60% homologous to the B-chain (Johnsson *et al.*, 1984), and each chain contains eight conserved cysteine residues (Giese *et al.*, 1987; Sauer and Donoghue, 1988). The predominant PDGF isoform purified from human platelets is AB, with the remainder mostly the BB form (Hammacher *et al.*, 1988). The PDGF isoform in porcine (Stroobant and Waterfield, 1984) and rat (Jawien *et al.*, 1992) plasma is largely the BB homodimer.

Although released from the α granules of platelets (Kaplan *et al.*, 1979), PDGF is also produced by vascular endothelial cells (Collins *et al.*, 1985), SMCs (Seifert *et al.*, 1984) and a wide array of other cell types (reviewed in Raines *et al.*, 1990). In this context, "PDGF" is a misnomer. PDGF is synthesized and dimerized in the endoplasmic reticulum and transferred to the Golgi stacks where it undergoes proteolytic processing. From the Golgi, it is transported to the cell surface *via* secretory vacuoles and released by exocytosis or remains cell-associated. Alternatively, it is returned to the ER or targetted to lysozomes for degradation (Igarashi *et al.*, 1997; Bywater *et al.*, 1988; Ostman *et al.*, 1988; Thyberg *et al.*, 1990; Ostman *et al.*, 1992). Once secreted, PDGF activity may be regulated by physical association with α_2-macroglobulin (Raines *et al.*, 1984) or the extracellular glycoprotein, SPARC (Raines *et al.*, 1992).

THE PDGF-A GENE

Somatic cell hybrid chromosome segregation analysis (Betsholtz *et al.*, 1986) and *in situ* hybridization (Bonthron *et al.*, 1988) mapped the PDGF A-chain gene to the proximal long arm of human chromosome 7 (7p21-p22). The gene consists of seven exons. Exon 1 contains an unusually long 5' untranslated region (UTR) and signal peptide. Exons 2 and 3 encode the amino-terminal propeptide. Exons 4 and 5 encode the majority of the mature protein. Exon 6, which is alternatively spliced, encodes the 18 amino acid carboxyl-terminus of the longer protein (Betsholtz *et al.*, 1986; Collins *et al.*, 1987a; Tong *et al.*,

1987; Rorsman *et al.*, 1988). Exon 7 encodes the three amino acids located at the carboxyl-terminus of the shorter chain (Collins *et al.*, 1987a; Tong *et al.*, 1987), as well as the 3′ UTR (Bonthron *et al.*, 1988).

Long A-chain cDNA (containing the exon 6 product) was originally cloned from a glioblastoma library (Betsholtz *et al.*, 1986; the shorter form was cloned from an endothelial library (Collins *et al.*, 1987a). This led to the suggestion that expression of the longer A-chain may be a feature of tumor cells. It has since been found that both long and short forms of PDGF-A mRNA are expressed in multiple tissues and cell types, and is clearly not tumor-specific (Matoskova *et al.*, 1989; Young *et al.*, 1990). Interestingly, alternative splicing of exon 6 has also been observed in *Xenopus* oocytes (Mercola *et al.*, 1988) suggesting conservation of function over millions of years. In contrast with the short A-chain, which is efficiently secreted, the long A-chain remains largely cell-associated (Rorsman *et al.*, 1988; Kelly *et al.*, 1993). The highly basic carboxyl-terminal region encoded by exon 6 serves to retain the longer PDGF-A protein at the cell surface and extracellular matrix *via* its interaction with heparan sulfate glycosaminoglycans (Khachigian *et al.*, 1992; Raines and Ross, 1992). A peptide bearing this sequence inhibits PDGF binding (Khachigian *et al.*, 1992), proliferation (Khachigian and Chesterman, 1992) and tumor growth (Khachigian *et al.*, 1995a). The exon 6 sequence is capable of directing cytoplasmic proteins to the nucleus (Maher *et al.*, 1989). Exon 2 of the PDGF-A RNA is also alternatively spliced, however, the product of this transcript is unlikely to be functional (Sanchez *et al.*, 1991).

Multiple mRNA species have been reported for PDGF-A (Betsholtz *et al.*, 1988). Three distinct transcripts arise from the one transcriptional start site (Bonthron *et al.*, 1988; Takimoto *et al.*, 1991). An alternate TATA-less promoter, with a transcriptional start site located 530bp downstream of the upstream start site, also generates three transcripts (Rorsman *et al.*, 1992). Heterogeneity of PDGF-A mRNA is due to differential utilization of polyadenylation sites (Bonthron *et al.*, 1988; Rorsman *et al.*, 1992).

THE PDGF-B GENE

The human PDGF-B gene occupies 24kb of genomic DNA on chromosome 22 (22q12.3-q13.1) (Dalla-Favera *et al.*, 1982; Swan *et al.*, 1982; Bartram *et al.*, 1984). Sequence analysis of the PDGF B-chain revealed striking homology (over 90%) with the transforming protein of the simian sarcoma virus, p28[sis] (Devare *et al.*, 1983; Doolittle *et al.*, 1983; Waterfield *et al.*, 1983). This provided the first link between an oncogene and a cellular growth factor.

The PDGF B-chain, like the A-chain, consists of seven exons. There is no evidence that PDGF-B undergoes alternative splicing. Exon 1 contains a long 5′ UTR and the signal sequence. The second and third exon encode the propeptide and the first two amino acids of the mature protein. Exons 4 and 5 encode the mature protein and the start of the carboxyl-terminal propeptide. Exon 6 codes for the remainder of the propeptide, as well as a nuclear targeting sequence (Lee *et al.*, 1987). Exon 7 encodes the 3′ UTR (Johnsson *et al.*, 1984). The size of the PDGF-B transcript is approximately 3.5kb (Barrett *et al.*, 1984; Collins *et al.*, 1985), and its half-life is dependent on the cell type (Majesky *et al.*, 1988; Daniel and Fen, 1988; Press *et al.*, 1988; Press *et al.*, 1989; Harsh *et al.*, 1989; Gay and Winkles, 1991). The 3′ UTR of PDGF-B mRNA is only distantly related to that of PDGF-A (Hoppe *et al.*, 1987).

Discrepancies in amounts of PDGF-B mRNA and protein have been observed in some cell types, suggesting that PDGF-B expression may be regulated at the level of translation

(Rao *et al.*, 1988). The 5′ UTR of PDGF-B mRNA inhibits efficient translation (Kozak, 1986). This region is highly G+C-rich and contains multiple open reading frames upstream of the translational start site (Rao *et al.*, 1986). Cell-free translational systems as well as reporter gene and transient transfection analysis have determined that the 5′ UTR contains potent translational inhibitory activity (Ratner *et al.*, 1987; Rao *et al.*, 1988) possibly by impeding ribosomal scanning. This region harbors an internal ribosomal entry site which relieves translational inhibition during megakaryocytic differentiation in a 5′ end-independent manner (Bernstein *et al.*, 1995).

RECEPTORS

A cDNA for a PDGF receptor was initially cloned from murine 3T3 cells cells by Lewis Williams and colleagues (Yarden *et al.*, 1986). Two separate PDGF receptor genes were subsequently cloned from human cells and designated the type α (Claesson-Welsh *et al.*, 1989) and β receptor (Claesson-Welsh *et al.*, 1988; Escobedo *et al.*, 1988; Gronwald *et al.*, 1988). Each receptor gene encodes five extracellular immunoglobulin-like domains and an intracellular split tyrosine kinase domain. The genes for the α and β subtypes have been mapped to human chromosome 4 (4q11-q12) (Matsui *et al.*, 1989) and 5 (5q23-31) (Yarden *et al.*, 1986), respectively. Each receptors spans over 1000 amino acid residues (reviewed in Hart and Bowen-Pope, 1990).

The PDGF isoforms bind to their receptors with different specificities (Hart *et al.*, 1988; Heldin *et al.*, 1988). The PDGF/receptor interaction results in autophosphorylation and receptor dimerization (Heldin *et al.*, 1989). PDGF-AA binds to $\alpha\alpha$, AB binds to $\alpha\alpha$ or $\alpha\beta$, while BB interacts with all three combinations. The number of autophosphorylation sites defined in the α receptor is three; there are nine sites in the β receptor (Yokote *et al.*, 1996). The tight complex (approximate K_d of 10^{-10}) between PDGF and its receptors involves one receptor subunit bound to one chain of the dimeric ligand (Seifert *et al.*, 1989). The receptors are internalized and down-regulated after ligand binding (Heldin *et al.*, 1982).

PDGF RECEPTORS MEDIATE MITOGENESIS AND CHEMOTAXIS

PDGF was originally discovered because of its ability to stimulate the growth of cells of mesenchymal origin in culture (Ross *et al.*, 1974). Receptors for PDGF appear on the surface of vascular SMCs (Ross *et al.*, 1974; Heldin *et al.*, 1981), certain endothelial cells (Smits *et al.*, 1989; Beitz *et al.*, 1992; Risau *et al.*, 1992; Battegay *et al.*, 1994), fibroblasts (Claesson-Welsh *et al.*, 1988; Gronwald *et al.*, 1988) and a wide variety of other normal and transformed cells in culture (reviewed in Raines *et al.*, 1990). Large vessel endothelial cells generally do not express PDGF receptors, unless activated (Lindner, 1995). Expression of PDGF receptors by microvascular endothelial cells suggests a role for PDGF in angiogenesis (Smits *et al.*, 1989).

Both receptor subtypes mediate mitogenic signals when bound; however, the chemotactic response to PDGF appears to be under tighter control. For example, Swiss 3T3 cells and human granulocytes, which each express α receptors, migrate towards PDGF-AA. In contrast, human foreskin fibroblasts, human monocytes, and rat vascular SMCs, which also express the α receptor, fail to undergo a chemotactic response to PDGF-AA. Regions in the α receptor important for chemotaxis have been determined (Yokote *et al.*, 1996). Responsiveness to PDGF may be regulated by different signaling cascades. Bornfeldt and collaborators have found that independent intracellular pathways mediate chemotaxis and proliferation of SMCs exposed to PDGF-BB (Bornfeldt *et al.*, 1994).

The PDGF ligand/receptor system is subject to cross-regulation. For example, PDGF-AA can inhibit chemotaxis induced by PDGF-AB or PDGF-BB (Siegbahn *et al.*, 1990; Koyama *et al.*, 1992), and α receptor activation in SMCs can inhibit chemotaxis through the β receptor (Koyama *et al.*, 1992; Koyama *et al.*, 1994). The mechanism underlying negative regulation of one receptor type by another is unclear.

MICE DEFICIENT IN PDGF OR ITS RECEPTORS

PDGF and its receptors are expressed during murine (Slamon and Cline, 1984; Mercola *et al.*, 1990; Orr-Urtreger and Lonai, 1992; Palmieri *et al.*, 1992) and rat (Richardson *et al.*, 1988; Pringle *et al.*, 1992; Sasahara *et al.*, 1992) embryogenesis. PDGF-B and the β receptor have been detected during glomerulogenesis (Alpers *et al.*, 1992) and placental development (Goustin *et al.*, 1985; Holmgren *et al.*, 1991) in humans. Mice have been generated with null mutations in the genes encoding the PDGF-B chain (Leveen *et al.*, 1994), the PDGF A-chain (Bostrom *et al.*, 1996) and the β receptor (Soriano, 1994). Animals deficient in PDGF-B and the β receptor die perinatally and exhibit similar phenotypes. Homozygous mice suffer cardiovascular, hematologic and renal abnormalities. These include anemia, thrombocytopenia, and a complete absence of mesangial cells in the kidney glomerulus (Leveen *et al.*, 1994; Soriano, 1994). PDGF-B null mice suffer dilatation of the heart and some large blood vessels. Severe hemorrhagic diathesis observed in PDGF-B deficient embryos may be due to the failure of vascular pericytes to develop (Lindahl *et al.*, 1996). Mice with a targetted mutation in the PDGF-A gene show defects in lung alveolar myofibroblast development and alveogenesis (Bostrom *et al.*, 1996). These studies demonstrate that PDGF signaling is critical for the early morphogenetic events during kidney and lung development.

PDGF IN ATHEROSCLEROTIC TISSUE

The PDGF ligand/receptor signaling system has been implicated in the pathogenesis of atherosclerosis, but as yet a direct causal role has not yet been proven. Barrett and Benditt used Northern blot and dot blot analysis to show that levels of PDGF B-chain mRNA were approximately five-fold greater in carotid plaques than normal aorta and carotid arteries (Barrett and Benditt, 1987). B-chain transcripts were associated with mRNA for cell type-specific markers for endothelial cells (von Willebrand's factor) and monocyte-macrophages (c-*fms*, the colony-stimulating factor type I receptor), implicating these cells as a source of PDGF-B in the plaque (Barrett and Benditt, 1988). *In situ* hybridization corroborated these findings by demonstrating that PDGF-B is associated with endothelium, interstitial cells (presumably SMCs) and capillary-like structures within the plaque (Wilcox *et al.*, 1988). There appeared to be little or no association of B-chain with macrophages (Wilcox *et al.*, 1988). The question of whether macrophages in the lesion were a source of PDGF-B was also addressed by Ross and coworkers, who used a double immunostaining technique with carotid endarterectomy specimens to locate PDGF-B protein in macrophages bearing little or no visible lipid in their cytoplasm (Ross *et al.*, 1990b).

PDGF-A transcripts are found in human carotid plaques as well as normal aortic media (Barrett and Benditt, 1988). Coexpression of PDGF-A with smooth muscle α-actin implicated SMCs in the plaque as a source of PDGF-A (Barrett and Benditt, 1988). *In situ* hybridization later localized the PDGF A-chain to smooth muscle-like cells, with little, if any, association with endothelial cells (Wilcox *et al.*, 1988). Libby and colleagues showed that SMCs cultured from human atherosclerotic plaques expressed PDGF-A transcripts and secreted PDGF-like binding and mitogenic activity (Libby *et al.*, 1988). None of these SMC isolates expressed PDGF-B transcripts (Libby *et al.*, 1988). Rekhter and Gordon (1994) used a triple immunolabeling approach within human carotid plaques to localize PDGF-A protein to smooth muscle-like cells and some endothelial cells. Regression analysis established a correlation between PDGF-A staining and SMC replication (Rekhter and Gordon, 1994). Schwartz and colleagues (Murry *et al.*, 1996) used competitive RT-PCR to show that normal aortic media contained several-fold more PDGF-A mRNA than plaque material. The precise role of PDGF-A in normal quiescent media is unclear, although as already mentioned, it does have the capacity to inhibit (Siegbahn *et al.*, 1990; Koyama *et al.*, 1992). PDGF-A in human atheroma appears to be mostly derived from SMCs, whereas macrophages and endothelial cells seemingly account for PDGF-B. Interestingly, both chains of PDGF are expressed 15–20-fold greater in circulating monocytes of hypercholesterolemic patients compared with normocholesterolemic individuals based on quantitative RT-PCR (Billet *et al.*, 1996). PDGF-A mRNA could not detected in thoracic and abdominal aortic segments from hypercholesterolemic non-human primates, whereas B-chain was observed in the same specimens (Ross *et al.*, 1990b).

The multifunctional glycoprotein, thrombospondin-1 (TSP-1), (Lawler *et al.*, 1978), has also been localized to vascular SMCs in proliferative lesions (Wight *et al.*, 1985; Raugi *et al.*, 1990). Majack and colleagues utilized TSP-1 antibodies to suggest that PDGF-inducible SMC migration and proliferation is dependent on endogenous TSP-1 (Majack *et al.*, 1988). PDGF-BB can interact tightly and specifically with TSP-1 (Hogg *et al.*, 1997). This interaction does not compromise the ability of PDGF-BB to bind to its receptor (Hogg *et al.*, 1997). Accordingly, TSP-1, glycosaminoglycans (Khachigian *et al.*, 1992; Raines and Ross, 1992; Kelly *et al.*, 1993), SPARC (Raines *et al.*, 1992) and α_2-macroglobulin (Bonner, 1994; Bonner and Osornio-Vargas, 1995) may sequester PDGF in the developing lesion, increasing its local concentration and accessibility to its receptors.

PDGF IN THE RESPONSE TO ARTERIAL INJURY

A frequently performed procedure in the treatment of occlusive vascular disease is mechanical dilatation of the artery, or percutaneous transluminal coronary angioplasty (Greuntzig *et al.*, 1987). Although angioplasty generally results in acute symptomatic improvement, its longer term success is complicated by restenosis which occurs in over 30% of patients within six months of the procedure (Califf *et al.*, 1991). Restenosis has been described as a "loss of gain" in luminal size following dilatation (Miller *et al.*, 1993) and is thought to be a culmination of many processes, including cell migration, proliferation and matrix deposition. These events may be due to chemotactic and mitogenic factors released from the stretched lesion into the local microenvironment (Jackson and Schwartz, 1992). PDGF-A and -B have been detected in SMCs, macrophages and endothelial cells, at sites of repair following angioplasty of human coronary arteries (Ueda *et al.*, 1996).

The principal animal model used to assess the molecular mechanisms mediating formation of the neointima after balloon injury is the rat carotid artery model (Clowes *et al.*, 1983). Although this model involves imparting physical injury to normal vessels and differs from balloon dilatation of atherosclerotic lesions, it provides a useful setting to investigate molecular and cellular events underlying the response to vascular injury. Majesky and colleagues (1990) used Northern blot analysis to demonstrate that PDGF-B is expressed at low levels in uninjured rat carotid arteries and that levels are unaffected by balloon injury. Miano *et al.* also reported the inability to induce PDGF-B mRNA by balloon injury (Miano *et al.*, 1993). However, using *in situ* hybridization with *en face* preparations of balloon-injured artery, a subsequent study observed that a small proportion of SMCs in the neointima inducibly expressed PDGF-B (Lindner *et al.*, 1995). Subculturing these SMCs yielded enriched populations that expressed or did not express PDGF-B (Lindner *et al.*, 1995). In contrast, PDGF-A mRNA is dramatically increased in the rat carotid artery after balloon injury (Majesky *et al.*, 1990; Miano *et al.*, 1993).

Intravenous administration of an antibody recognizing all three dimeric forms of PDGF after ballooning reduced intimal thickening without changing the thymidine labeling index (Ferns *et al.*, 1991). Continuous infusion of recombinant PDGF-BB into filament-denuded rat carotid arteries resulted in a marked increase in the size of the intimal lesion without significantly affecting the rate of SMC replication (Jawien *et al.*, 1992). In contrast, infusion of an FGF-2 antibody reduced proliferation (assessed 2d after injury) (Lindner and Reidy, 1991) while administration of FGF-2 increased the rate of intimal replication (Lindner *et al.*, 1991). Heparin, which modulates FGF-2 activity, inhibits injury-induced SMC proliferation when administered at the time of ballooning but not 6h subsequently (Lindner *et al.*, 1992; Olson *et al.*, 1992). These findings argue that PDGF stimulates SMC migration but not proliferation, in contrast to FGF-2 (Lindner and Reidy, 1991). *In vitro* studies have questioned the biologic role played by PDGF-A. PDGF A-chain is a poor mitogen for SMCs (Majack *et al.*, 1990) and can even inhibit migration (Koyama *et al.*, 1992; Koyama *et al.*, 1994). In contrast, PDGF B-chain is both a mitogen and chemoattractant for SMCs *in vitro* (Grotendorst *et al.*, 1982). It is possible that PDGF-A could play a stimulatory role in the injured vessel when it exists in dimeric form with the B-chain.

The predominant PDGF receptor expressed in luminal SMCs after balloon injury of the rat carotid artery is the β receptor (Majesky *et al.*, 1990), which only binds to the PDGF B-chain. Interestingly, this receptor is expressed basally in the rat arterial wall and the two-fold induction in transcript abundance one week after injury is preceded by a transient decrease (Majesky *et al.*, 1990). The α receptor is also expressed basally in the vessel wall, but transcript levels are not elevated by injury (Majesky *et al.*, 1990).

A second model used to investigate the expression of PDGF and its receptors after injury involves scraping the artery wall with an uninflated balloon catheter or filament loop (Lindner, 1995; Lindner *et al.*, 1989). "Gentle injury" leaves behind an intact subendothelial matrix and does not appear to damage cells of the media. *En face in situ* hybridization revealed that PDGF-A and -B transcripts are expressed at low or undetectable levels in unmanipulated endothelium. Both these genes are induced at the endothelial wound edge 4 h after partial denudation (Lindner, 1995). The α receptor is also induced at the wound edge in a manner coincident with PDGF-A. In contrast, the β receptor is not expressed by endothelium at any time. Coexpression of both chains of PDGF and the α receptor in replicating endothelial cells suggests an autocrine growth loop induced by injury. SMC expression of the β receptor suggests a possible paracrine role for endothelial cell-derived PDGF-B in SMC migration (Lindner *et al.*, 1995). Indeed, the first SMCs to migrate to the intima after endothelial denudation in the rat aorta do so at the endothelial wound edge where PDGF is expressed (Lindner, 1995).

Porcine models have also been used to establish a functional role for PDGF in the vessel wall. Nabel *et al.* observed increased intimal thickening when porcine iliofemoral arteries were transfected with a PDGF-B expression vector (Nabel *et al.*, 1993). Double-labeling immunohistochemical studies revealed that SMCs undergoing proliferation in the intima express PDGF-B, suggesting that PDGF-BB may stimulate neointimal thickening by chemotactic and mitogenic effects on SMCs (Pompili *et al.*, 1995). This apparent mitogenic role for PDGF-B may relate to the anatomy of the vessel wall in the rat and pig. Unlike normal rat arteries, porcine arteries contain intimal SMCs (Nabel *et al.*, 1993; Pompili *et al.*, 1995).

INVOLVEMENT OF PDGF IN OTHER VASCULAR PATHOLOGIC SETTINGS

PDGF may play a role in the pathogenesis of a number of other forms of vascular disease. These include hypertensive vascular hypertrophy, hypoxia-associated pulmonary vascular remodeling, hyperoxic lung injury, acute respiratory distress syndrome (ARDS) and diabetic retinopathy.

Vascular remodeling associated with chronically elevated blood pressure is characterized by vessel wall thickening due largely to SMC hypertrophy, hyperplasia and the accumulation of connective tissue in the media. PDGF A-chain mRNA levels are two-fold higher in SMCs of spontaneously hypertensive rats than normal rats (Negoro *et al.*, 1995). Pharmacologic treatment of these rats decreased both PDGF-A transcript levels and aortic hypertrophy (Negoro *et al.*, 1995). In addition, aortic endothelial cells from stroke-prone spontaneously hypertensive rats express higher levels of PDGF-B mRNA and mitogenic activity than control rats (Sasahara *et al.*, 1995). Increased PDGF-B expression was not due to altered mRNA stability but likely the result of new transcription (Sasahara *et al.*, 1995).

Lung disorders associated with hypoxia, such as chronic obstructive pulmonary disease and sleep apnea, are complicated by vascular remodeling. Thickening of the vessel wall in these hypoxic settings is mainly due to abnormal SMC and fibroblast growth, and increased extracellular matrix deposition. These remodeling events may be influenced by the release of inflammatory cytokines and growth factors such as PDGF. PDGF-B chain is secreted from cultured pulmonary artery endothelial cells exposed to hypoxia. Kourembanas and colleagues have demonstrated that exposure of human umbilical vein endothelial cells to hypoxic conditions (0–3% O_2) increases the rate of PDGF B-chain transcription and levels of steady-state mRNA (Kourembanas *et al.*, 1990; Kourembanas and Bernfield, 1994). This increase in PDGF-B expression was inversely proportional to the concentration of O_2 and was reversible upon restoration of normoxia. Dawes and colleagues determined that endothelial cells exposed to hypoxia for 24h release both chemotactic and mitogenic activity. This was partially blocked by an antibody to PDGF (Dawes *et al.*, 1994). Katayose *et al.*, have demonstrated a similar, albeit less dramatic effect using physiologic levels of hypoxia *in vivo*. Northern blot analysis revealed modest increases in PDGF-A and -B transcript levels in rats during hypoxia-associated pulmonary vascular remodeling (Katayose *et al.*, 1993).

Since hyperoxic lung injury and ARDS are both associated with altered vascular reactivity, permeability and structure, the remodeling and obliterative events that occur relatively late in these conditions may be partially mediated by PDGF. For example, hypoxia-induced remodeling in mouse lungs is preceded by the increased expression of both PDGF ligands and their receptors (Powell *et al.*, 1992). Similarly, PDGF-B mRNA and protein are elevated several-fold in human pulmonary artery endothelial cells exposed to endotoxin, an important cause of ARDS (Albelda *et al.*, 1989). These studies suggest that PDGF may play a role in the aberrant SMC and fibroblast growth which typifies remodeling pulmonary vessels and interstitium.

TRANSCRIPTIONAL REGULATION OF PDGF GENE EXPRESSION

Isolation of genomic clones for the PDGF ligand genes has allowed detailed investigation of the molecular mechanisms controlling transcription. This information is fundamental to our understanding of the regulatory pathways involved in activation. S1 nuclease and primer extension mapping localized a transcriptional start site in the PDGF-A gene 36 bp downstream of the TATA box (Bonthron *et al.*, 1988; Takimoto *et al.*, 1991). Serial 5′ deletion analysis defined minimal promoter regions in human HeLa epithelial cells (Lin *et al.*, 1992), African green monkey BSC-1 renal epithelial cells (Kaetzel *et al.*, 1994) and bovine aortic endothelial cells (Khachigian *et al.*, 1995b) to approximately 100 bp. Since identical promoter-reporter constructs were not used in these studies, it is difficult to predict whether differences in the 5′ endpoints of these "core promoters" are actually cell-type specific. The proximal region of the A-chain promoter is extremely G+C-rich and contains multiple consensus binding sites for the zinc finger transcription factor, Sp1 (Briggs *et al.*, 1986; Jones *et al.*, 1987), the related immediate-early gene product, Egr-1 (Sukhatme *et al.*, 1988; Tsai-Morris *et al.*, 1988), and activating protein-2 (AP-2) (Imagawa *et al.*, 1987).

Electrophoretic mobility shift assays using nuclear extracts from vascular endothelial cells (Khachigian *et al.*, 1995b) and recombinant sources (Lin *et al.*, 1992; Wang *et al.*, 1992a; Khachigian *et al.*, 1995b) determined that Sp1 binds to the −71 to −55 bp region in the proximal PDGF-A promoter. 5′ deletion and mutational analysis revealed that the Sp1 binding element is crucial for basal promoter activity in endothelial cells (Khachigian *et al.*, 1995b). This region (−69 to −60 bp) is hypersensitive to S1 nuclease (SHSII) (Wang *et al.*, 1992b). Additional functional elements have been defined in other cell types. For example, a region upstream of the Sp1 binding site, −102 to −82 bp, interacts with nuclear proteins from human mesangial cells exposed to PMA (Bhandari *et al.*, 1995). The −96 to −90 bp element in this region is hypersensitive (SHSI) (Wang *et al.*, 1992b). Additionally, a promoter element that mediates inducible PDGF-A promoter-dependent gene expression in human Hs68 foreskin fibroblasts exposed to PDGF-AB has been localized to the region between −477 and −468 bp (Lin *et al.*, 1992). This region shares sequence similarity with serum response elements in the human c-*fos* and murine *egr*-1 promoters (Lin *et al.*, 1992).

The PDGF-A promoter is subject to negative transcriptional regulation. The Wilms' tumor suppressor gene (*wt1*) encodes a zinc finger DNA binding protein that interacts with the proximal G+C-rich region (Gashler *et al.*, 1992; Wang *et al.*, 1992a). Transient transfection studies have determined that WT-1 functions as a potent repressor of PDGF-A transcription (Gashler *et al.*, 1992; Wang *et al.*, 1992a). Whether PDGF-A expression can be inversely correlated with the expression of WT-1 in normal and pathologic settings remains unresolved. In BSC-1 cells, two negative regulatory regions have been defined: the first between −1029 and −880 bp and the second between −1800 and −1029 bp (Kaetzel *et al.*, 1994). Further studies are required to identify specific elements as well as the nuclear proteins that interact with these sites.

The PDGF-B transcriptional unit contains a consensus TATA box and a single start site located 30 bp downstream (Rao *et al.*, 1986; van den Ouweland *et al.*, 1987). The minimal promoter has been defined in human K562 megakaryocytic cells (Pech *et al.*, 1989; Dirks *et al.*, 1995), U2OS osteosarcoma cells (Jin *et al.*, 1994; Liang *et al.*, 1996), JEG3 cytotro-phoblast-like cells, HeLa cells, PC3 prostate carcinoma cells, dermal fibroblasts (Dirks *et al.*, 1995) and bovine aortic endothelial cells (Khachigian *et al.*, 1994) also to the first 100bp. Several putative *cis*-acting elements regulating basal expression have been iden-tified within this region. In endothelial cells, two elements with structural similarity to consensus elements recognized by members of the ETS (−80 to −70bp) and AP-1 (−92 to −85bp) families of transcription factors are required for basal expression (Khachigian *et al.*, 1994). Recombinant AP1 and ETS family member, GABPα, do not bind to these sites *in vitro* (Khachigian *et al.*, 1994). Linker scanning mutational analysis revealed that these elements are not essential for promoter transactivation by the Tax protein of human T-cell leukemia virus type 1 (HTLV-1) in human Jurkat T-cells (Trejo *et al.*, 1996). In U2-OS osteosarcoma cells, two positive regulatory sites have been identified by linker scanning analysis designated the SIS distal (−102 to −95bp) and SIS proximal (−63 to −44bp) elements (Jin *et al.*, 1994). *In vivo* footprinting studies determined that the 5′-CCACCCAC-3′ sequence within the SIS proximal element is occupied in intact cells (Dirks *et al.*, 1995).

Sp1 was the first endogenous nuclear factor found to bind to the PDGF-B promoter (Khachigian *et al.*, 1994; Jin *et al.*, 1994). Sp1 binds to the 5′-CCACCC-3′ (−61 to −56bp) motif in the SIS proximal element and is required for the basal expression of the gene in endothelial cells (Khachigian *et al.*, 1994) and U2OS cells (Jin *et al.*, 1994; Liang *et al.*, 1996). The interaction of Sp1 with this region has since been demonstrated in human endothelial cells (Scarpati and DiCorleto, 1996), Jurkat T-cells (Trejo *et al.*, 1996) and SMCs (Silverman *et al.*, 1997). Increased levels of Sp1 may account for elevated PDGF A-chain gene expression in SMCs of spontaneously hypertensive rats (Negoro *et al.*, 1995).

Another member of the Sp transcription factor family, Sp3, has been found to interact with the 5′-CCACCC-3′ element. Overexpression studies in U2-OS cells indicate that Sp3, like Sp1, has a positive regulatory effect on PDGF-B promoter-dependent expression (Liang *et al.*, 1996). Sp1, Sp3 and Egr-1 also bind to this element in human Jurkat T-cells (Trejo *et al.*, 1996). Sp3 interacts with the PDGF-A promoter in SMCs and can activate reporter gene expression (Silverman *et al.*, 1997). Sp3 activation contrasts with previous reports that have shown that Sp3 can repress transcription (Hagen *et al.*, 1994; Majello *et al.*, 1994). Sp3 inhibition of Sp1 activity may be mediated by competition for common binding sites (Liang *et al.*, 1996).

PDGF-B gene expression is induced in vascular endothelial cells exposed to physiologic levels of fluid shear stress (Resnick *et al.*, 1993). Shear-inducible PDGF-B reporter expres-sion is mediated *via* a shear-stress response element (SSRE) located 150bp upstream of the core promoter (Resnick *et al.*, 1993). Binding studies determined that the p50–p65 hetero-dimeric form of NF-κB rapidly accumulates in the nuclei of endothelial cells exposed to fluid shear stress (Lan *et al.*, 1994; Khachigian *et al.*, 1995c) and interacts with the SSRE spanning the sequence 5′-GAGACC-3′ (Resnick *et al.*, 1993; Khachigian *et al.*, 1995c). Transfection experiments determined that the NF-κB binding site is crucial for shear-inducible reporter gene expression (Khachigian *et al.*, 1995c).

PDGF-B gene expression is also induced in endothelial cells exposed to thrombin (Dan-iel *et al.*, 1986, 1987; Shankar *et al.*, 1992), a multifunctional serine protease generated at sites of vascular injury. A thrombin-response element has been localized to the SIS prox-imal element (Scarpati and DiCorleto, 1996). Nuclear proteins induced by thrombin inter-act with this element in a specific and transient manner (Scarpati and DiCorleto, 1996). The precise nature of these putative regulatory factor(s) is not known, although supershift assays have excluded the involvement of Egr-1.

EGR-1/SP1 INTERPLAY AS A REGULATOR OF INDUCIBLE PDGF GENE EXPRESSION

PDGF-A transcripts are dramatically increased in endothelial cells exposed to PMA (Starksen *et al.*, 1987). Deletion analysis of the PDGF-A promoter localized the PMA-response element to the Sp1 binding site (Khachigian *et al.*, 1995b). Northern blot and gel shift analysis (Khachigian *et al.*, 1996) determined, however, that Sp1 levels are unchanged in cells exposed to PMA. Similarly, levels of Sp3 are not significantly altered by PMA (Silverman *et al.*, 1997). Since this region also contains consensus binding sites for Egr-1, we determined whether Egr-1 played a role in PMA-inducible PDGF-A expression. Egr-1 mRNA and protein are dramatically induced in endothelial cells (Khachigian *et al.*, 1995b) and SMCs (Silverman *et al.*, 1997) exposed to phorbol ester. Gel shift and *in vitro* DNase I footprint analysis determined that nuclear and recombinant Egr-1 binds to the Sp1 site in the proximal PDGF-A promoter (Khachigian *et al.*, 1995b). The ability of Egr-1 to interact with the promoter is crucial for PMA-inducible reporter gene expression (Khachigian *et al.*, 1995b).

The overlap in binding sites for Egr-1 and Sp1 in the PDGF-A promoter led to the hypothesis that inducible PDGF-A expression may involve the interplay of one transcription factor with the other. In this model, Sp1, which resides at the promoter in resting cells and directs basal expression of the gene, is displaced by increasing levels of Egr-1. Binding studies with recombinant proteins show that Egr-1 is capable of displacing prebound Sp1 from the promoter. Conversely, decreasing levels of Egr-1 in the presence of a fixed amount of Sp1 facilitates in reoccupation of the promoter by Sp1 (Khachigian *et al.*, 1995b).

Alterations in blood flow and shear stress in a baboon model can increase PDGF-A mRNA and protein levels in vascular endothelium (Kraiss *et al.*, 1996). Studies *in vitro* have shown that physiologic levels of fluid shear stress increase PDGF-A mRNA expression in vascular endothelial cells (Hsieh *et al.*, 1991; Khachigian *et al.*, 1997). Nuclear run-off experiments demonstrate that shear-induced PDGF-A expression occurs, at least in part, at the level of transcription (Khachigian *et al.*, 1997). Egr-1 is synthesized and accumulates in endothelial nuclei minutes after application of the biomechanical force. Transient transfection studies indicate that the ability of Egr-1 to interact with the promoter is crucial for shear-inducible gene PDGF-A expression (Khachigian *et al.*, 1997). Moreover, shear-induced Egr-1 displaces Sp1 from the promoter (Khachigian *et al.*, 1997). Shear-inducible expression of tissue factor, which also has a G + C-rich promoter, is mediated by Egr-1 in vascular endothelial cells (Braddock *et al.*, 1996).

IMPLICATIONS FOR EGR-1/SP1 INTERPLAY IN THE RESPONSE TO INJURY

A number of immediate-early genes have been reported to be induced in the rat arterial wall after balloon catheter injury. These include *c-myc* (Miano *et al.*, 1990; Miano *et al.*, 1993), c-*fos* (Miano *et al.*, 1990) and *c-jun* (Miano *et al.*, 1990). Upon endothelial denudation *egr-1* transcripts are rapidly expressed the endothelial wound edge (Khachigian *et al.*, 1996). The transient induction of Egr-1 precedes the earliest detection of PDGF-A or -B transcripts at the same location. When cultured vascular endothelial cells are injured by scraping, new *egr-1* transcription is initiated (L.M.K., unpublished data). Injury-induced Egr-1 protein interacts with a non-consensus binding element in the proximal PDGF-B promoter. Egr-1 fully protects this region from partial DNase I digestion. Mutational studies demonstrate that this interaction is required for injury-inducible activity of the PDGF-B promoter (Khachigian *et al.*, 1996). These findings established the first functional link between a transcription factor and the increased expression of a pathophysiologically-relevant gene in endothelial cells responding to injury.

Egr-1 also interacts with and can displace Sp1 from the proximal promoters (Khachigian *et al.*, 1996) of transforming growth factor β_1 (Kim *et al.*, 1989), urokinase-type plasminogen activator (Riccio *et al.*, 1985) and tissue factor (Mackman *et al.*, 1989; Cui *et al.*, 1996). These genes, like both chains of PDGF, are induced at the endothelial wound edge after injury (Khachigian *et al.*, 1996), spatially and temporally consistent with a regulatory role for Egr-1 (Khachigian *et al.*, 1996). Egr-1 could initiate a cascade of transcriptional events which lead to the inducible expression of genes whose products can influence the remodeling events assoicated with the response to injury. Sp1 displacement by Egr-1 may be a common theme in the inducible expression of these and other pathophysiologically-relevant genes.

Since its discovery almost two decades ago, it has become clear that PDGF is a ubiquitous molecule that can play important roles in normal processes such as embryogenesis and wound repair, and in pathologic conditions such as neoplasia and different manifestations of vascular disease. Further studies should provide further insights into the molecular mechanisms controlling the expression, structure and function of PDGF and its receptors, and better elucidate the precise biologic role(s) of the PDGF ligand/receptor system *in vivo*.

References

Albelda, S.M., Elias, J.A., Levine, E.M. and Kern, J.A. (1989). Endotoxin stimulates platelet-derived growth factor production from cultured human pulmonary endothelial cells. *American Journal of Physiology*, **257**:L65–70.

Alpers, C.E., Seifert, R.A., Hudkins, K.L., Johnson, R.J. and Bowen-Pope DF. (1992). Developmental patterns of PDGF B-chain, PDGF-receptor, and α-actin expression in human glomerulogenesis. *Kidney International*, **42**:390–399.

Andersson, M., Ostman, A., Backstrom, G., Hellman, U., George-Nascimento, C., Westermark B. *et al.* (1992). Assignment of interchain disulfide bonds in platelet-derived growth factor (PDGF) and evidence for agonist activity of monomeric PDGF. *Journal of Biological Chemistry*, **267**:11260–11266.

Antoniades, H.N., Scher, C.D. and Stiles, C.D. (1979). Purification of platelet-derived growth factor. *Journal of Biological Chemistry*, **76**:1809–1813.

Barrett, T.B., Gajdusek, C.M., Schwartz, S.M., McDougall, J.K. and Benditt, E.P. (1984). Expression of the sis gene by endothelial cells in culture and *in vivo*. *Proceedings of the National Academy of Sciences, USA*, **81**:6772–6774.

Barrett, T.B. and Benditt, E.P. (1987). Sis (platelet-derived growth factor B chain) gene transcripts are elevated in human atherosclerotic lesions compared to normal artery. *Proceedings of the National Academy of Sciences, USA*, **84**:1099–1103.

Barrett, T.B. and Benditt, E.W. (1988). Platelet-derived growth factor gene expression in human atherosclerotic plaques and normal artery wall. *Proceedings of the National Academy of Sciences, USA*, **85**:2810–2814.

Bartram, C.R., De Klein, A., Hagmeijer, A., Grosveld, G., Heisterkamp, N. and Groffen, J. (1984). Localization of the human c-sis oncogene in Ph(1)-negative myelocytic leukemia by *in situ* hybridization. *Blood*, **63**:223–225.

Battegay, E.J., Rupp, J., Iruela-Arispe, L., Sage, E.H. and Pech, M. (1994). PDGF-BB modulates endothelial proliferation and angiogenesis *in vitro via* PDGF β-receptors. *Journal of Cell Biology*, **125**:917–928.

Beitz, J.G., Kim, I.S., Calabresi, P. and Frackelton, A.R., Jr. (1992). Receptors for platelet-derived growth factor on microvascular endothelial cells. *EXS*, **61**:85–90.

Bernstein, J., Shefler, I. and Elroy-Stein, O. (1995). The translational repression mediated by the platelet-derived growth factor 2/c-sis mRNA leader is relieved during megakaryocytic differentiation. *Journal of Biological Chemistry*, **270**:10559–10565.

Betsholtz, C., Johnnson, A., Heldin, C.-H., Westermark, B., Lind, P., Urdea, M.S. *et al.* (1986). The human platelet-derived growth factor A-chain: complementary cDNA sequence, chromosomal localization and expression in tumour cell lines. *Nature*, **323**:6440–6444.

Bhandari, B., Wenzel, U.O., Marra, F. and Abboud, H.E. (1995). A nuclear protein in mesangial cells that binds to the promoter of the platelet-derived growth factor A-chain gene. *Journal of Biological Chemistry*, **270**:5541–5548.

Billet, M.A., Adbeish, I.S., Alrokayan, S.A.H., Bennett, A.J., Marenah, C.B. and White, D.A. (1996). Increased expression of genes for platelet-derived growth factor in circulating mononuclear cells of hypercholesterolemic patients. *Arteriosclerosis, Thrombosis, and Vascular Biology*, **16**:399–406.

Bonner, J.C. (1994). Regulation of platelet-derived growth factor (PDGF) and alveolar macrophage-derived PDGF by α_2-macroglobulin. *Annals of the New York Academy of Sciences*, **737**:324–338.

Bonner, J.C. and Osornio-Vargas, A.R. (1995). Differential binding and regulation of platelet-derived growth factor A and B chain isoforms by α_2-macroglobulin. *Journal of Biological Chemistry*, **270**:16236–16242.

Bonthron, D.T., Morton, C.C., Orkin, S.H. and Collins, T. (1988). Platelet-derived growth factor A chain: gene structure, chromosomal location, and basis for alternative mRNA splicing. *Proceedings of the National Academy of Sciences, USA*, **85**:1492–1496.

Bornfeldt, K.E., Raines, E.W., Nakano, T., Graves, L.M., Krebs, E.G. and Ross, R. (1994). Insulin-like growth factor-1 and platelet-derived growth factor-BB induce directed migration of human arterial smooth muscle cells *via* signaling pathways that are distinct from those of proliferation. *Journal of Clinical Investigation*, **93**:1266–1274.

Bostrom, H., Willetts, K., Pekny, M., Leveen, P., Lindahl, P., Hedstrand, H. *et al.* (1996). PDGF-A signaling is a critical event in lung alveolar myofibroblast development and alveogenesis. *Cell*, **85**:863–873.

Braddock, M., Houston, P., Dickson, M.C., McVey, J. and Campbell, C.J. (1996). Fluid shear stress activates the tissue factor promoter in vascular endothelial cells *via* up-regulation of the cellular transcription factor EGR-1. *Journal of Vascular Research*, **33** (Suppl. 1):11 (Abstract).

Briggs, M.R., Kadonaga, J.T., Bell, S.P. and Tjian, R. (1986). Purification and biochemical characterization of the promoter-specific transcription factor, Sp1. *Science*, **234**:47–52.

Bywater, M., Rorsman, F., Bongcam-Rudloff, E., Mark, G., Hammacher, A., Heldin, C.-H., Westermark, B. and Betsholtz, C. (1988). Expression of recombniant platelet-derived growth factor A-and B-chain homodimers in Rat-1 cells and human fibroblasts reveals differences in protein processing and autocrine effects. *Molecular and Cellular Biology*, **8**:2753–2762.

Califf, R.M., Fortin, D.F., Frid, D.J., Harlan, W.R., Ohman, E.M., Bengtson, J.R. *et al.* (1991). Restenosis after coronary angioplasty: an overview. *Journal of the American College of Cardiologists*, **17** (suppl. B):2B–13B.

Claesson-Welsh, L., Eriksson, A., Moren, A., Severinsson, L., Ek, B., Ostman, A., Betsholtz, C. *et al.* (1988). cDNA cloning and expression of a human platelet-derived growth factor (PDGF) receptor specific for a B-chain-containing PDGF molecule. *Molecular and Cellular Biology*, **8**:3476–3486.

Claesson-Welsh, L., Eriksson, A., Westermark, B. and Heldin, C.-H. (1989). cDNA cloning and expression of the human A-type platelet-derived growth factor (PDGF) receptor establishes structural similarity to the B-type receptor. *Proceedings of the National Academy of Sciences, USA*, **86**:4917–4921.

Cliff, W.J., Heathcote, C.R., Moss, N.S. and Reichenbach, D.D. (1988). The coronary arteries in cases of cardiac and noncardiac sudden death. *American Journal of Pathology*, **132**:319–329.

Clowes, A.W., Reidy, M.A. and Clowes, M.M. (1983). Kinetics of cellular proliferation after arterial injury. *Laboratory Investigation*, **49**:327–333.

Collins, T., Ginsburg, D., Boss, J.M., Orkin, S.H. and Pober, J.S. (1985). Cultured human endothelial cells express platelet-derived growth factor B chain: cDNA cloning and structural analysis. *Nature*, **316**:748–750.

Collins, T., Bonthron, D.T. and Orkin, S.H. (1987a). Alternative RNA splicing affects function of encoded platelet-derived growth factor A chain. *Nature*, **328**:621–624.

Collins, T., Pober, J.S., Gimbrone, M.A., Jr, Hammacher, A., Betsholtz, C., Westermark, B. *et al.* (1987b). Cultured human endothelial cells expression platelet-derived growth factor A chain. *American Journal of Pathology*, **127**:7–12.

Cui, M.-Z., Parry, G.C.N., Oeth, P., Larson, K.H., Smith, M., Huand, R.-P. *et al.* (1996). Transcriptional regulation of the tissue factor gene in human epithelial cells is mediated by Sp1 and EGR-1. *Journal of Biological Chemistry*, **271**:2731–2739.

Dalla-Favera, R., Gallo, R.C., Giallongo, A. and Croce, C.M. (1982). Chromosomal localization of the human homologue of the c-sis gene of the simian sarcoma virus onc gene. *Science*, **218**:686–688.

Daniel, T.O., Gibbs, V.C., Milfay, D.F., Garovoy, M.R. and Williams, L.T. (1986). Thrombin stimulates c-sis gene expression in mircrovascular endothelial cells. *Journal of Biological Chemistry*, **261**:9579–9582.

Daniel, T.O., Gibbs, V.C., Milfay, D.F. and Williams, L.T. (1987). Agents that increase cAMP accumulation block endothelial c-sis induction by thrombin and transforming growth factor-β. *Journal of Biological Chemistry*, **262**:11893–11896.

Daniel, T.O. and Fen, Z. (1988). Distinct pathways mediate transcriptional regulation of platelet-derived growth factor B/c-sis expression. *Journal of Biological Chemistry*, **263**:19815–19820.

Dawes, K.E., Peacock, A.J., Gray, A.J., Bishop, J.E. and Laurent, G.J. (1994). Characterization of fibroblast mitogens and chemoattractants produced by endothelial cells exposed to hypoxia. *American Journal of Respiratory Cell and Molecular Biology*, **10**:552–559.

Deuel, T.F., Huang, J.S., Proffit, R.T., Baenzinger, J.U., Chang, D. and Kennedy, B.B. (1981). Human platelet-derived growth factor: purification and resolution into two active protein fractions. *Journal of Biological Chemistry*, **256**:8896–8899.

Devare, S.G., Reddy, E.P., Law, D.J., Robbins, K.C. and Aaronson, S.A. (1983). Nucleotide sequence of the simian sarcoma virus genome: demonstration that its acquired cellular sequences encode the transforming gene product p28sis. *Proceedings of the National Academy of Sciences, USA,* **30**:731–735.

Dewey, C.F., Jr, Bussolari, S.R., Gimbrone, M.A., Jr and Davies, P.F. (1981). The dynamic response of endothelial cells to fluid shear stress. *Journal of Biomechanical Engineering,* **103**:177–185.

Dirks, R.P.H., Jansen, H.J., van Gerven, B., Onnekink, C. and Bloemers, H.P.J. (1995). *In vivo* footprinting and functional analysis of the human c-sis/PDGF B gene promoter provides evidence for two binding sites for transcriptional activators. *Nucleic Acids Research,* **23**:1119–1126.

Doolittle, R.F., Hunkapiller, M.W., Hood, L.E., Devare, S.G., Robbins, K.C. and Aaronson, S.A. (1983). Simian sarcoma virus onc gene, v-sis, is derived from the gene (or genes) encoding platelet-derived growth factor. *Science,* **221**:275–277.

Escobedo, J.A., Navanakasatussas, S., Coussens, L.S., Coughlin, S.R., Bell, G.I. and Williams, L.T. (1988). A common PDGF receptor is activated by homodimeric A and B forms of PDGF. *Science,* **240**:1532–1535.

Ferns, G.A.A., Raines, E.W., Sprugel, K.H., Motani, A.S., Reidy, M.A. and Ross, R. (1991). Inhibition of neointimal smooth muscle accumulation after angioplasty by an antibody to PDGF. *Science,* **253**:1129–1132.

Gashler, A.L., Bonthron, D.T., Madden, S.L., Rauscher III, F.J., Collins, T. and Sukhatme, V.P. (1992). Human platelet-derived growth factor A chain is transcriptionally repressed by the Wilms' tumor suppressor WT1. *Proceedings of the National Academy of Sciences, USA,* **89**:10984–10988.

Gay, C.G. and Winkles, J.A. (1991). The half-lives of platelet-derived growth factor A- and B-chain mRNAs are similar in endothelial cells and unaffected by heparin-binding growth factor-1 or cycloheximide. *Journal of Cellular Physiology,* **147**:121–127.

Giese, N.A., Robbins, K.C. and Aaronson, S.A. (1987). The role of individual cysteine residues in the structure and function of the v-sis gene product. *Science,* **236**:1315.

Goustin, A.S., Betsholtz, C., Pfeiffer-Ohlsson, S., Persson, H., Rydnert, J., Bywater, M. *et al.* (1985). Coexpression of the sis and myc proto-oncogenes in developing human placenta suggests autocrine control of trophoblast growth. *Cell,* **41**:301–312.

Greuntzig, A., King, S.I., Schlumpf, M. and Siegenthaler, W. (1987). Long-term follow-up after percutaneous transluminal coronary angioplasty: the early Zurich experience. *New England Journal of Medicine,* **316**:1127–1132.

Gronwald, R.G.K., Grant, E.J., Haldeman, B.A., Hart, C.E., O'Hara, P.J., Hagen, F.S. *et al.* (1988). Cloning and expression of a cDNA coding for the human platelet-derived growth factor receptor: evidence for more than one receptor class. *Proceedings of the National Academy of Sciences, USA,* **85**:3435–3439.

Grotendorst, G.R., Chang, T., Seppa, H.E.J., Kleinman, H.K. and Martin, G.R. (1982). Platelet-derived growth factor is a chemoattractant for vascular smooth muscle cells. *Journal of Clinical Investigation,* **113**:261–266.

Hagen, G., Muller, S., Beato, M. and Suske, G. (1994). Sp1-mediated transcriptional activation is repressed by Sp3. *EMBO Journal,* **13**:3843–3851.

Hammacher, A., Hellman, U., Johnsson, A., Ostman, A., Gunnarson, K., Westermark, B. *et al.* (1988). A major part of platelet-derived growth factor purified from human platelets is a heterodimer of one A and one B chain. *Journal of Biological Chemistry,* **263**:16493–16498.

Harsh, G.R., Kavanaugh, W.M., Starksen, N.F. and Williams, L.T. (1989). Cyclic AMP blocks expression of the c-sis gene in tumor cells. *Oncogene Research,* **4**:65–73.

Hart, C.E., Forstrom, J.W., Kelly, J.D., Seifert, R.A., Smith, R.A., Ross, R. *et al.* (1988). Two classes of PDGF receptor recognize different isoforms of PDGF. *Science,* **240**:1529–1531.

Hart, C.E. and Bowen-Pope, D.F. (1990). Platelet-derived growth factor receptor: current views of the two-subunit model. *Journal of Investigative Dermatolology,* **94**:53S–57S.

Heldin, C.-H., Westermark, B. and Wasteson, A. (1979). Platelet-derived growth factor: purification and partial characterization. *Proceedings of the National Academy of Sciences, USA,* **76**:3722–3726.

Heldin, C.-H., Westermark, B. and Wasteson, A. (1981). Specific receptors for platelet-derived growth factor on cells derived from connective tissue and glia. *Proceedings of the National Academy of Sciences, USA,* **78**:3664–3668.

Heldin, C.-H., Wasteson, A. and Westermark, B. (1982). Interaction of platelet-derived growth factor with its fibroblast receptor: demonstration of ligand degradation and receptor modulation. *Journal of Biological Chemistry,* **257**:4216–4221.

Heldin, C.-H., Backstrom, G., Ostman, A., Hammacher, A., Ronnstrand, L., Rubin, K. *et al.* (1988). Binding of diferrent dimeric forms of PDGF to human fibroblasts: evidence for two separate receptor types. *EMBO Journal,* **7**:1387–1393.

Heldin, C.-H., Ernlund, A., Rorsman, C. and Ronnstrand, L. (1989). Dimerization of B-type platelet-derived growth factor receptors occurs after ligand binding and is closely associated with receptor kinase activation. *Journal of Biological Science,* **264**:8905–8915.

Hogg, P.J., Hotchkiss, K.A., Jimenez, B.M., Stathakis, P. and Chesterman, C.N. (1997). Interaction of platelet-derived growth factor with thrombospondin 1. *Biochemical Journal,* **326**:709–16.

Holmgren, L., Claesson-Welsh, L., Heldin, C.H. and Ohlsson, R. (1992). The expression of PDGF α- and β-receptors in subpopulations of PDGF-producing cells implicates autocrine stimulatory loops in the control of proliferation in cytotrophoblasts that have invaded the maternal endometrium. *Growth Factors,* **6**:219–231.

Hoppe, J., Schumacher, L., Eichner, W. and Weich, H.A. (1987). The long 3′-untranslated regions of the PDGF-A and -B mRNAs are only distantly related. *FEBS Letters*, **223**:243–246.

Hsieh, H.-J., Li, N.-Q. and Frangos, J.A. (1991). Shear stress increases endothelial platelet-derived growth factor messenger RNA levels. *American Journal of Physiolology*, **260**:H642–H646.

Igarashi, H., Rao, C.D., Siroff, M., Leal, F., Robbins, K.C. and Aaronson, S.A. (1987). Detection of PDGF-2 homodimers in human tumor cells. *Oncogene*, **1**:79–85.

Imagawa, M., Chiu, R. and Karin, M. (1987). Transcription factor AP-2 mediates induction by two different signal-transduction pathways: protein kinase C and cAMP. *Cell*, **51**:251–60.

Jackson, C.L. and Schwartz, S.M. (1992). Pharmacology of smooth muscle replication. *Hypertension*, **20**:713–736.

Jawien, A., Bowen-Pope, D.F., Lindner, V., Schwartz, S.M. and Clowes, A.W. (1992). Platelet-derived growth factor promotes smooth muscle migration and intimal thickening in a rat model of balloon angioplasty. *Journal of Clinical Investigation*, **89**:507–511.

Jin, H.-M., Robinson, D.F., Liang, Y. and Fahl, W.E. (1994). SIS/PDGF-B promoter isolation and characterization of regulatory elements necessary for basal expression of the SIS/PDGF-B gene in U2-OS osteosarcoma cells. *Journal of Biological Chemistry*, **269**:28648–28654.

Johnsson, A., Betsholtz, C., Heldin, C.-H., Wasteson, A., Westermark, B., Deuel, T.F. *et al.* (1984). The c-sis gene encodes a precursor of the B chain of platelet-derived growth factor. *EMBO Journal*, **3**:921–928.

Jones, K.A., Kadonaga, J.T., Rosenfeld, P.J., Kelly, T.J. and Tjian, R. (1987). A cellular DNA-binding protein that activates eukaryotic transcription and DNA replication. *Cell*, **48**:79–89.

Kaetzel, D.M., Maul, R.S., Liu, B., Bonthron, D., Fenstermaker, R.A. and Coyne, D.W. (1994). Platelet-derived growth factor A-chain transcription is mediated by positive and negative regulatory regions in the promoter. *Biochemical Journal*, **301**:321–327.

Kaplan, D.R., Chao, F.C., Stiles, C.D., Antoniades, H.N. and Scher, C.D. (1979). Platelet α-granules contain a growth factor for fibroblasts. *Blood*, **53**:1043–1052.

Katayose, D., Ohe, M., Yamauchi, K., Ogata, M., Shirato, K., Fujita, H. *et al.* (1993). Increased expression of PDGF A- and B-chain genes in rat lungs with hypoxic pulmonary hypertension. *American Journal of Physiology*, **264**:L100–L106.

Kelly, J.L., Sanchez, A., Brown, G., Chesterman, C.N. and Sleigh, M.J. (1993). Accumulation of PDGF isoforms in the extracellular matrix. *Journal of Cell Biology*, **121**:1153–1163.

Khachigian, L.M. and Chesterman, C.N. (1992). Synthetic peptides representing the alternatively spliced exon of the PDGF A-chain modulate mitogenesis stimulated by normal human serum and several growth factors. *Journal of Biological Chemistry*, **267**:7478–7482.

Khachigian, L.M., Owensby, D.A. and Chesterman, C.N. (1992). A tyrosinated peptide representing the alternatively spliced exon of the PDGF A-chain binds specifically to cultured cells and interferes with binding of several growth factors. *Journal of Biological Chemistry*, **267**:1660–1666.

Khachigian, L.M. and Chesterman, C.N. (1994). Structural basis for extracellular retention of platelet-derived growth factor A-chain using a synthetic peptide corresponding to exon 6. *Peptides*, **15**:133–137.

Khachigian, L.M., Fries, J.W.U., Benz, M.W., Bonthron, D.T. and Collins, T. (1994). Novel *cis*-acting elements in the human platelet-derived growth factor B-chain core promoter that mediate gene expression in cultured vascular endothelial cells. *Journal of Biological Chemistry*, **269**:22647–22656.

Khachigian, L.M., Feild, S.E., Crouch, R. and Chesterman, C.N. (1995). Platelet-derived growth factor A-chain synthetic peptide inhibits human glioma xenograft proliferation in nude mice. *Anticancer Res.*, **15**:337–342.

Khachigian, L.M., Williams, A.J. and Collins, T. (1995b). Interplay of Sp1 and Egr-1 in the proximal PDGF-A promoter in cultured vascular endothelial cells. *Journal of Biological Chemistry*, **270**:27679–27686.

Khachigian, L.M., Resnick, N., Gimbrone, M.A., Jr. and Collins, T. (1995c). Nuclear factor-κB interacts functionally with the platelet-derived growth factor B-chain shear-stress response element in vascular endothelial cells exposed to fluid shear stress. *Journal of Clinical Investigation*, **96**:1169–75.

Khachigian, L.M., Lindner, V., Williams, A.J. and Collins, T. (1996). Egr-1-induced endothelial gene expression: a common theme in vascular injury. *Science*, **271**:1427–1431.

Khachigian, L.M., Anderson, K.A., Halnon, N.J., Resnick, N., Gimbrone, M.A., Jr. and Collins, T. (1997). Egr-1 is activated in endothelial cells exposed to fluid shear stress and interacts with a novel shear-stress-response element in the PDGF A-chain promoter. *Arteriosclerosis, Thrombosis and Vascular Biology*, **17**:2280–6.

Kim, S.-J., Glick, A., Sporn, M.B. and Roberts, A.B. (1989). Characterization of the promoter region of the transforming growth factor-β_1 gene. *Journal of Biological Chemistry*, **264**:402–408.

Kohler, N. and Lipton, A. (1974). Platelets as a source of fibroblast growth-promoting activity. *Experimental Cell Research*, **87**:297–301.

Kourembanas, S., Hannan, R.L. and Faller, D.V. (1990). Oxygen tension regulates the expression of the platelet-derived growth factor-B chain gene in human endothelial cells. *Journal of Clinical Investigation*, **86**:670–674.

Kourembanas, S. and Bernfield, M. (1994). Hypoxia and endothelial-smooth muscle cell interactions in the lung. *American Journal of Respiratory Cell and Molecular Biology*, **11**:373–374.

Koyama, N., Morisaki, N., Saito, Y. and Yoshida, S. (1992). Regulatory effects of platelet-derived growth factor-AA homodimer on migration of vascular smooth muscle cells. *Journal of Biological Chemistry*, **267**:22806–22812.

Koyama, N., Hart, C.E. and Clowes, A.W. (1994). Different functions of the platelet-derived growth factor-α and -β receptors for the migration and proliferation of cultured baboon smooth muscle cells. *Circulation Research*, **75**:682–691.

Kozak, M. (1986). Influences of mRNA secondary structure on initiation by eukaryotic ribosomes. *Proceedings of the National Academy of Sciences, USA*, **83**:2580–2584.

Kraiss, L.W., Geary, R.L., Mattson, E.J.R., Vergel, S., Au, Y.P.T. and Clowes, A.W. (1996). Acute reductions in blood flow and shear stress induce platelet-derived growth factor-A expression in baboon prosthetic grafts. *Circulation Research*, **79**:45–53.

Lan, Q., Mercurius, K.O. and Davies, P.F. (1994). Stimulation of transcription factors NF-κB and AP1 in endothelial cells subjected to shear stress. *Biochemical and Biophysical Research Communications*, **201**:950–956.

Lawler, J., Slayter, H.S. and Coligan, J.E. (1978). Isolation and characterization of a high molecular weight glycoprotein from human blood platelets. *Journal of Biological Chemistry*, **253**:8609–8616.

Lee, B.A., Maher, D.W., Hannick, M. and Donoghue, D.J. (1987). Identification of a signal for nuclear targeting in platelet-derived growth factor-related molecules. *Molecular and Cellular Biology*, **7**:3527–3537.

Leveen, P., Pekny, M., Gebre-Medhin, S., Swolin, B., Larsson, E. and Betsholtz, C. (1994). Mice deficient for PDGF B show renal, cardiovascular, and hematological abnormalities. *Genes & Development*, **8**:1875–1887.

Liang, Y., Robinson, D.F., Dennig, J., Suske, G. and Fahl, W.E. (1996). Transcriptional regulation of the SIS/PDGF-B gene in human osteosarcoma cells by the Sp family of transcription factors. *Journal of Biological Chemistry*, **271**:11792–11797.

Libby, P., Warner, S.J.C., Salomon, R.N. and Birinyi, L.K. (1988). Production of platelet-derived growth factor-like mitogen by smooth muscle cells from human atheroma. *New England Journal of Medicine*, **318**:1438–1438.

Lin, X., Wang, Z., Gu, L. and Deuel, T.F. (1992). Functional analysis of the human platelet-derived growth factor A-chain promoter region. *Journal of Biological Chemistry*, **267**:25614–25619.

Lindahl, P., Johansson, B.R., Leveen, P. and Betsholtz, C. (1996). Pericyte loss in PDGF B null mice leads to micro-aneurysm formation and rupture at late gestation. *Journal of Vascular Research*, **33** (Suppl.1):177 (Abstract)

Lindner, V., Reidy, M.A. and Fingerele, J. (1989). Regrowth of arterial endothelium: denudation with minimal trauma leads to complete endothelial cell regrowth. *Laboratory Investigation*, **61**:556.

Lindner, V. and Reidy, M.A. (1991). Proliferation of smooth muscle cells after vascular injury is inhibited by an antibody against basic fibroblast growth factor. *Proceedings of the National Academy of Sciences, USA*, **88**:3739–3743.

Lindner, V., Lappi, D.A., Baird, A., Majack, R.A. and Reidy, M.A. (1991). Role of basic fibroblast growth factor in vascular lesion formation. *Circulation Research*, **68**:106–113.

Lindner, V., Olson, N.E., Clowes, A.W. and Reidy, M.A. (1992). Inhibition of smooth muscle cell proliferation in injured rat arteries: interaction of heparin with basic fibroblast growth factor. *Journal of Clinical Investigation*, **90**:2044–2049.

Lindner, V. (1995). Expression of platelet-derived growth factor ligands and receptors by rat aortic endothelium *in vivo*. *Pathobiology*, **63**:257–264.

Lindner, V., Giachelli, C.M., Schwartz, S.M. and Reidy, M.A. (1995). A subpopulation of smooth muscle cells in injured rat arteries expresses platelet-derived growth factor-B chain mRNA. *Circulation Research*, **76**:951–957.

Mackman, N., Morrissey, J.H., Fowler, B. and Edgington, T.S. (1989). Complete sequence of the human tissue factor gene, a highly regulated cellular receptor that initiates the coagulation cascade. *Biochemistry*, **28**:1755–1762.

Maher, D.W., Lee, B.A. and Donoghue, D.J. (1989). The alternatively spliced exon of the platelet-derived growth factor A chain encodes a nuclear targeting signal. *Molecular and Cellular Biology*, **9**:2251–2253.

Majack, R.A., Goodman, L.V. and Dixit, V.M. (1988). Cell surface thrombospondin is functionally essential for vascular smooth muscle proliferation. *Journal of Cell Biology*, **106**:415–422.

Majack, R.A., Majesky, M.W. and Goodman, L.V. (1990). Role of PDGF-A expression in the control of vascular smooth muscle cell growth by transforming growth factor-β. *Journal of Cell Biology*, **111**:239–247.

Majello, B., Deluca, P., Hagen, G., Suske, G. and Lania, L. (1994). Different members of the Sp1 multigene family exert opposite transcriptional regulation of the long terminal repeat of HIV-1. *Nucleic Acids Research*, **22**:4914–4921.

Majesky, M.W., Benditt, E.P. and Schwartz, S.M. (1988). Expression and developmental control of platelet-derived growth factor A-chain and B-chain/Sis genes in rat aortic smooth muscle cells. *Proceedings of the National Academy of Sciences, USA*, **85**:1524–1528.

Majesky, M.W., Reidy, M.A., Bowen-Pope, D.F., Hart, C.E., Wilcox, J.N. and Schwartz, S.M. (1990). PDGF ligand and receptor gene expression during repair of arterial injury. *Journal of Cell Biology*, **111**:2149–2158.

Matoskova, B., Rorsman, F., Svensson, V. and Betsholtz, C. (1989). Alternative splicing of the platelet-derived growth factor A-chain transcript occurs in normal as well as tumor cells and is conserved among mammalian species. *Molecular and Cellular Biology*, **9**:3148–3150.

Matsui, T., Heidaran, M., Miki, T., Popescu, N., LaRochelle, W., Krauss, M. *et al.* (1989). Isolation of a novel receptor cDNA establishes the existence of two PDGF receptor genes. *Science*, **243**:1532–1535.

Mercola, M., Melton, D.A. and Stiles, C.D. (1988). Platelet-derived growth factor A chain is maternally encoded in Xenopus embryos. *Science*, **239**:1223–1225.

Mercola, M., Wang, C., Kelly, J., Brownlee, C., Jackson-Grusby, L., Stiles, C. *et al.* (1990). Selective expression of PDGF A and its receptor during early mouse embryogenesis. *Developmental Biology*, **138**:114–122.

Miano, J.M., Tota, R.R., Vlasic, N., Danishefsky, K.J. and Stemerman, M.B. (1990). Early proto-oncogene expression in rat aortic smooth muscle cells following endothelial removal. *American Journal of Pathology*, **137**:761–765.

Miano, J.M., Vlasic, N., Tota, R.R. and Stemerman, M.B. (1993). Smooth muscle cell immediate-early gene and growth factor activation follows vascular injury — a putative mechanism for autocrine growth. *Arteriosclerosis and Thrombosis*, **13**:211–219.

Miller, M.J., Kuntz, R.E., Friedrich, S.P., Leidig, G.A., Fishman, R.F., Schnitt, S.J. *et al.* (1993). Frequency and consequences of intimal hyperplasia in specimens retrieved by directional atherectomy of native primary coronary artery stenosis and subsequent restenosis. *American Journal of Cardiology*, **71**:652–658.

Murry, C.E., Bartosek, T., Giachelli, C.M., Alpers, C.E. and Schwartz, S.M. (1996). Platelet-derived growth factor-A mRNA expression in fetal, normal adult, and atherosclerotic human aortas. *Circulation*, **93**:1095–1106.

Nabel, E.G., Yang, Z., Liptay, S., San, H., Gordon, D., Haudenschild, C.C. *et al.* (1993). Recombinant platelet-derived growth factor B gene expression in porcine arteries induces intimal hyperplasia *in vivo*. *Journal of Clinical Investigation*, **91**:1822–1829.

Negoro, N., Kanayama, Y., Haraguchi, M., Umetani, N., Nishimura, M., Konishi, Y. *et al.* (1995). Blood pressure regulates platelet-derived growth factor A-chain gene expression in vascular smooth muscle cells *in vivo* — an autocrine mechanism promoting hypertensive vascular hypertrophy. *Journal of Clinical Investigation*, **95**:1140–1150.

Olson, N.E., Chao, S., Lindner, V. and Reidy, M.A. (1992). Intimal smooth muscle cell proliferation after balloon catheter injury — the role of basic fibroblast growth factor. *American Journal of Pathology*, **140**:1017–1023.

Orr-Urtreger, A. and Lonai, P. (1992). Platelet-derived growth factor-A and its receptor are expressed in separate, but adjacent cell layers of the mouse embryo. *Development*, **115**:1045–1058.

Ostman, A., Rall, L., Hammacher, A., Wormstead, M.A., Coit, D., Valenzuela, P. *et al.* (1988). Synthesis and assembly of a functionally active recombinant platelet-derived growth factor AB heterodimer. *Journal of Biological Chemistry*, **263**:16202–16208.

Ostman, A., Thyberg, J., Westermark, B. and Heldin, C.H. (1992). PDGF-AA and PDGF-BB biosynthesis: pro-protein processing in the Golgi complex and lysosomal degradation of PDGF-BB retained intracellularly. *Journal of Cell Biology*, **118**:509–519.

Palmieri, S.L., Payne, J., Stiles, C.D., Biggers, J.D. and Mercola, M. (1992). Expression of mouse PDGF-A and PDGF α-receptor genes during pre- and post-implantation development: evidence for a developmental shift from an autocrine mode of action. *Mechanisms of Development*, **39**:181–191.

Pech, M., Rao, C.D., Robbins, K.C. and Aaronson, S.A. (1989). Functional identification of regulatory elements within the promoter region of platelet-derived growth factor 2. *Molecular and Cellular Biology*, **9**:396–405.

Pompili, V.J., Gordon, D., San, H., Yang, Z., Muller, D.W.M., Nabel, G.J. *et al.* (1995). Expression and function of a recombinant PDGF B gene in porcine arteries. *Arteriosclerosis, Thrombosis, and Vascular Biology*, **15**:2254–2264.

Powell, P.P., Wang, C.C. and Jones, R. (1992). Differential regulation of the genes encoding platelet-derived growth factor receptor and its ligand in rat lung during microvascular and alveolar wall remodeling in hyperoxia. *American Journal of Respiratory Cell and Molecular Biology*, **7**:278–285.

Press, R.D., Samols, D. and Goldthwait, D.A. (1988). Expression and stability of c-sis mRNA in human glioblastoma cells. *Biochemistry*, **27**:5736–5741.

Press, R.D., Misra, A., Gillaspy, G., Samols, D. and Goldthwait, D.A. (1989). Control of the expression of c-sis mRNA in human glioblastoma cells by phorbol ester and transforming growth factor β_1. *Cancer Research*, **49**:2914–2920.

Pringle, N.P., Mudhar, H.S., Collarini, E.J. and Richardson, W.D. (1992). PDGF receptors in the rat CNS: during late neurogenesis, PDGF α-receptor expression appears to be restricted to glial cells of the oligodendrocyte lineage. *Development*, **115**:535–551.

Raines, E.W., Bowen-Pope, D.F. and Ross, R. (1984). Plasma binding proteins for platelet-derived growth factor that inhibit its binding to cell-surface receptors. *Proceedings of the National Academy of Sciences, USA*, **81**:3424–3428.

Raines, E.W., Bowen-Pope, D.F. and Ross, R. (1990). Platelet-derived growth factor. In *Handbook of Experimental Pharmacology: Peptide Growth Factors and their Receptors*, edited by M.B. Sporn and Roberts, A.B., pp. 173–262. Berlin: Springer-Verlag.

Raines, E.W., Lane, T.F., Iruela-Arispe, M.L., Ross, R. and Sage, E.H. (1992). The extracellular glycoprotein SPARC interacts with platelet-derived growth factor (PDGF)-AB and -BB and inhibits the binding of PDGF to its receptors. *Proceedings of the National Academy of Sciences, USA*, **89**:1281–1285.

Raines, E.W. and Ross, R. (1992). Compartmentalization of PDGF on extracellular binding sites dependent on exon 6-encoded sequences. *Journal of Cell Biology*, **116**:533–543.

Rao, C.D., Igarashi, H., Chiu, I.-M., Robbins, K.C. and Aaronson, S.A. (1986). Structure and sequence of the human c-sis/platelet-derived growth factor 2 (sis/PDGF-2) transcriptional unit. *Proceedings of the National Academy of Sciences, USA*, **83**:2392–2396.

Rao, C.D., Pech, M., Robbins, K.C. and Aaronson, S.A. (1988). The 5′ untranslated sequence of the c-sis/platelet-derived growth factor 2 transcript is a potent translational inhibitor. *Molecular and Cellular Biology*, **8**:284–292.

Ratner, L., Thielan, B. and Collins, T. (1987). Sequences of the 5′ portion of the human c-sis gene: characterization of the transcriptional promoter and regulation of expression of the protein product by 5′ untranslated mRNA sequences. *Nucleic Acids Research*, **15**:6017–6036.

Raugi, G.J., Mullen, J.S., Bark, D.H., Okada, T. and Mayberg, M.R. (1990). Thrombospondin deposition in rat carotid artery. *American Journal of Pathology*, **137**:179–185.

Rekhter, M. and Gordon, D. (1994). Does platelet-derived growth factor-A chain stimulate proliferation of arterial mesenchymal cells in human atherosclerostic plaques? *Circulation Research*, **75**:410–417.

Resnick, N., Collins, T., Atkinson, W., Bonthron, D.T., Dewey, C.F., Jr and Gimbrone, M.A., Jr. (1993). Platelet-derived growth factor B chain promoter contains a cis-acting fluid shear stress response element. *Proceedings of the National Academy of Sciences, USA*, **90**:4591–4595.

Riccio, A., Grimaldi, G., Verde, P., Sebastio, G., Boast, S. and Blasi, F. (1985). The human urokinase-plasminogen activator gene and its promoter. *Nucleic Acids Research*, **13**:2759–2771.

Richardson, W.D., Pringle, N., Mosley, M.J., Westermark, B. and Dubois-Dalcq, M. (1988). A role for platelet-derived growth factor in normal gliogenesis in the central nervous system. *Cell*, **53**:309–319.

Risau, W., Drexler, H., Mironov, V., Smits, A., Siegbahn, A., Funa, K. *et al.* (1992). Platelet-derived growth factor is angiogenic *in vivo*. *Growth Factors*, **7**:261–266.

Roberts, W.C. (1989). Qualitative and quantitative comparison of amounts of narrowing by atherosclerotic plaques in the major epicardial coronary arteries at necroscopy in sudden coronary death, transmural acute myocardial infarction, transmural healed myocardial infarction and unstable angina pectoris. *American Journal of Cardiology*, **64**:324–328.

Rorsman, F., Bywater, M., Knott, T.J., Scott, J. and Betsholtz, C. (1988). Structural characterization of the human platelet-derived growth factor A-chain cDNA and gene: alternative exon usage predicts two different precursor proteins. *Molecular and Cellular Biology*, **8**:571–577.

Rorsman, F., Leveen, P. and Betsholtz, C. (1992). Platelet-derived growth factor (PDGF) A-chain mRNA heterogeneity generated by the use of alternative promoters and alternative polyadenylation sites. *Growth Factors*, **7**:241–251.

Ross, R. (1986). The pathogenesis of atherosclerosis — an update. *New England Journal of Medicine*, **134**:488–500.

Ross, R. (1993). The pathogenesis of atherosclerosis: a perspective for the 1990s. *Nature*, **362**:801–809.

Ross, R. and Glomset, J.A. (1973). Arteriosclerosis and the arterial smooth muscle cell. *Science*, **180**:1332–1339.

Ross, R., Glomset, J.A., Kariya, B. and Harker, L. (1974). A platelet-dependent serum factor that stimulates the proliferation of arterial smooth muscle cells *in vitro*. *Proceedings of the National Academy of Sciences, USA*, **71**:1207–1210.

Ross, R., Masuda, J. and Raines, E.W. (1990a). Cellular interactions, growth factors, and smooth muscle proliferation in atherogenesis. *Annals of the New York Academy of Sciences*, **598**:102–112.

Ross, R., Masuda, J., Raines, E.W., Gown, A.M., Katsuda, S., Sasahara, M. *et al.* (1990b). Localization of PDGF-B protein in macrophages in all phases of atherogenesis. *Science*, **248**:1009–1012.

Sanchez, A., Chesterman, C.N. and Sleigh, M.J. (1991). Novel PDGF A gene transcripts derived by alternative RNA splicing. *Gene*, **98**:295–298.

Sasahara, A., Knott, J.N., Sasahara, M., Raines, E.W., Ross, R. and Westrum, L.E. (1992). Platelet-derived growth factor B-chain-like immunoreactivity in the developing and adult rat brain. *Developmental Brain Research*, **68**:41–53.

Sasahara, M., Hayase, Y., Yang, X.H., Iihara, K., Amano, S. and Hazama, F. (1995). Platelet-derived growth factor B-chain comprises the major part of enhanced released mitogen from aortic endothelial cells of stroke-prone spontaneously hypertensive rats. *Clinical Experimental Pharmacology and Physiology*, **22**:123–125.

Sauer, M.K. and Donoghue, D.J. (1988). Identification of nonessential disulfide bonds and altered conformations in the v-sis protein, a homologue of the B chain of platelet-derived growth factor. *Molecular and Cellular Biology*, **8**:1011–1018.

Scarpati, E.M. and DiCorleto, P.E. (1996). Identification of a thrombin reponse element in the human platelet-derived growth factor B-chain (c-sis) promoter. *Journal of Biological Chemistry*, **271**:3025–3032.

Seifert, R.A., Schwartz, S.M. and Bowen-Pope, D.F. (1984). Developmentally regulated production of platelet-derived growth factor-like molecules. *Nature*, **311**:669–671.

Seifert, R.A., Hart, C.E., Phillips, P.E., Forstrom, J.W., Ross, R., Murray, M.J. *et al.* (1989). Two different subunits associate to create isoform-specific platelet-derived growth factor receptors. *Journal of Biological Chemistry*, **264**:8771–8778.

Shankar, R., de la Motte, C. and DiCorleto, P.E. (1992). 3-Deazaadenosine inhibits thrombin-stimulated platelet-derived growth factor production and endothelial-leukocyte adhesion molecule-1-mediated monocytic cell adhesion in human aortic endothelial cells. *Journal of Biological Chemistry*, **267**:9376–9382.

Slamon, D.J. and Cline, M.J. (1984). Expression of cellular oncogenes during embryonic and fetal development of the mouse. *Proceedings of the National Academy of Sciences, USA*, **81**:7141–7145.

Siegbahn, A., Hammacher, A., Westermark, B. and Heldin, C.-H. (1990). Differential effects of the various isoforms of platelet-derived growth factor on chemotaxis of fibroblasts, monocytes, and granulocytes. *Journal of Clinical Investigation*, **85**:916–920.

Silverman, E.S., Khachigian, L.M., Lindner, V., Williams, A.J. and Collins, T. (1997). Inducible PDGF A-chain transcription in vascular smooth muscle cells is mediated by Egr-1 displacement of Sp1 and Sp3. *American Journal of Physiology*, **273**:H1415–H1426.

Smits, A., Hermansson, M., Nister, M., Karnushina, I., Heldin, C.-H., Westermark, B. *et al.* (1989). Rat brain capillary endothelial cells express functional PDGF B-type receptors. *Growth Factors*, **2**:1–8.

Soriano, P. (1994). Abnormal kidney development and hematological disorders in PDGF β-receptor mutant mice. *Genes & Development*, **8**:1888–1896.

Starksen, N.F., Harsh, G.R., Gibbs, V.C. and Williams, L.T. (1987). Regulated expression of the platelet-derived growth factor A chain gene in microvascular endothelial cells. *Journal of Biological Chemistry*, **262**:14381–14384.

Stary, H.C., Chandler, A.B., Dinsmore, R.E., Fuster, V., Glagov, S., Insull, W. *et al.* (1995). A definition of advanced types of atherosclerotic lesions and a histological classification of atherosclerosis: a report from the Committee on Vascular Lesions of the Council on Atherosclerosis, American Heart Association. *Arteriosclerosis, Thrombosis, and Vascular Biology*, **15**:1512–1531.

Stroobant, P. and Waterfield, M.D. (1984). Purification and properties of porcine platelet-derived growth factor. *EMBO Journal*, **3**:2963–2967.

Sukhatme, V.P., Cao, X., Chang, L.C., Tsai-Morris, C.-H., Stamenkovich, D., Ferreira, P.C.P. *et al.* (1988). A zinc finger-encoding gene coregulated with c-fos during growth and differentiation, and after cellular depolarization. *Cell*, **53**:37–43.

Swan, D.C., McBride, O.W., Robbins, K.C., Keithley, D.A., Reddy, E.P. and Aaronson, S.A. (1982). Chromosomal mapping of the simian sarcoma virus onc gene analogue in human cells. *Proceedings of the National Academy of Sciences, USA*, **79**:4691–4695.

Takimoto, Y., Wang, Z.Y., Kobler, K. and Deuel, T.F. (1991). Promoter region of the human platelet-derived growth factor A-chain gene. *Proceedings of the National Academy of Sciences, USA*, **88**:1686–1690.

Thyberg, J., Ostman, A., Backstrom, G., Westermark, B. and Heldin, C.-H. (1990). Localization of platelet-derived growth factor (PDGF) in CHO cells transfected with PDGF A- or B-chain cDNA: retention of PDGF-BB in the endoplasmic reticulum and Golgi complex. *Journal of Cell Science*, **97**:218–229.

Tong, B.D., Auer, D.E., Jaye, M., Kaplow, J.M., Ricca, G., McConathy, E. *et al.* (1987). cDNA clones reveal differences between human glial and endothelial cell platelet-derived growth factor A-chains. *Nature*, **328**:619–621.

Trejo, S.R., Fahl, W.E. and Ratner, L. (1996). c-sis/PDGF-B promoter transactivation by the Tax protein of human T-cell leukemia virus type 1. *Journal of Biological Chemistry*, **271**:14584–14590.

Tsai-Morris, C.-H., Cao, X. and Sukhatme, V.P. (1988). 5′-flanking sequence and genomic structure of Egr-1, a murine mitogen inducible zinc finger encoding gene. *Nucleic Acids Research*, **16**:8835–8846.

Ueda, M., Becker, A.E., Kasayuki, N., Kojima, A., Morita, Y. and Tanaka, S. (1996). *In situ* detection of platelet-derived growth factor-A and -B chain mRNA in human coronary arteries after percutaneous transluminal coronary angioplasty. *American Journal of Pathology*, **149**:831–843.

van den Ouweland, A.M.W., van Groningen, J.J.M., Schalken, J.A., van Neck, H.W., Bloemers, H.P.J. and van den Ven, W.J.M. (1987). Genomic structure of the c-sis transcriptional unit. *Nucleic Acids Research*, **15**:959–970.

Wang, Z.Y., Madden, S.L., Deuel, T.F. and Rauscher III, F.J. (1992a). The Wilms' tumor gene product, WT-1, represses transcription of the platelet-derived growth factor A-chain gene. *Journal of Biological Chemistry*, **267**:21999–22002.

Wang, Z., Lin, X.H., Qiu, Q.-Q. and Deuel, T.F. (1992b). Modulation of transcription of the platelet-derived growth factor A-chain gene by a promoter region sensitive to S1 nuclease. *Journal of Biological Chemistry*, **267**:17022–17031.

Waterfield, M.D., Scrace, G.T., Whittle, N., Stroobant, P., Johnsson, A., Wasteson, A. *et al.* (1983). Platelet-derived growth factor is structurally related to the putatively transforming protein p28sis of simian sarcoma virus. *Nature*, **304**:35–39.

Wight, T.N., Raugi, G.J., Mumby, S.M. and Bornstein, P. (1985). Light microscopic immunolocation of thrombospondin in human tissues. *Journal of Histochemistry and Cytochemistry*, **33**:295–302.

Wilcox, J.N., Smith, K.M., Williams, L.T., Schwartz, S.M. and Gordon, D. (1988). Platelet-derived growth factor mRNA detection in human atherosclerotic plaques by *in situ* hybridization. *Journal of Clinical Investigation*, **82**:1134–1143.

Yarden, Y., Escobedo, A., Kuang, W.-J., Tang-Feng, T.L., Daniel, T.O., Tremble, P.M. *et al.* (1986). Structure of the receptor for platelet-derived growth factor helps define a family of closely related growth factor receptors. *Nature*, **323**:226–234.

Yokote, K., Mori, S., Siegbahn, A., Ronnstrand, L., Wernstedt, C., Heldin, C.-H., *et al.* (1996). Structural determinants in the platelet-derived growth factor α-receptor implicated in modulation of chemotaxis. *Journal of Biological Chemistry*, **271**:5101–5111.

Young, R.M., Mendoza, A.E., Collins, T. and Orkin, S.H. (1990). Alternatively spliced platelet-derived growth factor A-chain transcripts are not tumor specific but encode normal cellular proteins. *Molecular and Cellular Biology*, **10**:6051–6054.

15 Antiphospholipid Antibodies

Tracy McNally and Michael C. Berndt

Hazel and Pip Appel Vascular Biology Laboratory, Baker Medical Research Institute,
P.O. Box 348, Prahran 3181, Victoria, Australia, Tel: 03-9522-4333, Fax: 03-9521-1362

INTRODUCTION

Laboratory identification of antiphospholipid antibodies (aPAs) began in 1907, when Wasserman introduced a diagnostic test for syphilis (Wasserman, 1907). The antigenic component of the Wasserman test was later identified as cardiolipin (Pangborn, 1941). Subsequently all syphilis tests were performed using a reagent containing cardiolipin extracted from bovine heart and, with the wide scale introduction of this test, it soon became apparent that there were a large number of patients whose sera contained antibodies that reacted with the reagent, but who had no clinical evidence of syphilis (Moore and Mohr, 1952). This phenomenon was termed the biological false positive serological test for syphilis (BFP-STS) and was identified in a large number of subjects with transient infections unrelated to syphilis. In the majority of cases, the positive reaction disappeared after resolution of the infection, however, in the remainder of cases, the positive test persisted over months or years. The two groups were classified as acute and chronic BFP-STS accordingly. With further study of patients with chronic BFP-STS, a high incidence of autoimmune disease, particularly systemic lupus erythematosus (SLE) was noted (Moore and Lutz, 1955).

In the early fifties, there were reports of three patients with SLE and BFP-STS whose plasma contained an acquired inhibitor of coagulation which prolonged the whole blood clotting time and prothrombin time. In two patients, an association with a haemorrhagic disorder was reported (Conley and Hartmann, 1952), and in the third, a history of thrombosis was reported (Mueller *et al.*, 1951). A direct association between the acquired inhibitor of coagulation and BFP-STS was later demonstrated when it was shown that the anticoagulant activity could be adsorbed by the reagent used in the syphilis test (Laurell and Nilsson, 1957). As a result of its frequent detection in patients with SLE, the atypical coagulation inhibitor was subsequently termed the lupus anticoagulant (LA) (Feinstein and Rappaport, 1972).

Contrary to the laboratory evidence of inhibition of *in vitro* coagulation, it soon became apparent that LAs were associated with clinical events of a thrombotic nature (Bowie *et al.*, 1963), and that if bleeding did occur in the face of this atypical inhibitor, it was almost always due to a specific prothrombin deficiency or thrombocytopenia (Feinstein and Rappaport, 1972). Subsequent studies have confirmed an association between the presence of aPAs and thromboembolic clinical events, and associations between the presence of aPAs and a history of recurrent fetal loss, intra-uterine death and intra-uterine growth retardation and thrombocytopenia have also been reported (reviewed in McNeil *et al.*, 1991).

Attempts to elucidate the immunological specificity of aPAs have yielded varying results that have highlighted the heterogeneous nature of this antibody population. Early studies focused on the charge and phase properties of the phospholipid target (Harris *et al.*, 1985a; Rauch*et al.*, 1986) but more recent studies have identified the requirement of a protein cofactor for the binding of aPAs (Galli *et al.*, 1990; McNeil *et al.*, 1990; Matsuura *et al.*, 1990; Bevers *et al.*, 1991). β_2 glycoprotein-I (β_2 GPI) was first identified as a cofactor for the binding of anticardiolipin antibodies (aCL) (McNeil *et al.*, 1990) and was later identified as a requirement for the expression of some LA activities (Galli *et al.*, 1992). However, the heterogeneous nature of these antibodies has been reflected in their cofactor requirements as additional protein cofactors have been identified (Bevers *et al.*, 1991; Oosting *et al.*, 1993).

LABORATORY IDENTIFICATION OF APAS

APAs are currently detected in plasma or serum by prolongation of phospholipid dependent coagulation tests (LA) or antibody binding in solid phase immunoassays employing phospholipid antigens, most commonly cardiolipin (aCL).

LA Antibodies

Recent guidelines (Exner *et al.*, 1991) recommend the following three step approach for LA identification: (i) Demonstration of prolongation of a phospholipid coagulation test; (ii) Demonstration that such prolongation is the result of an inhibitor rather than coagulation factor deficiency (no correction on addition of normal control plasma) and; (iii) Demonstration of the phospholipid dependent nature of the inhibitor, by neutralisation of the inhibitor by addition of lysed washed platelets or phospholipid liposomes containing phosphatidylserine or hexagonal phase phospholipids. Although no conclusive agreement has been reached about the phospholipid dependent coagulation tests that should be employed for LA identification, the current guidelines recommend at least one of the following: kaolin clotting time (KCT); dilute Russell's Viper venom time (DRVVT); tissue thromboplastin inhibition test (TTI) or a modified activated partial thromboplastin time (APTT).

ACL Antibodies

In 1983, Harris *et al.* developed a solid phase radioimmunoassay (RIA) to replace the syphilis screening test for the measurement of aCL. Two years later an enzyme-linked immunosorbent assay (ELISA) with comparable sensitivity to the RIA was introduced (Loizou *et al.*, 1985). Further developments in standardisation of the assay and the widespread availability of reference sera have continued to improve the inter-laboratory variability of results (Harris, 1990).

Of the different immunoglobulin isotypes, IgG aCL are the most closely associated with a history of thrombosis (Gharavi *et al.*, 1987). Positive LA results have been demonstrated to be more closely associated with thrombotic events than positive aCL results in a number of studies (McNeil *et al.*, 1991), which is perhaps a reflection of the inferior specificity of the aCL assay for antibodies associated with a significant thrombotic risk. Recently, solid phase assays employing β_2GPI as the antigenic component have been developed and have demonstrated good correlation with a history of thromboembolic disease (McNally *et al.*, 1995).

CLINICAL CONDITIONS ASSOCIATED WITH DEVELOPMENT OF APAS

Systemic Lupus Erythematosus And Other Autoimmune Disorders

Antiphospholipid antibodies were first recognised in patients with SLE, and it remains the most common condition in which aPAs are found (Derksen and Kater, 1985). SLE is a multi-system autoimmune disease with many potential clinical features. The aetiology of this disease, which predominantly affects women during child bearing years, is unknown. There is some evidence for a viral aetiology and there may be a genetically determined susceptibility to the development of SLE in appropriate environmental conditions, as there is an increased prevalence of the disease in families of patients with SLE.

SLE is perhaps best described as a clinical syndrome, with a complex multifactorial aetiology characterised by inflammation and involvement of most of the body's organs or systems (Tan *et al.*, 1982). It is subject to remissions and exacerbations and though the mucoskeletal system and skin are invariably affected, it also frequently gives rise to manifestations in the kidney, heart, lungs and central nervous system. The diversity of the clinical features of SLE is matched by an apparent diversity among the autoantibodies detectable in the serum from the patients. The various autoantibodies have specific associations with certain clinical findings and outcomes. The presence of antinuclear antibodies is detected by indirect immunofluoresence in approximately 95% of SLE patients. Antinuclear antibodies that are directed against double stranded DNA are closely associated with renal disease, and antibodies against Rho (a combination of protein and a small cytoplasmic RNA) are associated with a photosensitive subset of SLE patients. Antiphospholipid antibodies are detected in approximately 35% of SLE patients and are associated with a high prevalence of venous and arterial thrombosis and fetal loss in these patients (Love and Santorro, 1990).

Harris *et al.* (1985a, b) have reported associations between the presence of aPAs and other autoimmune diseases including Sjorgen's syndrome, mixed connective tissue disease, rheumatoid arthritis, Bechet's syndrome and immune thrombocytopenia. They also reported more tenuous links with multiple sclerosis and scleroderma.

Infection

The spectrum of infections associated with aPAs includes viral, bacterial, parasitic and treponemal infections (Moore and Mohr, 1952). The development of aPAs in patients with infection may represent a transient immune response to antigenic material of the organism, or alternatively the organism may cause host cell membrane damage that leads to exposure or modification of cell membrane phospholipids, such that they become antigenic. Generally, aPAs occurring in infection are not considered to be associated with thrombosis (Love and Santoro, 1990).

Drugs

Many drugs are associated with the development of aPAs, including hydralazine, procainamide, quinidine, amoxycillin and propanalol, although chlorpromazine is by far the most common cause (Derksen and Kater, 1985). Lupus anticoagulant and aCL have been detected in over 30% patients treated with chlorpromazine for more than one year (Canoso and Sise, 1982; Canoso and Oliviera, 1988). The rationale for development of aPAs in patients receiving chlorpromazine therapy is unclear but is probably related to the lipid binding properties of the compound (Mori *et al.*, 1980).

Normal Subjects

Antiphospholipid antibodies have also been detected in subjects with no known underlying disease, with reported incidences as high as 8% (Shi *et al.*, 1990). Interestingly, the majority of positive LA tests in this series were found in young females, the same subject group that predominates among SLE patients.

CLINICAL ASSOCIATIONS OF APAS

Since the early identification of an association between the presence of aPAs and thrombosis, large retrospective studies have confirmed these findings. Similar studies have also identified associations between the presence of aPAs and recurrent fetal loss and thrombocytopenia.

Analysis of the results of 21 studies of SLE patients between 1979 and 1989, showed that of a total of 1428 patients, thrombotic events were much more frequent in patients with aPAs (42%) than in those without (13%) (McNeil *et al.*, 1991). Analysis of patients identified as aPA positive and then investigated for previous history of thrombosis, showed that of 1147 patients with aPAs, 31% had a history of one or more thrombotic events (McNeil *et al.*, 1991). The use of assays for markers of thrombin generation such as the prothrombin fragment (F1+2) and fibrinopeptide A, have allowed the demonstration of hypercoagulability in patients with SLE and aPAs (Ginsberg *et al.*, 1993).

Retrospective studies of fetal loss in patients with SLE have also identified a strong association between aPAs and recurrent fetal loss. In a recent study of 42 consecutive patients with SLE, 122 pregnancies and a history of at least one pregnancy loss, pregnancy loss occurred in 24 of the 55 pregnancies in patients with persistently positive LA tests, compared with 8 of 67 pregnancies in LA negative patients (Ginsberg *et al.*, 1992). A similar trend was seen with aCL positivity in this group with 10 of 14 pregnancies in aCL positive patients resulted in pregnancy loss compared to 22 of 108 pregnancies in aCL negative patients. The cause of fetal wastage in patients with aPAs is thought to be due to thrombosis of placental vessels, with resultant ischaemia and infarction (de Wolf *et al.*, 1982).

Antiphospholipid antibodies are also associated significantly with mild thrombocytopenia, particularly in patients with SLE. In a review of 13 studies encompassing 869 SLE patients, it was shown that thrombocytopenia occurred more frequently in the 406 patients with aPAs (38%) than those without (11%) (Lechner *et al.*, 1985). Severe thrombocytopenia, however, is a less frequent finding and in a more recent study was reported to be present in 11% of patients with aPAs (Italian Registry of aPAs, 1993).

In addition to these highly significant clinical associations with aPAs, livedo reticularis and haemolytic anaemia have also been frequently reported in subjects with aPAs (Alarcon-Segovia and Sanchez-Guerrero, 1989; Asherson, 1991).

Antiphospholipid Syndrome

The term "antiphospholipid syndrome" (aPS) was proposed originally by Harris *et al.* (1987) in order to describe the concomitant occurrence of aPAs and clinical thrombotic events. Antiphospholipid syndrome is characterised by venous and/or arterial thrombotic events and/or recurrent fetal losses, which are accompanied frequently by mild thrombocyto-penia, in the presence of LA, elevated aCL levels, or both. The IgG aCL isotype is the most commonly detected in aPS, with IgM occurring less frequently and IgA detected occasionally (Gharavi *et al.*, 1987). APS is most commonly found in patients with greatly increased levels of IgG aCL levels (Harris *et al.*, 1986).

In addition to its frequent occurrence in patients with SLE, aPS is also encountered in patients with no clinical or laboratory evidence of SLE or any definable autoimmune disease. Such patients are classified as having primary antiphospholipid syndrome (PaPS) (Asherson, 1988). The original criteria for classification of PaPS are shown in Table 15.1. A few patients with PaPS have subsequently evolved probable or full blown SLE (Asherson *et al.*, 1991; Andrews *et al.*, 1993) and for this reason, annual year follow-up of PaPS patients is recommended in order to rule out subsequent development of autoimmune disease.

Table 15.1 Criteria for classification of PaPS (Asherson, 1988)

Clinical Events	Laboratory Findings	Conditions
venous thrombosis	IgG aCL (moderate/high level)	(i) patients should have at least one clinical and one laboratory finding during their illness
arterial thrombosis	IgM aCL (moderate/high level)	(ii) laboratory test must be positive on at least two occasions >3 months apart
recurrent fetal loss thrombocytopenia	positive LA test	(iii) follow-up of >5 years is recommended to rule out subsequent development of auto-immune disease

The criteria defined for the classification of SLE cannot satisfactorily distinguish between patients with PaPS and aPS secondary to SLE, as some of the SLE criteria could be fulfilled as a result of pathogenic mechanisms unrelated to SLE. For this reason, Piette *et al.* (1992) proposed a set of exclusion criteria to distinguish PaPS and aPS secondary to SLE (Table 15.2).

Table 15.2 Proposed empirical exclusion criteria to distinguish primary and Sle related APS (Piette *et al.*, 1992)

Criteria	Conditions
Malar rash Discoid rash	The presence of any of the SE criteria excludes the diagnosis of PaPS
Oral or pharyngeal ulceration (excluding nasal septum ulceration or perforation)	A follow-up of >5 years is necessary to rule out subsequent development of SLE
Frank arthritis Pleuritis in absence of heart failure or PE	
Pericarditis in absence of MI or uraemia	
Persistent proteinuria (0.5 g/day), due to biopsy proven immune complex related glomerulonephritis	
Lymphopenia (<×109/1)	
antibodies to native DNA by RIA or Crithidia fluorescence	
Anti-extractable nuclear antigen antibodies	
Anti-nuclear antibodies of more than 1:320	
Treatment with drugs known to induce aPAs	

TREATMENT OF APS

Arterial and Venous Thrombosis

Acute thrombotic episodes are treated by thrombolytic and/or anticoagulant in the conventional way. Prophylaxis in the face of a positive aPA result and absence of a previous thrombotic event is not indicated as only approximately 30% of patients with aPAs experience thrombotic complications (McNeil *et al.*, 1991). There is now strong evidence, however, that thrombotic events tend to reoccur in aPS (Rosove and Brewer, 1992; Derksen *et al.*, 1993) and thus the perceived risk:benefit ratio should be seriously considered for each individual patient. The decision is usually based on the nature and severity of the presenting thrombotic event. Although oral anticoagulant therapy is generally employed for the prophylaxis of venous thrombosis and anti-platelet drugs, such as aspirin and dipyrimadole, for prophylaxis of arterial thrombosis, the duration and intensity of prophylaxis remains ill defined. However, recent evidence has suggested that long-term high-intensity anticoagulation therapy is indicated and that low dose aspirin has little efficacy in the prevention of secondary thrombosis in aPS (Khamasta *et al.*, 1995).

Recurrent Fetal Loss

The management of recurrent fetal loss in aPS is largely based upon evidence from individual reports and to date there have been no reports of large prospective therapeutic trials. Good results have been claimed for the use of corticosteroids, aspirin, heparin and intravenous γ globulin (Greaves and Preston, 1991).

Thrombocytopenia

Treatment of thrombocytopenia in aPS is not necessary if the degree of thrombocytopenia is mild and no danger of bleeding exists. If treatment is required the strategies for treatment of immune thrombocytopenia are employed and include corticosteroid therapy, immunosuppressives, intravenous γ globulin and splenectomy.

IMMUNOLOGICAL SPECIFICITY OF APAS

The heterogeneity of aCL and LA activities has long been suggested by the disparity of activities in different patient plasmas and was confirmed when it was shown that in most cases, aCL could be separated from LA antibodies (Exner *et al.*, 1988; McNeil *et al.*, 1989). The two groups used similar methods for separation of aCL and LA IgGs, incorporating phospholipid affinity chromatography techniques. Analysis of the two IgG fractions showed that aCL IgG were able to bind to several anionic phospholipids in solid phase ELISAs but did not express anticoagulant activity, and LA antibodies could prolong phospholipid coagulation tests but showed no binding to phospholipids in the solid phase ELISA (McNeil *et al.*, 1989). Further evidence for the existence of separate antibody populations was provided when it was demonstrated that aCL and LA activities are the result of different immunoglobulin isotypes in some patients (McNeil *et al.*, 1989).

This concept was accepted widely and investigations of immunological specificities of aCL and LA antibodies were regarded as separate areas of study. Then in 1991, LA activity was demonstrated in aCL IgGs isolated from the plasma of five out of 16 patients with concomitant aCL and LA activities (Bevers *et al.*, 1991). Bevers proposed the terms aCL types A and B to denote aCL IgGs with and without anticoagulant activity, respectively. It is now considered that, whereas in approximately 60–70% patients LA and aCL activities are the result of different antibody populations, in the remainder a single antibody population can express both activities. Therefore at least three different aPAs have been recognised: aCL type A, which have intrinsic anticoagulant activity, aCL type B, which have no detectable anticoagulant activity, and LA antibodies, which possess anticoagulant activity but which show no binding in solid phase assays.

Phospholipid Specificities

Anticardiolipin antibodies

Early studies showed that aCL detected in patients with aPS cross-react with other anionic phospholipids but show no reaction with zwitterionic phospholipids (Harris *et al.*, 1985c). In some cases the cross-reactivity extended to DNA, and some anti-DNA antibodies were shown to bind to cardiolipin (Lafer *et al.*, 1981). For this reason, it was considered likely that aCL antibodies were polyreactive with anionic phospholipids and recognise epitopes consisting of repeating single negatively charged phosphodiester groups with neutral or anionic variable head groups. Interestingly, it was later shown that aCL cross react with oxidised low density lipoproteins (Vaarala *et al.*, 1993)

Lupus anticoagulant

The specificity of LA antibodies for anionic antiphospholipids was first demonstrated in 1980 when a monoclonal IgM with LA activity was purified from the plasma of a patient with Waldenstrom's macroglobulinaemia (Thiagarajan *et al.*, 1980). Reactivity of the immunoglobulin with several anionic phospholipids was demonstrated using immuno-precipitation techniques. The antibody was also shown to inhibit the calcium-dependent binding of coagulation factors II and X to an anionic procoagulant phospholipid surface.

Studies with human monoclonal antibodies with LA activity suggested that the biophysical configuration of the lipid may be important for LA specificity. Rauch's group demonstrated that LA activity of monoclonal antibodies could be neutralised by addition of hexagonal phase PE, but not lamellar PE (Rauch *et al.*, 1986) (Figure 15.1). This work was confirmed using plasma derived LA antibodies (Rauch *et al.*, 1989).

Protein Cofactors for APA Binding

In 1990, three groups reported independently the requirement of a protein cofactor for the binding of aCL antibodies from patients with autoimmune disorders to anionic phospholipids in solid phase ELISA (McNeil *et al.*, 1990; Galli *et al.*, 1990; Matsuura *et al.*, 1990). It was reported originally that aCL antibodies isolated by ion exchange chromatography or phospholipid affinity chromatography, failed to bind to phospholipid affinity columns in the absence of native or bovine plasma (McNeil *et al.*, 1989). This suggested the requirement of a cofactor present in human or bovine plasma and serum for binding to the phospholipid.

Low molecular weight substances such as phospholipids are generally considered to be poor antigens according to the classical hapten concept of Landsteiner, so the demonstration of a protein component in the antigenic determinant of aPAs went some way to explaining this paradox. The requirement of protein cofactor was also compatible with the previous findings of differences in binding associated with phase presentation of the phospholipid, as the phospholipid may assume an hexagonal phase on binding of protein.

β_2 Glycoprotein-I

McNeil *et al.* (1990) identified the cofactor required for aCL binding as β_2 Glycoprotein-I (β_2GPI). This new finding explained the previous observation that the use of 10% adult

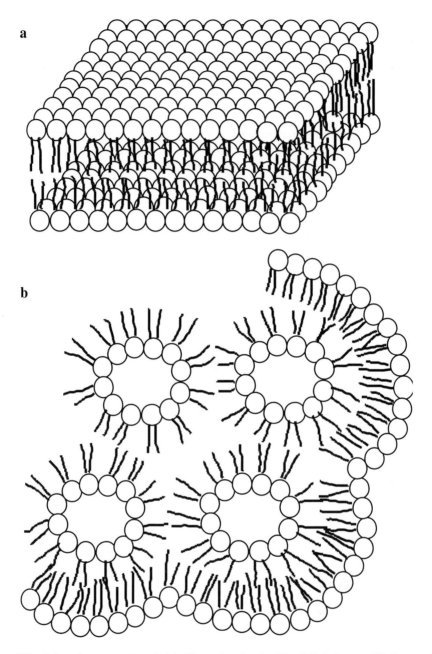

Figure 15.1   Schematic representation of (a) the bilayer phase forming PE and (b) the hexagonal II phase egg PE.

bovine serum improved the reliability of aCL assays (Loizou *et al.*, 1985), as the serum provides an exogenous source of β_2GPI. Although the β_2GPI requirement of aCL binding has been confirmed by several laboratories, some investigators have reported that β_2GPI is not an absolute requirement for, but enhances aCL binding (Pierangeli *et al.*, 1992; Sammaritano *et al.*, 1992). It is possible that the concentration of aCL IgG used in these studies was so high, or as the authors suggest, the antibody avidities for aCL so great, as to preclude the β_2GPI requirement.

Galli *et al.* (1992) identified the cofactor requirement of the anticoagulant activity of aCL type A IgG as β_2GPI. In a series of experiments, she showed that prolongation of the DRVVT by aCL type A IgG was abolished using β_2GPI deficient plasma, but could be restored upon addition of β_2GPI. In contrast, the LA antibodies caused prolongation of the DRVVT irrespective of the presence or absence of β_2GPI. Oosting later confirmed that β_2GPI was a cofactor requirement for LA activity in some patients when she demonstrated that the anticoagulant activity of four out of six plasmas with LA activity was abolished after depletion of β_2GPI from the plasma and restored after addition of exogenous β_2GPI (Oosting *et al.*, 1992).

β_2GPI remains the most commonly identified and closely studied of the reported aPA protein cofactors. This glycoprotein, which is normally present at concentrations of approximately 200 mg/l in plasma (Cohnen, 1969), was first described in 1961 (Schultze *et al.*, 1961), but as yet has no ascribed physiological role. β_2GPI has been reported in all the major lipoprotein fractions and accordingly, was assigned the name apolipoprotein H (Nakaya *et al.*, 1980). A role in lipoprotein metabolism has been proposed but remains unconfirmed. β_2GPI can bind several anionic macromolecular structures including DNA (Kroll *et al.*, 1976), heparin (Polz, 1979), platelets (Schousboe, 1980) and phospholipid vesicles (Wurm, 1984). A possible role in the regulation of haemostasis has been widely investigated since the demonstration of the following *in vitro* anticoagulant activities: inhibition of the contact phase of coagulation (Schousboe, 1985), inhibition of the pro-thrombinase and tenase coagulant reactions (Nimpf *et al.*, 1985) and inhibition of ADP induced platelet aggregation (Nimpf *et al.*, 1986).

β_2GPI is a single chain polypeptide of 326 amino acids, with an apparent molecular weight of 50 kDa. It is highly glycosylated, with the reported carbohydrate content being approximately 18% of the molecular weight (Schultze *et al.*, 1961). β_2GPI contains a sequence of five "sushi" domains and as such is a member of the short consensus repeat (SCR) superfamily. Whilst the first four β_2GPI domains are typical examples of this SCR superfamily, the fifth domain is unusual in that it contains two additional cysteine residues and a long C-terminal tail (Kato and Enjyoji, 1991; Figure 15.2).

Analysis of a β_2GPI preparation lacking aPA cofactor activity has facilitated investigation of the lipid binding area of this molecule (Hunt *et al.*, 1993). This molecule demonstrated a single band on SDS PAGE, but had a slightly lower molecular mass than that usually associated with β_2GPI. The preparation was unable to bind to anionic phospholipids. Amino acid analysis indicated that the molecule had been cleaved between Lys 317 and Thr 318 in the fifth domain. Investigation then focused on this part of the molecule for identification of the lipid binding site and work using synthetic peptides showed that the sequence Glu 274-Cys 288 inhibited both the binding of purified aCL antibodies to cardiolipin, and the binding of radio-labelled β_2GPI to cardiolipin, in a dose dependent manner (Hunt *et al.*, 1994). Using additional peptides, the lipid binding area of β_2GPI was further localised to Cys 281-Cys 288. This sequence is highly positively charged and computer generated models predict that it is surface exposed, while being linked to the main molecule by the flanking cysteines (Steinkasserer *et al.*, 1992).

Prothrombin

In some patients, LA activity requires prothrombin for its expression. Bevers *et al.* (1991) investigated the cofactor requirement of LA antibodies in patients with aCL type B antibodies. Using IgG fractions, they showed that prolongation of phospholipid dependent coagulation tests was only observed with normal plasma of human origin, not with bovine,

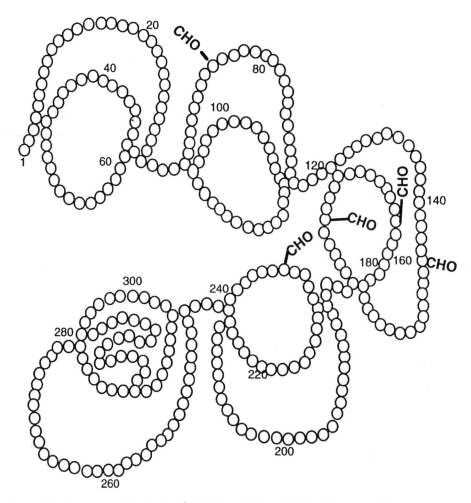

Figure 15.2 Proposed model of bovine β_2GPI (Kato & Enjyoji, 1991). CHO denotes N-linked glycosylation sites, 1 denotes N-terminal end and 326 denotes C-terminal end.

rat or sheep plasma. Whilst investigating thrombin generation using purified proteins and phospholipid vesicles, they then went on to show that a dose dependent inhibition by LA antibodies was observed only when human prothrombin was used as a substrate. No similar species specificity was demonstrated with factors V or X, indicating that these LA antibodies were most likely directed against an epitope exposed on binding of prothrombin to phospholipid. The involvement of prothrombin in LA activity was not a new concept as it had previously been postulated to be a cofactor for LA activity as early as 1959 (Loeliger, 1959). In a more recent study, Fleck *et al.* (1988) demonstrated anti-prothrombin antibodies in 74% of patients with LA, and showed that the LA activity could be removed by preincubation with prothrombin.

Other protein cofactors

Whilst investigating whether other phospholipid binding haemostatic proteins could serve as cofactors for aPAs, Oosting *et al.* (1993a) identified protein C and protein S as aPA cofactors. They demonstrated inhibition of the protein C mediated inactivation of factor Va by IgG fractions of seven out of 22 patients with aPAs and a history of thrombosis. This inhibitory activity could be adsorbed by protein C coated vesicles in three of these patients and by protein S coated cardiolipin vesicles in the remaining patients. Using a solid phase ELISA technique, antibodies directed against protein C were identified in 11% of patients with SLE (Ruiz-Arguelles *et al.*, 1993).

Phospholipase A_2 has also been implicated as a possible aPA cofactor. As early as 1981, aPAs were shown to inhibit prostacyclin production by endothelial cells and this inhibition was postulated to be the result of inhibition of phospholipase A_2 activity (Carreras *et al.*, 1981). The inhibition of phospholipase A_2 activity by IgG from patients with aPAs was later confirmed (Schorer *et al.*, 1992). As a result of these findings, it has been postulated that some aPAs are in fact directed towards phospholipid-phospholipase A_2 complexes (Vermylen and Arnout, 1992).

Investigations are continuing to identify additional protein cofactors involved in aPA activities, including annexin V (placental anticoagulant protein I), high molecular weight kininogen and platelet activating factor (Amiral *et al.*, 1994; Nakamura *et al.*, 1995; Sugi and McIntyre, 1995; and Barquinero *et al.*, 1993).

The Antigenic Target

Most of the recent work studying the antigenic target of aPAs has been performed using aCL antibodies and β_2GPI. Several groups have demonstrated binding of antibodies to β_2GPI in the absence of phospholipid in solid phase ELISA systems (Galli *et al.*, 1990; Arvieux *et al.*, 1991; Viard *et al.*, 1992; Keeling *et al.*, 1992; Matsuda *et al.*, 1994), whereas other groups have failed to show reactivity to β_2GPI alone (Gharavi *et al.*, 1993). Recently, Matsuura *et al.* (1994) reported that aCL antibodies reacted with β_2GPI coated on electron or γ-irradiated ELISA plates, but not standard polystyrene microtitre plates, and that the degree of binding depended on the irradiation dose. The finding that the type of microtitre plate affects antibody binding could therefore account for the discrepancies between results of different investigators.

γ irradiation acts to partially oxidise the surface, which renders the plate more anionic and enhances protein binding capacity significantly (Onyiriuka *et al.*, 1990). The requirement for γ irradiation of microtitre plates for demonstration of antibody binding to β_2GPI, therefore suggests that either the antibodies require a high density of immobilised antigen which cannot be achieved with standard polystyrene plates, or the antibodies are specific for a neo-epitope on β_2GPI that is only exposed after binding to an anionic surface. The work of Roubey *et al.* (1994) supports the first hypothesis as they demonstrated that the binding of antibodies was dependent on β_2GPI density, but the results of Matsuura *et al.* (1994) suggest that a conformational change of β_2GPI occurs and facilitates antibody binding.

The study of the specificities of aPAs, and in particular the identification of cofactor requirements for their binding, has highlighted the inadequacy of the nomenclature for their classification. Vermylen and Arnout (1992) have proposed that the term "antiphospholipid antibodies" should be replaced by "antiphospholipid-protein antibodies". However, because antibody binding to β_2GPI and prothrombin has been observed in the absence of phospholipid, the term "antibodies to phospholipid-binding plasma proteins" has also been suggested (Roubey, 1994).

Interestingly, the requirement for β_2GPI for binding seems to be limited to aPAs associated with autoimmune conditions and/or antiphospholipid syndrome. Conversely, β_2GPI appears to inhibit phospholipid binding of aPAs occurring in association with infectious disease, presumably by acting as a competitive inhibitor for the phospholipid (Hunt *et al.*, 1992).

Immunogenicity of APAs

The demonstration of reactivity of an antibody with phospholipid, or a phospholipid binding plasma protein, *in vitro*, does not prove that the antigen induces the production of autoantibodies, nor that it serves as the target antigen *in vivo*. In order to assess the nature of the true target antigen, examination of antigenicity or pathogenicity of the antigen is required. Accordingly, many groups have investigated the antigenicity of various phospholipids and phospholipid binding plasma proteins in animal models.

The original work of Rauch and Janoff (1990) demonstrated that hexagonal II phase phosphatidylethanolamine (PE) but not bilayer PE was immunogenic and induced the production of antibodies to PE in BALB/c mice. Interestingly, the antibodies were cross-reactive with cardiolipin and had LA activity. The same investigators later immunised the mice with cardiolipin alone, cardiolipin in combination with β_2GPI and β_2GPI alone. The results of these studies demonstrated that mice immunised with cardiolipin mixed with β_2GPI produced antibodies to cardiolipin and β_2GPI, and that mice immunised with cardiolipin alone did not produce any phospholipid antibodies (Rauch and Janoff, 1992). The mice immunised with β_2GPI produced antibodies to β_2GPI. These results suggest that cardiolipin undergoes a conformational change on binding to β_2GPI and that the resultant neoepitope(s) provide the antigenic stimulus.

In contrast to these results, Gharavi *et al.* (1992) reported the development of aPAs and anti-β_2GPI antibodies in mice and rabbits immunised with human and bovine β_2GPI. The binding of the aPAs was shown to be specific for anionic phospholipids and enhanced by β_2GPI. The authors state that possible contamination of the β_2GPI preparation had been excluded, and postulate that the aPAs were produced in response to the foreign β_2GPI bound to host phospholipid. This work has since been reproduced by Pierangeli and Harris (1993).

THE PATHOGENIC ROLE OF APAS IN APS

The close association between aPAs and thrombosis has suggested a direct pathogenic role of aPAs in inducing thrombotic events and studies using experimental animal models have provided direct evidence. Several animal models for aPS have been described. The MRL/lpr autoimmune strain of mice was the first reported animal model of aPS (Gharavi *et al.*, 1989). After the age of two months, MRL/lpr mice develop high levels of autoantibodies, including IgG aPAs, in association with thrombocytopenia. Recently, another mouse model for aPS has been reported. NZW X BXSB F_1 male mice develop systemic lupus-like disease and produce aCL with similar antigenic specificity to that observed in humans (Hashimoto *et al.*, 1992). These mice also develop myocardial infarctions and coronary vascular disease (Hang *et al.*, 1981).

In addition to animal models for spontaneous aPS, some investigators have induced aPS in normal mice by passive transfer or active immunisation with patient plasma derived aPAs. Branch *et al.* (1990) showed that pregnant BALB/c mice aborted within 24 hours following immunisation with IgG from patients with aPS and recurrent fetal loss. A reduced yield of live offspring has also been demonstrated in ICR mice following immunisation with aCL (Blank *et al.*, 1991).

The demonstration of a close association between the presence of aPAs and thromboembolic complications has prompted numerous studies to identify abnormal haemostatic regulation in such subjects. Although no single mechanism has yet been attributed to these events, these investigations have highlighted abnormalities of many different aspects of haemostasis in patients with aPAs. The recent identification of haemostatically active proteins as antigenic targets for aPAs has further intensified this area of study.

APAs and The Endothelium

The normal maintenance of an undamaged, and therefore non-thrombogenic, endothelial lining of blood vessels is dependent on the metabolic activity of the endothelial cells. Any mechanism causing disturbance of the normal functions of these cells may therefore predispose to a thrombotic process. Purified aPAs have been shown to bind to endothelial cells (Cines *et al.*, 1984) and modulate many aspects of endothelial cell function. However, aPA binding to endothelium is not a consistent finding. Adsorption studies with phospholipid micelles have shown that although adsorption completely eradicated aPA binding, endothelial binding was not always inhibited (Del Papa *et al.*, 1992). These results indicate that aPA and anti-endothelial activities may belong to different antibody populations in some patients.

Tissue factor synthesis

Tissue factor (TF) is a small transmembrane surface receptor that serves as the major activator of coagulation *in vivo* (Nemerson, 1988). The regulation of this system depends on control of expression of TF and activity of the tissue factor pathway inhibitor. Endothelial cells express very little TF on their surface under normal conditions. Inflammatory mediators such as interleukin-1 (IL-1), endotoxin and tumour necrosis factor (TNF) can induce endothelial cells to synthesise and express TF (Nawroth and Stern, 1986). Some workers have shown that whole IgG from patients with LA activity can induce endothelial cells to express TF (Tannenbaum *et al.*, 1986; Rustin *et al.*, 1988) in the manner observed with plasma from patients with heparin-induced thrombocytopenia (Cines *et al.*, 1987).

Prostacyclin release

Prostacyclin (PGI_2), a potent vasodilator and inhibitor of platelet aggregation (Weksler and Jaffe, 1986), is secreted by endothelial cells and is considered to play a major role in the maintenance of vessel patency *in vivo*. In 1981 it was reported that plasma with LA activity inhibited prostacyclin production of endothelial cells (Carreras *et al.*, 1981). This inhibition could be overcome by the addition of exogenous arachidonate, suggesting that the reduced production of prostacyclin was due to inhibition of arachidonate liberation. Inhibition of prostacyclin activity by aPAs is not a consistent finding (Hasselar *et al.*, 1988) and as yet, no significant correlation between the presence of a prostacyclin inhibitory activity and a history of thrombosis has been demonstrated.

Protein C activation

Protein C is activated by thrombin in association with the endothelial receptor thrombomodulin (TM). It has been shown that IgG fractions from patients with aPAs inhibit the generation of activated protein C by TM in the presence of phospholipid vesicles and endothelial cell cultures (Freyssinet *et al.*, 1986; Cariou *et al.*, 1988, Keeling *et al.*, 1993). However, other groups have failed to confirm the inhibitory activity of aPAs on protein C activation (Oosting *et al.*, 1991). The presence of anti-TM antibodies in patients with aPAs has been demonstrated recently (Oosting *et al.*, 1993) and the authors postulated that this may account for the reduced TM activity in the presence of whole IgG fractions from patients with aPAs.

Once activated, protein C inactivates factors V and VIII in the presence of protein S. Several groups have reported the inhibition of the activated protein C mediated inactivation of factor Va by aPAs (Marciniak and Romond, 1989; Malia *et al.*, 1990; Borrell *et al.*, 1992). A recent investigation reported that IgGs with activated protein C inhibitory activity were directed against a combination of protein C with phospholipids or protein S with phospholipids (Oosting *et al.*, 1993a).

Fibrinolysis

Fibrinolysis plays a crucial role in the maintenance of blood vessel patency. Plasmin is formed from the zymogen plasminogen by the action of the plasminogen activators (PA) tissue PA (tPA) and urokinase (uPA) (Collen and Lijnen, 1991). Both tPA and uPA are synthesised by endothelial cells, together with their main inhibitor, plasminogen activator inhibitor-1 (PAI-1). It has been reported that fibrinolysis, as measured by the euglobulin fraction lysis time is impaired in patients with SLE (Angles-Cano *et al.*, 1979). More specific assays have allowed demonstration that increased PAI-1 levels were associated with SLE (Glas-Greenwalt *et al.*, 1984). However, no correlation has been demonstrated between the presence of aPAs and increased PAI-1 activity in SLE patients, and investigation of the effect of aPA plasmas on tPA and PAI-1 release from cultured endothelial cells have failed to demonstrate an inhibitory effect (Francis and Neely, 1989). Increased PAI-1 in SLE may therefore reflect an acute phase response in patients with autoimmune disease.

von Willebrand factor

von Willebrand factor (vWF) is synthesised by endothelial cells and megakaryocytes and plays a prominent role in the procoagulant response of platelets. Increased vWF levels have been reported in patients with SLE and statistical associations between increased vWF, the presence of aPAs and thrombosis have been reported (Angles-Cano *et al.*, 1979). Although, increased plasma levels of vWF are indicative of endothelial cell injury, the demonstration of elevated levels of vWF in patients with aPAs does not in itself imply a causal relationship, the association has been demonstrated to be significant, and not merely part of a general acute phase response (Ames *et al.*, 1995).

APAs and Platelets

Increased platelet activation, as indicated by reduced platelet concentrations and increased plasma concentrations of β-thromboglobulin (βTG), has been reported in patients with LA activity (Galli *et al.*, 1988). The same group showed that binding of IgG from plasma with LA activity, induced secretion of βTG from normal platelets, although other investigators have since failed to confirm this (Shi *et al.*, 1993).

Thrombocytopenia in association with aPAs results from an autoimmune destruction of platelets by antiplatelet antibodies. While most anti-platelet antibodies bind recognisable platelet antigens and not phospholipids, antibodies with aPA specificity can bind to platelets (Rauch *et al.*, 1987). However, platelet antibodies directed against the GPIIb/IIIa and/ or GpIb/IX complexes have also been identified in approximately 40% patients with aPAs (Galli *et al.*, 1994).

β_2GPI has been shown to bind to activated platelets and to inhibit ADP induced aggregation (Nimpf, 1985), and it is possible that aPA binding results in perturbation of this anticoagulant property of β_2GPI. β_2GPI has been demonstrated to be a requirement for aPA IgG binding to platelets and aPAs have been shown to inhibit the anticoagulant effect of β_2GPI on the factor Xa generating activity of platelets (Shi *et al.*, 1993b).

Contact Activation

Activation of the contact system is believed to be initiated by the binding of plasma factor XII to a negatively charged surface, where autoactivation of the factor XII occurs, converting it to an active serine protease (Tankersley and Finlayson, 1984). The presence of a small amount of FXIIa can lead to the activation of its substrates, prekallikrein, factor XI and high molecular weight kininogen.

Recent evidence has suggested that contact activation may be important in conditions of hypercoagulability and that large lipoprotein particles carrying appropriate free fatty acid at a sufficient density of negative charge could provide the contact surface that induces autoactivation of factor XII (Mitropoulos *et al.*, 1992). β_2GPI has been shown to inhibit the autoactivation of factor XII on triglyceride rich lipoproteins and some aPAs have been shown to reduce this inhibitory action of β_2GPI (McNally *et al.*, 1996).

The recent identification of high molecular weight kininogen as an aPA protein cofactor (Sugi and McIntyre, 1995) has highlighted the possibility that perturbation of its function may contribute to the thrombotic process. It is possible that aPA binding reduces the high molecular weight kininogen-mediated inhibition of α-thrombin binding and activation of platelets (Puri *et al.*, 1991).

FUTURE DIRECTIONS

An interesting component of the aPS is the observation that recurring thrombotic events seem to be of the same nature, *i.e.* patients with a history of venous thrombosis will continue to present with venous thromboses and patients with recurrent fetal loss most often have no history of circulatory thrombosis. These findings suggest that particular aPAs may be linked with a particular clinical event and/or site. Previous attempts to correlate LA and aCL with specific clinical events have yielded disappointing results and studies are now underway in order to determine whether differing cofactor requirements are associated with different thrombotic risks. The identification of the protein cofactors has also stimulated further attempts to elucidate the pathogenic mechanisms involved in aPS and hopefully these studies will yield useful results that may ultimately improve therapeutic options.

The clinical management of patients with aPS is still based largely on individual case reports. Controlled perspective studies are now underway for the management of secondary prevention of thrombosis and fetal loss in patients with aPS and should hopefully soon yield valuable results.

References

Alarcon-Segovia, D. and Sanchez-Guerrero, J. (1989). Primary antiphospholipid syndrome. *Journal of Rheumatology*, **16**:482–488.
Ames, P.R. (1994). APAs, thrombosis and atherosclerosis in SLE: A unifying membrane stress syndrome. *Lupus*, **3**:371–377.
Amiral, J., Larrivaz, I., Cluzeau, D. and Adam, M. (1994). Standardisation of immunoassays for antiphospholipid antibodies with β_2 glycoprotein-I and the role of other phospholipid cofactors. *Haemostasis*, **24**: 191–203.

Andrews, P.A., Frampton, G. and Cameron, C.S. (1993). Antiphospholipid syndrome and systemic lupus erythematosus. *Lancet*, **342**:988–999.

Angles-Cano, E., Sultan, Y. and Clauvel, J.P. (1979). Predisposing factors to thrombosis in systemic lupus erythematosus: possible relation to endothelial cell damage. *Journal of Laboratory and Clinical Medicine*,
94:312–323.

Arvieux, J., Roussel, B., Jacob, M.C. and Colomb, M.G. (1991). Measurement of antiphospholipid antibodies by
ELISA using β_2 glycoprotein-I as an antigen. *Journal of Immunological Methods*, **143**:223–229.

Asherson, R.A. (1988). A "primary" antiphospholipid syndrome? *Journal of Rheumatology*, **15**:1742–1746.

Asherson, R.A. (1991). Antiphospholipid antibodies: clinical implications reported in medical literature. In *Phospholipid binding antibodies*, edited by E.N. Harris, T. Exner, G.R.V. Hughes and R.A. Asherson, pp. 387–
401. CRC Press.

Barquinero, J., Ordi-Ros, J., Selva, A., Perez-peman, P., Vilardell, M. and Khamasta, M. (1993). Antibodies
against platelet activating factor in patients with antiphospholipid antibodies. *Lupus*, **3**:55–58.

Bevers, E.M., Galli, M., Barbui, T., Comfurius P. and Zwaal RF. (1991). Lupus anticoagulant IgGs (LA) are not
directed to phospholipids only, but to a complex of lipid-bound prothrombin. *Thrombosis and Haemostasis*,
66:629–632.

Blank, M., Cohen, J., Toder, V. and Schoenfeld, Y. (1991). Induction of anti-phospholipid syndrome in naive
mice with mouse lupus monoclonal and human polyclonal anti-cardiolipin antibodies. *Proceedings of the
National academy of Sciences, USA*, **88**:3069–3073.

Borrell, M., Sala, N., de Castellarnau, C., Lopes, S., Gari, M. and Fontcuberta, J. (1992). Immunoglobulin fractions isolated from patients with antiphospholipid antibodies prevent the inactivation of factor Va by activated protein C on human endothelial cells. *Thrombosis and Haemostasis*, **68**:268–273.

Bowie, E.J.W., Thompson, J.H., Pascuzzi, C.A. and Owen, C.A. (1963). Thrombosis in systemic lupus erythematosus despite circulating anticoagulants. *Journal of Laboratory and Clinical Medicine*, **62**:416–430.

Branch, D.W., Dudley, D.J. and Mitchel, M.D. (1990). Immunoglobulin G fractions from patients with antiphospholipid antibodies cause fetal death in BALB/c mice: a model for autoimmune fetal loss. *American Journal
of Obstetrics and Gynaecology*, **163**:210–216.

Canoso, R.T. and Sise, H.S. (1982). Chlorpromazine-induced lupus anticoagulant and associated immunologic
abnormalities. *American Journal of Haematology*, **13**:121–129.

Canoso, R.T. and de Oliviera, R.M. (1988). Chlorpromazine-induced anticardiolipin antibodies and lupus anticoagulant: Absence of thrombosis. *American Journal of Haematology*, **27**:272–275.

Cariou, R., Toblem, G., Belluci, S., Soria, J., Sorai, C., Maclouf, J. and Caen, J. (1988). Effect of lupus anticoagulant on antithrombotic properties of endothelial cells. Inhibition of thrombomodulin-dependent protein C
activation. *Thrombosis and Haemostasis*, **60**:54–58.

Carreras, L.O., Defreyn, G., Machin, S.J., Vermylen, J., Deman, R., Spitz, B. and van Assche, A. (1981). Arterial
thrombosis, intrauterine death and "lupus anticoagulant". Detection of immunoglobulin interfering with
prostacyclin formation. *Lancet*, **1**:244–246.

Casu, B. (1991). Structural features and binding properties of chondroitin sulfates, dermatan sulfate and heparan
sulfate. *Seminars in Thrombosis and Haemostasis*, **17**:9–14.

Chong, B.H. (1995). Heparin-induced thrombocytopenia. *British Journal of Haematology*, **89**:431–439.

Cines, D.B., Lyss, A.P. and Reeber, M. (1984). Presence of complement-fixing anti-endothelial cell antibodies in
systemic lupus erythematosus. *Journal of Clinical Investigation*, **73**:611–625.

Cines, D.B., Tomask, A. and Tannenbaum, S. (1987). Immune endothelial cell injury in heparin-induced thrombocytopenia. *New England Journal of Medicine*, **316**:581–589.

Cohnen, G. (1970). Immunochemical quantitation of β_2-glycoprotein-I in various diseases. *Journal of Laboratory
and Clinical medicine*, **75**:212–216.

Collen, D. and Lijnen, H.R. (1991). Basic and clinical aspects of fibrinolysis and thrombolysis. *Blood*, **78**:3114–3124.

Conley, C.L. and Hartmann, R.C. (1952). Haemorrhagic disorder caused by circulating anticoagulant in patients
with disseminated lupus erythematosus. *Journal of Clinical Investigation*, **31**:621–622.

De Wolf, F., Carreras, L.O., Moerman, P., Vermylen, J., Van-Assche, A. and Ranaer, M. (1982). Decidual vasculo-
pathy and extensive placental infarction in a patient with repeated thromboembolic accidents, recurrent fetal
loss and a lupus anticoagulant. *American Journal of Obstetrics and Gynaecology*, **142**:829–834.

Del Papa, N., Meroni, P.L., Tincani, A. and Balestrieri, G. (1992). Relationship between antiphospholipid and
anti-endothelial antibodies: further characterisation of the reactivity on resting and cytokine activated
endothelial cells. *Clinical and Experimental Rheumatology*, **10**:37–42.

Derksen, R.H.W.M. and Kater, L. (1985). Lupus anticoagulant: revival of an old phenomenon. *Clinical and
Experimental Rheumatology*, **3**:349–357.

Derksen, R.H.W.M., De Groot, P.G., Kater, L. and Nieuwenhuis H.K. (1993). Patients with antiphospholipid antibodies and venous thrombosis should receive long-term anticoagulant treatment. *Annals of The Rheumatic
Diseases*, **52**:689–692.

Exner, T., Sahman, N. and Trudinger, B. (1988). Separation of anticardiolipin from lupus anticoagulant on a phospholipid polystyrene column. *Biochemical and Biophysical Research Communications*, **155**:1001–1007.

Exner, T., Triplett, D.A., Tabener, D. and Machin, S.J. (1991). Guidelines for testing and revised criteria for lupus
anticoagulants subcommittee for the standardisation of lupus anticoagulants. *Thrombosis and Haemostasis*,
65:320–322.

Feinstein, D.I. and Rappaport, S.I. (1972). Acquired inhibitors of coagulation. *Progress in Haemostasis and Thrombosis*, **1**:75–95.

Fleck, R.A., Rappaport, S.I. and Rao, L.V.M. (1988). Anti-prothrombin antibodies and the lupus anticoagulant. *Blood*, **72**:512–519.

Francis, R.B. and Neely, S. (1989). Effect of lupus anticoagulant on endothelial fibrinolytic activity *in vitro*. *Thrombosis and Haemostasis*, **61**:314–317.

Freyssinet, J.M., Gauchy, J. and Cazenave, J.P. (1986). The effect of phospholipids on the activation of protein C by human thrombin-thrombomodulin complex. *Biochemical Journal*, **238**:151–157.

Galli, M., Cortelazzo, S., Viero, P., Daldossi, M. and Barbui, T. (1988). Interaction between platelets and lupus anticoagulant. *European Journal of Haematology*, **41**:88–94.

Galli, M., Comfurius, P., Maassen, C., Hemker, H.C., De Baets, H.M., Van Breda-Vriesman, P.J.C. *et al.* (1990). Anticardiolipin antibodies (ACA) are directed not to a cardiolipin but to a plasma protein co-factor. *Lancet*, **335**:1544–1547.

Galli, M., Comfurius, P., Barbui, T., Zwaal, R.F.A. and Bevers, E.M. (1992). Anticoagulant activity of β_2 glyco-protein-I is potentiated by a distinct subgroup of anticardiolipin antibodies. *Thrombosis and Haemostasis*, **68**:297–300.

Galli, M., Daldossi, M. and Barbui T. (1994). Anti-glycoprotein-Ib/IX and IIb/IIIa antibodies in patients with antiphospholipid antibodies. *Thrombosis and Haemostasis*, **71**:571–575.

Gharavi, A.E., Harris, E.N., Asherson, R.A. and Hughes, G.R.V. (1987). Anticardiolipin antibodies: isotype distribution and phospholipid specificity. *Annals of The Rheumatic Diseases*, **46**:1–6.

Gharavi, A.E., Mellors, R.C. and Elkon, K.B. (1989). IgG anticardiolipin antibodies in murine lupus. *Clinical and Experimental Immunology*, **78**:233–238.

Gharavi, A.E., Sammaritano, L.R., Wen, J. and Elkon, K.B. (1992). Induction of antiphospholipid antibodies by immunisation with β_2 glycoprotein-I (apolipoprotein H). *Journal of Clinical Investigation*, **90**:1105–1109.

Gharavi, A.E., Harris, E.N., Sammaritano, L.R., Pierangeli, S.S. and Wen, J. (1993). Do patients with antiphospholipid syndrome have autoantibodies to β_2 glycoprotein-I? *Journal of Laboratory and Clinical medicine*, **122**:426–428.

Ginsberg, J.S., Brill-Edwards, P., Johnston, M., Denburg, J.A., Andrew, M., Burrows, R.F. *et al.* (1992). Relationship of antiphospholipid antibodies to pregnancy loss in patients with systemic lupus erythematosus: A cross sectional study. *Blood*, **80**:975–980.

Ginsberg, J.S., Demers, C., Brill-Edwards, P., Johnston, M., Bona, R., Burrows, R.F. *et al.* (1993). Increased thrombin generation and activity in patients with systemic lupus erythematosus and anticardiolipin antibodies: evidence for a prothrombotic state. *Blood*, **81**:2958–2963.

Glas-Greenwalt, P., Kant, K.S., Allen, C. and Pollack, V.E. (1984). Fibrinolysis in health and disease: severe abnormalities in systemic lupus erythematosus. *Journal of Laboratory and Clinical Medicine*, **104**:962–976.

Greaves, M. and Preston, F.E. (1991). Clinical and laboratory aspects of thrombophilia. In *Recent Advances In Blood Coagulation V*, edited by L. Poller, pp. 119–140.

Hang, L.M., Izui, S. and Dixon, F.J. (1981). (NZW x BXSB) F_1 hybrid, a model of acute lupus and coronary vascular disease with myocardial infarction. *Journal of Experimental Medicine*, **154**:216–221.

Harris, E.N., Gharavi, A.E., Boey, M.L., Patel, B.M., Mackworth-Young, C.G., Loizou, S. and Hughes, G.R.V. (1983). Anticardiolipin antibodies: detection by radioimmunoassay and association with thrombosis in systemic lupus erythemaotosus. *Lancet*, **2**:1211–1214.

Harris, E.N., Gharavi, A.E., Loizou, S., Derue, G., Chan, J.K., Patel, B.M. *et al.* (1985a). Cross reactivity of antiphospholipid antibodies. *Journal of Clinical and Laboratory Immunology*, 16:1–6.

Harris, E.N., Asherson, R.A., Gharavi, A.E. and Hughes, G.R.V. (1985b). Thrombocytopenia in SLE and related disorders: Association with anticardiolipin antibody. *British Journal of Haematology*, **59**:227–230.

Harris, E.N., Gharavi, A.E., Mackworth-Young, C.G., Asherson, R.A. and Hughes, G.R.V. (1985c). Lupoid sclerosis: A possible pathogenic role for antiphospholipid antibodies. *Annals Of The Rheumatic Diseases*, **44**:281–283.

Harris, E.N., Chan, J.K.H., Asherson, R.A., Gharavi, A.E. and Hughes, G.R.V. (1986). Thrombosis, recurrent fetal loss and thrombocytopenia: predictive value of the anticardiolipin antibody test. *Archives of Internal Medicine*, **146**:2153–2156.

Harris, E.N. (1987). Syndrome of the black swan. *British Journal of Rheumatology*, **26**:324–326.

Harris, E.N. (1990). The second international anticardiolipin standardisation workshop/The Kingston antiphospholipid study (KAPS) group. *American Journal of Clinical Pathology*, **94**:476–484.

Hashimoto, Y., Kawamura, M., Ichikawa, K., Suzuki, T., Sumida, T., Yoshida, S. *et al.* (1992). Anticardiolipin antibodies in NZW x BXSB F_1 mice: a model of antiphospholipid syndrome. *Journal of Immunology*, **149**:1063–1069.

Hasselaar, P., Derksen, R.H.W.M., Blokzijl, L. and de Groot, P.H.G. (1988). Thrombosis associated with antiphospholipid antibodies cannot be explained by effects on endothelial and platelet prostanoid synthesis. *Thrombosis and Haemostasis*, **59**:80–85.

Hunt, J.E., McNeil, H.P., Morgan, G.J., Crameri, R.M., and Krilis, S.A. (1992). A phospholipid-β_2 glycoprotein I complex is an antigen for anticardiolipin antibodies occurring in autoimmune disease but not with infection. *Lupus*, **1**:83–90.

Hunt, J.E., Simpson, R.J. and Krilis, S.A. (1993). Identification of a region of β_2-glycoprotein-I critical for lipid binding and anticardiolipin cofactor activity. *Proceedings of the National academy of Science, USA,* **90**:2141–2145.

Hunt, J.E. and Krilis, S.A. (1994). The fifth domain of β_2 glycoprotein I contains a phospholipid binding site (Cys 281-288) and region recognised by anticardiolipin antibodies. *Journal of Immunology,* **152**:653–659.

Italian registry of Antiphospholipid Antibodies (IR-APA) (1993). Thrombosis and thrombocytopenia in antiphospholipid syndrome (idiopathic and secondary to SLE): first report from the Italian Registry. *Haematologica,* **78**:3131–318.

Kato, H. and Enjyoji, K. (1991). Amino acid sequence and location of the disulphide bonds in bovine β_2-glycoprotein-I: The presence of five Suchi domains. *Biochemistry,* **30**:11687–11694.

Keeling, D.M., Wilson, A.J.G., Mackie, I.J., Machin, S.J. and Isenberg, D.A. (1992). Some "antiphospholipid antibodies" bind to β_2 glycoprotein-I in the absence of phospholipid. *British Journal of Haematology,* **82**:571–574.

Keeling, D.M., Wilson, A.J.G., Mackie, I.J., Machin, S.J. and Isenberg, D.A. (1993). Role of β_2 glycoprotein-I and antiphospholipid antibodies in activation of protein C *in vitro. Journal of Clinical Pathology,* **46**:908–911.

Khamasta, M.A., Guadrado, M.J., Mujic, F., Taub, N.A., Hunt, B.J. and Hughes, G.R.V. (1995). The management of thrombosis in the antiphospholipid syndrome. *New England Journal of Medicine,* **332**:993–997.

Kroll, J., Larsen, J.K., Loft, H., Ezbou, M., Wallevik, K. and Faber, M. (1976). DNA-binding protein in Yoshida ascites tumour fluid. *Biochimica Biophysical Acta,* **434**:490–501.

Lafer, E.M., Rauch, J., Andrezejewski, C,. Mudd, D., Furie, B., Schwartz, R.S. and Stollar, B.D. (1981). Polyspecific monoclonal lupus autoantibodies reactive with both polynucleotides and phospholipids. *Journal of Experimental Medicine,* **153**:897–909.

Laurell, A. and Nilsson, I.M. (1957). Hypergammaglobulinaemia, circulating anticoagulant and biologic false positive Wasserman reaction. *Journal of Laboratory and Clinical Medicine,* **49**:694–707.

Lechner, K. and Pabinger-Faching, I. (1985). Lupus anticoagulant and thrombosis. a study of 25 cases and a review of the literature. *Haemostasis,* **15**:254–262.

Loeliger, A. (1959). Prothrombin for cofactor for the circulating anticoagulant in systemic lupus erythematosus? *Thrombosis et Diathesis Haemorrhagica,* **3**:237–256.

Loizou, S., McCrea, J.D., Rudge, A.C., Reynolds, R., Boyle, C.C. and Harris, E.N. (1985). Measurement of anticardiolipin antibodies by enzyme-linked immunosorbent assay: standardisation and quantitation of results. *Clinical and Experimental Immunology,* **62**:739–744.

Love, P.E. and Santoro, S.A. (1990). Antiphospholipid antibodies: anticardiolipin and the lupus anticoagulant in systemic lupus erythematosus (SLE) and non-SLE disorders. *Annals of Internal Medicine,* **112**:682–698.

Malia, R.G., Kitchen, S., Greaves, M. and Preston, F.E. (1990). Inhibition of activated protein C and its cofactor protein S by antiphospholipid antibodies. *British Journal of Haematology,* **76**:101–107.

Marciniak, E. and Romond, E.H. (1989). Impaired catalytic function of activated protein C: A new *in vitro* manifestation of lupus anticoagulant. *Blood,* **74**:2426–2432.

Matsuda, J., Saitoh, N., Gohchi, K., Gotoh, M. and Tsukamoto, M. (1994). Detection of β_2 glycoprotein-I-dependent antiphospholipid antibodies and anti-β_2 glycoprotein-I antibody in patients with systemic lupus erythematosus and in patients with syphilis. *Internal Archives of Allergy and Applied Immunology,* **103**:239–244.

Matsuura, E., Igarashi, Y., Fujimoto, M., Ichikawa, K. and Koike, T. (1990). Anticardiolipin cofactor(s) and differential diagnosis of autoimmune disease. *Lancet,* **336**:177–178.

Matsuura, E., Igarashi, Y., Yasuda, T., Triplett, D.A. and Koike, T. (1994). Anticardiolipin antibodies recognise β_2 glycoprotein-I structure altered by interacting with an oxygen modified solid phase surface. *Journal of Experimental Medicine,* **179**:457–462.

McNally, T., Mackie, I.J., Machin, S.J. and Isenberg, D.A. (1995). Increased levels of β_2 glycoprotein-I antigen and β_2 glycoprotein-I binding antibodies are associated with a history of thromboembolic complications in patients with SLE and primary antiphospholipid syndrome. *British Journal of Rheumatology,* **34**:1031–1036.

McNally, T., Mackie, I.J., Isenberg, D.A. and Machin, S.J. (1996). β_2 glycoprotein-I inhibits factor XII activation on triglyceride rich lipoproteins: the effect of antibodies from plasma of patients with antiphospholipid syndrome. *Thrombosis and Haemostasis,* in press.

McNeil, H.P., Chesterman, C.N. and Krilis, S.A. (1989). Anticardiolipin antibodies and lupus anticoagulants comprise separate antibody subgroups with different binding characteristics. *British Journal of Haematology,* **73**:506–513.

McNeil, P.H., Simpson, R.J. Chesterman, C. and Krilis, S.A. (1990). Anti-phospholipid antibodies are directed against a complex antigen that includes a lipid binding inhibitor of coagulation: β_2 glycoprotein-I, *Proceedings of the National Academy of Science, USA,* **87**:4120–4124.

McNeil, H.P., Chesterman, C.N. and Krilis, S.A. (1991). Immunology and clinical importance of antiphospholipid antibodies. *Advances in Immunology,* **49**:193–280.

Mitropoulos, K.A., Miller, G.J., Watts, G.F. and Durrington, P.N. (1992). Lipolysis of triglyceride rich lipoproteins activates coagulant factor XII: a study in familial lipoprotein lipase deficiency. *Atherosclerosis,* **95**:119–125.

Moore, J.E. and Mohr, C.F. (1952). Biologically false positive serologic tests for syphilis. *Journal of The American Medical Association,* **150**:467–473.

Moore, J.E. and Lutz, W.B. (1955). The natural history of systemic lupus erythematosus: An approach to its study through chronic biologic false positive reactions. *Journal of Chronic Diseases*, **1**:297–316.

Mori, T., Takai, Y., Minakuchi, R., Yu, B. and Nishizuka, Y. (1980). Inhibitory action of chlorpromazine, dibucaine and other phospholipid interacting drugs on calcium-activated phospholipid dependent protein kianse. *Journal of Biological Chemistry*, **255**:8378–8380.

Mueller, J.F., Ratnoff, O. and Heinle, R.W. (1951). Observations on the characteristics of an unusual circulating anticoagulant. *Journal of Laboratory and Clinical Medicine*, **38**:254–261.

Nakamura, N., Kuragaki, C., Shidara, Y., Yamaji, K. and Wada, Y. (1995). Antibody to annexin V has antiphospholipid and lupus anticoagulant properties. *American Journal of Haematology*, **49**:347–348.

Nakaya, Y., Schafer, E.J. and Brewer, B.H. (1980). Activation of human post heparin lipoprotein lipase by Apolipoprotein H (β_2 glycoprotein-I). *Biochemical Biophysical Research Communications*, **95**:1168–172.

Nawroth, P.P. and Stern, D.M. (1986). Modulation of endothelial cell haemostatic properties by tumour necrosis factor. *Journal of Experimental Medicine*, **163**:740–745.

Nemerson Y. (1988). Tissue factor and haemostasis. *Blood*, **71**:1–8.

Nimpf, J., Wurm, H. and Kostner, G.M. (1985). Interaction of β_2 glycoprotein-I with human blood platelets: Influence upon ADP induced aggregation. *Thrombosis and Haemostasis*, **54**:397–401.

Nimpf, J., Bevers, E.M., Bomans, P.H.H., Till, U., Wurm, H., Kostner, G.M. and Zwaal, R.F.A. (1986). Prothrombinase activity of human platelets is inhibited by β_2 glycoprotein-I. *Biochimica et Biophysica Acta*, **884**:142–149.

Onyiriuka, E.C., Hersh, I.S. and Hert, W.I. (1990). Surface modification of polystyrene by γ-radiation. *Applied Spectrophotometry*, **44**:808.

Oosting, J.D., Derksen, R.H.W.M., Hackeng, T.M., Van Vliet, M., Preissner, K.T., Bouma, B.N. and De Groot, P.G. (1991). *In vitro* studies of antiphospholipid antibodies and its cofactor β_2 glycoprotein-I, show no effect on endothelial cell mediated protein C activation. *Thrombosis and Haemostasis*, **66**:666–671.

Oosting, J.D., Derksen, R.H.W.M., Entjes, H.T.I., Bouma, B.N. and De Groot, P.G. (1992). Lupus anticoagulant activity is frequently dependent upon the presence of β_2 glycoprotein-I. *Thrombosis and Haemostasis*, **67**:499–502.

Oosting, J.D., Derksen, R.H.W.M., Bobbink, T.W.G., Hackeng, T.M., Bouma, B.N. and De Groot, P.G. (1993a). Antiphospholipid antibodies directed against a combination of phospholipids with prothrombin, protein C or protein S: An explanation for their pathogenic mechanism. *Blood*, **81**:2618–2625.

Oosting, J.D., Preissner, K.T., Derksen, R.H.W.M. and De Groot, P.G. (1993b). Autoantibodies directed against the epidermal growth factor-like domains of thrombomodulin inhibit protein C activation *in vitro*. *British Journal of Haematology*, **85**:761–768.

Pangborn, M.C. (1941). A new serologically active phospholipid from beef heart. *Proceedings of The Society for Experimental Biology and Medicine*, **48**:484–486.

Pierangeli, S.S., Harris, E.N., Davis, S.A. and DeLorenzo, G. (1992). β_2 glycoprotein-I enhances cardiolipin binding activity but is not the antigen for antiphospholipid antibodies. *British Journal of Haematology*, **82**:565–570.

Pierangeli, S.S. and Harris, E.N. (1993). Induction of antiphospholipid binding antibodies in mice and rabbits by immunisation with human β_2 glycoprotein-I or anticardiolipin antibodies alone. *Clinical and Experimental Immunology*, **93**:268–272.

Piette, J.C., Wechsler, B., Frances, C. and Godeay, P. (1992). Systemic lupus erythematosus and the antiphospholipid syndrome: reflections about the relevance of ARA criteria. *Journal of Rheumatology*, **19**:1835–1837.

Polz, E. (1979). Isolation of a specific lipid binding protein from human serum byaffinity chromatography using heparin sepharose. In *Proteins of Biological Fluids*, edited by L. Peters, pp. 817–820. Oxford: Pergammon Press.

Puri, R.N., Zhou, F., Hu, C-J., Colman, R.F. and Colman, R.W. (1991). High molecular weight kininogen inhibits thrombin induced platelet aggregation and cleavage of aggregin by inhibiting binding of thrombin to platelets. *Blood*, **77**:500–507.

Rauch, J., Tannenbaum, M., Tannenbaum, H., Ramelson, H., Cullis, P.R. Tilcock, C.P. *et al.* (1986). Human hybridome lupus anticoagulants distinguish between lamellar and hexagonal phase lipid systems. *Journal Of Biological Chemistry*, **261**:9672–9677.

Rauch, J., Meng, Q. and Tannenbaum, H. (1987). Lupus anticoagulant and anti-platelet properties of human hybridoma autoantibodies. *Journal of Immunology*, **139**:2598–2604.

Rauch, J., Tannenbaum, M. and Janoff, A.S. (1989). Distinguishing plasma lupus anticoagulants from anti-factor antibodies using hexagonal II phase phospholipids. *Thrombosis and Haemostasis*, **62**:892–896.

Rauch, J. and Janoff, A.S. (1990). Phospholipid in the hexagonal phase is immunogenic: Evidence for immunorecognition of non-bilayer lipid phases *in vivo*. *Proceedings of the National Academy Of Science, USA*, **87**:4112–4114.

Rauch, J. and Janoff, A.S. (1992). The nature of antiphospholipid antibodies. *Journal of Rheumatology*, **19**:1782–1785.

Rosove, M.H. and Brewer, P.M.C. (1992). Antiphospholipid antibodies: clinical course after the first thrombotic event in 70 patients. *Annals of Internal Medicine*, **117**:303–308.

Roubey, R.A. (1994). Autoantibodies to phospholipid-binding plasma proteins: A new view of lupus anticoagulants and other "antiphospholipid" autoantibodies. *Blood*, **84**:2854–2867.

Ruiz-Arguelles, A., Vazquez-Prado, J., Deleze, M., Perez-Romano, B., Drenkard, C., Alarcon-Segovia, D. and Ruiz-Arguelles, G.J. (1993). Presence of serum antibodies to coagulation protein C in patients with systemic lupus erythematosus is not associated with antigenic or functional protein C deficiencies. *American Journal of Haematology*, **44**:58–59.

Rustin, M.H.A., Bull, H.A., Machin, S.J. and Dowd, P. (1988). Effects of lupus anticoagulant in patients with systemic lupus erythematosus on endothelial cells, prostacyclin release and procoagulant activity. *Journal of Investigative Dermatology*, **90**:744–748.

Sammaritano, L.R., Lockshin, M.D. and Gharavi, A.E. (1992). Antiphospholipid antibodies differ in aPL cofactor requirement. *Lupus*, **1**:83–90.

Schorer, A.E., Wickam, N.R.W. and Watson, K.V. (1989). Lupus anticoagulant induces a selective defect in thrombin mediated endothelial prostacyclin release and platelet aggregation. *British Journal of Haematology*, **71**:399–407.

Schousboe, I. (1980). Binding of β_2 glycoprotein-I to platelets: effect on adenylate cyclase activity. *Thrombosis Research*, **19**:225–237.

Schousboe, I. (1983). Characterisation of the interaction between β_2 glycoprotein-I and mitochondria, platelets, liposomes and bile acids. *International Journal Of Biochemistry*, **15**:1393–1401.

Schousboe, I. (1985). β_2 glycoprotein-I: A plasma inhibitor of the contact activation of the intrinsic blood coagulation pathway. *Blood*, **66**:1086–1091.

Schultze, H.E., Heide, K. and Haupt, H. (1961). Uber Einbisher unbekanntes niedermole kulares β_2 globulin des human serums. *Naturwissen Schaftern*, **48**:719.

Shi, W., Chong, B.H. and Chesterman, C.N. (1993). β_2 glycoprotein-I is a requirement for anticardiolipin antibodies binding to activated platelets: differences with lupus anticoagulants. *Blood*, **81**:1255–1262.

Shi, W. Chong, B.H., Hogg, P.J. and Chesterman, C.N. (1993b). Anticardiolipin antibodies block the inhibition by β_2 glycoprotein-I of the FXa generating activity of platelets. *Thrombosis and Haemostasis*, **70**:342–345.

Shi, W., Krilis, S.A., Chong, B.H. Gordon, S. and Chesterman C.N. (1990). Prevalence of lupus anticoagulant in a healthy population: Lack of correlation with anticardiolipin antibodies. *Australian and New Zealand Journal of Medicine*, **20**:231–236.

Steinkasser, A., Barlow, P., Willis, A., Kertesz, Z., Campbell, I., Sim, R. and Norman, D. (1992). Activity, disulphide mapping and structural modelling of the fifth domain of human β_2 glycoprotein-I. *FEBS Letters*, **313**:193–197.

Sugi, T. and McIntyre, J.A. (1995). Autoantibodies to phosphatidylethanolamine (PE) recognise a kininogen-PE complex. *Blood*, **86**:3083–3089.

Tan, E.M., Cohen, A.S., Fries, F.J., Masi, A.T., McShane, D.J., Rothfield, N.F. *et al.* (1982). Revised criteria for the classification of systemic lupus erythematosus. *Arthritis and Rheumatism*, **25**:1271–1277.

Tankersley, D.L. and Finlayson, J.S. (1984). Kinetics of activation and autoactivation of human factor XII. *Biochemistry*, **23**:273–279.

Tannenbaum, S.H., Finko, R. and Cines, D.B. (1986). Antibody and immune complexes induce tissue factor production by human endothelial cells. *Journal of Immunology*, **137**:1532–1537.

Thiagarajan, P., Shapiro, S.S. and de Marco, L. (1980). A monoclonal immunoglobulin M coagulation inhibitor with phospholipid specificity: mechanism of a lupus anticoagulant. *Journal Of Clinical Investigation*, **66**:397–405.

Vaarala, U., Alfthan, G., Jauhiainen, M., Leirisal-Repo, M. Aho, K. and Palosuo, T. (1993). Crossreaction between antibodies to oxidised low density lipoprotein and to cardiolipin in systemic lupus erythematosus. *Lancet*, **341**:920–923.

Vermylen, J. and Arnout, J. (1992). Is the antiphospholipid syndrome caused by antibodies directed against physiologically relevant phospholipid-protein complexes? *Journal of Laboratory and Clinical Medicine*, **120**:10–14.

Viard, J.P., Amoura, Z. and Bach, J.F. (1992). Association of anti-β_2 glycoprotein-I antibodies with lupus-type circulating anticoagulant and thrombosis in systemic lupus erythematosus. *The American Journal of Medicine*, **93**:181–186.

Wasserman, A. (1907). Uber die Entwicklung und den gegenwartigen Stand der Serodiagnostik gegenuber syphilis. *Berl Klin Wochenschr*, **44**:1599.

Weksler, B.B. and Jaffe, E.A. (1986). Prostacyclin and the endothelium. In *Vascular Endothelium In Haemostasis and Thrombosis*, edited by M.A. Gimbrone. Edinburgh: Churchill Livingstone.

Wurm, H. (1984). β_2 glycoprotein-I (apolipoprotein H) interactions with phospholipid vesicles. *International Journal of Biochemistry*, **16**:511–515.

16 Resistance to Activated Protein C and Inherited Thrombosis — Molecular Mechanism, Diagnosis and Clinical Management

Ross Baker and John Eikelboom

Clinical Thrombosis Unit, Department of Haematology, Royal Perth Hospital, University of Western Australia, GPO Box X2213 Perth WA 6001, Tel: 61-9-224-2897, Fax: 61-9-224-3449, e-mail: ross.baker@rph.health.wa.gov.au.

INTRODUCTION

Venous thromboembolism is a common clinical problem accounting for more than 250 000 hospital admissions per year in the United States. It has an estimated incidence of 1 event per 1000 persons each year, approximately two-thirds of which are first-time episodes and one-third of which are recurrent episodes (Anderson *et al.*, 1991; Goldhaber, 1994). However, despite the use of optimal therapies, this disorder is frequently associated with long-term morbidity and mortality. Up to 50% of patients develop a post-phlebitic syndrome (lower limb oedema, ulceration, dermatitis, varicose veins and pain), and there is a 7% risk of death from pulmonary embolism (Marder, 1995).

Central to the process of clot formation is the generation of thrombin through a series of proteolytic cleavage events (Davie *et al.*, 1991). Thrombin is the pivotal serine protease, exerting its pro-thrombotic effect *via* the conversion of fibrinogen to insoluble fibrin, stimulation of platelet aggregation and provision of a positive feedback loop on itself by activation of factors V and VIII (Hemker, 1994). After binding to thrombomodulin on the endothelial surface, thrombin also initiates a major physiological mechanism for the down regulation of thrombosis, which is mediated by the protein C anticoagulant pathway (Davie *et al.*, 1991; Dahlback and Stenflo, 1994).

Since the early 1980's, hereditary deficiencies of the vitamin K-dependent glycoproteins (protein C and protein S) involved in the protein C anticoagulant pathway, have been linked to an increased risk of venous thrombosis (Griffin *et al.*, 1981; Comp *et al.*, 1984). However, the combination of these two defects plus antithrombin III deficiency, were identified in only up to 15% of patients with a personal or family history of recurrent venous thromboembolism, and less than 7% of consecutive unselected patients with venous thrombosis (Heijboer *et al.*, 1990; Allaart and Briet, 1994; Bertina *et al.*, 1995). As a result, there remained a substantial number of patients with unexplained familial or recurrent deep vein thrombosis, although a large proportion of thrombotic events in these individuals occurred in the setting of exposure to environmental risk factors (major surgery, pregnancy, oral contraception).

In 1993, a new genetic abnormality in the protein C anticoagulant pathway was described in patients with a family history of venous thrombosis, who were found to have a

diminished anticoagulant response to activated protein C (APC) in the activated partial
thromboplastin time (APTT) clotting assay (Dahlback *et al.*, 1993). The term APC resistance was used to describe this defect, and in consecutive patients presenting with deep
venous thrombosis, it was subsequently found to be at least ten times more common than
deficiencies of all the other known genetic defects which cause thrombophilia (Svensson
and Dahlback, 1994). Further studies have found APC resistance in up to 60% of patients
presenting with venous thromboembolism (Zoller and Dahlback, 1995; Bertina *et al.*, 1995),
with thrombotic events occurring in association with exposure to enviromental risk factors
in approximately half of these cases.

   This recent discovery of APC resistance has led to a new understanding of the pathophysiology of the protein C anticoagulant pathway, and has had a major impact on the
management of individual patients and families with venous thromboembolism, particularly
with regards to genetic testing, risk stratification and prophylactic therapy. Ultimately this
may substantially impact upon the morbidity and mortality resulting from this common and
frequently devastating clinical problem.

THE DISCOVERY THAT ACTIVATED PROTEIN C RESISTANCE IS A NOVEL MECHANISM OF FAMILIAL THROMBOSIS

The initial report of APC resistance was of a patient with recurrent deep venous thrombosis
whose plasma, unlike the normal response, did not produce a prolongation in the APTT

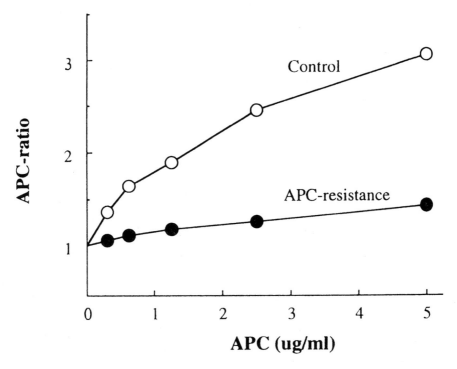

Figure 16.1   The discovery of APC resistance. The addition of activated protein C (APC) prolongs the activated
partial thromboplastin time (APTT) in a dose dependent fashion in normal plasma (○). However in a patient with
recurrent thrombosis there is a poor response in the anticoagulant activity to APC (●). The APC ratio is the ratio of
the APTT with and without APC. (Adapted from Dahlback *et al.*, 1993 with permission).

based clotting assay when purified APC was added (Figure 16.1) (Dahlback *et al.*, 1993). An underlying genetic basis was suspected because several family members who had also experienced thrombosis, were likewise found to be APC resistant. Early experiments suggested that the abnormality was either a factor V mutation in one of the known APC binding or cleavage sites, or an unknown plasma APC cofactor (Dahlback *et al.*, 1993). *In-vitro* APC resistance was shown to be corrected by a crude protein extract of normal plasma, and after further purification, this protein was found to be identical to the intact form of factor V (Dahlback and Hildebrand, 1994). In subsequent mixing experiments, APC resistance could be corrected by all known factor deficient plasmas except for factor V deficient plasma (Bertina *et al.*, 1994). Affinity purified factor V from normal plasma also corrected the phenotype of APC resistance (Bertina *et al.*, 1994), while factor V isolated from a patient with homozygous APC resistance could cause the same abnormality when added to normal plasma (Heeb *et al.*, 1995).

The association of APC resistance with a mutation involving the factor V gene was independently demonstrated by linkage studies in several large families. Co-segregation with the APC resistance phenotype was found using microsatellite markers for loci in chromosome 1q21-25 region (Bertina *et al.*, 1994) and restriction fragment length polymorphism of exon 13 in the factor V gene (Zoller and Dahlback, 1994). Subsequent investigation concentrated on detecting mutations either in the factor V gene at the putative APC binding site corresponding to amino acid residues 1865–1874, or the APC cleavage site at arginine 506. In two patients with homozygous APC resistance, direct sequencing after PCR amplification revealed a substitution (guanine to adenine) at the APC cleavage site, involving nucleotide position 1691 (Bertina *et al.*, 1994). This point mutation predicts the replacement of arginine with glutamine (factor V:Q^{506}). A simple PCR genotyping screening method was then designed by utilizing specific primers and recognition of the digestion pattern by the enzyme M*nl* 1 of the 267 base pair product. It was confirmed that all carriers of the mutation had the phenotype of APC resistance, and that this genetic defect was common in the Dutch general population with a 2% prevalence (Bertina *et al.*, 1994). Linkage of APC resistance to the factor V:Q^{506} mutation was independently confirmed by several other groups of investigators (Voorberg *et al.*, 1994; Beufe *et al.*, 1995).

MOLECULAR MECHANISM OF ACTIVATED PROTEIN C RESISTANCE

Normal Factor V:R^{506} Regulation By Activated Protein C

Factor V is a single chain high molecular weight (330 000 kD) pro-cofactor, circulating in the plasma at a concentration of 10 mg/l (Jenny *et al.*, 1994). It has little procoagulant activity, but during the process of vascular injury, it is converted on the membrane surface to active factor V (factor Va) *via* limited proteolysis by the activated clotting factors α thrombin and factor Xa (Monkovic and Tracy, 1990). Most of the whole blood factor V circulates in plasma (80%), but a significant proportion (20%) is also found in α granules of platelets. At the site of vascular damage, activated platelets not only provide an appropriate membrane surface for coagulation activity, but also explosively release and locally target factor V/Va at concentrations which are 600-fold higher than plasma levels (Tracy *et al.*, 1982). Generated factor Va in the presence of platelet phospholipid, divalent cations and factor Xa forms the enzyme complex prothrombinase, which proteolytically converts prothrombin to α-thrombin. Factor Va is crucial to the activity of this complex by not only binding the substrate prothrombin, but also acting as a receptor for Xa at the phospholipid surface. Without

NORMAL FACTOR Va:A506

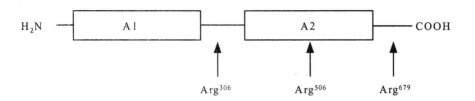

MUTATED FACTOR Va:Q506

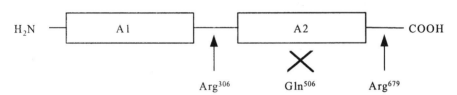

Figure 16.2   Inactivation of normal and mutated heavy chain (A_1 and A_2 domains) of factor Va by APC.
The peptide bonds located at Arg^{306}, Arg^{506} and Arg^{679} are susceptible to proteolytic cleavage by APC. However
in mutated factor Va where Arg^{506} is replaced by Gln $(Q)^{506}$, the preferential cleavage sites by APC are only at
Arg^{306} and Arg^{679}. The inactivation rate is ten times less efficient without the initial priming cleavage at Arg^{506}.

factor Va, prothrombinase complex activity is reduced ten thousand fold (Nesheim *et al.*,
1979). Therefore any alteration in factor Va activity profoundly influences the rate of
α-thrombin generation and modulates the amount of cross-linked insoluble fibrin deposited
in the thrombus at the site of vascular injury.

The physiological down regulation of plasma factor Va occurs as a result of proteolytic
inactivation by APC. Plasma factor Va loses its cofactor activity by ordered and sequential
peptide bond cleavages in the A_1 and A_2 regions of the 105 kD heavy chain (see Figure
16.2). The initial reaction involves rapid cleavage at Arg^{506} resulting in a heavy chain portion of 75 kD, which has reduced but still significant cofactor activity in the prothrombinase
complex. Subsequently, this reaction intermediate is fully inactivated by slow cleavage at
Arg^{306} and Arg^{679} (Kalafatis *et al.*, 1994; Rosing *et al.*, 1995).

Recent reports suggest that platelet derived factor Va behaves very differently to inactivation by APC compared to plasma derived factor Va. Experiments have shown that
platelet derived factor Va still has 20% cofactor activity remaining at 2 hours after incubation with APC, compared to negligible residual activity (less than 5%) of normal plasma
Va incubated for the same period of time (Camire *et al.*, 1995). The explanation for this
resistance of platelet factor Va to the inactivation of APC is unclear, although it is apparent
that there is a difference in the sequence of events resulting in the inactivation of platelet
factor Va, compared to plasma derived factor Va. Residual platelet factor Va function
appears to be due to the presence of a partially cleaved but active 60 kD heavy chain molecule resulting from an initial cleavage by APC at Arg^{306} rather than at Arg^{506} (Camire *et al.*,
1995). This 60 kD intermediate is slowly inactivated by subsequent cleavage at Arg^{679} and
Arg^{506} (Camire *et al.*, 1995). It is speculated that this change in the APC inactivation
sequence is an inherent property of platelet factor Va rather than being due to its binding to
a particular membrane surface (phospholipid vesicles or thrombin activated platelets)

(Camire *et al.*, 1995). It is likely that the apparent substrate differences between plasma and platelet factor Va are caused either by undefined structural and post-translational modifications, or alternatively by modulation of APC inactivation through previously unrecognized platelet release molecules. Further study is required to clarify these issues in view of the high concentration of relatively APC resistant platelet derived factor Va which is found at the site of vascular damage.

Mutated Factor V:R^{506} Regulation by Activated Protein C

The factor V:Q^{506} mutation is found in over 90% of cases of APC resistance (Zoller *et al.*, 1994). The reason why patients with APC resistance have only a mild predisposition to thrombosis is that only one aspect of the protein C anticoagulant pathway is defective, and even then, this results in a decrease in the rate but does not totally prevent factor Va inactivation. Both the rate of procoagulant activation of factor V:Q^{506} and its anticoagulant function are normal (Shen and Dahlback, 1994; Lu *et al.*, 1996), but there is disturbance of the ordered sequence of proteolytic downregulation of the mutated plasma and platelet factor Va by APC as a result of the loss of the crucial initial cleavage event at Arg^{506} (Figure 16.2). Instead, inactivation occurs *via* cleavage at Arg^{306} and Arg^{679}, which is 10 times slower in the absence of the prior priming cleavage at Arg^{506} (Heeb *et al.*, 1995; Kalafatis *et al.*, 1995). Similarly to plasma factor Va:Q^{506}, platelet derived factor Va:Q^{506} is resistant to inactivation by APC, as demonstrated by the detection of 40% (compared to 20% of factor Va:R^{506}) of initial cofactor activity after two hours incubation with APC (Camire *et al.*, 1995).

It appears that carriers of the factor Va:Q^{506} mutation, even in the absence of acute thrombosis, have markers of increased *in-vivo* thrombin generation as measured by increased levels of D-dimer, prothrombin fragments 1+2 and thrombin-antithrombin complexes (Simioni *et al.*, 1996; Leroy-Matheron *et al.*, 1996). This suggests that inefficient degradation of mutated factor V by APC is an important mechanism of thrombin generation, predisposing to venous thrombosis even in asymptomatic family members.

The APC resistance clotting assay requires appropriate concentrations of reagents to detect the factor Va:Q^{506} mutation. Excessive activation of coagulation or high APC concentrations (greater than 5.5 nm) will produce clotting times independent of the factor Va:Q^{506} mutation, with most of the APC inactivation occurring rapidly at the Arg^{306} site rather than at position 506 (Kalafatis *et al.*, 1996). The reasons for this are twofold: Firstly, at low concentrations, APC preferentially cleaves factor V at position Arg^{679}. The resultant procoagulant heavy chain of factor Va is dependent on subsequent APC cleavage at position 506 for its inactivation, a step which does not occur in patients with the factor Va:Q^{506} mutation. As a result, the ultimate inactivation step at Arg^{306} is delayed, and there is accumulation of the procoagulant heavy chain which in turn causes the phenotype of APC resistance. Higher concentrations of APC will directly bypass this process and inactivate factor Va at Arg^{306} independent of the mutation (Kalafatis *et al.*, 1996). Secondly, strong procoagulant forces cause enhanced factor Xa generation which not only increases the level of factor Va, but also prevents APC inactivation of factor Va by blocking its cleavage site at Arg^{506} (Rosing *et al.*, 1995). Therefore, with overwhelming factor Xa activity, the dependence of APC inactivation on the presence or absence of the factor Va:Q^{506} mutation is negated. This complex molecular balance of pro and anticoagulant forces may explain why the choice of the APTT reagent and the concentration of APC is crucial for the reliable detection of APC resistance (factor Va:Q^{506}) by clotting based methods.

Recently, a new mutation on factor V (Arg 306 to Thr) was found in one patient from a group of 17 patients who had DVT with APC resistance in the absence of the factor

Va:Q^{506} mutation (Williamson *et al.*, 1998). However in the majority of these cases no new mutations have been found, a situation which may occur in up to 10% of familial venous thrombosis cases (Zoller *et al.*, 1994). It is not caused by a mutation in the factor VIII cleavage site (Roelse *et al.*, 1996). Further examination of these events may uncover the abnormality causing APC resistance in this subset of patients with hereditary thrombosis who do not have the factor V:Q^{506} mutation.

GEOGRAPHICAL VARIATION IN THE PREVALENCE OF APC RESISTANCE (FACTOR V:Q^{506})

Venous thrombosis is common in Europeans but is rarely reported in patients from other racial backgrounds. It is unclear whether this is due to heredity, environmental factors or reporting bias (Nathwani and Tuddenham, 1992). One explanation may be the substantial variation in the prevalence of the factor V:Q^{506} mutation amongst different populations. The factor V:Q^{506} mutation appears exclusively in the Indo-European population with the reported frequency ranging from between 2 and 15% (Bertina *et al.*, 1994; Beuchamp *et al.*, 1994; van Bockxmeer *et al.*, 1995; Kontula *et al.*, 1995; Rees *et al.*, 1995; Ridker *et al.*, 1995a; Holm *et al.*, 1996). In contrast, this mutation is very rarely if at all found in individuals of indigenous populations from Asian, African, Australasian, Greenland and Americas origin (Rees *et al.*, 1995; Takamiya *et al.*, 1995; Arrunda *et al.*, 1996; Chan *et al.*, 1996; Maat *et al.*, 1996). In particular regions of Southern Sweden (Holm *et al.*, 1996) and Greece (Rees *et al.*, 1995), 15% of the population are heterozygous for the factor V:Q^{506} mutation. However even within the same country, significant variations in prevalence can occur. For example, in France the prevalence of the factor V:Q^{506} mutation in control subjects from the north, east and southwest respectively, ranged from 0.7 to 8.2% (Emmerich *et al.*, 1995). This highlights the importance of establishing the gene frequency of the factor V mutation in common ethnically derived populations rather than using political boundaries. In future, geographical differences in the prevalence of the factor V:Q^{506} mutation may influence decisions regarding the clinical usefulness and cost-benefit of population screening for APC resistance in asymptomatic individuals.

The low prevalence of the factor V:Q^{506} mutation outside of European derived populations suggests that a founding effect from a single ancient mutation has occurred. This was originally suggested by Bertina *et al.* (1994) who studied a cohort of patients with venous thrombosis, and found that the common H*inf* 1 allele of the factor V gene was significantly over-represented in those with the factor V:Q^{506} mutation compared to controls (p<0.001). This hypothesis is further supported by the results of a recent study, in which the haplotype frequency amongst homozygous and heterozygous carriers of the factor V:Q^{506} mutation was significantly different (p=0.046) compared to the non-mutation carrying population (Cox *et al.*, 1996). Recent migration from Europe probably accounts for the distribution in the Western World of the factor V:Q^{506} mutation rather than recurrent mutations occurring at the well established hot spot for mutation at the CpD dinucleotide.

LABORATORY DIAGNOSIS AND IMPACT OF ACTIVATED PROTEIN C RESISTANCE

The balance between the pro and anticoagulant actions of APC is complex, and reliably reproducing this physiological activity in the laboratory with clotting-based assays has been problematic. As APC resistance is commonly found in the general population, a clinically robust screening test is required that will not only recognize the well defined factor V:Q^{506} mutation but also identify patients with "acquired" APC resistance, either associated with thrombosis or with an as yet unrecognized abnormality of the protein C anticoagulant pathway.

Early Experience with APTT Based Clotting Assays and APC Resistance

The original principle of the APC resistance clotting assay was based on the performance of two simultaneous APTTs with and without the addition of a fixed concentration of APC (Dahlback *et al.*, 1993). The ratio of the APTT with and without APC was then calculated for each patient. In the original commercial assay (Coatest APC Resistance, Chromogenix, Sweden), a ratio of less than two was considered to be diagnostic of APC resistance. However, when analyzed as a continuous variable, the APC resistance ratio was shown to be inversely related to the risk of familial venous thrombosis (Svensson and Dahlback, 1994).

After the discovery of the factor V:Q^{506} mutation as the molecular basis for over 90% of cases of APC resistance (Bertina *et al.*, 1994; Zoller *et al.*, 1994), it became apparent in the clinical laboratory that the specificity and sensitivity of the APTT based clotting method for the point mutation varied considerably, ranging from 81 to 94% and 70 to 100% respectively (Baker *et al.*, 1994). In addition, repeat testing in the same patient did not always confirm the original phenotype despite similar sample collection and processing conditions (Alhenc-Gelas *et al.*, 1994; Legnani *et al.*, 1994; Baker *et al.*, 1994). A number of patient, reagent and pre-analytic factors that alter the APC resistance ratio have been identified, including platelet contamination (Shizuka *et al.*, 1995; Sidelmann *et al.*, 1995), prolonged baseline APTT (de Ronde and Bertina, 1994), source and concentration of APTT activator and APC (de Ronde and Bertina 1994), calcium chloride concentration (de Ronde and Bertina, 1994), freeze/thawing (Trossaert *et al.*, 1995a) and anticoagulant concentration and sample handling (de Ronde and Bertina, 1994). The choice of instrument that identifies the clot end-point in the assay also appears to affect the intra-laboratory reproducibility of clotting times, probably because of differences in the sensitivity of the electromagnetic versus optical clot-detection methods (Rosen *et al.*, 1994; de Stefano *et al.*, 1996).

Initial modifications to APC resistance test (after thorough standardization of pre-analytic variables) included increasing the APC concentration in particular instruments to allow clear discrimination between normal and factor V:Q^{506} patients, and use of an internal quality control procedure to adjust or "normalize" the APC ratio of the patient. The latter process involved re-calculating the patient's APC ratio against the APC ratio of a standard (comprised of pooled normal plasma) that is performed with each batch of tests (Bertina *et al.*, 1994). This was then termed the APC sensitivity ratio (APC-SR), a measure which may reduce the intra- and inter-assay variation of different batches of reagents (de Ronde and Bertina, 1994) and minimize instrument bias (de Stefano *et al.*, 1996).

Despite the utmost care and the implementation of strict standardization criteria, there was still overlap between the APC resistance ratios in heterozygote factor V:Q^{506} patients and normal controls (Voorberg *et al.*, 1994; Baker *et al.*, 1994; Aillaud *et al.*, 1995).

Recent Modifications to Clotting Based Assays to Detect APC Resistance

The appreciation that the majority of patients with APC resistance have the point mutation at the APC cleavage site on factor V, has led to more specific clotting based assays using dilution of the patients plasma in factor V deficient plasma (Jorquera *et al.*, 1994; Trossaert *et al.*, 1996). This procedure diminishes potentially confounding changes in the other coagulation factors involved in the APTT based APC resistance test, in particular the effect of warfarin and elevations of factor VIII and fibrinogen levels which occur during acute thrombosis, infection, use of oral contraception and pregnancy (Henkens *et al.*, 1995; Sakato *et al.*, 1996).

The approach of using a one in five dilution of the patient's plasma in factor V deficient plasma has allowed for a clear discrimination between factor V:R^{506} and the factor V:Q^{506} mutation using the APC resistance ratio alone, reducing the need for diagnostic molecular analysis. This has been confirmed in patients with a history of thrombosis or being invest-igated for possible thrombophilia (Trossaret *et al.*, 1996; Svensson *et al.*, 1996), stable oral anticoagulation patients (Jorquera *et al.*, 1994; Trossaert *et al.*, 1994; Denson *et al.*, 1995; Cadroy *et al.*, 1995; Jorquera *et al.*, 1996), neonates and infant with septicaemia (Nowak-Gottl *et al.*, 1996), and during pregnancy (Cumming *et al.*, 1996).

Another clotting method developed to avoid the potential pitfalls of the APTT based APC resistance assay, involves the use of venom from a species of the *Vipera russelli* snake (Russell Viper Venom). In this technique, intact factor X is directly cleaved by the venom to activated Xa, thereby initiating the coagulation cascade *via* factor Va and the pro-thrombinase complex. This forms the basis of the dilute Russell Viper venom time (DRVVT) clotting assay for APC resistance, which similarly to the one in five plasma dilution APTT method, clearly differentiates normal from the factor V:Q^{506} allele. Other methods which have been reported to be useful in the identification of patients with the factor V:Q^{506} allele include a tissue factor-dependent factor V assay (Le *et al.*, 1995) and a modified chromo-genic assay (Varadi *et al.*, 1995). Whichever clotting method is used to detect APC resist-ance, each clinical laboratory must establish its own reference range (confirmed by molecular analysis) for the patient population that it examines.

Various modifications to the clotting based assays have been used to overcome the prob-lem of testing patients who have a prolonged baseline APTT as a result of heparin therapy or a lupus anticoagulant. Patients receiving low molecular weight heparin or standard hep-arin at therapeutic levels can have the heparin activity neutralized by adding polybrene in the one in five dilution APTT (Hall *et al.*, 1996) or the DRVVT APC resistance assays (unpublished observations). Alternatively, pre-treatment of plasma with heparinases (Hep-zyme, Dade) will effectively neutralize both unfractionated and low molecular weight heparin activity in the unmodified APTT based assay, with the exception of the heparinoid Orgaran (de Ronde and Bertina, 1994; Gilmore *et al.*, 1996). Adding phospholipid in patients with lupus anticoagulants either as a platelet extract or in the form of the hexagonal phase, corrects the functional APC resistance which has been reported in these patients using an APTT based assay (Martorell *et al.*, 1995).

Molecular Diagnosis of the Factor V:Q^{506} Mutation in the Clinical Laboratory

Unlike deficiencies of protein C, protein S and antithrombin III, where numerous different gene mutations produce a similar clinical phenotype, activated protein C resistance is caused in the vast majority of patients by a single point mutation on the factor V molecule where Arginine^{506} is replaced by glutamine^{506}. It is usually detected by PCR to amplify the sequence of interest followed by M*nl* I restriction site digestion (Zoller *et al.*, 1994; Zoller and Dahlback, 1994). The major advantage of detecting the factor V mutation by gene ana-lysis is that it gives a distinct result with the patient unequivocally being classified as nor-mal, heterozygote or homozygote for the mutation. Whole blood samples regardless of anticoagulant may be sent intact in the mail for DNA extraction and molecular analysis. The disadvantages of DNA testing are that it is labour intensive, time consuming and requires expertise and dedicated equipment to complete the assays. In addition, PCR techniques are subject to contamination and are generally more expensive than clotting based assays.

Various modification to the M*nl* I restriction site digest method have been proposed in an attempt to overcome specific problems with this technique relating to the cost and stability of M*nl* I, the specificity of the PCR reaction for the mutation, DNA extraction methods and the means of detection of final PCR product. Each individual modification has been reported to increase specificity for detection of the mutation without loss of sensitivity, and achieves an improved turnaround time for results. Several laboratories have described increased economy, greater efficiency and enhanced resolution for the mutation by using site directed mutagenesis with different primers that introduce the cheaper H*ind* III or T*aq* I restriction digest sites (Gandrille *et al.*, 1995; Rabes *et al.*, 1995; Guillerm *et al.*, 1996). Other groups have avoided the requirement to perform the endonuclease digestion of the final PCR fragment by designing sequence specific primers (PCR-SSP) that are complementary to the normal or mutated factor V. This technique, also known as allele specific PCR or amplification refractory mutation system (ARMS), requires two instead of one PCR procedures per sample, with an internal control and strict conditions to ensure similar PCR efficiency in each reaction (Kirshbaum and Foster, 1995; van de Locht *et al.*, 1995; Bellissimo *et al.*, 1996; Engel *et al.*, 1996; Blasczyk *et al.*, 1996). Other researchers have concentrated on the problem of DNA extraction from whole blood which is expensive, time consuming and a rate limiting step in achieving timely PCR results for clinical laboratories. Diluting whole blood and boiling without further preparation has been shown to reliably detect the factor V genotype by M*nl* I or T*aq* I restriction site digest from the PCR product (Rabes *et al.*, 1995; Rees *et al.*, 1996). Lastly, semi-automated analysis of samples and storage of results can be achieved by detecting the final PCR product by capillary electrophoresis. This also eliminates the possibility of observer misinterpretation of results when using conventional gel electrophoresis (van de Locht *et al.*, 1995). Clearly each step in the process of DNA extraction, PCR reaction and product detection for the diagnosis of the factor V mutation will be targets for improvement using new technologies. As yet no one group has published a technique which combines all the above modifications into a single, reliable, efficient and cost-effective method suitable for use by the clinical laboratory in the detection of the factor V mutation.

The Detection and Significance of Acquired APC Resistance

The term acquired APC resistance is used to describe the laboratory finding of a reduced APC resistance ratio (usually less than 2) as defined by the undiluted plasma APTT based assay, in the absence of the factor V:Q[506] mutation. It has been reported to occur in families with venous thrombosis (Zoller and Dahlback, 1994), in association with acute thrombosis (Sakata *et al.*, 1996), pregnancy (Cumming *et al.*, 1995; Bokarewa *et al.*, 1996; Mathonnet *et al.*, 1996), oral contraception (Osterud *et al.*, 1994; Henkens *et al.*, 1995; Oliveri *et al.*, 1995), antiphospholipid antibodies (Matsuda *et al.*, 1995; Ruiz-Arguelles *et al.*, 1996) and the lupus anticoagulant (Hampton *et al.*, 1994; Halbmayer *et al.*, 1994a; Gschwandtner *et al.*, 1995; Potzsch *et al.*, 1995; Villa *et al.*, 1995; Gschwandtner *et al.*, 1996). Abnormally increased levels of factor VIII, factor VIIa, ristocetin cofactor and whole blood viscosity have been associated with acquired APC resistance (Freyburger *et al.*, 1996; Sakata *et al.*, 1996). Whether the reduced APC resistance ratio in these individuals reflects a hypercoagulable state and whether this may have diagnostic or prognostic significance is yet to be established, and remains an area of great interest for future clinical research.

APC RESISTANCE AND ITS IMPACT ON VENOUS THROMBOEMBOLISM

Numerous studies have examined the association of APC resistance to venous thrombosis using either a functional APC resistance test (Table 16.1a) or DNA-based assays (Table 16.1b). An overall analysis of the patients with venous thrombosis reveals a prevalence of APC resistance of 21% compared to 4.9% in controls. Reports from among selected patients with idiopathic or familial thrombosis, have found a prevalence of APC resistance ranging from 0–64%, however this wide range is most likely explained by the differences in recruitment conditions including age, presence of other risk factors for thrombosis (pregnancy, oral contraceptives, malignancy) and study population (wide variation in frequency of the factor $V:Q^{506}$ mutation in different geographic areas and among different ethnic groups). The previously outlined variations in laboratory techniques employed to measure functional APC resistance may also contribute to this wide range.

Table 16.1a Prevalence of functional APC resistance in healthy control and in patients with venous thromboembolism

Study	Design and Age (years)	Controls		Venous Thromboembolism	
		Number	APCR	Number	APCR
Faioni *et al.*, 1993	Prospective selected cases	–	–	106	35%
Griffin *et al.*, 1993	Retrospective selected cases mean age 39	–	–	25	64%
Koster *et al.*, 1993	Prospective population based consecutive cases <70	301	5%	301	21%
Cadroy *et al.*, 1994	Prospective selected cases	75	1%	48	19%
Cushman *et al.*, 1994	Retrospective selected cases <55	39	5%	37	24%
Halbmayer *et al.*, 1994b	Retrospective selected cases mean age 46	50	2%	40	17%
Svensson *et al.*, 1994	Prospective consecutive cases 14–74	130	7%	104	40%
de Stefano *et al.*, 1995	Retrospective selected cases	–	–	118	28%
Fujimura *et al.*, 1995	Retrospective selected cases	–	–	22	18%
Kambayashi *et al.*, 1995	Retrospective selected cases	–	–	43	12%
Samaha *et al.*, 1995	Retrospective selected cases	–	–	183	13%
Trossaert *et al.*, 1995b	Prospective selected cases 15–77	50	4%	175	17%
Voorberg *et al.*, 1995	Prospective consecutive cases	–	–	27	30%
Svensson *et al.*, 1996	Prospective consecutive cases 20–89	288	11%	99	28%
Total		933	6.6%	1328	22.7%

Table 16.1b Prevalence of the factor $V:Q^{506}$ mutation in healthy controls and in patients with venous thromboembolism

Study	Design and Age (years)	Controls		Venous Thromboembolism	
		Number	Factor $V:Q^{506}$ Mutation	Number	Factor $V:Q^{506}$ Mutation
Alhenc-Gelas *et al.*, 1994	Prospective consecutive cases	–	–	87	16%
Bertina *et al.*, 1994	Prospective consecutive cases <70	301	1%	301	17.6%
Voorberg *et al.*, 1994	Prospective consecutive cases 23–79	–	–	27	37%
Arruda *et al.*, 1995	Retrospective cases <50	100	2%	40	18%
Fujimura *et al.*, 1995	Retrospective selected cases	–	–	18	0
Ma *et al.*, 1995	Prospective consecutive cases 20–86	–	–	45	27%
Ridker *et al.*, 1995a	Prospective cohort selected males 40–84	704	6.0%	121	11.6%
Rosendaal *et al.*, 1995	Prospective consecutive cases <70	474	2.9%	471	19.5%
Total		1579	3.9%	1110	18%

Estimated Risk of Venous Thromboembolism in Patients with APC Resistance

Several studies have shown that APC resistance is associated with an increased relative risk of venous thrombosis, with episodes occurring spontaneously in around half of cases, and more commonly and at a younger age in homozygotes compared to heterozygotes for the factor V:Q^{506} mutation. In the Leiden Thrombophilia Study, a large population based case-control study, APC resistance was found in 21% of otherwise unselected consecutive patients aged less than 70 years with a first ever deep vein thrombosis compared to 5% of an equal number of healthy age and sex matched controls, resulting in a relative risk of thrombosis in those with impaired response to APC of 6.6 (95% confidence limits 3.6–12) (Koster *et al.*, 1993). Zoller *et al.* (1994) studied 50 thrombosis prone families with APC resistance and found that by 33 years of age, 8% of those not carrying the mutation, 20% of heterozygotes and 40% of homozygotes had experienced at least one thrombotic complication (Figure 16.3). Thrombotic complications were more common among homozygotes than among heterozygotes (44% vs 30%) and the first thrombotic complication also occurred at an earlier age in homozygotes compared to heterozygotes (25 vs 36 years). Deep vein thrombosis was the most common clinical manifestation, but pulmonary embolism and superficial thrombophlebitis also occurred. First thrombotic events were found to have occurred in conjunction with a circumstantial risk factor (oral contraception, pregnancy, trauma and surgery) in 63% of symptomatic carriers of the factor V:Q^{506} mutation. Overall, heterozygosity for the factor V:Q^{506} mutation is associated with a 5- to 10-fold increased risk of venous thrombosis, and homozygosity with 50- to 100-fold increased risk (Bertina *et al.*, 1994 and Rosendaal *et al.*, 1995).

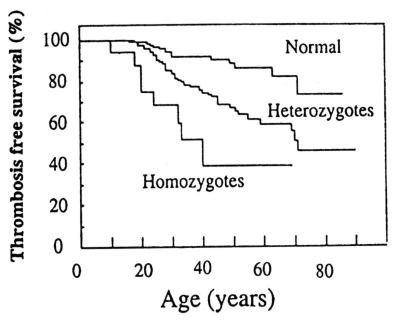

Figure 16.3   Thrombosis free survival curves from 50 families with APC resistance. It shows the age dependent relationship between genotype and frequency of thrombosis. The probability to be free from thrombosis according to age is shown in the Kaplin-Meier analysis of 146 normals, 144 heterozygotes and 18 homozygotes. Difference between normals and heterozygotes, and heterozygotes and homozygotes were highly significant (p < 0.001 and p = 0.01 respectively). (Adapted from The Journal of Clinical Investigation, 1994, **94**:2521–2524, by copyright permission of the American Society for Clinical Investigation).

Somewhat surprisingly, Desmaris *et al.* (1996) found a low prevalence of resistance to APC of 5.5% in 146 consecutive unselected patients with confirmed pulmonary embolism, compared to 4% in 348 controls also referred with possible pulmonary embolism but in whom the diagnosis was ruled out (odds ratio 1.36; 95% confidence limits 0.56–3.32). However, in this study, a total of 136 patients were excluded from analysis because of an elevated APTT, and an unspecified number received heparin prior to blood sampling and required the addition of polybrene (which has an uncertain influence on testing for APC resistance), before testing for the defect could be completed. In addition, the presence of APC resistance was not correlated with testing for factor $V:Q^{506}$ mutation, and it therefore difficult to draw definitive conclusions from this study.

There are only limited studies specifically addressing the question of recurrent thrombosis risk in patients with APC resistance. Ridker *et al.* (1995b) prospectively followed 77 participants of the Physicians Health Study with a first deep vein thrombosis and calculated that the relative risk of recurrent thrombosis in heterozygotes for the factor $V:Q^{506}$ mutation was 4.6 (p = 0.04) compared to unaffected patients. Rintelen *et al.* (1996b) determined the risk of recurrent venous thrombosis as being 9.5% per patient per year for homozygotes and 4.8% per patient per year for both unaffected controls and heterozygotes for the factor $V:Q^{506}$ mutation, although these results did not achieve statistical significance. There are no studies which have examined whether the high (20%) rate of recurrence following an episode of pulmonary embolism (Monreal *et al.*, 1989) may be related to APC resistance and large prospective studies are required before definitive conclusions can be made regarding the risk of recurrent venous thromboembolism in patients who are APC resistant.

APC resistance has been linked to childhood thromboembolism in a small number of studies. In a case report, Zenz *et al.* (1995) describe APC resistance in a boy with multiple thromboses, and a small case series has found a 21% prevalence of APC resistance in 14 children with thrombosis during childhood (Sifontes *et al.*, 1996).

Clinical Presentation of APC Resistance

The clinical manifestations of APC resistance are similar to those of the natural anticoagulant deficiencies, with venous thrombosis being the most common and accounting for >90% of all thrombotic episodes (de Stefano *et al.*, 1996). Almost half of thrombotic complications occur in association with circumstantial risk factors (surgery, pregnancy, immobilization), while thrombosis at unusual sites accounts for <5% of all episodes of thromboembolism (de Stefano *et al.*, 1996). The age related incidence of venous thrombosis in these disorders is compared in Table 16.2.

Table 16.2   The risk of first thrombotic episodes associated with natural anticoagulant defects and APC resistance. (Data obtained from Zoller *et al.*, 1994 and Allaart and Briet *et al.*, 1994)

Risk of Thrombosis (% of individuals who have experienced a thrombotic event)		
	Up to 24 years	*Up to 50 years*
Antithrombin III	50	85
Protein S	40	80
APC (homozygous)	30	62
Protein C	15	50
APC (heterozygous)	10	33

In selected study populations, heterozygous individuals appear to have a lifelong increased risk of thrombosis (Rosendaal *et al.*, 1995 and Ridker *et al.*, 1995a), however a significant proportion will remain asymptomatic, and there appears to be no survival disadvantage in affected individuals. This is confirmed by the finding of a 2% prevalence of APC resistance among 124 healthy centenarians (Mari *et al.*, 1995). None of the APC resistant centenarians in this study had a past history of thrombosis despite having all been exposed (without prophylaxis) on at least one occasion to circumstantial risk factors (trauma, pregnancy, major surgery) during their lifetime. Most individuals with the homozygous factor $V:Q^{506}$ mutation will experience at least 1 thrombotic episode in their lifetime, and are more likely to experience thrombosis at an earlier age than those with heterozygosity (Zoller *et al.*, 1994; Rosendaal *et al.*, 1995) (Figure 16.3). It is however clear that the homozygous factor $V:Q^{506}$ mutation is considerably more benign than other homozygous defects of the protein C pathway (homozygous protein C and protein S deficiency), which most commonly present with neonatal purpura fulminans (Branson *et al.*, 1983; Mahasandana *et al.*, 1990), and not all homozygous individuals will necessarily be affected by thrombosis during their lifetime (Greengard *et al.*, 1994; Zoller *et al.*, 1994).

APC RESISTANCE AND OTHER INHERITED THROMBOPHILIA

It has been recognized for some time that the presence of a particular mutation causing inherited thrombophilia does not predict the severity of the thrombotic phenotype in different families with the same mutation (Miletich *et al.*, 1987; Reitsma *et al.*, 1991), and this has led to speculation that thrombophilia is a polygenic disorder. More recently, this variable thrombotic phenotype has also been seen in families with APC resistance, and consequently there have been a number of studies examining the prevalence of co-segregation of this mutation with other forms of inherited thrombophilia (Table 16.3). The recently described molecular influence of protein S on the rate of factor Va cleavage may partially explain why the combined deficiency of protein S and the factor $V:Q^{506}$ mutation causes a substantial additive increased risk of venous thrombosis compared to either one defect alone (Koeleman *et al.*, 1995; Zoller *et al.*, 1995). Protein S has been found to accelerate by twenty fold the efficiency of the Arg^{306} cleavage of factor Va by APC, but has no effect on rate of inactivation at the Arg^{506} site (Rosing *et al.*, 1995). The anticoagulant function of protein S is a particularly important as a downregulator of thrombosis in patients with the factor $V:Q^{506}$ mutation, where the rate of Arg^{306} cleavage even without protein S deficiency is already substantially diminished and inefficient (Heeb *et al.*, 1995; Kalafatis

Table 16.3 Prevalence of thrombotic complications in members of families with combined APC resistance and natural anticoagulant deficiencies

Study	Thrombosis Risk (% with thrombotic complications)		
	Natural Anticoagulant Deficiency Alone	*Factor V:Q^{506} Mutation Alone*	*Combined Natural Anticoagulant Deficiency and Factor V: Q^{506} Mutation*
Protein C Koeleman *et al.*, 1994	0%	0%	45%
Brenner *et al.*,1996	31%	13%	73%
Protein S Koeleman *et al.*, 1995	66%	43%	80%
Zoller *et al.*, 1995	19%	19%	72%
Antithrombin III van Boven *et al.*, 1996	54%	20%	92%

et al., 1995). A similar pattern of increased thrombotic risk has been reported in members of families with combined APC resistance and protein C deficiency (Koeleman *et al.*, 1994; Brenner *et al.*, 1996) and antithrombin III deficiency (van Boven *et al.*, 1996) (Table 16.3). Other studies have likewise demonstrated a high prevalence of the factor V:Q^{506} mutation in patients with natural anticoagulant deficiencies (Gandrille *et al.*, 1995; Hallam *et al.*, 1995), although this has not been a universal finding (Simioni *et al.*, 1994; Martinelli *et al.*, 1996a).

Hyperhomocysteinemia is another common form of thrombophilia which occurs both as an inherited and acquired defect, and constitutes an independent risk factor for venous thromboembolism (de Stefano *et al.*, 1996). However there have only been limited studies examining the possible association between this defect and APC resistance. In the Leiden Thrombophilia Study, high plasma homocysteine levels were confirmed to be an independent risk factor for venous thrombosis, however there was no increase in risk of thrombosis in patients with combined mild hyperhomocysteinemia and the factor V:Q^{506} mutation (Den Heijer *et al.*, 1996). In contrast, Mandel *et al.* (1996) in a study of 45 members of seven unrelated consanguineous kindreds of whom at least one member had homozygous homocystinuria, found that major thrombotic events occurred only in those with combined hyperhomocysteinemia and the factor V:Q^{506} mutation. These conflicting results illustrate the need for further clinical studies to explore the possible link between hyperhomocysteinemia and the factor V:Q^{506} mutation.

Overall, these results support the concept of inherited thrombophilia as a multigene disorder. However, within families with genetic thrombophilia there are still unexplained episodes of thrombosis involving individuals in whom no genetic defect has been found, and it is therefore clear that there are as yet undiscovered genetic abnormalities causing thrombophilia (Koeleman *et al.*, 1994; Zoller *et al.*, 1994; Gandrille *et al.*, 1995; Zoller *et al.*, 1995). One such recent discovery is the prothrombin 20210 mutation that is hereditary, common and associated with venous thrombosis (Poort *et al.*, 1996)

APC RESISTANCE, ORAL CONTRACEPTION AND PREGNANCY

Pregnancy and oral contraception appear to be two of the most common pre-disposing factors for thrombosis in APC resistant women who have experienced a venous thromboembolic event (Zoller *et al.*, 1994). Oral contraceptives are an independent risk factor for venous thrombosis, with an odds ratio for a first venous thromboembolism in patients taking levonorgestrel-containing oral contraception of 4.2 in Europe and 3.3 in developing countries (Poulter *et al.*, 1995). This risk is further increased by a factor of between 2.1 and 2.6 with the introduction of new low-dose oral contraception containing a third generation prostagen (Jick *et al.*, 1995; Farley *et al.*, 1995). A series of studies have looked specifically at the risk of thrombosis in patients with both APC resistance and oral contraceptive use. Hellgren *et al.* (1995) reported a higher prevalence of APC resistance in women with thromboembolic complications during oral contraceptive use compared to healthy controls (30% vs 10%, p<0.05). In an extension of the population-based Leiden Thrombophilia Study, Vandenbroucke *et al.* (1994) found that the relative risk of a first venous thrombosis in consecutive female patients aged 15–49 years was increased 4-fold amongst oral contraceptive users, 8-fold amongst carriers of the factor V:Q^{506} mutation and greater than 30-fold (relative risk 34.7, 95% confidence interval 7.8–154) in those with both risk factors. The absolute risk of first venous thrombosis was calculated at 0.8 per 10000 person-years, and in those taking oral contraception who had the factor V mutation, this increased to 28.5 per

10 000 person-years. Data from the same case-control study was subsequently re-analyzed to focus on the third generation progestagen-containing oral contraceptives (Bloemenkamp *et al.*, 1995), with the findings being of relative risk of a first venous thrombosis of almost 50-fold in users of this form of oral contraception who also had the factor V:Q^{506} mutation. Rintelen *et al.* (1996a) studied 29 patients (17 females) with the homozygous factor V:Q^{506} mutation of whom 25 had experienced a thrombotic complication, and found that thrombosis occurred significantly earlier in females (median age 26) compared to males (median age 38). The most common triggering factor for thrombosis in females was use of oral contraception (80% of females had taken oral contraception prior to thrombosis), while in males 80% of thrombotic events were spontaneous.

Thromboembolism complicates 0.07–2.6% of pregnancies (Greer *et al.*, 1994) and is a major cause of maternal mortality (Sachs *et al.*, 1987), however deficiencies of the natural anticoagulants have been found in only a minority of these patients (de Boer *et al.*, 1992). APC resistance has now been clearly demonstrated to cause an increased risk of thromboembolic complications during pregnancy. In a retrospective case-control study, Hellgren *et al.* (1995) reported a 59% prevalence of APC resistance in 34 women with previous thromboembolic complications during pregnancy compared to 10% in 57 non-pregnant controls (p < 0.001) and 11% in 18 pregnant controls, although the factor V mutation was not investigated in this study. Similarly, de Stefano *et al.* (1995) found that 55% of women with thrombosis relating to pregnancy had APC resistance, while Bokarewa *et al.* (1996), using a molecular technique to identify the factor V:Q^{506} mutation, found a 46% prevalence of APC resistance in 70 women who had experienced thrombosis relating to pregnancy. Carriers of the mutation were more prone to the development of thrombosis during their first pregnancy (odds ratio 3.4) and had a higher probability of recurrent thrombosis (odds ratio 3.8) compared to non-carriers (Bokarewa *et al.*, 1996). In this study, women with the mutation were more likely to develop thrombosis in the first trimester of pregnancy or immediately post-partum compared to those without the mutation.

Other complications of pregnancy have been linked to APC resistance, including second-trimester miscarriage (Rai *et al.*, 1996), pre-eclampsia and abruptio placentae (Grisaru *et al.*, 1996), however these observations require confirmation in larger studies.

APC RESISTANCE AND LUPUS ANTICOAGULANTS/ ANTIPHOSPHOLIPID ANTIBODIES

Anti-phospholipid antibodies (APA) and lupus anticoagulants (LA) are commonly associated with arterial and venous thrombosis, although the pathogenesis of thrombotic events in these patients remains poorly understood. Given the high frequency of the heterozygous factor V:Q^{506} mutation in the general population, it is possible that the combination of APC resistance and APA/LA accounts for an increased thrombotic risk in these patients. However published studies have failed to demonstrate a relationship between APA/LA patients and inherited APC resistance caused by the factor V:Q^{506} mutation (Bertina *et al.*, 1994; Biousse *et al.*, 1995; Davies *et al.*, 1995; Dizon-Townsend *et al.*, 1995; Ruiz-Arguelles *et al.*, 1996).

As previously discussed, acquired APC resistance is commonly seen in patients with APA/LA, and there have been reports of APC resistance in the absence of the factor V:Q^{506} mutation in up to 70% of patients with LA and thrombosis (Potzsch *et al.*, 1995). Several investigators (Marciniak and Romond, 1989; Malia *et al.*, 1990 and Borrell *et al.*, 1992) have also demonstrated that the IgG fraction from patients with APA inhibits the degradation of factor Va by activated protein C. However, further studies are clearly needed to elucidate the mechanism of thrombosis and the possible role of acquired APC resistance in the pathogenesis of thrombosis in these patients.

APC RESISTANCE AND MYOCARDIAL INFARCTION

Only one of ten studies (Table 16.4) found an increased prevalence of the factor V:Q^{506} mutation in myocardial infarction (Marz et al., 1995). A second report did show a significantly increased prevalence of the factor V:Q^{506} mutation in a male subgroup of survivors of myocardial infarction compared to controls (23% vs 10%, p = 0.03), however when males and females were considered together there was no significant difference (Holm et al., 1996). An overall analysis of patients with myocardial infarction reveals a prevalence of the factor V:Q^{506} mutation of 6.5% compared to 5.4% of controls. A further case control study of 134 survivors of myocardial infarction and 100 healthy controls also failed to demonstrate an association with APC resistance, although the factor V:Q^{506} mutation was not analyzed in this report (Biasiutti et al., 1995). The majority of these studies were retrospective and confined to survivors of myocardial infarction, and it therefore remains possible that APC resistance affects clinical outcome, and in particular early mortality, in myocardial infarction by influencing the extent of ischaemic damage, response to therapy and rate of re-thrombosis. However, such an effect was not evident in the prospective case-control study by Ridker et al. (1995a) although this report was confined to males, half of whom were receiving Aspirin.

Table 16.4 Prevalence of the factor V:Q^{506} mutation in healthy controls and in patients with myocardial infarction

Study	Design and Age (years)	Controls		Myocardial Infarction	
		Number	% Factor V:Q^{506} Mutation	Number	% Factor V:Q^{506} Mutation
Kontula et al., 1994	Retrospective selected survivors <60	137	2.9%	122	5.7%
Samani et al., 1994	Prospective consecutive cases <50	–	–	60	3%
Emmerich et al., 1995	Retrospective selected survivors 24–64	726	4.6%	643	5.1%
* Marz et al., 1995	Retrospective selected survivors 28–78	196	4%	224	9%
* Prohaska et al., 1995	Retrospective selected survivors 28–85	190	9.5%	317	9.1%
Ridker et al., 1995a	Prospective selected males 40–84	704	6%	374	6.1%
van Bockxmeer et al., 1995	Prospective consecutive cases <50	126	4.0%	222	5.0%
Ardissino et al., 1996	Retrospective selected survivors <45	100	2.0%	100	1%
Holm et al., 1996	Retrospective selected survivors <50	101	11%	101	18%
van der Bom et al., 1996	Retrospective Survivors	222	5%	114	4%
Total		2502	5.4%	2277	6.5%

* Marz et al. and Prohaska et al. studied patients with documented coronary artery disease, some of whom were selected survivors of myocardial infarction.

No association has been found between APC resistance and graft occlusion after coronary artery bypass graft surgery (Eritsland et al., 1995). However, isolated case reports of homozygous APC resistance in myocardial infarction (Holm *et al.*, 1994) raise the possibility that homozygosity for the factor V:Q^{506} mutation may play a more important role in the pathogenesis of arterial thrombosis, and myocardial infarction in particular, than heterozygosity. As there are certain geographic areas with an extremely high prevalence of the mutant factor V gene, and hence also an increased prevalence of the homozygous form of this mutation (Zoller *et al.*, 1996), screening for APC resistance in patients with myocardial infarction may prove to be valuable in these populations.

APC RESISTANCE AND ISCHAEMIC STROKE

APC resistance in ischaemic stroke has been investigated using both a functional assay (Table 16.5a) and DNA-based techniques (Table 16.5b). One of six studies examining the factor V:Q^{506} mutation found a higher prevalence of this defect in patients with stroke compared to controls (Albucher *et al.*, 1995), however an overall analysis of the factor V:Q^{506} mutation in stroke revealed a 3.0% prevalence of this defect in patients compared to 5.1% in controls (Table 16.5b). This suggests that the genetic mutation causing APC resistance does not play a significant role in the pathogenesis of the majority of cases of ischaemic stroke.

Table 16.5a Prevalence of functional APC resistance in healthy controls and patients with ischaemic stroke

Study	Design and Age (years)	Controls		Stroke	
		Number	% APCR	Number	% APCR
Cushman *et al.*, 1994	Retrospective selected cases <55	39	5%	15	0%
Halbmayer *et al.* 1994b	Retrospective selected survivors mean age 46	50	2%	30	20%
Forsyth *et al.*, 1995	Uncontrolled selected cases <60	–	–	45	2.2%

Table 16.5b Prevalence of the factor V:Q^{506} mutation in healthy controls and in patients with ischaemic stroke

Study	Design and Age (years)	Controls		Stroke	
		Number	% Factor V:Q^{506} Mutation	Number	% Factor V:Q^{506} Mutation
Albucher *et al.*, 1995	Prospective consecutive cases <45	75	1.3%	30	10%
Catto *et al.*, 1995	Prospective selected cases 65–80	247	5.6%	368	4.1%
Kontula *et al.*, 1995	Retrospective selected cases <60	137	2.9%	236	3.8%
Ridker *et al.*, 1995a	Prospective selected males 40–84	704	6%	209	4.3%
Press *et al.*, 1996	Prospective selected cases mean age 64	170	4.7%	161	2.5%
van der Bom *et al.*, 1996	Retrospective survivors >55	222	5%	107	6%
Total		1555	5.1%	1111	3.0%

Functional APC resistance has been linked to ischaemic stroke in isolated case reports (Bachmeyer *et al.*, 1995; Simioni *et al.*, 1995a) and one (Halbmayer *et al.*, 1994b) of the three case series which have examined this association (see Table 16.5a). A separate study (van der Bom *et al.*, 1996) reported APC resistance in patients with a history of stroke as a continuous variable in addition to performing genetic studies for the factor V:Q^{506} mutation, and found that the prevalence of cerebrovascular disease, but not of myocardial infarction, was inversely proportional to the APC ratio, and was independent of the factor V:Q^{506} mutation. The majority of reports in the literature have not analyzed the APC ratio as a continuous variable, and this finding raises the possibility that response to APC independent of the factor V:Q^{506} mutation may serve as a risk indicator for cerebrovascular disease. Further studies are required to confirm this finding and to elucidate the mechanism of acquired APC resistance in these patients.

Conflicting reports in two small case series of childhood stroke concluded that there was no association (Riikonen *et al.*, 1996) and an increased prevalence (Ganesan *et al.*, 1996) of APC resistance in these patients. Again, further data is required to clarify this question.

APC RESISTANCE AND PERIPHERAL VASCULAR DISEASE

In a small case series, Lindblat *et al.* (1994) suggested an association between APC resistance and occlusion of peripheral arterial vascular reconstructions. This has subsequently been supported in a larger case series by Ouriel *et al.* (1996) who followed 74 patients undergoing reconstructive lower limb vascular surgery and found a significantly increased risk of graft occlusion in patients with APC resistance compared to patients without APC resistance (6/10 vs 16/64, p = 0.02). Meanwhile, Cushman *et al.* (1994) found no increase in prevalence of APC resistance in 44 cases of arterial thrombosis including 17 patients with peripheral vascular disease.

Although an association between peripheral vascular disease and APC resistance has not been demonstrated, these studies do raise the possibility that APC resistance may be predictive of an increased risk of graft occlusion in patients with peripheral vascular disease who are undergoing reconstructive surgery with the placement of arterial grafts.

APC RESISTANCE AND UNUSUAL SITES OF THROMBOSIS

There have been isolated case reports linking APC resistance to Budd-Chiari syndrome (Denninger *et al.*, 1995; Mahmoud *et al.*, 1995; Mambrini *et al.*, 1996), heparin-induced thrombocytopenia (Gardyn *et al.*, 1995), relapsing nephrotic syndrome with recurrent thrombosis (Petaja *et al.*, 1995), priapism (de Prost *et al.*, 1995), systemic embolism in non-rheumatic atrial fibrillation (Eikelboom *et al.*, 1996), Warfarin induced skin necrosis (Makris *et al.*, 1996) and thrombotic complications in a factor V deficient family (Zehnder and Jain, 1996). Single case-control studies have also reported an association between APC resistance and venous leg ulceration (Munkvad and Jorgensen, 1996) and thrombosis in the setting of myeloproliferative disorders (Buscalossi *et al.*, 1996).

The potential role of the factor V:Q^{506} mutation in amelioration of the bleeding phenotype in patients with severe haemophilia A and B has been examined in a small case series however no association was demonstrated (Arbin *et al.*, 1995). Likewise, in single case series, the factor V:Q^{506} mutation could not be implicated in the thrombotic complications which occur in acute childhood lymphoblastic leukaemia (Nowak-Gottl *et al.*, 1995) and Sickle Cell Disease (Laffan *et al.*, 1996).

APC resistance appears to be an important risk factor in cases thrombosis involving the cerebral venous circulation, including both cerebral sinus thrombosis and retinal vein thrombosis. Martinelli *et al.* (1996b) in a retrospective case-control study found a 20% prevalence of the heterozygous factor V:Q^{506} mutation in 25 patients with cerebral venous thrombosis compared to 2.7% of 75 controls (p = 0.01). In the setting of central retinal vein thrombosis, Larsson *et al.* (1996), in an uncontrolled retrospective case series which included 31 patients aged <50 years, found a 26% prevalence of APC resistance, while Williamson *et al.* (1996) in a prospective case-control study, found a 12% prevalence of APC resistance in 35 patients compared to 5% of control subjects. There has also been a case report of APC resistance in central retinal vein occlusion (Dhote *et al.*, 1995).

WHO SHOULD BE TESTED FOR APC RESISTANCE?

Venous thromboembolism is a common cause of in-hospital mortality accounting for an estimated 5–10% of all deaths (Anderson and Wheeler, 1995), and between 2.5% and 5% of the population have been estimated to suffer from at least one thrombotic episode during their lifetime (Zoller and Dahlback, 1995). However, despite long-standing and convincing evidence of the efficacy of a number of prophylactic agents for the prevention of venous thromboembolism, studies have shown that less than one third of high risk patients receive appropriate prophylaxis (Anderson and Wheeler, 1995). As APC resistance is now known to be the single most common underlying cause of venous thrombosis, identification of those affected by this defect will distinguish a large proportion of those who are at highest risk of thrombosis, and thereby improve our ability to prevent this complication. This realization has led to suggestions that it might be more cost-effective to screen for APC resistance in the general population, than to screen for and treat hypertension (Zoller *et al.*, 1996). However the determination as to who should be tested is still not clearly resolved and awaits the results of prospective population studies.

Testing of The General Population

Before any large population-based screening program can be considered, certain specific and often difficult criteria need to be fulfilled (Hodgkin, 1978). While the majority of these criteria would appear to apply readily to screening for APC resistance, a positive test for this defect is not always associated with serious thrombotic complications, and there is clearly a lack of knowledge regarding the natural history of this defect in asymptomatic individuals. Further information will only become available when the results of large, controlled, prospective, population studies are published, and until such times, it is not possible to recommend wide-scale testing for APC resistance in the general population. These studies may also provide the necessary information which will allow a comprehensive cost-benefit analysis of any proposed screening program for APC resistance.

Testing of Asymptomatic Persons in High Risk Situations

An emerging issue is whether or not to test otherwise asymptomatic patients without a past or family history of thrombosis for APC resistance, prior to exposure to common circumstantial risk factors (surgery, pregnancy, oral contraception).

Prior to surgery

In the pre-operative setting there is no current evidence that screening for APC resistance should be performed in asymptomatic persons without other risk factors for venous thrombosis, except in the setting of a personal or family history of thrombosis.

Prior to commencement of oral contraception

Women with a family history of venous thrombosis could consider screening for the factor $V:Q^{506}$ mutation to help the decision making processs regarding the risk versus the benefits of the oral contraceptive pill. The increased relative risk to 35 fold of venous thrombosis in patients with the combination of the oral contraceptive pill and the factor $V:Q^{506}$ mutation would appear to provide a basis for screening of all women contemplating oral contraception. However, the absolute baseline annual prevalence of thrombosis at a young age is small, approximately one per 10 000 (range 0.6–1.9 per 10 000 for persons aged 0–49) (Rosendaal *et al.*, 1995) or about 0.01% per person per year. The absolute risk in those with the factor $V:Q^{506}$ mutation increases to around 0.3% per year or 3% over a 10 year period of oral contraceptive use. Rosendaal *et al.* (1996) estimated that 2.25 million women requesting the pill would need to be tested (assuming a prevalence of 4% of the factor $V:Q^{506}$ mutation in the general population) to prevent one death from fatal thromboembolism, although such calculations do not take into consideration the morbidity caused by thromboembolic episodes including acute symptoms of pain and discomfort, need for hospitalization and diagnostic testing, the hazards of anticoagulation, the significant probability of post phlebitic syndrome and increased risks of recurrent thrombosis.

Prior to pregnancy

There is currently no clear evidence to support a policy of testing for APC resistance prior to pregnancy in asymptomatic patients without a past or family history of thrombosis. Unfortunately there is no data available which allows an estimate of absolute risk of thrombosis in asymptomatic patients with APC resistance who become pregnant, and as prophylactic therapy with heparin during pregnancy and the post-partum period is both inconvenient and associated with potential hazards (bleeding, osteoporosis).

Testing for APC Resistance in Patients with A Personal or Family History of Venous Thrombosis

Patients with venous thrombosis

The identification of APC resistance in symptomatic individuals may have implications with regards to management of the acute thrombotic episode, duration and intensity of oral anticoagulation, future prophylactic therapy when exposed to situations of increased thrombotic, decisions regarding use of hormonal contraception and the identification of affected family members. Therefore, it would be reasonable to test all individuals with venous thrombosis for this defect. In particular, testing is indicated in young patients with venous thrombosis, those with a history of recurrent or familial thrombosis and cases of unusual thrombosis (e.g. involving the cerebral venous circulation, retinal veins, portal veins).

Patients with other forms of inherited thrombophilia

The high frequency of co-segregation of APC resistance with other forms of inherited thrombophilia, and the severe thrombotic phenotype in individuals with more than one genetic defect make it important to screen all persons with a previously identified natural anticoagulant deficiency for APC resistance.

Family members of symptomatic patients

It seems reasonable to test all relatives of symptomatic APC resistant patients for this defect. Half of family members will be expected to have the mutation, and the implications concerning thrombosis prophylaxis in high risk situations will apply in a similar fashion to both APC resistant probands and to asymptomatic APC resistant family members.

PRACTICAL MANAGEMENT OF APC RESISTANCE

Treatment of APC Resistant Patients Who Experience Venous Thrombosis

It appears that the management of acute thrombosis in patients with APC resistance is no different from that of patients without inherited thrombophilia and consists of standard intravenous unfractionated or low molecular weight Heparin followed by Warfarin therapy. Patients who have experienced one episode of thrombosis are, as a group, at increased risk of further thrombosis, and the presence of APC resistance may confer an additional risk of recurrence. However, there is no data that can answer the question of optimal duration of anticoagulation in these patients. In making this clinical decision, due consideration should be given to individual patient characteristics including the number, site and severity of thrombosis, whether the thrombotic episode was spontaneous or whether precipitating factors were present, the gender and life-style of the individual and the family history of thromboembolism (Bauer, 1995). First episodes of thrombosis should generally receive standard duration and intensity of oral anticoagulation, and long-term Warfarin therapy should be restricted to patients with recurrent thrombosis, spontaneous life-threatening thrombosis, thrombosis at unusual sites or patients with more than one genetic defect (Bauer, 1995; de Stefano *et al.*, 1996).

Secondary Prevention of Thrombosis in APC Resistant Patients

As almost half of thrombotic complications in patients with inherited thrombophilia occur in the setting of known risk exposure, appropriate anti-thrombotic prophylaxis is important in this patient group. Options include the use of both pharmacological treatments (e.g. Heparin, Warfarin) and non-pharmacological therapies (e.g. graduated compression stockings, early mobilization post surgery), with the choice being dictated by the level of thrombosis risk (low vs high), potential bleeding risk and individual patient characteristics (e.g. co-existent disease).

As previously alluded to, pregnancy poses a unique problem particularly in terms of risks to the foetus as a result of anticoagulant therapy. It would be reasonable to treat patients with APC resistance and a past history of thrombosis with prophylactic subcutaneous unfractionated or low molecular weight Heparin during pregnancy and for at least four to six weeks post-partum. In view of the increased risk of thrombosis in APC resistant individuals who who have already experienced a thrombotic event, the oral contraception should generally be avoided.

Treatment of Patient with Homozygous APC Resistance

In view of the greater risk of thrombotic complications in homozygous APC resistance compared to heterozygotes, these patients may warrant extended and/or more intense anticoagulation after a single episode of thrombosis.

CONCLUSION

The recent discovery of APC resistance has substantially altered our understanding of the importance of the protein C anticoagulant pathway in patients with a personal or family history of venous thrombosis. This abnormality is caused by a single point mutation involving the factor V gene which can be detected readily in the clinical laboratory by either clotting-based or molecular techniques. It is commonly found in Western populations and explains the majority of cases of thrombophilia. The impact of screening for this mutation to prevent thrombosis in the general population or in selected patients prior to high risk environmental exposure remains to be established. The discovery of APC resistance offers an improved understanding of the relationship between genetic and environmental factors in the predisposition to venous thrombosis and confirms the polygenic nature of this common clinical problem.

References

Aillaud, M.F., Succo, E., Alessi, M.C., Gandois, J.M., Gallian, P., Morange, P. *et al.* (1995). Resistance to activated protein C — diagnostic strategy in a laboratory of haemostasis. *Thrombosis and Haemostasis*, **74**:1197–1198.

Albucher, J.F., Guiraud-Chaumeil, B., Chollet, F., Vadroy, Y. and Sie, P. (1996). Frequency of resistance to activated protein C due to factor V mutation in young patients with ischaemic stroke. *Stroke*, **27**:766–777.

Alhenc-Gelas, M., Gandrille, S., Aubry, M., Emmerich, J., Fiessinger, J. and Aiach, M. (1994). Unexplained thrombosis and factor V Leiden mutation. *Lancet*, **343**:1535–1536.

Allaart, C.F. and Briet, E. (1994). Familial venous thrombophilia. In *Haemostasis and Thrombosis* (volume 1), edited by A.L. Bloom, C.D. Forbes, D.P Thomas, E.G.D. Tuddenham, pp. 1349–1360. New York: Churchill Livingstone.

Anderson, F.A. and Wheeler, H.B. (1995). Venous thromboembolism: Risk Factors and Prophylaxis. *Clinics in Chest Medicine*, **16**:235–251.

Anderson, F.A., Wheeler, H.B., Goldberg, R.J., Hosmer, D.W., Patwardhan, N.A., Jovanovic, B. *et al.* (1991). A population-based perspective of the incidence and case-fatality rates of venous thrombosis and pulmonary embolism: The Worcester DVT Study. *Archives of Internal Medicine*, **151**:933–938.

Arbini, A.A., Mannucci, P.M. and Bauer, K.A. (1995). Low prevalence of the factor V Leiden mutation among "severe" hemophiliacs with a "milder" bleeding diathesis. *Thrombosis and Haemostasis*, **74**:1255–1258.

Ardissino, D., Peyvandi, F., Merlini, P.A., Colombi, E. and Mannucci, P.M. (1996). Factor V (Arg^{506}→Gln) mutation in young survivors of myocardial infarction. *Thrombosis and Haemostasis*, **75**:701–702.

Arruda, V.R., Annichino-Bizzacchi, J.M., Costa, F.F. and Reitsma, P.H., (1995). Factor V Leiden (factor VQ506) is common in a Brazilian population. *American Journal of Hematology*, **49**:242–243.

Arruda, V.R., von Zuben, P.M., Soares, M.C.P., Menezes, R., Annichino-Bizzacchi, J.M. and Costa, F.F. (1996). Very low incidence of Arg^{506}→Gln mutation in the factor V gene among the Amazonian Indians and the Brazilian Black population. *Thrombosis and Haemostasis*, **75**:859–860.

Bachmeyer, C., Toulon, P., Dhote, R., Christoforov, B., Zuber, M. and Mas, J. (1995). Ischemic stroke and activated protein C resistance. *Journal of the American Medical Association*, **274**:1266.

Baker, R., Thom, J. and van Bockxmeer, F. (1994). Diagnosis of Activated Protein C Resistance (Factor V Leiden). *Lancet*, **344**:1162.

Bauer, K.A. (1994). Hypercoagulability — A new cofactor in the protein C anticoagulant pathway. *New England Journal of Medicine*, **330**:566–567.

Bauer, K.A. (1995). Management of patients with hereditary defects predisposing to thrombosis including pregnant women. *Thrombosis and Haemostasis*, **74**:94–100.

Beauchamp, N.J., Daly, M.E., Cooper, P.C., Preston, F.E. and Peake, I.R. (1994). Rapid two-stage PCR for detecting factor V G1691A mutation. *Lancet*, **344**:694–695.

Bellissimo, D.B., Kirschbaum, N.E. and Foster, P.A. (1996). Improved method for factor V Leiden typing by PCR-SSP. *Thrombosis and Haemostasis*, **75**:520.

Bertina, R.M., Koeleman, B.P.C., Koster, T., Rosendaal F.R., Dirven R.J., de Ronde, H. *et al.* (1994). Mutation in blood coagulation factor V associated with resistance to activated protein C. *Nature*, **369**:64–67.

Bertina, R.M., Reitsma, P.H., Rosendaal, F.R. and Vandenbroucke J.P. (1995). Resistance to activated protein C and factor V Leiden as risk factors for venous thrombosis. *Thrombosis and Haemostasis*, **74**:449–453.

Beufe, S., Borg, J., Vasse, M., Charbonnier, F., Moreau, V., Monconduit M. *et al.* (1995). Co-segregaton of thrombosis with factor V Q^{506} mutation in an extended family with resistance to activated protein C. *British Journal of Haematology*, **89**:659–662.

Biasiutti, F.D., Merlo, C., Furlan, M., Sulzer, I., Binder, B.R. and Lammle, B. (1995). No association of APC resistance with myocardial infarction. *Blood Coagulation and Fibrinolysis*, **6**:456–459.

Biousse, V., Piette, J., Frances, C., Bletry, O., Papo, T. and Tournier-Lasserve, E. (1995). Primary antiphospholipid syndrome is not associated with activated protein C resistance caused by factor V Arg^{506}→Gln mutation. *Journal of Rheumatology*, **22**:1215.

Blaszczk, R., Ritter, M., Thiede, C., Wehling, J., Hintz, G. and Neubauer, A. (1996). Simple and rapid detection of factor V Leiden by allele-specific PCR amplification. *Thrombosis and Haemostasis*, **75**:757–759.

Bloemenkamp, K.W.M., Rosendaal, F.R., Helmerhorst, F.M., Buller, H.R. and Vandenbroucke, J.P. (1995). Enhancement by factor V Leiden mutation of risk of deep-vein thrombosis associated with oral contraceptives containing a third-generation progestagen. *Lancet*, **346**:1593–1596.

Bokarewa, M.I., Bremme, K. and Blomback, M. (1996). Arg^{506}Gln mutation in factor V and risk of thrombosis during pregnancy. *British Journal of Haematology*, **92**:473–478.

Borrell, M., Sala, N., de-Castellarnau, C., Lopex, S. and Gari, M., Fontcuberta, J. (1992). Immunoglobulin fractions isolated from patients with antiphospholipid antibodies prevent the inactivation of factor Va by activated protein C on human endothelial cells. *Thrombosis and Haemostasis*, **68**:268–272.

Branson, H.E., Marble, R. and Griffin, J.H. (1983). Inherited protein C deficiency and coumarin-responsive chronic relapsing purpura fulminans in a newborn infant. *Lancet*, **2** (8360):1165–1168.

Brenner, B., Zivelin, A., Lanir, N., Greengard, J.S., Griffin, J.H. and Selingsohn, U. (1996). Venous thromboembolism associated with double heterozygosity for R^{506}Q mutation of factor V and for T298M mutation of protein C in a large family of a previously described homozygous protein C-deficient newborn with massive thrombosis. *Blood*, **88**:877–880.

Bucalossi, A., Marotta, G., Bigazzi, C., Galieni, P. and Dispensa, R. (1996). Reduction of antithrombin III, protein C, and protein S levels and activated protein C resistance in polycythemia vera and essential thrombocythemia patients with thrombosis. *American Journal of Hematology*, **52**:14–19.

Cadroy, Y., Sie, P., Alhenc-Gelas, M. and Aiach, M. (1995). Evaluation of APC resistance in the plasma of patients with Q^{506} mutation of factor V (Factor V Leiden) and treated by oral anticoagulants. *Thrombosis and Haemostasis*, **73**:734–735.

Cadroy, Y., Sie, P. and Boneu, B. (1994). Frequency of a defective response to activated protein C in patients with a history of venous thrombosis. *Blood*, **83**:2008–2009.

Camire, R.M., Kalafatis, M., Cushman, M., Tracy, R.P., Mann, K.G. and Tracy P.B. (1995). The mechanism of inactivation of human platelet factor Va from normal and activated protein C-resistant individuals. *Journal of Biological Chemistry*, **270**:20794–20800.

Catto, A., Carter, A., Ireland, H., Bayston, T.A., Philippou, H., Barrett, J. *et al.* (1995). Factor V Leiden gene mutation and thrombin generation in relation to the development of acute stroke. *Arteriosclerosis, Thrombosis, and Vascular Biology*, **15**:783–785.

Chan, L.C., Bourke, C., Lam, C.K., Liu, H.W., Brookes, S., Jenkins, V. *et al.* (1996). Lack of activated protein C resistance in healthy Hong Kong Chinese blood donors — correlation with absence of Arg^{506}-Gln mutation of factor V gene. *Thrombosis and Haemostasis*, **75**:522–523.

Comp, P.C., Nixon, R.R., Cooper, M.R. and Esmon, C.T. (1984). Familial protein S deficiency is associated with recurrent thrombosis. *Journal of Clinical Investigation*, **74**:2082–2088.

Cox, M.J., Rees, D.C., Martinson, J.J. and Clegg, J.B. (1996). Evidence for a single origin of factor V Leiden. *British Journal of Haematology*, **92**:1022–1025.

Cumming, A.M., Tait, R.C., Fildes, S. and Hay, C.R.M. (1996). Diagnosis of APC resistance during pregnancy. *British Journal of Haematology*, **92**:1026–1027.

Cumming, A.M., Tait, R.C., Fildes, S., Yoong, A., Keeney, S. and Hay, C.R.M. (1995). Development of resistance to activated protein C during pregnancy. *British Journal of Haematology*, **90**:725–727.

Cushman, M., Bhushan, F., Bovill, E. and Tracy, R. (1994). Plasma resistance to activated protein C in venous and arterial thrombosis. *Thrombosis and Haemostasis*, **72**:647.

Dahlback, B, Carlsson, M. and Svensson, P.J. (1993). Familial thrombophilia due to a previously unrecognized mechanism characterized by poor anticoagulant response to activated protein C; Prediction of a cofactor to activated protein C. *Proceedings of the National Academy of Sciences of the United States of America*, **90**:1004–1008.

Dahlback, B. and Hildebrand, B. (1994). Inherited resistance to activated protein C is corrected by anticoagulant cofactor activity found to be property of factor V. *Proceedings of the National Academy of Sciences of the United States of America*, **91**:3396–1400.

Dahlback, B. and Stenflo, J. (1994). A natural anticoagulant pathway: Proteins C, S, C4b-binding protein and thrombomodulin. In *Haemostasis and Thrombosis* (volume 1), edited by A.L. Bloom, C.D. Forbes, D.P Thomas, E.G.D. Tuddenham, pp. 671–698. New York: Churchill Livingstone.

Davie, E.W., Fujikawa, K. and Kisiel, W. (1991). The coagulation cascade: Initiation, maintenance, and regulation. *Biochemistry*, **30**:10363–103703.

Davies, K.A., Ireland, H., Athanassiou, P., Loizou, S., Lane, D. and Walport, M.J. (1994). Factor V Leiden mutation and venous thrombosis. *Lancet*, **345**:132–133.

de Boer, K., Buller, H.R., tenCate, J.W. and Levi, M. (1992). Deep vein thrombosis in obstetric patients: Diagnosis and risk factors. *Thrombosis and Haemostasis*, **67**:4–7.

de Maat, M.P.M., Kluft, C., Jespersen, J. and Gram, J. (1996). World distribution of factor V Leiden mutation. *Lancet*, **347**:58.

de Prost, D., Delmas, V., Lefebvre, M., Lacombe, C. and Bridey, F. (1996). Priapism revealing Arg^{506} to Gln factor V mutation. *Journal of Urology*, **155**:1392.

de Ronde, H. and Bertina, R.M., (1994). Laboratory diagnosis of APC-resistance: A critical evaluation of the test and the development of diagnostic criteria. *Thrombosis and Haemostasis*, **72**:880–886.

de Stefano, V., Mastrangelo, S., Paciaroni, K., Ireland, H., Lane, D.A., Scirpa, P. *et al.* (1995). Thrombotic risk during pregnancy and puerperium in women with APC-resistance. Effective subcutaneous heparin prophylaxis in a pregnant patient. *Thrombosis and Haemostasis*, **74**:793–794.

de Stefano, V., Paciaroni, K., Mastrangelo, S., Rutella, S., Bizzi, B. and Leone, G. (1996). Instrument effect on the activated protein C resistance plasma assay performed by a commercial kit. *Thrombosis and Haemostasis*, **75**:752–756.

den Heijer, M., Koster, T., Blom, H.J., Bos, G.M.J., Briet, E., Reitsma, P.H. *et al.* (1996). Hyperhomocysteinemia as a risk factor for deep vein thrombosis. *New Zealand Journal of Medicine*, **334**:759–762.

Denninger, M., Beldjord, K., Durand, F., Denie, C., Valla, D. and Guillin, M. (1995). Budd-Chiari syndrome and factor V Leiden mutation. *Lancet*, **345**:525–525.

Desmarais, S., de Moerloose, P., Reber, G., Minazio, P. and Perrier, A., Bounameaux, H. (1996). Resistance to activated protein C in an unselected population of patients with pulmonary embolism. *Lancet*, **347**:1374–1375.

Dhote, R., Bachmeyer, C., Horellou, M.H., Toulon, P. and Christoforov, B. (1995). Central retinal vein thrombosis associated with resistance to activated protein C. *American Journal of Ophthalmology*, **120**:388–389.

Dizon-Townson, D., Hutchison, C., Silver, R., Branch, D.W. and Ward, K. (1995). The factor V Leiden mutation which predisposes to thrombosis is not common in patients with antiphospholipid syndrome. *Thrombosis and Haemostasis*, **74**:1029–1031.

Eikelboom, J., Thom, J., van Bockxmeer, F. and Baker, R. (1996). Does activated protein C resistance increase the risk of systemic embolism in non rheumatic atrial fibrillation? *Australian and New Zealand Journal of Medicine*, **26**:243–244.

Emmerich, J., Poirier, O., Evans, A., Marques-Vidal, P., Arveiler, D., Luc, G. *et al.* (1995). Myocardial infarction, Arg^{506} to Gln factor V mutation, and activated protein C resistance. *Lancet*, **345**:321.

Engel, H., Zwang, L., van Vliet, H.H.D.M., Michiels, J.J., Stibbe, J. and Lindemans, J. (1996). Phenotyping and genotyping of coagulation factor V Leiden. *Thrombosis and Haemostasis*, **75**:267–269.

Eritsland, J., Gjonnes, G., Sandset, P.M., Seljeflot, I. and Arnesen, H. (1995). Activated protein C resistance and graft occlusion after coronary artery bypass surgery. *Thrombosis Research*, **79**:223–226.

Farley, T.M.M., Meirik, O., Chang, C.L., Marmot, M.G. and Poulter, N.R. (1995). Effect of different progestagens in low oestrogen oral contraceptives on venous thromboembolic disease. *Lancet*, **346**:1582–1588.

Forsyth, P.D. and Dolan, G. (1995). Activated protein C resistance in cases of cerebral infarction. *Lancet*, **345**:795.

Freyburger, G., Bilhou-Nabera, C., Dief, S., Javorschi, S., Labrouche, S., Lerebeller, M.J. *et al.* (1996). Technical and biological conditions influencing the functional APC resistance test. *Thrombosis and Haemostasis*, **75**:460–465.

Fujimura, H., Kambayashi, J. and Monden, M. (1995). Coagulation factor V Leiden may have a racial background. *Thrombosis and Haemostasis*, **74**:1379–1387.

Ganesan, V., Kelsey, H., Cookson, J., Osborn, A. and Kirkham, F.J. (1996). Activated protein C resistance in childhood stroke. *Lancet*, **347**:260.

Gandrille, S., Alhenc-Gelas, M. and Aiach, M. (1995). A rapid screening method for the factor V Arg^{506}→Gln mutation. *Blood Coagulation and Fibrinolysis*, **6**:245–248.

Gandrille, S., Greengard, J.S., Alhenc-Gelas, M., Juhan-Vague, I., Abgrall, J.F., Jude, B. *et al.* (1995). Incidence of activated protein C resistance caused by the ARG^{506} GLN mutation in factor V in 113 unrelated symptomatic protein C - deficient patients. *Blood*, **86**:219–224.

Gardyn, J., Sorkin, P., Kluger, Y., Kabili, S., Klausner, J.M., Zivelin, A. and Eldor, A. (1995). Heparin-induced thrombocytopenia and fatal thrombosis in a patient with activated protein C resistance. *American Journal of Hematology*, **50**:292–295.

Gilmore, G., Thom, J. and Baker, R. (1996). Diagnosis of APC resistance in patients on standard or low molecular weight heparin. *Thrombosis and Haemostasis*, **75**:372–373.

Goldhaber, S.Z., (1994). Epidemiology of pulmonary embolism and deep vein thrombosis. In *Haemostasis and Thrombosis* (volume 1), edited by A.L. Bloom, C.D. Forbes, D.P Thomas, E.G.D. Tuddenham, pp. 1327–1333. New York, Churchill Livingstone.

Greengard, J.S., Eichinger, S., Griffin, J.H. and Bauer, K.A. (1994). Brief report: Variability of thrombosis among homozygous siblings with resistance to activated protein C due to an Arg→Gln mutation in the gene for factor V. *New England Journal of Medicine*, **23**:1559–1562.

Greer, I.A. (1994). Haemostasis and thrombosis in pregnancy. *Haemostasis and Thrombosis*, edited by A.L. Bloom, C.D. Forbes, D.P Thomas, E.G.D. Tuddenham, pp. 1349–1360. New York: Churchill Livingstone.

Griffin, J.H., Evatt, B., Wideman, C. and Fernandez, J.A. (1993). Anticoagulant protein C pathway defective in majority of thrombophilic patients. *Blood*, **8**:1989–1993.

Griffin, J.H., Evatt, B., Zimmerman, T.S., Kleiss, A.J. and Wideman, C. (1981). Deficiency of protein C in congenital thrombotic disease. *Journal of Clinical Investigation*, **68**:1370.

Grisaru, D., Fait, G. and Eldor, A. (1996). Activated protein C resistance and pregnancy complications (letter). *American Journal of Obstetrics and Gynecology*, **174**:801–802.

Gschwandtner, M.E., Eichinger, S., Hutter, D., Korninger, L., Lechner, K., Panzer, S. *et al.* (1996). Lupus anti-coagulant and thromboembolism: Evaluation of fibrinogen, natural inhibitors and molecular markers of thrombosis. *Blood Coagulation and Fibrinolysis,* **7**:325–330.

Gschwandtner, M.E., Lechner, K. and Pabinger, I. (1995). Erroneously low APC ratio in patients with lupus anti-coagulant. *Annals of Hematology,* **70**:169–170.

Guillerm, C., Lellonche, F., Darnige, L., Schandelong, A., Geffroy, F. and Dorval, I. (1996). Rapid detection of the G1691A mutation of coagulation factor V by PCR-mediated site-directed mutagenesis. *Clinical Chemistry,* **42**:329.

Halbmayer, W.M., Haushofer, A., Schon, R. and Fischer, M. (1994a). Influence of lupus anticoagulant on a commercially available kit for APC-resistance. *Thrombosis and Haemostasis,* **72**:645–646.

Halbmayer, W.M., Haushofer, A., Schon, R. and Fischer, M. (1994b). The prevalence of poor anticoagulant response to activated protein C (APC resistance) among patients suffering from stroke or venous thrombosis and among healthy subjects. *Blood Coagulation and Fibrinolysis,* **5**:51–58.

Hall, C.M., Andersseon, N.E., Andras, M., Rosen, S. and Zetterberg, U. (1996). Complete discrimination for factor V: Q^{506} in plasma from OAC-patients, heparin patients and non-treated individuals using the COAT-EST® APC™ resistance assay after predilution in V-DEF plasma. *Blood Coagulation and Fibrinolysis,* **7**:59.

Hallam, P.J., Millar, D.S., Krawczak, M., Kakkar, V.V. and Cooper, D.N. (1995). Population differences in the frequency of the factor V Leiden variant among people with clinically symptomatic protein C deficiency. *Journal of Medical Genetics,* **32**:543–545.

Hampton, K.K., Preston, F.E. and Greaves, M. (1994). Resistance to activated protein C. *New England Journal of Medicine,* **331**:129–130.

Heeb, M.J., Kojima, Y., Greengard, J.S. and Griffin, J.H. (1995). Activated protein C resistance: Molecular mechanisms based on studies using purified Gln^{506}-factor V. *Blood,* **85**:3405–3411.

Heijboer, H., Brandjes, D.P., Buller, H.R., Sturk, A. and ten Cate J.W. (1990). Deficiencies of coagulation-inhibiting and fibrinolytic proteins in outpatients with deep-vein thrombosis. *New England Journal of Medicine,* **323**:1512–1518.

Hellgren, M., Svensson, P.J. and Dahlback, B. (1995). Resistance to activated protein C as a basis for venous thromboembolism associated with pregnancy and oral contraceptives. *American Journal of Obstetrics and Gynecology,* **173**:210–213.

Hemker, H.C. (1994). Thrombin generation, an essential step in haemostasis and thrombosis. In *Haemostasis and Thrombosis* (volume 1), edited by A.L. Bloom, C.D. Forbes, D.P Thomas, E.G.D. Tuddenham, pp. 477–490. New York: Churchill Livingstone.

Henkens, C.M.A., Bom, V.J.J., Seinen, A.J. and van der Meer, J. (1995). Sensitivity to activated protein C; influence of oral contraceptives and sex. *Thrombosis and Haemostasis,* **73**:402–404.

Henkens, C.M.A., Bom, V.J.J. and van der Meer, J. (1995). Lowered APC-sensitivity ratio related to increased factor VIII-clotting activity. *Thrombosis and Haemostasis,* **74**:1198–1199.

Hodgkin, K. (1978). Early Diagnosis and Prevention. *Towards Earlier Diagnosis in Primary Care,* pp. 32–35. Edinburgh: Churchhill Livingstone.

Holm, J., Zoller, B., Berntorp, E., Erhardt, L. and Dahlback, B. (1996). Prevalence of factor V gene mutation amongst myocardial infarction patients and healthy controls is higher in Sweden than in other countries. *Journal of Internal Medicine,* **239**:221–226.

Holm, J., Zoller, B., Svensson, P.J., Berntorp, E., Erhardt, L. and Dahlback, B. (1994). Myocardial infarction associated with homozygous resistance to activated protein C. *Lancet,* **344**:952–953.

Hunt, B. (1996). Activated protein C and retinal vein occlusion. *British Journal of Ophthalmology,* **80**:194.

Jenny, R.J., Tracy, P.B. and Mann, K.G. (1994). The physiology and biochemistry of factor V. In *Haemostasis and Thrombosis* (volume 1), edited by A.L. Bloom, C.D. Forbes, D.P Thomas, E.G.D. Tuddenham, pp. 465–474. New York: Churchill Livingstone.

Jorquera, J.I., Montoro, J.M., Fernandez, M.A., Aznar, J.A. and Aznar, J. (1994). Modified test for activated protein C resistance. *Lancet,* **344**:1162–1163.

Jorquera, J.I., Aznar, J., Fernandez, M.A., Montoro, J.M., Curats, R. and Casana, P. (1996). A modification of the APC resistance test and its application to the study of patients on coumarin therapy. *Thrombosis Research,* **82**:217–224.

Jick, H., Jick, S.S., Gurewich, V., Myers, M.W. and Vasilakis, C. (1995). Risk of idiopathic cardiovascular death and nonfatal venous thromboembolism in women using oral contraceptives with differing progestagen components. *Lancet,* **346**:1589–1592.

Kalafatis, M., Haley, P.E., Rogier, D.L., Bertina, R.M., Long, G.L. and Mann K.G. (1996). Proteolytic events that regulate factor V activity in whole plasma from normal and activated protein C (APC)-resistant individuals during clotting: An insight into the APC-resistance assay. *Blood,* **87**:4695–4707.

Kalafatis, M., Lu, D., Bertina, R.M., Long, G.L. and Mann, K.G. (1995). Biochemical prototype for familial thrombosis: A study combining a functional protein C mutation and factor V Leiden. *Arteriosclerosis, Thrombosis, and Vascular Biology,* **15**:2181–2187.

Kalafatis, M., Rand, M.D. and Mann, K.G. (1994). The mechanism of inactivation of human factor V and human factor Va by activated protein C. *Journal of Biological Chemistry,* **269**:31869–31880.

Kambayashi, J., Fujimura, H., Kawasaki, T. *et al.* (1995). The incidence of activated protein C resistance among patients with deep vein thrombosis and healthy subjects in Osaka. *Thrombosis Research,* **79**:227–229.

Kirschbaum, N.E. and Foster, P.A. (1995). The polymerase chain reaction with sequence specific primers for the detection of the factor V mutation associated with activated protein C resistance. *Thrombosis and Haemostasis*, **74**:874–878.

Koeleman, B.P.C., Reitsma, P.H., Allaart, C.F. and Bertina, R.M. (1994). Activated protein C resistance as an additional risk factor for thrombosis in protein C deficient families. *Blood*, **84**:1031–1035.

Koeleman, B.P.C., van Rumpt, D., Hamulyak, K., Reitsma, P.H. and Bertina R.M. (1995). Factor V Leiden: An additional risk factor for thrombosis in protein S deficient families? *Thrombosis and Haemostasis*, **74**:580–583.

Kontula, K., Ylikorkala, A., Miettinen, H., Vuorio, A., Kauppinen-Makelin, R., Hamalainen, L. *et al.* (1995). $Arg^{506}Gln$ factor V mutation (factor V Leiden). in patients with ischaemic cerebrovascular disease and survivors of myocardial infarction. *Thrombosis and Haemostasis*, **73**:558–560.

Laffan, M.A., Vulliamy, T., Schmitz, E. and Swirsky, D. (1996). Factor V Leiden and sickle cell disease. *Thrombosis and Haemostasis*, **75**:859–860.

Larsson, J., Olafsdottir, E. and Bauer, B. (1996). Activated protein C resistance in young adults with central retinal vein occlusion. *British Journal of Ophthalmology*, **80**:200–202.

Le, D.T., Griffin, J.H., Greengard, J.S., Mujumdar, V. and Rapaport, S.I. (1995). Use of a generally applicable tissue factor — dependent factor V assay to detect activated protein C - resistant factor Va in patients receiving warfarin and in patients with a lupus anticoagulant. *Blood*, **85**:1704–1711.

Legnani, C., Palareti, G., Biagi, R. and Coccheri, S. (1994). Activated protein C resistance in deep-vein thrombosis. *Lancet*, **343**:541–542.

Leroy-Matherson, C., Levent, M., Pignon, J., Mendonca, C. and Gouault-Heilmann, M. (1996). The 1691 G→A mutation in the factor V gene: Relationship to activated protein C (APC). resistance and thrombosis in 65 patients. *Thrombosis and Haemostasis*, **75**:4–10.

Lindblad, B., Svensson, P.J. and Dahlback, B. (1994). Arterial and venous thromboembolism with fatal outcome and resistance to activated protein C. *Lancet*, **343**:917.

Lu, D., Kalafatis, M., Mann, K.G. and Long, G.L. (1996). Comparison of activated protein C, protein S - mediated inactivation of human factor VIII and factor V. *Blood*, **87**:4708–4717.

Ma, D.D.F., Williams, B.G., Aboud, M.R. and Isbister, J.P. (1995). Activated protein C resistance (APC) and inherited factor V (factor V) mis-sense mutation in patients with venous and arterial thrombosis in a haematology clinic. *Australian and New Zealand Journal of Medicine*, **25**:151–154.

Mahasandana, C., Suvatte, V., Marlar, R.A., Manco-Johnson, M.J., Jacobson, L.J. and Hathaway, W.E. (1990). Neonatal purpura fulminans associated with homozygous protein S deficiency (letter). *Lancet*, **335**:61–62.

Mahmoud, A.E.A., Wilde, J.T. and Elias, E. (1995). Budd-Chiari syndrome and factor V Leiden mutation *Lancet*, **345**:525–526.

Makris, M., Bardhan, G. and Preston, F.E. (1996). Warfarin induced skin necrosis associated with activated protein C resistance. *Thrombosis and Haemostasis*, **75**:523–524.

Malia, R.G., Kitchen, S., Greaves, M. and Preston, F.E. (1990). Inhibition of activated protein C and its cofactor protein S by antiphospholipid antibodies. *British Journal of Haematology*, **76**:101–107.

Mambrini, P., Mallet, D., O'Callaghan, T., Sebahoun, G., Salducci, J. and Grimaud, J.C. (1996). Budd-Chiari syndrome and activated protein C resistance. *Journal of Hepatology*, **24**:246.

Mandel, H., Brenner, B., Berant, M., Rosenberg, N., Lanir, N., Jakobs, C. *et al.* (1996). Coexistence of hereditary homocystinuria and factor V Leiden — effect on thrombosis. *The New England Journal of Medicine*, **334**:763–768.

Marciniak, E. and Romond, E.H. (1989). Impaired catalytic function of activated protein C: A new *in vitro* manifestation of lupus anticoagulant. *Blood*, **74**:2426–2432.

Marder, V.J. (1995). Thrombolytic Therapy: Overview of results in major vascular occlusions. *Thrombosis and Haemostasis*, **74**:101–105.

Mari, D., Mannucci, P.M. and Duca, F., Bertolini, S., Franceschi, C. (1996). Mutation factor V ($Arg^{506}Gln$) in healthy centenarians. *Lancet*, **347**:1044.

Martorell, J.R., Munoz-Castillo, A. and Gil, J.L. (1995). False positive activated protein C resistance test due to anti-phospholipid antibodies is corrected by platelet extract. *Thrombosis and Haemostasis*, **74**:796–797.

Martinelli, I., Magatelli, R., Cattaneo, M. and Mannucci, P.M. (1996a). Prevalence of mutant factor V in Italian patients with hereditary deficiencies of antithrombin, protein C or protein S. *Thrombosis and Haemostasis*, **75**:694–695.

Martinelli, I., Landi, G., Merati, G., Cella, R., Tosetto, A. and Mannucci, P.M. (1996b). Factor V gene mutation is a risk factor for cerebral venous thrombosis. *Thrombosis and Haemostasis*, **75**:393–394.

Marz, W., Seydewitz, H., Winkelmann, B., Chen, M., Nauck, M. and Witt, I. (1995). Mutation in coagulation factor V associated with resistance to activated protein C in patients with coronary artery disease. *Lancet*, **345**:526.

Matsuda, J., Gotoh, M., Gohchi, K., Kawasugi, K., Tsukamoto, M. and Saitoh, N. (1995). Resistance to activated protein C activity of an anti-β_2-glycoprotein I antibody in the presence of β_2-glycoprotein I. *British Journal of Haematology*, **90**:204–206.

Mathonnet, F., de Mazancourt, P., Bastenaire, B., Morot, M., Benattar, N. and Beufe, S. (1996). Activated protein C sensitivity ratio in pregnant women at delivery. *British Journal of Haematology*, **92**:244–246.

Monkovic, D. and Tracy, P.B. (1990). Activation of human factor V by factor Xa and thrombin. *Biochemistry*, **29**:1118–1128.

Monreal, M., Ruiz, J., Salvador, R., Morera, J. and Arias, A. (1989). Recurrent pulmonary embolism: A Prospective study. *Chest*, **95**:976–979.

Munkvad, S. and Jorgensen, M. (1996). Resistance to activated protein C: A common anticoagulant deficiency in patients with venous leg ulceration. *British Journal of Dermatology*, **134**:296–298.

Nathwani, A.C. and Tuddenham, E.G.D., (1992). Epidemiology of coagulation disorders. *Bailliere's Clinical Haematology*, **5**:383–349.

Nowak-Gottl, U., Aschka, I., Koch, H.G., Boos, J., Dockhorn-Dworniczak, B., Deufel, T. *et al.* (1995). Resistance to activated protein C (APCR) in children with acute lymphoblastic leukaemia - the need for a prospective multicentre study. *Blood Coagulation and Fibrinolysis*, **6**:761–764.

Nowak-Gottl, U., Kohlhase, B.K., Vielhaber, H., Aschka, I., Schneppenheim, R. and Jurgens, H. (1996). APC resistance in neonates and infants: Adjustment of the APTT - based method. *Thrombosis Research*, **81**:665–670.

Olivieri, O., Friso, S., Manzato, F., Guella, A., Bernardi, F., Lunghi, B. *et al.* (1995). Resistance to activated protein C in healthy women taking oral contraceptives. *British Journal of Haematology*, **91**:465–470.

Osterud, B., Robertsen, R., Asveng, G.B. and Thijssen, F. (1994). Resistance to activated protein C is reduced in women using oral contraceptives. *Blood Coagulation and Fibrinolysis*, **5**:853–854.

Ouriel, K., Green, R.M., De Weese, J.A. and Cimino, C. (1996). Activated protein C resistance: Prevalence and implications in peripheral vascular disease. *Journal of Vascular Surgery*, **23**:46–52.

Petaja, J., Jalanko, H., Holmberg, C., Kinnunen, S. and Syrala, M. (1995). Resistance to activated protein C as an underlying cause of recurrent venous thrombosis during relapsing nephrotic syndrome. *Journal of Pediatrics*, **127**:103–105.

Poort S.R., Rosendaal F.R., Reitsma P.H. and Bertina R.M., A common genetic variation in the 3′ untranslated region of the prothrombin gene is associated with elevated plasma prothrombin levels and an increase in venous thrombosis. *Blood*, **88**:3698–703

Potzsch, B., Kawamura, H., Preissner, K.T., Schmidt, M., Seelig, C. and Muller-Berghaus, G. (1994). Acquired protein C dysfunction but not decreased activity of thrombomodulin is a possible marker of thrombophilia in patients with lupus anticoagulant. *Journal of Laboratory and Clinical Medicine*, **125**:56–65.

Poulter, N.R., Chang, C.L., Farley, T.M.M., Meirik, O. and Marmot M.G. (1995). Venous thromboembolic disease and combined oral contraceptives: Results of international multicentre case-control study. *Lancet*, **346**:1575–1581.

Press, R.D., Liu, X., Beamer, N. and Coull, B.M. (1996). Ischemic stroke in the elderly. Role of the common factor V mutation causing resistance to activated protein C. *Stroke*, **27**:44–48.

Prohaska, W., Mannebach, H., Schmidt, M., Gleichmann, U. and Kleesiek, K. (1995). Evidence against heterozygous coagulation factor V 1691G→A mutation with resistance to activated protein C being a risk factor for coronary artery disease and myocardial infarction. *Journal of Molecular Medicine*, **73**:521–524.

Rabes, J.P., Trossaert, M., Conard, J., Samama, M., Giraudet, P. and Boileau, C. (1995). Single point mutation at Arg[506] of factor V associated with APC resistance and venous thromboembolism: Improved detection by PCR-mediated site-directed mutagenesis. *Thrombosis and Haemostasis*, **74**:1379–1380.

Rai, R., Regan, L., Hadley, E., Dave, M. and Cohen, H. (1996). Second-trimester pregnancy loss is associated with activated protein C resistance. *British Journal of Haematology*, **92**:489–490.

Rees, D.C., Cox, M. and Clegg, J.B. (1995). World distribution of factor V Leiden. *Lancet*, **346**:1133.

Rees, D.C., Cox, M. and Clegg, J.B. (1996). Detection of the factor V Leiden mutation using whole blood PCR. *Thrombosis and Haemostasis*, **75**:520–521.

Ridker, P.M., Hennekens, C.H., Lindpaintner, K., Stampfer, M.J., Eisenber, P.R. and Miletich, J.P. (1995a). Mutation in the gene coding for coagulation factor V and the risk of myocardial infarction, stroke, and venous thrombosis in apparently healthy men. *New England Journal of Medicine*, **332**:912–917.

Ridker, P.M., Miletich, J.P., Stampfer, M.J., Goldhaber, S.Z., Lindpaintner, K. and Hennekens, C.H. (1995b). Factor V Leiden and risks of recurrent idiopathic venous thromboembolism. *Circulation*, **92**:2800–2802.

Riikonen, R.S., Vahtera, E.M. and Kekomaki, R.M. (1996). Physiological anticoagulants and activated protein C resistance in childhood stroke. *Acta Paediatrica*, **85**:2542–244.

Rintelen, C., Mannhalter, C., Ireland, H., Lane, D.A., Knobl, P., Lechner, K. *et al.* (1996a). Oral contraceptives enhance the risk of clinical manifestation of venous thrombosis at a young age in females homozygous for factor V Leiden. *British Journal of Haematology*, **93**:487–490.

Rintelen, C., Pabinger, I., Knobl, P., Lechner, K. and Mannhalter, C.H. (1996b). Probability of recurrence of thrombosis in patients with and without factor V Leiden. *Thrombosis and Haemostasis*, **75**:229–232.

Roelse, J.C., Koopman, M.M., Buller, H.R., ten Cate, J.W., Montaruli, B., van Mourik, J.A. *et al.* (1996). Absence of mutations at the activated protein C cleavage sites of factor VIII in 125 patients with venous thrombosis. *British Journal of Haematology*, **92**:740–743.

Rosen, S., Johansson, K., Lindberg, K. and Dahlback, B. (1994). Multicenter evaluation of a kit for activated protein C resistance on various coagulation instruments using plasmas from healthy individuals. *Thrombosis and Haemostasis*, **72**:255–260.

Rosendaal, F.R., (1996). Oral contraceptives and screening for factor V Leiden. *Thrombosis and Haemostasis*, **74**:524–525.

Rosendaal, F.R., Koster, T., Vandenbroucke, J.P. and Reitsma, P.H. (1995). High risk of thrombosis in patients homozygous for factor V Leiden (activated protein C resistance). *Blood*, **85**:1504–1508.

Rosing, J., Hoekema, L., Nicolaes, G.A, Thomassen, M.C., Hemker, H.C., Varadi, K. *et al.* (1995). Effects of protein S and factor Xa on peptide bond cleavages during inactivation of factor Va and factor Va^{R506Q} by
activated protein C. *Journal of Biological Chemistry*, **270**:27852–27858.

Ruiz-Arguelles, G.J., Garces-Eisele, J., Alarcon-Segovia, D. and Ruiz-Arguelles, A. (1996). Activated protein C
resistance phenotype and genotype in patients with primary antiphospholipid syndrome. *Blood Coagulation
and Fibrinolysis*, **7**:344–348.

Sachs, B.P., Brown, D.A.J., Driscoll, S.G. *et al.* (1987). Maternal mortality in Massachusetts: Trends and prevention. *New England Journal of Medicine*, **316**:667–672.

Sakata, T., Kario, K., Katayama, Y., Matsuyama, T., Kato, H. and Miyata, T. (1996). Clinical significance of
activated protein C resistance as a potential marker for hypercoagulable state. *Thrombosis Research*,
82:235–244.

Samaha, M., Trossaert, M., Conard, J., Horellou, M., Elalamy, I. and Samama, M. (1995). Prevalence and patient
profile in activated protein C resistance. *American Journal of Clinical Pathology*, **104**:450–454.

Samani, N.J., Lodwick, D., Martin, D. and Kimber, P. (1994). Resistance to activated protein C and risk of premature myocardial infarction. *Lancet*, **344**:1709–1710.

Shen, L. and Dahlback, B. (1994). Factor V and protein S as synergistic cofactors to activated protein C in degradation of factor VIIIa. *Journal of Biological Chemistry*, **269**:18735–18738.

Shizuka, R., Kanda, T., Amagai, H. and Kobayashi, I. (1995). False-positive activated protein C (APC) sensitivity ratio caused by freezing and by contamination of plasma with platelets. *Thrombosis Research*,
78:189–190.

Sidelmann, J., Gram, J., Pedersen, O.D. and Jespersen, J. (1995). Influence of plasma platelets on activated protein C resistance assay. *Thrombosis and Haemostasis*, **74**:993–994.

Sifontes, M.T., Nuss, R. and Jacobson, L.J. (1996). Thrombosis in otherwise well children with the factor V Leiden mutation. *Journal of Pediatrics*, **128**:324–328.

Simioni, P., de Ronde, H., Prandoni, P., Saladini, M., Bertina, R.M. and Girolami, A. (1995a). Ischemic stroke in
young patients with activated protein C resistance. A report of three cases belonging to three different kindreds. *Stroke*, **26**:885–890.

Simioni, P., Gavasso, S., Luni, S., Invidiato, S. and Girolami, A. (1995b). A protein S functional assay yields
unsatisfactory results in patients with activated protein C resistance. *Blood Coagulation and Fibrinolysis*,
6:286–287.

Simioni, P., Prandoni, P. and Girolami, A. (1994). Patients with AT III, protein C or protein S defects show no associated hereditary APC-resistance. *Thrombosis and Haemostasis*, **72**:481–482.

Simioni, P., Scarano, L., Gavasso, S., Sardella, C., Girolami, B., Scudeller, A. *et al.* (1996). Prothrombin fragment
1+2 and thrombin-antithrombin complex levels in patients with inherited APC resistance due to factor V
Leiden mutation. *British Journal of Haematology*, **92**:435–441.

Svensson, P. and Dahlback, B, (1994). Resistance to activated protein C as a basis for venous thrombosis. *New
England Journal of Medicine*, **330**:517–522.

Svensson, P.J., Zoller, B. and Dahlback, B. (1997) Evaluation of original and modified APC-resistance tests in
unselected outpatients with clinically suspected thrombosis and in healthy controls. *Thrombosis and Haemostasis*, **77**:332–5.

Takamiya, O., Ishida, F., Kodaira, H. and Kitano, K. (1995). APC-Resistance and Mnl I genotype (Gln^{506}) of coagulation factor V are rare in Japanese population. *Thrombosis and Haemostasis*, **74**:996.

Tracy, P.B., Eide, L.L., Bowie, E.J. *et al.* (1982). Radioimmunoassay of factor V in human plasma and platelets.
Blood, **60**:59–63.

Trossaert, M., Conard, J., Horellou, M.H., Elalamy, I. and Samama, M.M. (1996). The modified APC resistance
test in the presence of factor V deficient plasma can be used in patients without oral anticoagulant. *Thrombosis and Haemostasis*, **75**:521–522.

Trossaert, M., Conard, J., Horellou, M.H. and Samama, M.M. (1995a). Influence of storage conditions on activated protein C resistance assay. *Thrombosis and Haemostasis*, **73**:162–166.

Trossaert, M., Conard, J., Horellou, Samaha, M., Elalamy, I. and Samama, M.M. (1995b). Resistance a la proteine
C activee dans les accidents thrombo-emboliques veineux frequence et manifestations cliniques. *La Presse
Medicale*, **24**:209.

Trossaert, M., Conard, J., Horellou, M.H., Samama, M.M., Ireland, H. and Bayston, T.A. (1994). Modified APC
resistance assay for patients on oral anticoagulants. *Lancet*, **344**:1709.

van Bockxmeer, F., Baker, R.I. and Taylor, R.R. (1995). Premature ischaemic heart disease and the gene for coagulation factor V. *Nature Medicine*, **1**:185.

van Boven, H.H., Reitsma, P.H., Rosendaal, F.R., Bayston, T.A., Chowdhury, V., Bauer, K.A. *et al.* (1996). Factor V Leiden (factor V R^{506}Q) in families with inherited antithrombin deficiency. *Thrombosis and Haemostasis*, **75**:417–421.

van de Locht, L.T.F., Kuypers, A.W.H.M., Verbruggen, B.W., Linssen, P.C.M., Novakova, I.R.O. and Mensink,
E.J.B.M. (1995). Semi-automated detection of the factor V mutation by allele specific amplification and
capillary electrophoresis. *Thrombosis and Haemostasis*, **74**:1276–1279.

van der Bom, J.G., Bots, M.L., Haverkate, F., Slagboom, P.E., Meijer, P., de Jong, P.T.V.M. *et al.* (1996).
Reduced response to activated protein C is associated with increased risk for cerebrovascular disease.
Annals of Internal Medicine, **125**:265–269.

Vandenbroucke, J.P., Koster, T., Briet, E., Reitsma, P.H., Bertina, R.M. and Rosendaal, F.R. (1994). Increased risk of venous thrombosis in oral-contraceptive users who are carriers of factor V Leiden mutation. *Lancet*, **344**:1453–1457.

Varadi, K., Moritz, B., Lang, H., Bauer, K., Preston, E., Peake, I. *et al*. (1995). A chromogenic assay for activated protein C resistance. *British Journal of Haematology*, **90**:884–891.

Villa, P., Aznar, J., Jorquera, J.I., Mira, Y., Vaya, A. and Fernandez, M.A. (1995). Inherited homozygous resistance to activated protein C. *Thrombosis and Haemostasis*, **74**:794–795.

Viskup, R.W., Tracy, P.B. and Mann, K.G. (1987). The isolation of human platelet factor V. *Blood*, **69**:1185–1189.

Voorberg, J, Roelse, J., Koopman, R., Buller, H., Berends, F., ten Cate, J.W. *et al*. (1994). Association of idiopathic venous thromboembolism with single point mutation at Arg[506] of factor V. *Lancet*, **343**:1535–1536.

Williamson, T.H., Rumley, A. and Lowe, G.D.O. (1996). Blood viscosity, coagulation, and activated protein C resistance in central retinal vein occlusion: A population controlled study. *British Journal of Ophthalmology*, **80**:203–208.

Williamson, D., Brown, K., Luddington, R., Baglin, C. and Baglin, C. (1998). Factor V Cambridge: a new mutation (Arg306 to Thr) associated with resistance to activated protein C. *Blood*, **91**:1140–4.

Zehnder, J.L. and Jain, M. (1996). Recurrent thrombosis due to compound heterozygosity for factor V Leiden and factor V deficiency. *Blood Coagulation and Fibrinolysis*, **7**:361.

Zenz, W., Muntean, W., Gallistl, S., Leschnik, B. and Beitzke, A. (1995). Inherited resistance to activated protein C in a boy with multiple thromboses in early infancy. *European Journal of Pediatrics*, **154**:285–288.

Zoller, B., Berntsdotter, A., de Frutos, P.G. and Dahlback, D. (1995). Resistance to activated protein C as an additional genetic risk factor in hereditary deficiency of protein S. *Blood*, **85**:3518–3523.

Zoller, B. and Dahlback, B, (1994). Linkage between inherited resistance to activated protein C and factor V gene mutation in venous thrombosis. *Lancet*, **343**:1536–1538.

Zoller, B. and Dahlback, B, (1995). Resistance to activated protein C caused by a factor V gene mutation. *Current Opinions In Haematology*, **2** (5):358–364.

Zoller, B., Norlund, L., Leksell, H., Nilsson, J., von Schenck, H., Rosen, U. *et al*. (1996). High prevalence of the factor V[R506Q] mutation causing APC-resistance in a region of Southern Sweden with a high incidence of venous thrombosis. *Thrombosis Research*, **83**:475–7.

Zoller, B., Svensson, P.J., He, X. and Dahlback, B. (1994). Identification of the same factor V gene mutation in 47 out of 50 thrombosis-prone families with inherited resistance to activated protein C. *Journal of Clinical Investigation*, **94**:2521–2524.

SUBJECT INDEX